Springer Proceedings in Mathematics & Statistics

Volume 319

Springer Proceedings in Mathematics & Statistics

This book series features volumes composed of selected contributions from workshops and conferences in all areas of current research in mathematics and statistics, including operation research and optimization. In addition to an overall evaluation of the interest, scientific quality, and timeliness of each proposal at the hands of the publisher, individual contributions are all refereed to the high quality standards of leading journals in the field. Thus, this series provides the research community with well-edited, authoritative reports on developments in the most exciting areas of mathematical and statistical research today.

More information about this series at http://www.springer.com/series/10533

Shigeru Kuroda · Nobuharu Onoda ·
Gene Freudenburg
Editors

Polynomial Rings and Affine Algebraic Geometry

PRAAG 2018, Tokyo, Japan, February 12–16

Springer

Editors
Shigeru Kuroda
Department of Mathematical Sciences
Tokyo Metropolitan University
Hachioji, Tokyo, Japan

Nobuharu Onoda
Faculty of Applied Physics
University of Fukui
Fukui, Japan

Gene Freudenburg
Department of Mathematics
Western Michigan University
Kalamazoo, MI, USA

ISSN 2194-1009 ISSN 2194-1017 (electronic)
Springer Proceedings in Mathematics & Statistics
ISBN 978-3-030-42135-9 ISBN 978-3-030-42136-6 (eBook)
https://doi.org/10.1007/978-3-030-42136-6

Mathematics Subject Classification (2010): 13A02, 13A50, 13B25, 13N15, 14E07, 14E25, 14J70, 14L30, 14R20

This Springer imprint is published by the registered company Springer Nature Switzerland AG
The registered company address is: Gewerbestrasse 11, 6330 Cham, Switzerland

Preface

Affine algebraic geometry studies algebraic subvarieties of complex affine space $\mathbb{C}^n$. Nascent forms of the subject can be traced in many directions, but one prominent series of developments is the description of the automorphism group of the plane due to Jung and Van der Kulk (1942, 1953), together with the classification of its $\mathbb{C}^*$-actions due to Gutwirth (1962) and its $\mathbb{C}^+$-actions due to Ebey and Rentschler (1962, 1968). Affine algebraic geometry emerged as an independent subject in the decade of the 1970s at the appearance of several celebrated results, including the topological characterization of the affine plane given by Ramanujam (1971), the cancelation theorem for curves given by Abhyankar, Eakin and Heinzer (1972), the Abhyankar-Moh Suzuki Theorem (1975), and the cancelation theorem for the plane given by Fujita, Miyanishi, and Sugie (1979). In addition, the all-important Quillen-Suslin Theorem (1976), coming from a larger algebraic framework, supplied a powerful new tool for research in the subject. In this same decade, Nagata (1972) published his famous automorphism of $\mathbb{C}^3$ with the conjecture that it was not tame, and it was discovered that published solutions to Keller's Problem, first posed in 1939, were incorrect. This problem, nowadays called the Jacobian Conjecture, remains open and continues to inspire a great deal of interest and new insight in the field.

Developments in the following decades continue apace. Lin and Zaidenberg (1983) characterized contractible plane curves, Danielewski (1989) gave counterexamples to cancelation for complex affine surfaces, and Schwarz (1989) gave counterexamples to linearization for reductive group actions on affine space. For $\mathbb{C}^+$-actions on affine space, Roberts (1990) showed that the ring of invariants is not generally of finite type, and Winkelmann (1990) constructed free actions which are not translations. Koras and Russell (1997) proved linearization of $\mathbb{C}^*$-actions on $\mathbb{C}^3$ with the aid of Makar-Limanov's work on absolute constants. Their work led to new examples of exotic affine spaces, and the Makar-Limanov invariant is now an important tool in classifying affine varieties. Kaliman showed that polynomial maps on $\mathbb{C}^3$ with general $\mathbb{C}^2$-fibers are variables (2002), and that every free $\mathbb{C}^+$-action on

$\mathbb{C}^3$ is a translation (2004). Shestakov and Umirbaev (2004) showed that Nagata's automorphism is not tame, thus confirming Nagata's conjecture.

One can study affine algebraic geometry over fields other than $\mathbb{C}$, in particular, over an algebraically closed field k of positive characteristics. Russell (1980) showed that cancelation holds for the affine plane over k, while Asanuma and Gupta (1987, 2014) showed that cancelation does not hold for higher dimensional affine spaces over k. Their constructions are based on exotic line embeddings in the plane due to Segre (1956/57), and their proof makes use of the Makar-Limanov invariant.

The foregoing description of significant developments is far from complete. For an extended survey, see the article of M. Miyanishi, *Recent developments in affine algebraic geometry*, in the volume *Affine Algebraic Geometry*, published by Osaka University Press (2007). Areas of ongoing research include cancelation and embedding problems, the Abhyankar Sathaye Conjecture, flexible varieties, automorphism groups and invariant theory, classification of affine varieties, linearization of $\mathbb{C}^*$-actions, characterization of affine spaces, and the Dolgachev-Weisfeiler Conjecture.

This volume presents recent advances in the field of affine algebraic geometry which were featured in the talks of the conference, *Polynomial Rings and Affine Algebraic Geometry*, held at Tokyo Metropolitan University in 2018. In organizing such a conference, we sought not only to bring together established researchers in this area, but also to invite students, early career researchers, and those less familiar with the area to explore and contribute to its rich content and history.

Hachioji, Tokyo, Japan *The Editors*
Fukui, Japan
Kalamazoo, USA

CONFERENCE PHOTO (Left to right)

FIRST ROW: Zbigniew Jelonek
SECOND ROW: Nobuharu Onoda, Arno van den Essen, David L. Wright, Lucy Moser-Jauslin, Shigeru Kuroda, Gene Freudenburg
THIRD ROW: Xiaosong Sun, Vladimir Popov, Toru Sugie, Masayoshi Miyanishi, Karol Palka, Tomasz Pelka
FOURTH ROW: Susanna Zimmermann, Michiel DeBondt, Ivan Arzhantsev, Jean-Phillipe Furter, Drew Lewis
FIFTH ROW: Teruo Asanuma, Pierre-Marie Poloni, Mitsuyasu Hashimoto, Daniel Daigle, Eric Edo
SIXTH ROW: Takashi Kishimoto, Takeshi Kawasaki, Junzo Watanabe, Charlie Petitjean, Yuri Prokhorov, Alexander Perepechko, Frank Kutzschebauch
STANDING: [participant], Takanori Nagamine, Riku Kudo, Ryuji Tanimoto, Yoshifumi Takeda, [participant], Chinami Yamaoka, Yu Yasufuku (seated), Motoki Kuroda, Tatsuki Sato, Hiro-o Tokunaga, Shihoko Ishii, Shrikant Bhatwadekar, Kentaro Mitsui

Contents

On Fano Schemes of Complete Intersections

C. Ciliberto and M. Zaidenberg

Abstract We provide enumerative formulas for the degrees of varieties parameterizing hypersurfaces and complete intersections which contain projective subspaces and conics. Besides, we find all cases where the Fano scheme of the general complete intersection is irregular of dimension at least 2, and for the Fano surfaces we deduce formulas for their holomorphic Euler characteristic.

Keywords Hypersurfaces · Complete intersections · Fano schemes

2010 Mathematics Subject Classification Primary: 14J70 · 14M10 · 14N10 · 14N15 · Secondary: 14C05 · 14C15

1 Introduction

The study of hypersurfaces in projective space, or more generally, of complete intersection, and specifically of varieties contained in them, is a classical subject in algebraic geometry [14]. The present paper is devoted to this subject, and in particular to some enumerative aspects of it.

Recall that the *Fano scheme* $F_k(X)$ of a projective variety $X \subset \mathbb{P}^r$ is the Hilbert scheme of k-planes (that is, linear subspaces of dimension k) contained in X; see [1] or [24, 14.7.13]. For a hypersurface $X \subset \mathbb{P}^r$ of degree d the integer

$$\delta(d, r, k) = (k+1)(r-k) - \binom{d+k}{k}$$

C. Ciliberto
Dipartimento di Matematica, Università degli Studi di Roma "Tor Vergata", Via della Ricerca Scientifica, 00133 Rome, Italy
e-mail: cilibert@mat.uniroma2.it

M. Zaidenberg (✉)
CNRS, Institut Fourier, Univ. Grenoble Alpes, 38000 Grenoble, France
e-mail: Mikhail.Zaidenberg@univ-grenoble-alpes.fr

S. Kuroda et al. (eds.), *Polynomial Rings and Affine Algebraic Geometry*, Springer Proceedings in Mathematics & Statistics 319,
https://doi.org/10.1007/978-3-030-42136-6_1

is called the *expected dimension* of $F_k(X)$. Let $\Sigma(d, r)$ be the projective space of dimension $\binom{d+r}{r} - 1$ which parameterizes the hypersurfaces of degree $d \geq 3$ in $\mathbb{P}^r$. If either $\delta(d, r, k) < 0$ or $2k \geq r$ then $F_k(X) = \emptyset$ for a general $X \in \Sigma(d, r)$. Otherwise $F_k(X)$ has dimension δ for a general $X \in \Sigma(d, r)$ [10, 18, 33, 42]. Setting

$$\gamma(d, r, k) = -\delta(d, r, k) > 0$$

the general hypersurface of degree $d \geq 3$ in $\mathbb{P}^r$ contains no k-plane. Let $\Sigma(d, r, k)$ be the subvariety of $\Sigma(d, r)$ of points corresponding to hypersurfaces which carry k-planes. Then $\Sigma(d, r, k)$ is a nonempty, irreducible, proper subvariety of codimension $\gamma(d, r, k)$ in $\Sigma(d, r)$ (see [40]), and its general point corresponds to a hypersurface of degree d which carries a unique k-plane (see [4]). The degree of this subvariety of the projective space $\Sigma(d, r)$ was computed in [36]. In Sect. 2 we reproduce this computation. This degree $\deg(\Sigma(d, r, k))$ is the total number of k-planes in the members of the general linear system $\mathcal{L}$ of degree d hypersurfaces, provided $\dim(\mathcal{L}) = \gamma(d, r, k)$. It can be interpreted also as the top Chern number of a vector bundle. Having in mind the further usage, we explore three different techniques for computing it:

- the Schubert calculus;
- a trick due to Debarre–Manivel;
- the Bott residue formula and the localization in the equivariant Chow rings.

In Sect. 3 we extend these computations to the Fano schemes of complete intersections in $\mathbb{P}^r$.

In Sects. 4–6 we turn to the opposite case $\gamma(d, r, k) < 0$, that is, the expected dimension of the Fano scheme is positive. In Sect. 4 we compute certain Chern classes related to the Fano scheme. In Sect. 5 we apply these computations in the case where the Fano scheme is a surface, and provide several concrete examples. The main result of Sect. 6 describes all the cases where the Fano scheme of the general complete intersection has dimension ≥ 2 and a positive irregularity. This happens only for the general cubic threefolds in $\mathbb{P}^4$ ($k = 1$), the general cubic fivefolds in $\mathbb{P}^6$ ($k = 2$), and the general intersections of two quadrics in $\mathbb{P}^{2k+3}$, $k \geq 1$; see Theorem 6.1.

In the final Sect. 7 we turn to the conics in degree d hypersurfaces in $\mathbb{P}^r$. Let

$$\epsilon(d, r) = 2d + 2 - 3r\,.$$

Let $\Sigma_c(d, r)$ be the subvariety of $\Sigma(d, r)$ consisting of the degree d hypersurfaces which contain conics. We show that $\Sigma_c(d, r)$ is irreducible of codimension $\epsilon(d, r)$ in $\Sigma(d, r)$, provided $\epsilon(d, r) \geq 0$. Then we prove that the general hypersurface in $\Sigma_c(d, r)$ contains a unique (smooth) conic if $\epsilon(d, r) > 0$. Our main results in this section are formulas (38)–(39) which express the degree of $\Sigma_c(d, r)$ via Bott's residue formula. Notice that there exists already a formula for $\deg(\Sigma_c(d, r))$ in the case $r = 3$, $d \geq 5$, that is, for the surfaces in $\mathbb{P}^3$, see [35, Proposition 7.1]. It expresses this degree as a polynomial in d.

Let us finish with a few comments on the case $\epsilon(d,r) < 0$. It is known (see [27]) that for $2d \le r+1$, given a general hypersurface $X \subset \mathbb{P}^r$ of degree d and any point $x \in X$, there is a family of dimension $e(r+1-d)-2 \ge ed$ of degree e rational curves containing x. In particular, X carries a $2(r-d)$-dimensional family of smooth conics through an arbitrary point. Moreover (see [7]), for $3d \le 2r-1$ the Hilbert scheme of smooth rational curves of degree e on a general X is irreducible of the *expected dimension* $e(r-d+1)+r-4$. In particular, the Hilbert scheme of smooth conics in X is irreducible of dimension $3r-2d-2 = -\epsilon(d,r)$. Analogs of the latter statements hold as well for general complete intersections (see [7]). See also [6, 9] for enumerative formulas counting conics in complete intersections.

2 Hypersurfaces Containing Linear Subspaces

The results of this section are known, see [36], except maybe for formula (4). Our aim is rather didactic, we introduce here the techniques that will be explored in the subsequent sections.

Recall that the *Fano scheme* $F_k(X)$ of a projective variety $X \subset \mathbb{P}^r$ is the Hilbert scheme of linear subspaces of dimension k contained in X; see [1] or [24, 14.7.13]. For a hypersurface $X \subset \mathbb{P}^r$ of degree d the integer

$$\delta(d,r,k) = (k+1)(r-k) - \binom{d+k}{k}$$

is called the *expected dimension* of $F_k(X)$. Let $\Sigma(d,r)$ be the projective space of dimension $\binom{d+r}{r}-1$ which parameterizes the hypersurfaces of degree $d \ge 3$ in $\mathbb{P}^r$. If either $\delta(d,r,k) < 0$ or $2k \ge r$ then $F_k(X) = \emptyset$ for a general $X \in \Sigma(d,r)$. Otherwise $F_k(X)$ has dimension δ for a general $X \in \Sigma(d,r)$ [10, 18, 33, 42]. We assume in the sequel that

$$\gamma(d,r,k) := -\delta(d,r,k) > 0\,. \tag{1}$$

Then the general hypersurface of degree $d \ge 3$ in $\mathbb{P}^r$ contains no linear subspace of dimension k. Let $\Sigma(d,r,k)$ be the subvariety of $\Sigma(d,r)$ of points corresponding to hypersurfaces which do contain a linear subspace of dimension k. The following statement, proven first in [36, Theorem (1)-(2)] in a slightly weaker form, is a particular case of Theorem 1.1 in [4]; see Proposition 3.1 below.

Proposition 2.1 *Assume $\gamma(d,r,k) > 0$. Then $\Sigma(d,r,k)$ is a nonempty, irreducible and rational subvariety of codimension $\gamma(d,r,k)$ in $\Sigma(d,r)$. The general point of $\Sigma(d,r,k)$ corresponds to a hypersurface which contains a unique linear subspace of dimension k and has singular locus of dimension* $\max\{-1, 2k-r\}$ *along its unique k-dimensional linear subspace (in particular, it is smooth provided $2k < r$).*

For instance, take $d=3, r=5$, and $k=2$. Then $\Sigma(3,5)$ parameterizes the cubic fourfolds in $\mathbb{P}^5$, and $\Sigma(3,5,2)$ parameterizes those cubic fourfolds which contain a plane. Since $\gamma(3,5,2)=1$, we conclude that $\Sigma(3,5,2)$ is a divisor in $\Sigma(3,5)$, and the general point of $\Sigma(3,5,2)$ corresponds to a smooth cubic fourfold which contains a unique plane.

Our aim is to compute the degree of $\Sigma(d,r,k)$ in the projective space $\Sigma(d,r)$ in the case $\gamma(d,r,k)>0$.

On the Grassmannian $\mathbb{G}(k,r)$ of k–subspaces of $\mathbb{P}^r$, consider the dual $\mathcal{S}^*$ of the tautological vector bundle $\mathcal{S}$ of rank $k+1$. Let $\Pi\in\mathbb{G}(k,r)$ correspond to a k-subspace of $\mathbb{P}^r$. Then the fiber of $\mathcal{S}^*$ over Π is $H^0(\Pi,\mathcal{O}_\Pi(1))$. It is known ([22, Sect. 5.6.2], [24]) that

$$c(\mathcal{S}^*)=1+\sum_{i=1}^{k+1}\sigma_{(1^i)},$$

where (1^i) stays for the vector $(1,\dots,1)$ of length i, and $\sigma_{(1^i)}$ is the (Poincaré dual of the) class of the Schubert cycle $\Sigma_{(1^i)}$. This cycle has codimension i in $\mathbb{G}(k,r)$, therefore, $\Sigma_{(1^i)}\in A^i(\mathbb{G}(k,r))$ in the Chow ring $A^*(\mathbb{G}(k,r))$.

The *splitting principle* (see [22, Sect. 5.4]) says that any relation among Chern classes which holds for all split vector bundles holds as well for any vector bundle. So, we can write formally

$$\mathcal{S}^*=L_0\oplus\ldots\oplus L_k\,,$$

the L_is being (virtual) line bundles. In terms of the *Chern roots* $x_i=c_1(L_i)$ one can express

$$c(\mathcal{S}^*)=1+c_1(\mathcal{S}^*)+\ldots+c_{k+1}(\mathcal{S}^*)=(1+x_0)\cdots(1+x_k)\,.$$

Hence $\sigma_{(1^i)}$ is the i–th elementary symmetric polynomial in $x_0,\dots,x_k$, i.e.,

$$\sigma_{(1)}=x_0+\ldots+x_k,\quad \sigma_{(1^2)}=\sum_{0\leqslant i<j\leqslant k}x_ix_j,\quad \ldots,\quad \sigma_{(1^{k+1})}=x_0\ldots x_k.$$

Consider further the vector bundle $\mathrm{Sym}^d(\mathcal{S}^*)$ on $\mathbb{G}(k,r)$ of rank

$$\binom{d+k}{k}>(k+1)(r-k)=\dim(\mathbb{G}(k,r)).$$

To compute the Chern class of $\mathrm{Sym}^d(\mathcal{S}^*)$ one writes

$$\mathrm{Sym}^d(\mathcal{S}^*)=\bigoplus_{v_0+\ldots+v_k=d}L_0^{v_0}\cdots L_k^{v_k}\,.$$

Since $c_1(L_0^{v_0}\cdots L_k^{v_k})=v_0x_0+\ldots+v_kx_k$ one obtains

$$c(\mathrm{Sym}^d(\mathcal{S}^*))=\prod_{v_0+\ldots+v_k=d}(1+v_0x_0+\ldots+v_kx_k)\,. \tag{2}$$

The following lemma is standard, see, e.g., [36, Theorem (3)]. For the reader's convenience we include the proof. As usual, the integral of the top degree cohomology class stands for its value on the fundamental cycle. The integral of the dual of a zero cycle α coincides with the degree of α.

Lemma 2.2 *Suppose* (1) *holds. Then one has*

$$\deg(\Sigma(d,r,k)) = \int_{\mathbb{G}(k,r)} c_{(k+1)(r-k)}(\mathrm{Sym}^d(\mathcal{S}^*)).$$

Proof Let $p : V(k,r) \to \mathbb{G}(k,r)$ be the tautological $\mathbb{P}^k$-bundle over the Grassmannian $\mathbb{G}(k,r)$. Consider the composition

$$\varphi : V(k,r) \overset{\phi}{\hookrightarrow} \mathbb{P}^r \times \mathbb{G}(k,r) \overset{\pi}{\longrightarrow} \mathbb{P}^r\,,$$

where ϕ is the natural embedding and π stands for the projection to the first factor. Letting $\mathcal{T} = \mathcal{O}_{\mathbb{P}^r}(-1)$ and $\mathcal{S}_d = \mathrm{Sym}^d(\mathcal{T}^*)$ one obtains

$$\mathcal{S}^* = R^0 p_*\varphi^*(\mathcal{T}^*) \quad \text{and} \quad \mathrm{Sym}^d(\mathcal{S}^*) = R^0 p_*\varphi^*(\mathcal{S}_d)\,.$$

Any $F \in H^0(\mathbb{P}^r, \mathcal{O}_{\mathbb{P}^r}(d))$ defines a section σ_F of $\mathrm{Sym}^d(\mathcal{S}^*)$ such that $\sigma_F(\Pi) = F|_\Pi \in H^0(\Pi, \mathcal{O}_\Pi(d))$. Consider the hypersurface X_F of degree d on $\mathbb{P}^r$ with equation $F = 0$. The support of X_F contains a linear subspace $\Pi \in \mathbb{G}(k,r)$ if and only if $\sigma_F(\Pi) = 0$, i.e., the subspaces $\Pi \in \mathbb{G}(k,r)$ lying in $\mathrm{Supp}(X_F)$ correspond to the zeros of σ_F in $\mathbb{G}(k,r)$, which have a natural scheme structure.

Let $\rho = \dim(\mathbb{G}(k,r)) = (k+1)(r-k)$. By our assumption one has

$$\mathrm{rk}\,(\mathrm{Sym}^d(\mathcal{S}^*)) - \rho = \gamma(r,k,d) > 0\,.$$

Choose a general linear subsystem $\mathcal{L} = \langle X_0, \ldots, X_\gamma \rangle$ in $\Sigma(d,r) = |\mathcal{O}_{\mathbb{P}^r}(d)|$ of dimension $\gamma = \gamma(r,k,d)$, where $X_i = \{F_i = 0\}$. By virtue of Proposition 2.1, $\mathcal{L}$ meets $\Sigma(d,r,k) \subset \Sigma(d,r)$ transversally in $\deg(\Sigma(d,r,k))$ simple points, and to any such point $X \in \Sigma(d,r,k) \cap \mathcal{L}$ corresponds a unique k-dimensional subspace $\Pi \in \mathbb{G}(k,r)$ such that $\Pi \subset X$.

Consider now the sections $\sigma_i := \sigma_{F_i}, i = 0, \ldots, \gamma$, of $\mathrm{Sym}^d(\mathcal{S}^*)$. The intersection of $\mathcal{L}$ with $\Sigma(d,r,k)$ is exactly the scheme of points $\Pi \in \mathbb{G}(k,r)$ where there is a linear combination of $\sigma_0, \ldots, \sigma_\gamma$ vanishing on Π. This is the zero dimensional scheme of points of $\mathbb{G}(k,r)$ where the sections $\sigma_0, \ldots, \sigma_\gamma$ are linearly dependent. This zero dimensional scheme represents the top Chern class $c_\rho(\mathrm{Sym}^d(\mathcal{S}^*))$ (see [22, Theorem 5.3]). Its degree (which is equal to $\deg(\Sigma(d,r,k))$) is the required Chern number $\int_{\mathbb{G}(k,r)} c_{(k+1)(r-k)}(\mathrm{Sym}^d(\mathcal{S}^*))$. □

Let us explain now three methods for computing $\deg(\Sigma(d,r,k))$.

2.1 Schubert Calculus

In order to compute $c_{(r-k)(k+1)}(\mathrm{Sym}^d(\mathcal{S}^*))$, one computes the polynomial in $x_0, \ldots, x_k$ appearing in (2) and extracts the homogeneous component $\mathcal{T}_{(d,r,k)}$ of degree $(k+1)(r-k)$. The latter homogeneous polynomial in $x_0, \ldots, x_k$ is symmetric, hence it can be expressed via a polynomial in the elementary symmetric functions $\sigma_{(1^i)}$, $i = 0, \ldots, k+1$:

$$\mathcal{T}_{(d,r,k)} = \sum_{j_1+2j_2+\ldots+(k+1)j_{k+1}=(k+1)(r-k)} \phi_{d,r}(j_1, j_2, \ldots, j_{k+1})\sigma_{(1)}^{j_1}\sigma_{(1^2)}^{j_2}\cdots\sigma_{(1^{k+1})}^{j_{k+1}} \quad (3)$$

with suitable coefficients $\phi_{d,r}(j_1, j_2, \ldots, j_{k+1})$. In this way the top Chern number

$$\mathcal{T}_{(d,r,k)} = c_{(k+1)(r-k)}\left(\mathrm{Sym}^d(\mathcal{S}^*)\right)$$

in Lemma 2.2 is expressed in terms of the Chern numbers

$$\sigma_{(1)}^{j_1}\sigma_{(1^2)}^{j_2}\cdots\sigma_{(1^{k+1})}^{j_{k+1}}$$

appearing in (3). By computing the intersection products among Schubert classes and plugging these in (3) one obtains the desired degree

$$\deg(\Sigma(d, r, k)) = \mathcal{T}_{(d,r,k)}.$$

2.2 Debarre–Manivel's Trick

This trick (applied for a similar purpose by van der Waerden [50]) allows to avoid passing to the elementary symmetric polynomials, which requires to compute the coefficients in (3). Let us recall the basics on the Chow ring of the Grassmannian $\mathbb{G}(k, r)$ following [37].

A *partition* λ of length $k+1$ is a (non-strictly) decreasing sequence of non-negative integers $(\lambda_0, \ldots, \lambda_k)$. To such a partition λ there corresponds a homogeneous symmetric *Schur polynomial* $s_\lambda \in \mathbb{Z}[x_0, \ldots, x_k]$ of degree $|\lambda| = \lambda_0 + \ldots + \lambda_k$. These polynomials form a base of the $\mathbb{Z}$-module Λ_{k+1} of symmetric polynomials in $x_0, \ldots, x_k$. One writes $\lambda \subset (k+1) \times (r-k)$ if $r - k \geq \lambda_0 \geq \ldots \geq \lambda_k \geq 0$. This inclusion means that the corresponding *Ferrers diagram* of λ is inscribed in the rectangular matrix of size $(k+1) \times (r-k)$ occupying λ_{i-1} first places of the ith line for $i = 1, \ldots, k+1$. To any $\lambda \subset (k+1) \times (r-k)$ there correspond:

- a Schubert variety $\Sigma_\lambda \subset \mathbb{G}(k, r)$ of codimension $|\lambda|$;
- the corresponding Schubert cycle $[\Sigma_\lambda]$ in the Chow group $A_*(\mathbb{G}(k, r))$;
- the corresponding dual Schubert class σ_λ of degree $|\lambda|$ in the Chow ring $A^*(\mathbb{G}(k, r))$.

The nonzero Schubert classes form a base of the free $\mathbb{Z}$-module $A^*(\mathbb{G}(k,r))$ ([37, Cor. 3.2.4]). There is a unique partition $\lambda_{\max} = (r-k, \ldots, r-k) \subset (k+1) \times (r-k)$ of maximal weight $|\lambda_{\max}| = (k+1)(r-k)$. Its Ferrers diagram coincides with the whole rectangle $(k+1) \times (r-k)$. The corresponding Schur polynomial is $s_{\lambda_{\max}} = (x_0 \cdots x_k)^{r-k}$. The corresponding Schubert cycle is a reduced point, and the corresponding Schubert class $\sigma_{\lambda_{\max}}$ generates the $\mathbb{Z}$-module $A^{(k+1)(r-k)}(\mathbb{G}(k,r)) \simeq \mathbb{Z}$.

Let as before $x_0, \ldots, x_k \in A^1(\mathbb{G}(k,r))$ be the Chern roots of the vector bundle $\mathcal{S}^*$ over $\mathbb{G}(k,r)$. A homogeneous symmetric polynomial $\tau \in \mathbb{Z}[x_0, \ldots, x_k]$ of degree $(k+1)(r-k)$ admits a unique decomposition as an integral linear combination of Schur polynomials s_λ of the same degree. The corresponding Schubert classes σ_λ vanish except for $\sigma_{\lambda_{\max}}$. If τ corresponds to an effective zero cycle on $\mathbb{G}(k,r)$, then the degree of this cycle equals the coefficient of $s_{\lambda_{\max}} = (x_0 \cdots x_k)^{r-k}$ in the decomposition of τ as a linear combination of Schur polynomials. Multiplying τ by the Vandermonde polynomial

$$V = V(x_0, \ldots, x_k) = \prod_{0 \leqslant i < j \leqslant k} (x_i - x_j),$$

this coefficient becomes the coefficient of the monomial $x_0^r x_1^{r-1} \cdots x_k^{r-k}$ in the product $\tau \cdot V$, see the proofs of [18, Theorem 4.3] and [37, Theorem 3.5.18].

Let $P(x_0, \ldots, x_k)$ be a polynomial, and let $x_0^{i_0} \cdots x_k^{i_k}$ be a monomial, which we identify with the lattice vector $\mathbf{i} = (i_0, \ldots, i_k) \in \mathbb{Z}^{k+1}$. We write $\psi_{\mathbf{i}}(P)$ for the coefficient of $x_0^{i_0} \cdots x_k^{i_k}$ in $P = \sum_{\mathbf{i}} \psi_{\mathbf{i}}(P) x_0^{i_0} \cdots x_k^{i_k}$. Summarizing the preceding discussion and taking into account Lemma 2.2 one arrives at the following conclusion.

Proposition 2.3 ([36, pp. 311–312]) *One has*

$$\deg(\Sigma(d,r,k)) = \psi_{(r,r-1,\ldots,r-k)}(V \cdot \tau_{(d,r,k)}),$$

that is, the degree $\deg(\Sigma(d,r,k))$ *equals the coefficient of* $x_0^r x_1^{r-1} \cdots x_k^{r-k}$ *in the product* $\tau \cdot V$, *where* $\tau = \tau_{(d,r,k)}(x_0, \ldots, x_k)$ *is as in* (3).

See also the table in [36, p. 312] of the values of $\deg(\Sigma(d,r,k))$ for several different values of (d,r,k), and an explicit formula in [36, Cor. on p. 312] which expresses $\deg(\Sigma(d,3,1))$ as a polynomial in d of degree 8.

2.3 *Bott's Residue Formula*

Bott's residue formula [11, Thms. 1, 2] says, in particular, that one can compute the degree of a zero–dimensional cycle class on a smooth projective variety X in terms of local contributions given by the fixed point loci of a torus action on X. Here we follow the treatment in [28] based on [12, 21, 39] and adapted to our setting.

We consider the diagonal action of $T = (\mathbb{C}^*)^{r+1}$ on $\mathbb{P}^r$ given in coordinates by

$$(t_0, \ldots, t_r) \cdot (x_0 : \ldots : x_r) = (t_0 x_0 : \ldots : t_r x_r).$$

This induces an action of T on $\mathbb{G}(k, r)$, with $\binom{r+1}{k+1}$ isolated fixed points corresponding to the coordinate k–subspaces in $\mathbb{P}^r$, which are indexed by the subsets I of order $k+1$ of the set $\{0, \ldots, r\}$. We let $\mathcal{I}_{k+1}$ denote the set of all these subsets, and $\Pi_I \in \mathbb{G}(k, r)$ denote the subspace which corresponds to $I \in \mathcal{I}_{k+1}$. Bott's residue formula, applied in our setting, has the form

$$\deg(\Sigma(d, r, k)) = \int_{\mathbb{G}(k,r)} c_{(r-k)(k+1)}(\mathrm{Sym}^d(\mathcal{S}^*)) = \sum_{I \in \mathcal{I}_{k+1}} \frac{c_I}{e_I},$$

where c_I results from the local contribution of $c_{(r-k)(k+1)}(\mathrm{Sym}^d(\mathcal{S}^*))$ at Π_I, and e_I is determined by the torus action on the tangent space to $\mathbb{G}(k, r)$ at Π_I.

As for the computation of e_I, this goes exactly as in [28, p. 116], namely

$$e_I = (-1)^{(k+1)(r-k)} \prod_{i \in I} \prod_{j \notin I} (t_i - t_j).$$

Also the computation of c_I is similar to the one made in [28, p. 116]. Recalling (2), for a given $I \in \mathcal{I}_{k+1}$, consider the polynomial

$$\prod_{v_0 + \ldots + v_k = d} (1 + \sum_{i \in I} v_i t_i)$$

and extract from this its homogeneous component $\tau^I_{(d,r,k)}$ of degree $(k+1)(r-k)$. Then

$$c_I = \tau^I_{(d,r,k)}(-t_i)_{i \in I} = (-1)^{(r-k)(k+1)} \tau^I_{(d,r,k)}(t_i)_{i \in I}.$$

In conclusion we have

$$\deg(\Sigma(d, r, k)) = \sum_{I \in \mathcal{I}_{k+1}} \frac{\tau^I_{(d,r,k)}(t_i)_{i \in I}}{\prod_{i \in I} \prod_{j \notin I} (t_i - t_j)}. \tag{4}$$

As in [28, p. 111], we notice that the right hand side of this formula is, a priori, a rational function in the variables $t_0, \ldots, t_k$. As a matter of fact, it is a constant and a positive integer.

3 Fano Schemes of Complete Intersections

In this section we extend the considerations of Sect. 2 to complete intersections in projective space. We consider the case in which a general complete intersection X of type $\mathbf{d} := (d_1, \ldots, d_m)$ in $\mathbb{P}^r$, where $\prod_{i=1}^m d_i > 2$, does not contain any linear subspace of dimension k. Like in the case of hypersurfaces, the latter happens if and only if either $2k > r - m = \dim(X)$, or

$$\gamma(\mathbf{d}, r, k) := \sum_{j=1}^{m} \binom{d_j + k}{k} - (k+1)(r-k) > 0,$$

see [10, 18, 33, 40–42].

Let $\Sigma(\mathbf{d}, r)$ be the parameter space for complete intersections of type $\mathbf{d}$ in $\mathbb{P}^r$. This is a tower of projective bundles over a projective space, hence a smooth variety. Consider the subvariety $\Sigma(\mathbf{d}, r, k)$ of $\Sigma(\mathbf{d}, r)$ parameterizing complete intersections which contain a linear subspace of dimension k. One has ([4, Theorem 1.1]):

Proposition 3.1 *Assume $\gamma(\mathbf{d}, r, k) > 0$. Then $\Sigma(\mathbf{d}, r, k)$ is a nonempty, irreducible and rational subvariety of codimension $\gamma(\mathbf{d}, r, k)$ in $\Sigma(\mathbf{d}, r)$. The general point of $\Sigma(\mathbf{d}, r, k)$ corresponds to a complete intersection which contains a unique linear subspace of dimension k and has singular locus of dimension* $\max\{-1, 2k + m - 1 - r\}$ *along its unique k-dimensional linear subspace (in particular, it is smooth provided $r \geq 2k + m$).*

Next we would like to make sense of, and to compute, the *degree* of $\Sigma(\mathbf{d}, r, k)$ inside $\Sigma(\mathbf{d}, r)$ when $\gamma(\mathbf{d}, r, k) > 0$. To do this we consider the general complete intersection X of type $(d_1, \ldots, d_{m-1})$, and the complete linear system

$$\Sigma(d_m, X) = |\mathcal{O}_X(d_m)|.$$

We assume that the Fano scheme $F_k(X)$ of linear subspaces of dimension k contained in X is non–empty. This implies that

$$\dim(F_k(X)) = (k+1)(r-k) - \sum_{j=1}^{m-1} \binom{d_j + k}{k} = \binom{d_m + k}{k} - \gamma(\mathbf{d}, r, k) \geqslant 0 \tag{5}$$

(see [10, 33, 42]). Moreover, assume

$$\dim(\Sigma(d_m, X)) > \gamma(\mathbf{d}, r, k) \geqslant 0. \tag{6}$$

Notice that (5) and (6) do hold if $\gamma(\mathbf{d}, r, k)$ is sufficiently small, e.g., if $\gamma(\mathbf{d}, r, k) = 1$.

Let now $\Sigma(d_m, X, k)$ be the set of points in $\Sigma(d_m, X)$ corresponding to complete intersections of type $\mathbf{d} = (d_1, \ldots, d_m)$ contained in X and containing a subspace of dimension k. As an immediate consequence of Proposition 3.1, we have

Corollary 3.2 *Assume $\gamma(\mathbf{d}, r, k) > 0$ and (5) holds. Let X be a general complete intersection of type $(d_1, \ldots, d_{m-1})$ verifying (6). Then $\Sigma(d_m, X, k)$ is irreducible of codimension $\gamma(\mathbf{d}, r, k)$ in $\Sigma(d_m, X)$. The general point of $\Sigma(d_m, X, k)$ corresponds to a complete intersection of type $\mathbf{d} = (d_1, \ldots, d_m)$ which contains a unique subspace of dimension k.*

Next we would like to compute the degree of $\Sigma(d_m, X, k)$ inside the projective space $\Sigma(d_m, X)$.

Consider the vector bundle $\mathrm{Sym}^{d_m}(\mathcal{S}^*)$ on $\mathbb{G}(k, r)$ and set

$$\rho := \binom{d_m + k}{k} - \gamma(\mathbf{d}, r, k) = \dim(F_k(X)).$$

Similarly as in Lemma 2.2, one sees that

$$\deg(\Sigma(d_m, X, k)) = \int_{\mathbb{G}(k,r)} c_\rho(\mathrm{Sym}^{d_m}(\mathcal{S}^*)) \cdot [F_k(X)], \tag{7}$$

where $[F_k(X)]$ stands for the dual class of $F_k(X)$ in the Chow ring $A^*(\mathbb{G}(k, r))$.

3.1 Schubert Calculus

The Chern class $c_\rho(\mathrm{Sym}^{d_m}(\mathcal{S}^*))$ is the homogeneous component θ of degree ρ of the polynomial

$$\prod_{v_0+\ldots+v_k=d_m} (1 + v_0 x_0 + \ldots + v_k x_k).$$

As usual, θ can be written as a polynomial in the elementary symmetric functions of the Chern roots $x_0, \ldots, x_k$, which can be identified with the $\sigma_{(1^i)}$s. Eventually, one has a formula of the form

$$c_\rho(\mathrm{Sym}^{d_m}(\mathcal{S}^*)) = \sum_{j_1+2j_2+\ldots+(k+1)j_{k+1}=\rho} \phi_{d_m,r}(j_1, j_2, \ldots, j_{k+1}) \sigma_{(1)}^{j_1} \sigma_{(1^2)}^{j_2} \cdots \sigma_{(1^{k+1})}^{j_{k+1}}.$$

In conclusion one has

$$\deg(\Sigma(d_m, X, k)) = \\ = \int_{\mathbb{G}(k,r)} [F_k(X)] \cdot \sum_{j_1+2j_2+\ldots+(k+1)j_{k+1}=\rho} \phi_{d_m,r}(j_1, \ldots, j_{k+1}) \sigma_{(1)}^{j_1} \cdots \sigma_{(1^{k+1})}^{j_{k+1}}. \tag{8}$$

3.2 Debarre–Manivel's Trick

Formula (8) is rather unpractical, since both, the computation of the coefficients and of the intersection products appearing in it are rather complicated, in general. A better result can be gotten using again Debarre–Manivel's idea as in Sect. 2.2. Taking into account (7) one sees that $\deg(\Sigma(d_m, X, k))$ equals the coefficient of the monomial $x_0^r x_1^{r-1} \cdots x_k^{r-k}$ in the product of the following polynomials in $x_0, \ldots, x_k$:

(i) the product $Q_{k,\mathbf{d}} = \prod_{i=1}^{m-1} Q_{k,d_i}$ of the polynomials

$$Q_{k,d_i} = \prod_{v_0+\ldots+v_k=d_i} (v_0 x_0 + \cdots + v_k x_k);$$

(ii) the polynomial θ;
(iii) the Vandermonde polynomial $V(x_0, \ldots, x_k)$.

Notice ([24, 14.7], [37, 3.5.5]) that $Q_{k,\mathbf{d}}$ in (i) corresponds to the class $[F_k(X)]$ of degree $(k+1)(r-k) - \rho$ in the Chow ring $A^*(\mathbb{G}(k, r))$, whereas θ in (ii) corresponds to the class of $c_\rho(\mathrm{Sym}^{d_m}(\mathcal{S}^*))$ of degree ρ. In conclusion,

$$\deg(\Sigma(d_m, X, k)) = \psi_{(r,r-1,\ldots,r-k)}(Q \cdot \theta \cdot V).$$

The Bott residue formula does not seem to be applicable in this case.

4 Numerical Invariants of Fano Schemes

In this section we consider the complete intersections whose Fano schemes have positive expected dimension

$$\delta(\mathbf{d}, r, k) := -\gamma(\mathbf{d}, r, k) = (k+1)(r-k) - \sum_{j=1}^{m} \binom{d_j + k}{k} > 0 \qquad (9)$$

where $\mathbf{d} = (d_1, \ldots, d_m)$. We may and we will assume $d_i \geq 2$, $i = 1, \ldots, m$. If also $r \geq 2k + m + 1$ then, for a general complete intersection X of type $\mathbf{d}$ in $\mathbb{P}^r$, the Fano variety $F_k(X)$ of linear subspaces of dimension k contained in X is a smooth, irreducible variety of dimension $\delta(\mathbf{d}, r, k)$ (see [10, 17, 18, 33, 41, 42]). We will compute some numerical invariants of $F_k(X)$. If $\delta(\mathbf{d}, r, k) = 1$ then $F_k(X)$ is a smooth curve; its genus was computed in [29]. In the next section we treat the case where $F_k(X)$ is a surface, that is, $\delta(\mathbf{d}, r, k) = 2$; our aim is to compute the Chern numbers of this surface. Actually, we deduce formulas for $c_1(F_k(X))$ and $c_2(F_k(X))$ for the general case $\delta(\mathbf{d}, r, k) > 0$. To simplify the notation, we set in the sequel $F = F_k(X)$, $\mathbb{G} = \mathbb{G}(k, r)$, $\delta = \delta(\mathbf{d}, r, k)$, and we let $\mathfrak{h}$ be the hyperplane section class of $\mathbb{G}$ in the Plücker embedding.

Recall the following fact (cf. Proposition 2.3).

Proposition 4.1 ([18, Theorem 4.3]) *In the notation and assumptions as before, one has*

$$\deg(F) = \psi_{(r,r-1,\dots,r-k)}(Q_{k,\mathbf{d}} \cdot e^{\delta} \cdot V) \quad \textit{where} \quad e(\mathbf{x}) := x_0 + \cdots + x_k\,,$$

that is, the degree of the Fano scheme F under the Plücker embedding equals the coefficient of the monomial $x_0^r x_1^{r-1} \cdots x_k^{r-k}$ of the product of $Q_{k,\mathbf{d}} \cdot e^{\delta} \cdot V$ where V stands for the Vandermonde polynomial (see Sect. 3.2 for the notation).

Remark 4.2 An alternative expression for $\deg(F)$ based on the Bott residue formula can be found in [28, Theorem 1.1] and [29, Theorem 2]; cf. also [24, Ex. 14.7.13] and [37, Sect. 3.5].

The next lemma is known in the case of the Fano scheme of lines on a general hypersurface, that is, for $k = m = 1$, see, [1, 32, Ex. V.4.7].

Lemma 4.3 *In the notation and assumptions as before, one has*

$$c_1(T_F) = \left(r + 1 - \sum_{i=1}^{m} \binom{d_i + k}{k + 1}\right) \mathfrak{h}_{|F} \tag{10}$$

and

$$K_F \sim \mathcal{O}_F\left(\sum_{i=1}^{m} \binom{d_i + k}{k + 1} - (r + 1)\right) \tag{11}$$

where $\mathcal{O}_F(1)$ corresponds to the Plücker embedding. In particular, F is a smooth Fano variety provided $\sum_{i=1}^{m} \binom{d_i+k}{k+1} \le r$.

Proof From the exact sequence

$$0 \to T_F \to T_{\mathbb{G}|F} \to N_{F|\mathbb{G}} \to 0$$

one obtains

$$c(T_{\mathbb{G}|F}) = c(T_F) \cdot c(N_{F|\mathbb{G}})\,.$$

Expanding one gets

$$c_1(T_F) = c_1(T_{\mathbb{G}|F}) - c_1(N_{F|\mathbb{G}}) \tag{12}$$

and, for the further usage,

$$c_2(T_F) = c_2(T_{\mathbb{G}|F}) - c_2(N_{F|\mathbb{G}}) - c_1(T_{\mathbb{G}|F}) \cdot c_1(N_{F|\mathbb{G}}) + c_1(N_{F|\mathbb{G}})^2. \tag{13}$$

Notice ([22, Theorem 3.5]) that $T_{\mathbb{G}} = \mathcal{S}^* \otimes \mathcal{Q}$, where, as usual, $\mathcal{S} \to \mathbb{G}$ is the tautological vector bundle of rank $k+1$ and $\mathcal{Q} \to \mathbb{G}$ is the tautological quotient bundle. Furthermore ([29, Lemma 3]), F is the zero scheme of a section of the vector bundle $\oplus_{i=1}^m \mathrm{Sym}^{d_i}(\mathcal{S}^*)$ on $\mathbb{G}$. It follows that

$$N_{F|\mathbb{G}} \simeq \oplus_{i=1}^m \mathrm{Sym}^{d_i}(\mathcal{S}^*)_{|F}\,. \tag{14}$$

By [29, Lemma 2] one has

$$c_1(T_{\mathbb{G}}) = (r+1)\mathfrak{h}. \tag{15}$$

Taking into account (14), [29, Lemma 1] (see also Lemma 4.6 below), and the fact that $c_1(\mathcal{S}^*) = \mathfrak{h}$ (see [22, Sect. 4.1]), one gets

$$c_1(N_{F|\mathbb{G}}) = \sum_{i=1}^m c_1(\mathrm{Sym}^{d_i}(\mathcal{S}^*)_{|F}) = \Big(\sum_{i=1}^m \binom{d_i+k}{k+1}\Big)\mathfrak{h}_{|F}. \tag{16}$$

Plugging (15) and (16) in (12) we find (10) and then (11). □

Corollary 4.4 *One has*

$$K_F^\delta = \Big(\sum_{i=1}^m \binom{d_i+k}{k+1} - (r+1)\Big)^\delta \deg(F), \tag{17}$$

where $\deg(F)$ *is computed in Proposition 4.1.*

Next we proceed to compute $c_2(T_F)$. Recalling (13), we need to compute $c_2(N_{F|\mathbb{G}})$ and $c_2(T_{\mathbb{G}})$. This requires some preliminaries. First of all, we need the following auxiliary combinatorial formula.

Lemma 4.5 *For any integers* n, m, k *where* $n \geq m \geq 1$ *and* $k \geq 0$ *one has*

$$\sum_{i=1}^n \binom{i-1}{m-1}\binom{n-i+k}{k} = \binom{n+k}{m+k}.$$

Proof[1] The choice of $m+k$ integers $i_1, \ldots, i_{m+k}$ among $\{1, \ldots, n+k\}$, where $1 \leq i_1 < \ldots < i_m < \ldots < i_{m+k} \leq n+k$, can be done in two steps. At the first step one fixes the choice of $i_m = i$, where, clearly, $i \in \{1, \ldots, n\}$. It remains to choose $i_1, \ldots, i_{m-1}$ among $\{1, \ldots, i-1\}$ and $i_{m+1}, \ldots, i_{m+k}$ among $\{i+1, \ldots, n+k\}$. □

[1] The authors are grateful to Roland Basher for communicating this beautiful, elementary argument.

Lemma 4.6 *Let E be a vector bundle of rank $k+1$. Then*

$$c_2(\mathrm{Sym}^n(E)) = \alpha c_1(E)^2 + \beta c_2(E) \quad \textit{and} \quad c_1(\mathrm{Sym}^n(E)) = \gamma c_1(E) \tag{18}$$

where[2]

$$\begin{aligned} \alpha &= \frac{1}{2}\binom{n+k}{k+1}^2 - \frac{1}{2}\binom{n+k}{k+1} - \binom{n+k}{k+2}, \\ \beta &= \binom{n+k+1}{k+2}, \quad \textit{and} \quad \gamma = \binom{n+k}{k+1}. \end{aligned} \tag{19}$$

Proof We use the splitting principle. Write E as a formal direct sum of line bundles $E = L_0 \oplus \ldots \oplus L_k$, with $c_1(L_i) = x_i$, for $0 \leqslant i \leqslant k$. From the equality

$$c(E) = (1 + x_0) \cdots (1 + x_k)$$

one deduces

$$c_1(E) = x_0 + \cdots + x_k \quad \text{and} \quad c_2(E) = \sum_{0 \leqslant i < j \leqslant k} x_i x_j. \tag{20}$$

Since

$$\mathrm{Sym}^n(E) = \sum_{v_0 + \cdots + v_k = n} L_0^{v_0} \cdots L_k^{v_k}$$

one has

$$c(\mathrm{Sym}^n(E)) = \prod_{v_0 + \cdots + v_k = n} (1 + v_0 x_0 + \cdots + v_k x_k) = \prod_{|\mathbf{v}|=n} (1 + \langle \mathbf{v}, \mathbf{x} \rangle),$$

where $\mathbf{x} = (x_0, \ldots, x_k)$, $\mathbf{v} = (v_0, \cdots, v_k)$, and $|\mathbf{v}| = v_0 + \cdots + v_k$. Therefore,

$$c_1(\mathrm{Sym}^n(E)) = \sum_{|\mathbf{v}|=n} \langle \mathbf{v}, \mathbf{x} \rangle \tag{21}$$

and

$$c_2(\mathrm{Sym}^n(E)) = \frac{1}{2} \sum_{|\mathbf{v}|=|\mathbf{w}|=n, \mathbf{v} \neq \mathbf{w}} \langle \mathbf{v}, \mathbf{x} \rangle \langle \mathbf{w}, \mathbf{x} \rangle. \tag{22}$$

The right hand sides of (21) and (22) are symmetric homogeneous polynomials in $x_0, \ldots, x_k$ of degree 1 and 2, respectively. Using (20) one deduces

$$c_1(\mathrm{Sym}^n(E)) = \sum_{|\mathbf{v}|=n} \langle \mathbf{v}, \mathbf{x} \rangle = \gamma(x_0 + \cdots + x_k) = \gamma c_1(E)$$

---

[2] See also [29, Lemma 1] for γ.

and

$$c_2(\mathrm{Sym}^n(E)) = \frac{1}{2} \sum_{|\mathbf{v}|=|\mathbf{w}|=n, \mathbf{v}\neq \mathbf{w}} \langle \mathbf{v}, \mathbf{x}\rangle \langle \mathbf{w}, \mathbf{x}\rangle$$
$$= \alpha(x_0 + \cdots + x_k)^2 + \beta \sum_{0\leqslant i<j\leqslant k} x_i x_j = \alpha c_1(E)^2 + \beta c_2(E),$$

cf. (18). In order to compute α, β and γ, we let in these relations $x_0 = 1, x_1 = \ldots = x_k = 0$, so that the coefficient of β vanishes and the coefficients of α and γ become 1. Similarly, for $x_0 = x_1 = 1, x_2 = \ldots = x_k = 0$ the coefficient of β in the decomposition of $c_2(\mathrm{Sym}^n(E))$ is 1 and the coefficient of α is 4. So, one gets

$$\alpha = \frac{1}{2} \sum_{|\mathbf{v}|=|\mathbf{w}|=n, \mathbf{v}\neq \mathbf{w}} v_0 w_0, \quad \beta + 4\alpha = \frac{1}{2} \sum_{|\mathbf{v}|=|\mathbf{w}|=n, \mathbf{v}\neq \mathbf{w}} (v_0 + v_1)(w_0 + w_1),$$
$$\gamma = \sum_{|\mathbf{v}|=n} v_0 \,. \tag{23}$$

For $k = 1$, (23) yields

$$\alpha = \frac{1}{2}\Big(\sum_{i,j=1}^{n} ij - \sum_{i=1}^{n} i^2\Big) = \frac{1}{2}\Big(\Big(\sum_{i=1}^{n} i\Big)^2 - \sum_{i=1}^{n} i^2\Big)$$
$$= \frac{1}{2}\left(\frac{n^2(n+1)^2}{4} - \frac{n(n+1)(2n+1)}{6}\right) = \frac{(3n+2)}{4}\binom{n+1}{3},$$

and

$$\beta + 4\alpha = \frac{1}{2} \sum_{v_0+v_1=w_0+w_1=n} n^2 - \frac{1}{2} \sum_{v_0+v_1=n} n^2$$
$$= \frac{1}{2}n^2(n+1)^2 - \frac{1}{2}n^2(n+1) = \frac{1}{2}n^3(n+1)\,.$$

Plugging in the value of α gives

$$\beta = \frac{1}{2}n^3(n+1) - 4\alpha = \frac{1}{2}n^3(n+1) - (3n+2)\binom{n+1}{3} = \binom{n+2}{3}.$$

Similarly, if $k = 2$ one has

$$\alpha = \frac{1}{2}\left(\sum_{i,j=1}^{n} ij(n-i+1)(n-j+1) - \sum_{i=1}^{n} i^2(n-i+1)\right) = \frac{5}{3}(n+1)\binom{n+3}{5}$$

and

$$\beta = \frac{1}{2}\left(\sum_{i,j=1}^{n} i(i+1)j(j+1) - \sum_{i=1}^{n} i^2(i+1)\right) - 4\alpha = \binom{n+3}{4}.$$

In the general case, applying Lemma 4.5 with a suitable choice of parameters we find

$$\gamma = \sum_{i=1}^{n} i\binom{n-i+k-1}{k-1} = \binom{n+k}{k+1}$$

and

$$\begin{aligned}
\alpha &= \frac{1}{2}\sum_{i,j=1}^{n} ij\binom{n-i+k-1}{k-1}\binom{n-j+k-1}{k-1} - \frac{1}{2}\sum_{i=1}^{n} i^2\binom{n-i+k-1}{k-1}\\
&= \frac{1}{2}\Big(\sum_{i=1}^{n} i\binom{n-i+k-1}{k-1}\Big)^2 - \sum_{i=1}^{n}\binom{i+1}{2}\binom{n-i+k-1}{k-1} +\\
&+ \frac{1}{2}\sum_{i=1}^{n} i\binom{n-i+k-1}{k-1}\\
&= \frac{1}{2}\binom{n+k}{k+1}^2 - \binom{n+k+1}{k+2} + \frac{1}{2}\binom{n+k}{k+1} = \frac{1}{2}\binom{n+k}{k+1}^2 - \frac{1}{2}\binom{n+k}{k+1} -\\
&- \binom{n+k}{k+2},
\end{aligned}$$

where at the last step one uses the standard identity

$$\binom{N+1}{k+1} = \binom{N}{k+1} + \binom{N}{k}. \tag{24}$$

Applying Lemma 4.5 and the identity

$$i^2(i+1) = 2\binom{i+1}{2} + 6\binom{i+1}{3},$$

for $k \geq 3$ we find:

$$\begin{aligned}\beta+4\alpha &= \frac{1}{2}\sum_{i,j=1}^{n} i(i+1)j(j+1)\binom{n-i+k-2}{k-2}\binom{n-j+k-2}{k-2}\\ &-\frac{1}{2}\sum_{i=1}^{n} i^2(i+1)\binom{n-i+k-2}{k-2} = 2\Big(\sum_{i=1}^{n}\binom{i+1}{2}\binom{n-i+k-2}{k-2}\Big)^2\\ &-\sum_{i=1}^{n}\binom{i+1}{2}\binom{n-i+k-2}{k-2} - 3\sum_{i=1}^{n}\binom{i+1}{3}\binom{n-i+k-2}{k-2}\\ &= 2\binom{n+k}{k+1}^2-\binom{n+k}{k+1}-3\binom{n+k}{k+2}.\end{aligned}$$

Using the formula for α and (24) we deduce

$$\beta=\binom{n+k}{k+1}+\binom{n+k}{k+2}=\binom{n+k+1}{k+2}.$$

□

Remark 4.7 The proof shows that for $k = 1, 2$, (19) can be simplified as follows:

$$(\alpha,\beta)=\begin{cases}\left(\frac{3n+2}{4}\binom{n+1}{3},\ \binom{n+2}{3}\right), & k=1\,,\\ \left(\frac{5(n+1)}{3}\binom{n+3}{5},\ \binom{n+3}{4}\right), & k=2\,.\end{cases} \tag{25}$$

One can readily check that the expressions for α in these formulas agree with the one in (19).

Lemma 4.8 *One has*

$$\begin{aligned} &c_1(\mathcal{Q}) = c_1(\mathcal{S}^*) = \mathfrak{h} \quad \textit{and}\\ &c_2(\mathcal{Q}) = c_1(\mathcal{S}^*)^2 - c_2(\mathcal{S}^*) = \mathfrak{h}^2 - c_2(\mathcal{S}^*).\end{aligned}$$

Proof One has $c(\mathcal{Q})\cdot c(\mathcal{S}) = 1$. By expanding and taking into account that $c_i(\mathcal{S}) = (-1)^i c_i(\mathcal{S}^*)$ for all positive integers i and $c_1(\mathcal{S}^*) = \mathfrak{h}$ ([22, Sect. 4.1]), the assertion follows. □

Lemma 4.9 *One has*

$$c_2(T_{\mathbb{G}}) = \left(\binom{r+1}{2}+k\right)\mathfrak{h}^2 + (r-2k-1)\,c_2(\mathcal{S}^*)\,. \tag{26}$$

Proof We use again the splitting principle. Write

$$\mathcal{S}^* = L_0\oplus\cdots\oplus L_k,\quad \mathcal{Q} = M_1\oplus\cdots\oplus M_{r-k}$$

with $c_1(L_i) = x_i$, $c_1(M_j) = y_j$, for $0 \leqslant i \leqslant k$ and $1 \leqslant j \leqslant r-k$. Since $T_{\mathbb{G}} = \mathcal{Q} \otimes \mathcal{S}^*$, see [22, Theorem 3.5], one obtains

$$c(T_{\mathbb{G}}) = c(\mathcal{Q} \otimes \mathcal{S}^*) = \prod_{i=0}^{k} \prod_{j=1}^{r-k} (1 + x_i + y_j),$$

whence

$$c_2(T_{\mathbb{G}}) = \frac{1}{2} \sum_{\substack{\lambda,\mu=0,\dots,k \\ \sigma,\rho=1,\dots,r-k \\ (\lambda,\sigma)\neq(\mu,\rho)}} (x_\lambda + y_\sigma)(x_\mu + y_\rho)\,.$$

By expanding, we see that in $c_2(T_{\mathbb{G}})$ appear the following summands:

- $\xi = \sum_{i=0}^{k} x_i^2$ and $\eta = \sum_{i=1}^{r-k} y_j^2$, the former appearing $\binom{r-k}{2}$ times, the latter $\binom{k+1}{2}$ times;
- $c_2(\mathcal{S}^*) = \sum_{0 \leqslant i < j \leqslant k} x_i x_j$, $c_2(\mathcal{Q}) = \sum_{1 \leqslant i < j \leqslant r-k} y_i y_j$, the former appearing $(r-k)^2$ times, the latter $(k+1)^2$ times;
- $c_1(\mathcal{Q})c_1(\mathcal{S}^*) = \sum_{i=0}^{k} \sum_{j=1}^{r-k} x_i y_j$ appearing $(k+1)(r-k) - 1$ times.

Using Lemma 4.8 one obtains

$$\xi = \sum_{i=0}^{k} x_i^2 = (x_0 + \ldots + x_k)^2 - 2 \sum_{0 \leqslant i < j \leqslant k} x_i x_j = c_1(\mathcal{S}^*)^2 - 2c_2(\mathcal{S}^*) = \mathfrak{h}^2 - 2c_2(\mathcal{S}^*),$$

and similarly

$$\eta = c_1(\mathcal{Q})^2 - 2c_2(\mathcal{Q}) = \mathfrak{h}^2 - 2(\mathfrak{h}^2 - c_2(\mathcal{S}^*)) = 2c_2(\mathcal{S}^*) - \mathfrak{h}^2.$$

Collecting these formulas and taking into account Lemma 4.8 one arrives at:

$$\begin{aligned}
c_2(T_{\mathbb{G}}) &= \binom{r-k}{2}\xi + \binom{k+1}{2}\eta + (r-k)^2 c_2(\mathcal{S}^*) + (k+1)^2 c_2(\mathcal{Q}) \\
&\quad + ((k+1)(r-k) - 1)\, c_1(\mathcal{Q})c_1(\mathcal{S}^*) \\
&= \left[\binom{r-k}{2} - \binom{k+1}{2}\right] \left(\mathfrak{h}^2 - 2c_2(\mathcal{S}^*)\right) + (r-k)^2 c_2(\mathcal{S}^*) \\
&\quad + (k+1)^2 \left(\mathfrak{h}^2 - c_2(\mathcal{S}^*)\right) + ((k+1)(r-k) - 1)\mathfrak{h}^2 \\
&= \left[\binom{r-k}{2} - \binom{k+1}{2} + (k+1)^2 + (k+1)(r-k) - 1\right] \mathfrak{h}^2 \\
&\quad + \left[2\binom{k+1}{2} - 2\binom{r-k}{2} + (r-k)^2 - (k+1)^2\right] c_2(\mathcal{S}^*)
\end{aligned}$$

$$= \left(\binom{r+1}{2} + k \right) \mathfrak{h}^2 + (r - 2k - 1) c_2(\mathcal{S}^*) . \qquad \square$$

Now we can deduce the following formulas.

Lemma 4.10 *Let α_i, β_i, and γ_i be obtained from α, β, and γ in* (19) *by replacing n by d_i, $i = 1, \ldots, m$. Then one has*

$$c_2(F) = c_2(T_F) = \left(A\mathfrak{h}^2 + Bc_2(\mathcal{S}^*) \right) \cdot [F] \tag{27}$$

where $[F]$ is the class of F in the Chow ring $A^*(\mathbb{G})$, and[3]

$$\begin{aligned} A = \binom{r+1}{2} + k - \sum_{i=0}^{m} \alpha_i - \sum_{1 \le i < j \le m} \gamma_i \gamma_j \\ - (r+1) \cdot \sum_{i=1}^{m} \binom{d_i + k}{k+1} + \Big(\sum_{i=1}^{m} \binom{d_i + k}{k+1} \Big)^2 , \end{aligned} \tag{28}$$

and

$$B = r - 2k - 1 - \sum_{i=1}^{m} \beta_i . \tag{29}$$

Proof Using (15) and (16) we deduce

$$c_1(T_{\mathbb{G}})|_F \cdot c_1(N_{F|\mathbb{G}}) = \left((r+1) \sum_{i=1}^{m} \binom{d_i + k}{k+1} \right) \mathfrak{h}^2 \cdot [F] \tag{30}$$

and

$$c_1(N_{F|\mathbb{G}})^2 = \Big(\sum_{i=1}^{m} \binom{d_i + k}{k+1} \Big)^2 \mathfrak{h}^2 \cdot [F] . \tag{31}$$

Furthermore, the Whitney formula and Lemma 4.6 yield

$$\begin{aligned} c_2(N_{F|\mathbb{G}}) &= \sum_{i=1}^{m} c_2\big(\mathrm{Sym}^{d_i}(\mathcal{S}^*)|_F\big) + \sum_{1 \le i < j \le m} c_1\big(\mathrm{Sym}^{d_i}(\mathcal{S}^*)|_F\big) \cdot c_1\big(\mathrm{Sym}^{d_j}(\mathcal{S}^*)|_F\big) \\ &= \Big(\sum_{i=1}^{m} \big(\alpha_i c_1(\mathcal{S}^*)^2 + \beta_i c_2(\mathcal{S}^*)\big) + \sum_{1 \le i < j \le m} \gamma_i \gamma_j c_1(\mathcal{S}^*)^2 \Big) \cdot [F] \\ &= \Big(\sum_{i=1}^{m} \alpha_i + \sum_{1 \le i < j \le m} \gamma_i \gamma_j \Big) \mathfrak{h}^2 \cdot [F] + \Big(\sum_{i=1}^{m} \beta_i \Big) c_2(\mathcal{S}^*) \cdot [F] . \end{aligned}$$

[3]The sum $\sum_{1 \le i < j \le m} \gamma_i \gamma_j$ disappears if $m = 1$.

Plugging this in (13) together with the values of the Chern classes from (26), (30), and (31) gives (27), (28), and (29). □

Remark 4.11 The cycle F on $\mathbb{G}$ is the reduced zero scheme of a section of the vector bundle $\mathcal{E}_F := \oplus_{i=1}^m \mathrm{Sym}^{d_i}(\mathcal{S}^*)$ on $\mathbb{G}$ of rank

$$\mathrm{rk}(\mathcal{E}_F) = \binom{\mathbf{d}+k}{k} := \sum_{i=1}^m \binom{d_i+k}{k}.$$

The Poincaré dual $[F] \in A^{\binom{\mathbf{d}+k}{k}}(\mathbb{G})$ of the class of F in $A_\delta(\mathbb{G})$ is the top Chern class $c_{\binom{\mathbf{d}+k}{k}}(\mathcal{E}_F)$. The latter can be expressed in terms of the Chern roots as

$$[F] = Q_{k,\mathbf{d}}(x_0,\ldots,x_k) = \prod_{i=1}^m Q_{k,d_i}(x_0,\ldots,x_k) \in A^{\binom{\mathbf{d}+k}{k}}(\mathbb{G}),$$

see Sect. 3.2.

5 The Case of Fano Surfaces

Let us turn to the case where the Fano scheme $F = F_k(X)$ of a general complete intersection $X \subset \mathbb{P}^r$ of type $\mathbf{d}$ is an irreducible surface, that is, $\delta = 2$ and $r \geq 2k+m$. Let us make the following observations.

In the surface case, $\int_F c_2(F) = e(F)$ is the Euler–Poincaré characteristic of F. By Lemma 4.10 one can compute $e(F)$ once one knows $\int_{\mathbb{G}} \mathfrak{h}^2 \cdot [F]$ and $\int_{\mathbb{G}} c_2(\mathcal{S}^*) \cdot [F]$. As for the former, one can use the Debarre–Manivel formula for the degree $\deg(F) = \int_{\mathbb{G}} \mathfrak{h}^2 \cdot [F]$, see Proposition 4.1; cf. also Remark 4.2.

As for the latter, recall that $c_2(\mathcal{S}^*) = \sigma_{(1^2)}$ is the class of the Schubert cycle of the $\mathbb{P}^k$s in $\mathbb{P}^r$ intersecting a fixed $\mathbb{P}^{r-k-1}$ in a line. Computing $\int_{\mathbb{G}} c_2(\mathcal{S}^*) \cdot [F]$ geometrically is difficult. However, one can compute it using Debarre–Manivel's trick. Indeed, arguing as in the proof of [18, Theorem 4.3], cf. Sect. 2.2, one can see that $\int_{\mathbb{G}} c_2(\mathcal{S}^*) \cdot [F]$ equals the coefficient of $x_0^r x_1^{r-1} \cdots x_k^{r-k}$ in the product of the three factors:

- $Q_{k,\mathbf{d}} = \prod_{i=1}^m Q_{k,d_i}$, see Sect. 3.2;
- $c_2(\mathcal{S}^*) = \sum_{0\leqslant i<j\leqslant k} x_i x_j$;
- the Vandermonde polynomial $V(x_0,\ldots,x_k) = \prod_{i<j}(x_i - x_j)$.

Notice that for $\delta = 2$ one has

$$\deg\left(Q_{k,\mathbf{d}} \cdot \sum_{0\leqslant i<j\leqslant k} x_i x_j\right) = \binom{\mathbf{d}+k}{k} + 2 = (k+1)(r-k) = \dim(\mathbb{G}).$$

Putting together (27), (28) and (29) one finds a formula for the Euler characteristic $e(F) = \int_F c_2(F)$. Then, using (17) and the Noether formula

$$\chi(\mathcal{O}_F) = \frac{1}{12}\left(K_F^2 + e(F)\right) = \frac{1}{12}\left(c_1(F)^2 + c_2(F)\right)$$

one can compute the holomorphic Euler characteristic $\chi(\mathcal{O}_F)$, the arithmetic genus $p_a(F) = \chi(\mathcal{O}_F) - 1$, and the signature $\tau(F) = 4\chi(\mathcal{O}_F) - e(F)$.

Example 5.1 Let us apply these recipes to the well known case of the Fano surface $F = F_1(X)$ of lines on the general cubic threefold in $\mathbb{P}^4$. Letting $r = 4$, $k = m = 1, d = 3$ one gets $\delta = 2$ and

$$Q_{1,(3)}(x_0, x_1) = 9x_0x_1(2x_0 + x_1)(x_0 + 2x_1), \quad V(x_0, x_1) = x_0 - x_1\,.$$

Therefore,

$$\deg(F) = \int_{\mathbb{G}(1,4)} \mathfrak{h}^2 \cdot [F] = \int_{\mathbb{G}(1,4)} c_1(\mathcal{S}^*)^2 \cdot [F] = \psi_{4,3}(Q_{1,(3)} \cdot (x_0 + x_1)^2 \cdot V) = 45$$

and

$$\int_{\mathbb{G}(1,4)} c_2(\mathcal{S}^*) \cdot [F] = \psi_{4,3}(Q_{1,(3)} \cdot x_0x_1 \cdot V) = 27\,.$$

Applying (25) and (27) one obtains

$$\alpha = 11, \quad \beta = 10, \quad A = 6, \quad \text{and} \quad B = -9\,.$$

Using the Noether formula and (17) one arrives at the classical values (see [1, 34])

$$e(F) = c_2(F) = 6\deg(F) - 9\int_{\mathbb{G}(1,4)} c_2(\mathcal{S}^*) \cdot [F] = 6 \cdot 45 - 9 \cdot 27 = 27$$

and

$$c_1(F)^2 = K_F^2 = \left(\binom{4}{2} - 5\right)^2 \deg(F) = 45, \quad \chi(\mathcal{O}_F) = \frac{1}{12}(45 + 27) = 6\,.$$

Example 5.2 More generally, one can consider the Fano surface $F = F_1(X)$ of lines on a general hypersurface X of degree $d = 2r - 5$ in $\mathbb{P}^r$, $r \geq 4$. Plugging in (27)–(29) the values of α and β from (25) one obtains

$$\begin{aligned} A &= \binom{2r-4}{2}^2 + \binom{r+1}{2} + 1 - \frac{6r-13}{4}\binom{2r-4}{3} - (r+1)\binom{2r-4}{2} \\ &= \frac{1}{3}(6r^4 - 56r^3 + 177r^2 - 211r + 78)\,, \\ B &= r - 3 - \binom{2r-3}{3}. \end{aligned}$$

Furthermore,

$$e(F) = c_2(F) = A\deg(F) + B\int_{\mathbb{G}(1,r)} c_2(\mathcal{S}^*)\cdot[F],\quad c_1^2(F) = \left(\binom{2r-4}{2} - (r+1)\right)^2,$$

and

$$\chi(\mathcal{O}_F) = \frac{1}{12}\left(c_1^2(F) + c_2(F)\right),$$

where

$$\deg(F) = \psi_{r,r-1}\left(Q_{1,(d)}\cdot(x_0+x_1)^2(x_0-x_1)\right)$$

and

$$\int_{\mathbb{G}(1,r)} c_2(\mathcal{S}^*)\cdot[F] = \psi_{r,r-1}\left(Q_{1,(d)}\cdot x_0x_1(x_0-x_1)\right)$$

with

$$Q_{1,(d)} = \prod_{v_0+v_1=d}(v_0x_0 + v_1x_1)\,.$$

Consider, for instance, the Fano scheme F of lines on a general quintic fourfold in $\mathbb{P}^5$. One has

$$r=5,\ d=5,\ k=m=1,\ \delta=2\,.$$

One gets

$$\alpha = 85,\quad \beta = 35,\quad A = 66,\quad B = -33\,,$$

and further (cf. [18, Table 1] and [50])

$$\deg(F) = \psi_{5,4}\left(Q_{1,(5)}\cdot(x_0+x_1)^2(x_0-x_1)\right) = 25\cdot 245 = 6125$$

and

$$c_2(\mathcal{S}^*)\cdot[F] = \psi_{5,4}\left(Q_{1,(5)}\cdot x_0x_1(x_0-x_1)\right) = 25\cdot 115 = 2875\,.$$

Hence

$$e(F) = c_2(F) = 25\cdot 33\cdot 375 = 309375,\quad c_1^2(F) = 25\cdot 81\cdot 245 = 496125\,.$$

Finally,

$$\chi(\mathcal{O}_F) = \frac{1}{12}\left(c_1^2(F) + c_2(F)\right) = 25 \cdot 15 \cdot 179 = 67125\,.$$

Example 5.3 Consider further the Fano surface $F = F_1(X)$ of lines on the intersection X of two general quadrics in $\mathbb{P}^5$. We have

$$r = 5,\ \ m = 2,\ \ \mathbf{d} = (2, 2),\ \ k = 1,\ \ \delta = 2\,,$$

$$Q_{1,(2,2)}(x_0, x_1) = 16x_0^2x_1^2(x_0 + x_1)^2, \quad \text{and} \quad V(x_0, x_1) = x_0 - x_1\,.$$

Hence

$$\deg(F) = \psi_{5,4}(Q_{1,(2,2)} \cdot (x_0 + x_1)^2 \cdot V) = 32$$

and

$$\int_{\mathbb{G}(1,5)} c_2(\mathcal{S}^*) \cdot [F] = \psi_{5,4}(Q_{1,(2,2)} \cdot x_0x_1 \cdot V) = 16\,.$$

Furthermore,

$$\alpha_1 = \alpha_2 = 2, \quad \beta_1 = \beta_2 = 4, \quad \gamma_1 = \gamma_2 = 3, \quad A = 3, \quad \text{and} \quad B = -6\,.$$

Therefore,

$$e(F) = 3\deg(F) - 6\int_{\mathbb{G}(1,5)} c_2(\mathcal{S}^*) \cdot [F] = 3 \cdot 32 - 6 \cdot 16 = 0\,,$$

$$c_1(F)^2 = \left(2\binom{3}{2} - 6\right)^2 \deg(F) = 0, \quad \text{and so,} \quad \chi(\mathcal{O}_F) = 0\,.$$

In fact, F is an abelian surface ([43]).

Example 5.4 Let now $F = F_2(X)$ be the Fano scheme of planes on a general cubic fivefold X in $\mathbb{P}^6$. Thus, one has

$$r = 6,\ \ m = 1,\ \ d = 3,\ \ k = 2, \quad \text{and} \quad \delta = 2\,.$$

Letting

$$Q_{2,(3)} = 27x_0x_1x_2(2x_0 + x_1)(2x_0 + x_2)(x_0 + 2x_1)(x_0 + 2x_2)(2x_1 + x_2)(x_1 + 2x_2)(x_0 + x_1 + x_2)$$

and

$$V(x_0, x_1, x_2) = (x_0 - x_1)(x_0 - x_2)(x_1 - x_2)\,,$$

the Wolfram Alpha gives (cf. [18, Table 2])

$$\deg(F) = \psi_{6,5,4}(Q_{2,(3)} \cdot (x_0 + x_1 + x_2)^2 \cdot V) = 27 \cdot 105 = 2835$$

and

$$\int_{\mathbb{G}(2,6)} c_2(\mathcal{S}^*) \cdot [F] = \psi_{6,5,4}(Q_{2,(3)} \cdot (x_0x_1 + x_0x_2 + x_1x_2) \cdot V) = 27 \cdot 63 = 1701\,.$$

Standard calculations yield

$$\alpha = 40, \quad \beta = 15, \quad \gamma = 10, \quad A = 13, \quad \text{and} \quad B = -14\,.$$

So, one obtains

$$e(F) = 13\deg(F) - 14c_2(\mathcal{S}^*) \cdot [F] = 13041\,,$$

$$K_F^2 = \left(\binom{5}{3} - 7\right)^2 \deg(F) = 9\deg(F) = 9 \cdot 27 \cdot 105 = 25515\,,$$

and

$$\chi(\mathcal{O}_F) = \frac{13041 + 25515}{12} = 3213\,.$$

6 Irregular Fano Schemes

In this section we study the cases in which the Fano scheme F of a general complete intersection is irregular, that is, $q(F) = h^1(\mathcal{O}_F) > 0$. As follows from the next proposition, for the Fano surfaces F this occurs only if F is one of the surfaces in Examples 5.1 (or, which is the same, in 5.2 for $r = 4$), 5.3, and 5.4. In all these cases one has $r = 2k + m + 1$.

Theorem 6.1 *Let X be a general complete intersection of type $\mathbf{d} = (d_1, \ldots, d_m)$ in $\mathbb{P}^r$. Suppose that the Fano scheme $F = F_k(X)$ of k-planes in X, $k \geq 1$, is irreducible of dimension $\delta \geq 2$. Then F is irregular if and only if one of the following holds:*

(i) F is the variety of lines on a general cubic threefold in $\mathbb{P}^4$ ($\dim(F) = 2$);
(ii) F is the variety of planes on a general cubic fivefold in $\mathbb{P}^6$ ($\dim(F) = 2$);
(iii) F is the variety of k-planes on the intersection of two general quadrics in $\mathbb{P}^{2k+3}$, $k \in \mathbb{N}$ ($\dim(F) = k + 1$).

Proof By our assumption, $\delta \geq 2$. By [18, Theorem 3.4] one has $q(F) = 0$ if $r \geqslant 2k + m + 2$. By [18, Theorem 2.1], F being nonempty implies $r \geqslant 2k + m$. Therefore, $q(F) > 0$ leaves just two possibilities:

$$r = 2k + m \quad \text{and} \quad r = 2k + m + 1\,.$$

We claim that the first possibility is not realized. Indeed, let $r = 2k + m$. We may assume that $d_i \geq 2$ for all $i = 1, \dots, m$. From (9) one deduces:

$$(k+1)(k+m) = (k+1)(r-k) = \delta + \sum_{i=1}^{m} \binom{d_i + k}{k} \geq 2 + m\binom{k+2}{2}. \quad (32)$$

This implies the inequality

$$4 \leq k(k+1)(2-m), \quad (33)$$

and so, $m = 1$, that is, X is a hypersurface in $\mathbb{P}^{2k+1}$. Letting $d = d_1$, (32) reads

$$\delta = (k+1)^2 - \binom{d+k}{k} \geq 2.$$

This inequality holds only when $d = 2$, that is, X is a smooth quadric of dimension $2k$. However, in the latter case $F = F_k(X)$ consists of two components ([20, Lemma 1.1]), contrary to our assumption. This proves our claim.

In the case $r = 2k + m + 1$, (32) and (33) must be replaced, respectively, by

$$(k+1)(k+m+1) = (k+1)(r-k) = \delta + \sum_{i=1}^{m} \binom{d_i + k}{k} \geq \delta + m\binom{k+2}{2}$$

and

$$4 \leq 2\delta \leq (k+1)[2(k+m+1) - m(k+2)] = (k+1)[(2-m)k+2]. \quad (34)$$

It follows from (34) that either $m = 1$ and $r = 2k + 2$, or $m = 2$ and $r = 2k + 3$.

In the hypersurface case (i.e., $m = 1$) one has

$$2 \leq \delta = (k+1)(k+2) - \binom{d+k}{k}.$$

This holds only if either $d = 2$, or $d \geq 3$ and $k \in \{1, 2\}$.

If $d = 2$, that is, X is a smooth quadric in $\mathbb{P}^{2k+2}$, then $\delta = \binom{k+2}{2}$, cf. [20, Lemma 1.1]. However, by [20, Lemma 1.2], in this case F is unirational, hence $q(F) = 0$, contrary to our assumption.

The possibility $d \geq 3$ realizes just in the following two cases:

(i) $(d, r, k) = (3, 4, 1)$, that is, F is the Fano surface of lines on a smooth cubic threefold in $\mathbb{P}^4$;
(ii) $(d, r, k) = (3, 6, 2)$, that is, F is the Fano surface of planes on a smooth cubic fivefold in $\mathbb{P}^6$.

If further $m = 2$ then $r = 2k + 3$ and

$$2 \le \delta = (k+1)(k+3) - \binom{d_1+k}{k} - \binom{d_2+k}{k}.$$

This inequality holds only for $\mathbf{d} = (2, 2)$, that is, only in the case where

(iii) $F = F_k(X)$ is the Fano scheme of k-planes in a smooth intersection of two quadrics in $\mathbb{P}^{2k+3}$.

Notice that F as in (iii) is smooth, irreducible, of dimension $\delta = k+1$, cf. [43, Chap. 4] and Remarks 6.2 below.

It remains to check that $q(F) > 0$ in (i)–(iii) indeed.

The Fano surface $F = F_1(X)$ of lines on a smooth cubic threefold $X \subset \mathbb{P}^4$ in (i) was studied by Fano ([23]) who found, in particular, that $q(F) = 5$. From Example 5.1, we deduce that $p_g(F) = 10$ (cf. also [5, Theorem 4], [8, 15, 26, 34, 43, Sect. 4.3], [44, 46, 47]). There is an isomorphism $\mathrm{Alb}(F) \simeq J(X)$ where $J(X)$ is the intermediate Jacobian (see [15]). The latter holds as well for $F = F_2(X)$ where $X \subset \mathbb{P}^6$ is a smooth cubic fivefold as in (ii), see [16]. Thus, $q(F) > 0$ in (i) and (ii).

By a theorem of Reid [43, Theorem 4.8] (see also [19, Theorem 2], [49]), the Fano scheme $F = F_k(X)$ of k-planes on a smooth intersection X of two quadrics in $\mathbb{P}^{2k+3}$ as in (iii) is isomorphic to the Jacobian $J(C)$ of a hyperelliptic curve C of genus $g(C) = k+1$ (of an elliptic curve if $k = 0$). Hence, one has $q(F) = \dim(F) = k+1 > 0$ for $k \ge 0$. Notice that there are isomorphisms $F \simeq J(C) \simeq J(X)$ where $J(X)$ is the intermediate Jacobian, see [20]. □

Remarks 6.2 1. The complete intersections in (i)–(iii) are Fano varieties. The ones in (i) are the Fano threefolds of index 2 with a very ample generator of the Picard group. The complete intersections Fano threefolds of index 1 with a very ample anticanonical divisor are the varieties $V_3^{2g-2} \subset \mathbb{P}^{g+1}$ of genera $g = 3, 4, 5$, that is, the smooth quartics V_3^4 in $\mathbb{P}^4$ ($g = 3$), the smooth intersections V_3^6 of a quadric and a cubic in $\mathbb{P}^5$ ($g = 4$), and the smooth intersections V_3^8 of three quadrics in $\mathbb{P}^6$ ($g = 5$), see [30, Chap. IV, Proposition 1.4]. The Fano scheme of lines $F = F_1$ on a general such Fano threefold V_3^{2g-2} is a smooth curve of a positive genus $g(F) > 0$. In fact, $g(F) = 801$ for $g = 3$, $g(F) = 271$ for $g = 4$, and $g(F) = 129$ for $g = 5$, see [38] and [29, Examples 1–3]. For these $X = V_3^{2g-2}$, the Abel-Jacobi map $J(F) \to J(X)$ to the intermediate Jacobian is a surjection, and $J(X)$ coincides with the Prym variety of X, see [30] and [48, Lect. 4, Sect. 1, Ex. 1 and Sect. 3].

2. Notice that the complete intersections whose Fano schemes of lines are curves of positive genera are not exhausted by the above Fano threefolds V_3^{2g-2}. The same holds, for instance, for a general hypersurface of degree $2r-4$ in $\mathbb{P}^r$, $r \ge 4$, and for general complete intersections of types $\mathbf{d} = (r-3, r-2)$ and $\mathbf{d} = (r-4, r-4)$ in $\mathbb{P}^r$ for $r \ge 5$ and $r \ge 6$, respectively, see [29, Examples 1–3], etc. One can find in [29] a formula for the genus of the curve F.

3. Let X be a smooth intersection of two quadrics in $\mathbb{P}^{2k+2}$. Then the Fano scheme $F_k(X)$ is reduced and zero-dimensional, and X contains exactly 2^{2k+2} k-planes, see [43, Chap. 3], whereas $F_{k-1}(X)$ is a rational Fano variety of dimension $2k$ and index 1, whose Picard number is $\rho = 2k+4$, see [2, 13], and the references therein.

As for the Picard numbers of the Fano schemes of complete intersections, one has the following result (cf. also [18]).

Theorem 6.1 ([31, Theorem 03]) *Let X be a very general complete intersection in $\mathbb{P}^r$. Assume $\delta(\mathbf{d}, r, k) \geq 2$. Then $\rho(F_k(X)) = 1$ except in the following cases:*

- *X is a quadric in $\mathbb{P}^{2k+1}$, $k \geq 1$. Then $F_k(X)$ consists of two isomorphic smooth disjoint components, and the Picard number of each component is 1;*
- *X is a quadric in $\mathbb{P}^{2k+3}$, $k \geq 1$. Then $\rho(F_k(X)) = 2$;*
- *X is a complete intersection of two quadrics in P^{2k+4}, $k \geq 1$. Then $\rho(F_k(X)) = 2k + 6$.*

Recall that a complete intersection is 'very general' if it varies in the complement to a certain countable union of proper subvarieties in the parameter space. The very generality assumption of the theorem cannot be replaced by generality; one can find corresponding examples in [31].

7 Hypersurfaces Containing Conics

Recall (see [25, Theorem 1.1]) that for the general hypersurface X of degree d in $\mathbb{P}^r$, the variety $R_2(X)$ of smooth conics in X is smooth of the expected dimension $\mu(d, r) = 3r - 2d - 2$ provided $\mu(d, r) \geq 0$, and is empty otherwise. It is connected provided $\mu(d, r) \geq 1$ and X is not a smooth cubic surface in $\mathbb{P}^3$. In this section we concentrate on the case where $\mu(d, r) < 0$.

7.1 *The Codimension Count and Uniqueness*

Set

$$\epsilon(d, r) = 2d + 2 - 3r\,.$$

Consider the subvariety $\Sigma_c(d, r)$ of $\Sigma(d, r)$ whose points correspond to hypersurfaces containing plane conics. By abuse of language, in the sequel we say "conic" meaning "plane conic"; thus, a pair of skew lines does not fit in our terminology. A conic is smooth if it is reduced and irreducible.

Theorem 7.1 *Assume $d \geq 2$, $r \geq 3$, and $\epsilon(d, r) \geq 0$. Then the following hold.*

(a) $\Sigma_c(d, r)$ is irreducible of codimension $\epsilon(d, r)$ in $\Sigma(d, r)$.
(b) If $\epsilon(d, r) > 0$ and $(d, r) \neq (4, 3)$ then the hypersurface corresponding to the general point of $\Sigma_c(d, r)$ contains a unique conic, and this conic is smooth. In the case $(d, r) = (4, 3)$ it contains exactly two distinct conics, and these conics are smooth and coplanar.

Proof (a) Let $\mathcal{H}_{c,r}$ be the component of the Hilbert scheme whose points parameterize conics in $\mathbb{P}^r$. There is an obvious morphism

$$\pi : \mathcal{H}_{c,r} \to \mathbb{G}(2, r)$$

sending a conic Γ to the plane $\Pi = \langle \Gamma \rangle$. The fibers of π are projective spaces of dimension 5, hence $\mathcal{H}_{c,r}$ is a $\mathbb{P}^5$-bundle over $\mathbb{G}(2, r)$. Therefore $\mathcal{H}_{c,r}$ is a smooth, irreducible projective variety of dimension $3r - 1$.

Consider the incidence relation

$$I = \{(\Gamma, X) \in \mathcal{H}_{c,r} \times \Sigma(d, r) \mid \Gamma \subset X\}$$

and the natural projections

$$p \colon I \to \mathcal{H}_{c,r} \quad \text{and} \quad q \colon I \to \Sigma(d, r)\,.$$

It is easily seen that $q(I) = \Sigma_c(d, r)$ and that, for any $\Gamma \in \mathcal{H}_{c,r}$, $p^{-1}(\Gamma)$ is a linear subspace of $\{\Gamma\} \times \Sigma(d, r)$ of codimension $2d + 1$. Indeed, Γ being a complete intersection, it is projectively normal, hence the restriction map

$$H^0(\mathbb{P}^r, \mathcal{O}_{\mathbb{P}^r}(d)) \to H^0(\Gamma, \mathcal{O}_\Gamma(d)) \simeq \mathbb{C}^{2d+1}$$

is surjective. It follows that I and $\Sigma_c(d, r)$ are irreducible proper schemes. Moreover, one has

$$\begin{aligned} \dim(I) &= \dim(p^{-1}(\Gamma)) + \dim(\mathcal{H}_{c,r}) = \\ &= \binom{d+r}{d} - 1 - (2d + 2 - 3r) = \dim(\Sigma(d, r)) - \epsilon(d, r)\,. \end{aligned}$$

Letting $\kappa(d, r)$ be the dimension of the general fiber of $q \colon I \to \Sigma_c(d, r)$, one obtains

$$\dim(\Sigma_c(d, r)) = \dim(I) - \kappa(d, r) = \dim(\Sigma(d, r)) - \epsilon(d, r) - \kappa(d, r)\,,$$

and therefore

$$\operatorname{codim}(\Sigma_c(d, r), \Sigma(d, r)) = \epsilon(d, r) + \kappa(d, r)\,.$$

Next we prove that $\kappa(d, r) = 0$, which will accomplish the proof of part (a). To do this, we imitate the argument in [10, p. 29].

First of all, consider again the surjective morphism $q : I \to \Sigma_c(d, r)$. Since I is irreducible, the general element $(\Gamma, X) \in I$ maps to the general element $X \in \Sigma_c(d, r)$. Since $(\Gamma, X) \in I$ is general and $p \colon I \to \mathcal{H}_{c,r}$ is surjective, then Γ is smooth. Hence the general $X \in \Sigma_c(d, r)$ contains some smooth conic Γ. Moreover, the general fibre of q could be reducible, but, by Stein factorization, all components of it are of the same dimension and exchanged by monodromy. This implies that the general element (Γ, X) of any component of $q^{-1}(X)$ with $X \in \Sigma_c(d, r)$ general, is such that Γ is smooth (cf. Claim 7.2 below for an alternative argument).

By choosing appropriate coordinates, we may assume that if (Γ, X) is the general element of a component of $q^{-1}(X)$ with $X \in \Sigma_c(d, r)$ general, then Γ has equations

$$x_0x_1 - x_2^2 = x_3 = \cdots = x_r = 0$$

and X has equation $F = 0$ with

$$F = A(x_0x_1 - x_2^2) + B_3x_3 + \cdots + B_rx_r + R$$

where

$$A = \sum_{v_0+v_1+v_2=d-2} \alpha_{\mathbf{v}} x_0^{v_0} x_1^{v_1} x_2^{v_2}, \; B_i = \sum_{w_0+w_1+w_2=d-1} \beta_{\mathbf{w}} x_0^{w_0} x_1^{w_1} x_2^{w_2}, \quad \text{for} \quad i = 3, \ldots r$$

and $R \in I_\Gamma^2$. By Bertini's theorem we may assume that X is smooth. We have the normal bundles sequence

$$0 \to N_{\Gamma|X} \to N_{\Gamma|\mathbb{P}^r} \simeq \mathcal{O}_\Gamma(2) \oplus \mathcal{O}_\Gamma(1)^{\oplus r-2} \to N_{X|\mathbb{P}^r}{}_{|\Gamma} \simeq \mathcal{O}_\Gamma(d) \to 0.$$

We want to show that $h^0(N_{\Gamma|X}) = 0$, which implies that $\kappa(d, r) = 0$, as desired. In order to prove this, we will show that the map

$$\varphi : H^0(N_{\Gamma|\mathbb{P}^r}) \to H^0(N_{X|\mathbb{P}^r}{}_{|\Gamma})$$

is injective. Notice that $h^0(N_{\Gamma|\mathbb{P}^r}) = 3r - 1$ and $h^0(N_{X|\mathbb{P}^r}{}_{|\Gamma}) = 2d + 1$, and so, the assumption $\epsilon(d, r) \geq 0$ reads $h^0(N_{\Gamma|\mathbb{P}^r}) \leq h^0(N_{X|\mathbb{P}^r}{}_{|\Gamma})$.

We can interpret a section in $H^0(N_{\Gamma|\mathbb{P}^r})$ as the datum of $(f, f_3, \ldots, f_r)$, where $f \in H^0(\mathcal{O}_\Gamma(2)))$ is a homogeneous polynomial

$$f = \sum_{0 \leq i \leq j \leq 2} b_{ij} x_i x_j$$

taken modulo $x_0x_1 - x_2^2$, and $f_i \in H^0(\mathcal{O}_\Gamma(1)))$ is a linear form

$$f_i = a_{i0}x_0 + a_{i1}x_1 + a_{i2}x_2, \quad \text{for} \quad i = 3, \ldots, r.$$

Notice that the parameters on which $(f, f_3, \ldots, f_r)$ depends are indeed $3r - 1$, namely the $3(r - 2)$ coefficients a_{ij}s plus the 5 coefficients b_{ij}s. The map φ sends $(f, f_3, \ldots, f_r)$ to the restriction of $Af + B_3f_3 + \cdots + B_rf_r$ to Γ. By identifying Γ with $\mathbb{P}^1$ via the map sending $t \in \mathbb{P}^1$ to

$$x_0 = 1, x_1 = t^2, x_2 = t, x_3 = \cdots = x_r = 0 \tag{35}$$

the restriction of $Af + B_3f_3 + \cdots + B_rf_r$ to Γ identifies (after the substitution (35)) with a polynomial $P(t)$ of degree $2d$ in t. Let us order the $3(r-2)$ coefficients a_{ij}s and the 5 coefficients b_{ij}s in such a way that the b_{ij}s come before the a_{ij}s, and inside each group they are ordered lexicographically. Then we can consider the matrix Φ of the map φ, which is of type $(2d+1) \times (3r-1)$. Indeed, each one of the $2d+1$ coefficients of the polynomial $P(t)$ of degree $2d$ is in turn a polynomial on the b_{ij} and a_{ij}. Given b_{ij} or a_{ij}, its coefficients in those polynomials form the corresponding column of Φ.

Notice that the latter coefficients (that is, the entries of Φ) are linear functions of the $\alpha_{\mathbf{v}}$s and the $\beta_{\mathbf{w}}$s. Moreover, in each row and in each column of Φ a given $\alpha_{\mathbf{v}}$ and a given $\beta_{\mathbf{w}}$ appear at most once.

The map φ is injective if and only if Φ has rank $3r-1$ for sufficiently general values of the $\alpha_{\mathbf{v}}$s and the $\beta_{\mathbf{w}}$s. We will in fact consider the $\alpha_{\mathbf{v}}$s and the $\beta_{\mathbf{w}}$s as indeterminates and prove that there is a maximal minor of Φ, e.g., the one Φ' determined by the first $3r-1$ rows, which is a non–zero polynomial in these variables. This will finish our proof.

Consider, for example, the order of the $\alpha_{\mathbf{v}}$s and the $\beta_{\mathbf{w}}$s in which the former come before the latter and in each group they are ordered lexicographically. Let us order the monomials appearing in the expression of Φ' according to the following rule: the monomial m_1 comes before the monomial m_2 if for the smallest variable appearing in m_1 and in m_2 with different exponents, the exponent in m_1 is larger than the exponent in m_2. The greatest monomial in this ordering will have coefficient ± 1 in Φ', since in each row, the choice of the $\alpha_{\mathbf{v}}$s and the $\beta_{\mathbf{w}}$s entering in it is prescribed. This proves that $\Phi' \neq 0$.

(b) We have to show that, if $\epsilon(d, r) > 0$ and, except for $(d, r) = (4, 3)$, the hypersurface X corresponding to the general point of $\Sigma_c(d, r)$ contains a unique conic. To do this we use counts of parameters, which show that the codimension in $\Sigma(d, r)$ of the locus of hypersurfaces X containing at least two distinct conics is strictly larger than $\epsilon(d, r)$. The proof is a bit tedious, since it requires to consider a number of different possibilities, namely that two conics on X do not intersect, or they intersect in one, two or in four points (counting with multiplicity). We will not treat in detail all the cases, but only the former and the latter, leaving some easy details in the remaining two cases to the reader, which could profit from similarity with the dimension count we made at the beginning of this proof.

We start with the following two claims.

Claim 7.1 *The subset $\Sigma_{2l}(d, r)$ of all the $X \in \Sigma_c(d, r)$ such that X contains a double line is a proper subvariety of $\Sigma_c(d, r)$.*

Proof of Claim 7.1 Consider the closed subset $\mathcal{H}_{2l,r} \subset \mathcal{H}_{c,r}$ whose general point corresponds to a double line in $\mathbb{P}^r$. There is a natural $\mathbb{P}^2$-fibration $\mathcal{H}_{2l,r} \to \mathbb{G}(2, r)$. Hence one has $\dim(\mathcal{H}_{2l,r}) = 3r - 4$. Consider further the incidence relation

$$I_{2l} = \{(\Gamma, X) \in \mathcal{H}_{2l,r} \times \Sigma(d, r) \,|\, \Gamma \subset X\}$$

with projections p_{2l}, q_{2l} to the first and the second factors, respectively. The general fiber F_{2l} of p_{2l} is a linear subspace of $\Sigma(d,r)$ of codimension $2d+1$. It follows that I_{2l} is an irreducible projective variety of dimension

$$\dim(I_{2l}) = \dim(\Sigma(d,r)) - (2d+5-3r).$$

Therefore, the image $\Sigma_{2l}(d,r) = q_{2l}(I_{2l})$ is an irreducible proper subvariety of $\Sigma(d,r)$ of codimension at least

$$2d+5-3r = \epsilon(r,d)+3 = \operatorname{codim}(\Sigma_c(d,r), \Sigma(d,r))+3,$$

see (a). □

We know by (a) that the general $X \in \Sigma_c(d,r)$ contains only finitely many conics (recall that the general fiber of $q : I \to \Sigma_c(d,r)$ has dimension $\kappa(d,r) = 0$). Our next claim is the following.

Claim 7.2 *The conics contained in the general $X \in \Sigma_c(d,r)$ are all smooth.*

Proof of Claim 7.2 The incidence variety I being irreducible, the monodromy group of the generically finite morphism $q : I \to \Sigma_c(d,r)$ acts transitively on the general fiber $q^{-1}(X)$. Its action on I lifts to the universal family of conics over I. The latter action by homeomorphisms of the general fiber (which consists of a finite number of conics) preserves the Euler characteristic. We know already that the general X contains a smooth conic. Due to Claim 7.1, X does not carry any double line. Since the Euler characteristic (equal 3) of the union of two crossing lines is different from the one of a smooth conic, all the conics in X are smooth. □

Suppose now the general $X \in \Sigma_c(d,r)$ contains more than one conic, and assume first it contains two conics which do not intersect. We will see this leads to a contradiction.

Let $\mathcal{H}_{cc,r}$ be the component of the Hilbert scheme whose general point corresponds to a pair of conics in $\mathbb{P}^r$ which do not meet. It is easy to see that $\mathcal{H}_{cc,r}$ is an irreducible projective variety of dimension $6r-2$.

Consider the incidence relation

$$I = \{(\Gamma, X) \in \mathcal{H}_{cc,r} \times \Sigma(d,r) \mid \Gamma \subset X\}$$

and the natural projections

$$p\colon I \to \mathcal{H}_{cc,r} \quad \text{and} \quad q\colon I \to \Sigma(d,r).$$

Claim 7.3 *For any $\Gamma \in \mathcal{H}_{cc,r}$ which corresponds to a pair of disjoint smooth conics, $p^{-1}(\Gamma)$ is a linear subspace of $\{\Gamma\} \times \Sigma(d,r)$ of codimension $4d+2$.*

Proof of Claim 7.3 By our assumption, $r \geq 3$. Then the hypothesis $\epsilon(d,r) > 0$ implies $d \geq 4$. So, we have to prove that the restriction map

$$\rho : H^0(\mathbb{P}^r, \mathcal{O}_{\mathbb{P}^r}(d)) \to H^0(\Gamma, \mathcal{O}_\Gamma(d)) \simeq \mathbb{C}^{4d+2},$$

where $\Gamma = \Gamma_1 \cup \Gamma_2$ with Γ_1, Γ_2 disjoint smooth conics, is surjective as soon as $d \geq 4$. Actually we will prove it for $d \geq 3$. By projecting generically into $\mathbb{P}^3$, it suffices to prove the assertion if $r = 3$.

The restriction map

$$H^0(\mathbb{P}^3, \mathcal{O}_{\mathbb{P}^r}(d)) \to H^0(\Gamma_1, \mathcal{O}_{\Gamma_1}(d)) \simeq \mathbb{C}^{2d+1}$$

is surjective for all $d \geq 1$ because Γ_1 is projectively normal. Hence the kernel of this map, i.e., $H^0(\mathbb{P}^3, \mathcal{I}_{\Gamma_1|\mathbb{P}^3}(d))$ has codimension $2d + 1$ in $H^0(\mathbb{P}^3, \mathcal{O}_{\mathbb{P}^3}(d))$. Consider now the restriction map

$$\rho' : H^0(\mathbb{P}^3, \mathcal{I}_{\Gamma_1|\mathbb{P}^3}(d)) \to H^0(\Gamma_2, \mathcal{O}_{\Gamma_2}(d)) \simeq \mathbb{C}^{2d+1}.$$

We will prove that this map is also surjective. This will imply that its kernel, i.e., $H^0(\mathbb{P}^3, \mathcal{I}_{\Gamma|\mathbb{P}^3}(d))$ has codimension $2d + 1$ in $H^0(\mathbb{P}^3, \mathcal{I}_{\Gamma_1|\mathbb{P}^3}(d))$, hence it has codimension $4d + 2$ in $H^0(\mathbb{P}^3, \mathcal{O}_{\mathbb{P}^r}(d))$, which proves that ρ is surjective.

Let $\Pi_i = \langle \Gamma_i \rangle$ be the plane spanned by Γ_i, for $i = 1, 2$. Consider the intersection scheme $\mathfrak{D}$ of Γ_1 with Π_2, so that $\mathfrak{D}$ is a zero dimensional scheme of length 2, and $\mathfrak{D}$ is not contained in Γ_2. To prove that ρ' is surjective, notice that it is composed of the following two restriction maps

$$\rho_1 : H^0(\mathbb{P}^3, \mathcal{I}_{\Gamma_1|\mathbb{P}^3}(d)) \to H^0(\Pi_2, \mathcal{I}_{\mathfrak{D}|\Pi_2}(d)),$$

$$\rho_2 : H^0(\Pi_2, \mathcal{I}_{\mathfrak{D}|\Pi_2}(d)) \to H^0(\Gamma_2, \mathcal{O}_{\Gamma_2}(d)).$$

The map ρ_1 is surjective, because its cokernel is $H^1(\mathbb{P}^3, \mathcal{I}_{\Gamma_1|\mathbb{P}^3}(d-1))$ which is zero because Γ_1 is projectively normal. So, we are left to prove that the map ρ_2 is surjective. The kernel of ρ_2 is $H^0(\Pi_2, \mathcal{I}_{\mathfrak{D}|\Pi_2}(d-2))$, whose dimension is

$$h^0(\Pi_2, \mathcal{I}_{\mathfrak{D}|\Pi_2}(d-2)) = \binom{d}{2} - 2$$

as soon as $d \geq 3$. Similarly

$$h^0(\Pi_2, \mathcal{I}_{\mathfrak{D}|\Pi_2}(d)) = \binom{d+2}{2} - 2$$

for any $d \geq 1$. Hence the dimension of the image of ρ_2 is

$$h^0(\Pi_2, \mathcal{I}_{\mathfrak{D}|\Pi_2}(d)) - h^0(\Pi_2, \mathcal{I}_{\mathfrak{D}|\Pi_2}(d-2)) = \binom{d+2}{2} - \binom{d}{2} = 2d + 1$$

which proves that ρ_2 is surjective. □

By Claim 7.3, I has a unique component I' which dominates $\mathcal{H}_{cc,r}$, and

$$\dim(I') = \dim(\mathcal{H}_{cc,r}) + \dim(\Sigma(d,r)) - (4d+2) = \\ = \dim(\Sigma(d,r)) - (4d - 6r + 4).$$

Since we are assuming $q(I') = \Sigma_c(d,r)$, we have

$$\dim(\Sigma(d,r)) - (4d - 6r + 4) = \dim(I') \geq \\ \geq \dim(q(I')) = \dim(\Sigma(d,r)) - \epsilon(d,r) = \dim(\Sigma(d,r)) - (2d - 3r + 2),$$

whence $\epsilon(d,r) = 2d + 2 - 3r \leq 0$, contrary to the assumption $\epsilon(d,r) > 0$.

A similar argument works also in the cases where the general $X \in \Sigma_c(d,r)$ contains two smooth conics which meet in one or two points, counting with multiplicity. The corresponding closed subset $\mathcal{H}^{(i)}_{cc,r} \subset \mathcal{H}_{cc,r}$ whose general point corresponds to a pair of smooth conics which meet in i points, where $i = 1, 2$, is an irreducible proper scheme of dimension $5r$ for $i = 1$ and $4r + 2$ for $i = 2$. Letting $I^{(i)} \subset \mathcal{H}^{(i)}_{cc,r} \times \Sigma_c(d,r)$ be the corresponding incidence relation and arguing as in the proof of Claim 7.3 one can easily show that any fiber of the projection $I' \to \mathcal{H}^{(i)}_{cc,r}$ over a point $\Gamma \in \mathcal{H}^{(i)}_{cc,r}$ representing a pair of conics with exactly i places in common counting with multiplicity, is a linear subspace of $\Sigma(d,r)$ of codimension $4d + 1$ if $i = 1$ and $4d$ if $i = 2$, where I' is the unique component of $I^{(i)}$ which dominates $\Sigma_c(d,r)$. Proceeding as before, this leads in both cases to the inequality $r \leq 2$, which contradicts the assumption $r \geq 3$.

Consider finally the remaining (extremal) case in which the general $X \in \Sigma_c(d,r)$ contains two conics which are coplanar, i.e., they intersect (counting with multiplicity) at 4 points.

We denote by $\mathcal{F}(= \mathcal{H}^{(4)}_{cc,r})$ the subvariety of the Hilbert scheme whose general point corresponds to a pair of coplanar conics in $\mathbb{P}^r$. It is easy to see that $\mathcal{F}$ is an irreducible projective variety of dimension $3r + 4$.

Consider the incidence relation

$$I = I^{(4)} = \{(\Gamma, X) \in \mathcal{F} \times \Sigma(d,r) \mid \Gamma \subset X\}$$

and the projections

$$p : I \to \mathcal{F} \quad \text{and} \quad q : I \to \Sigma(d,r).$$

For any $\Gamma \in \mathcal{F}$, $p^{-1}(\Gamma)$ is a linear subspace of $\{\Gamma\} \times \Sigma(d,r)$ of codimension $4d - 2$. Indeed, since Γ is a complete intersection, it is projectively normal. Hence the restriction map

$$H^0(\mathbb{P}^r, \mathcal{O}_{\mathbb{P}^r}(d)) \to H^0(\Gamma, \mathcal{O}_\Gamma(d))$$

is surjective for all $d \geq 1$. On the other hand, Γ is a curve of arithmetic genus 3, and the dualizing sheaf of Γ is $\mathcal{O}_\Gamma(1)$. Hence, $h^0(\Gamma, \mathcal{O}_\Gamma(d)) = 4d - 3 + 1 = 4d - 2$, as soon as $d \geq 2$.

Thus, I is irreducible, and

$$\begin{aligned}\dim(I) &= \dim(\mathcal{F}) + \dim(\Sigma(d,r)) - (4d-2) = \\ &= \dim(\Sigma(d,r)) - (4d-3r-6).\end{aligned}$$

Since we are assuming $q(I) = \Sigma_c(d,r)$, we have

$$\begin{aligned}&\dim(\Sigma(d,r)) - (4d-3r-6) = \dim(I) \geq \\ &\geq \dim(q(I)) = \dim(\Sigma(d,r)) - \epsilon(d,r) = \dim(\Sigma(d,r)) - (2d+2-3r),\end{aligned}$$

whence $d \leq 4$. Since

$$0 < \epsilon(d,r) = 2d+2-3r \leq 10-3r$$

we see that the only possibility is $d = 4$, $r = 3$. In this case a similar argument proves that the general $X \in \Sigma_c(4,3)$ contains exactly two coplanar, smooth conics.

7.2 *The Degree Count*

Next we compute the degree of $\Sigma_c(d,r)$ in $\Sigma(d,r)$, provided $\epsilon(d,r) > 0$.

Let $f : \mathcal{C} \to \mathcal{H}_{c,r}$ be the universal family over $\mathcal{H}_{c,r}$, which is endowed with a map $g : \mathcal{C} \to \mathbb{P}^r$. We denote by $\mathcal{E}_d$ the vector bundle $f_*(g^*(\mathcal{O}_{\mathbb{P}^r}(d))$ over $\mathcal{H}_{c,r}$. If Γ is a conic, the fiber $\mathcal{E}_{d,\Gamma}$ of $\mathcal{E}_d$ at (the point corresponding to) Γ is $H^0(\Gamma, \mathcal{O}_\Gamma(d))$. We set $\mathcal{E} = \mathcal{E}_1$. Notice that $\mathcal{E}_d$ is a vector bundle of rank $2d+1 > 3r-1 = \dim(\mathcal{H}_{c,r})$; in particular, $\mathrm{rk}(\mathcal{E}) = 3$.

Lemma 7.4 *If $\epsilon(d,r) > 0$ and $(d,r) \neq (4,3)$, then*

$$\deg(\Sigma_c(d,r)) = \int_{\mathcal{H}_{c,r}} c_{3r-1}(\mathcal{E}_d)\,.$$

Moreover

$$\deg(\Sigma_c(4,3)) = \frac{1}{2}\int_{\mathcal{H}_{c,3}} c_8(\mathcal{E}_4)\,.$$

Proof Any homogeneous form F of degree d in $r+1$ variables defines a section σ_F of $\mathcal{E}_d$ such that $\sigma_F(\Gamma) = F|_\Gamma \in H^0(\Gamma, \mathcal{O}_\Gamma(d))$. Consider the effective divisor X_F of degree d on $\mathbb{P}^r$ of zeros of F. The support of X_F contains Γ if and only if $\sigma_F(\Gamma) = 0$. Counting the conics $\Gamma \in \mathcal{H}_{c,r}$ lying in $\mathrm{Supp}(X_F)$ is the same as counting the zeros of σ_F in $\mathcal{H}_{c,r}$ with their multiplicities.

Let further $\rho = \dim(\mathcal{H}_{c,r}) = 3r-1$. By our assumption one has

$$\mathrm{rk}\,(\mathcal{E}_d) - \rho = \epsilon(d,r) > 0\,.$$

Choose a general linear subsystem $\mathcal{L} = \langle X_0, \ldots, X_\epsilon \rangle$ in $|\mathcal{O}_{\mathbb{P}^r}(d)|$ of dimension $\epsilon = \epsilon(d, r)$, where $X_i = \{F_i = 0\}$. By virtue of Theorem 7.1, $\mathcal{L}$ meets $\Sigma(d, r, k) \subset \Sigma(d, r)$ transversally in $\deg(\Sigma_c(d, r))$ simple points.

Consider now the sections $\sigma_i := \sigma_{F_i}$, $i = 0, \ldots, \epsilon$, of $\mathcal{E}_d$ and assume $(d, r) \neq (4, 3)$. By Theorem 7.1, the intersection of $\mathcal{L}$ with $\Sigma(d, r, k)$ is exactly the scheme of points $\Gamma \in \mathcal{H}_{c,r}$ where there is a linear combination of $\sigma_0, \ldots, \sigma_\epsilon$ vanishing on Γ. This is the zero dimensional scheme of points of $\mathcal{H}_{c,r}$ where the sections $\sigma_0, \ldots, \sigma_\epsilon$ are linearly dependent. This zero dimensional scheme represents the top Chern class $c_\rho(\mathcal{E}_d)$ (see [22, Theorem 5.3]). Its degree is the top Chern number $\int_{\mathcal{H}_{c,r}} c_{3r-1}(\mathcal{E}_d)$.

The case $(d, r) = (4, 3)$ is similar: one has to take into account again Theorem 7.1, which says that the general quartic surface in $\mathbb{P}^3$ contains exactly two smooth conics, and these conics are coplanar. □

To compute $c_{3r-1}(\mathcal{E}_d)$ we proceed as follows. For a positive integer d consider $\mathrm{Sym}^d(\mathcal{E})$. Note that the universal family $\mathcal{C}$ over $\mathcal{H}_{c,r}$ is the zero set of a section ξ of $\mathrm{Sym}^2(\mathcal{E})$. For any $d \geqslant 2$ one has the exact sequence

$$0 \to \mathrm{Sym}^{d-2}(\mathcal{E}) \xrightarrow{\cdot \xi} \mathrm{Sym}^d(\mathcal{E}) \to \mathcal{E}_d \to 0\,.$$

Hence

$$c(\mathcal{E}_d) = c(\mathrm{Sym}^d(\mathcal{E})) \cdot c(\mathrm{Sym}^{d-2}(\mathcal{E}))^{-1}\,. \tag{36}$$

To compute Chern classes, as usual, we use the splitting principle. We write formally

$$\mathcal{E} = L_1 \oplus L_2 \oplus L_3,$$

the L_is being (virtual) line bundles. Consider the Chern roots $x_i = c_1(L_i)$ of $\mathcal{E}$. One has

$$c(\mathcal{E}) = (1 + x_1)(1 + x_2)(1 + x_3)\,,$$

that is,

$$c_1(\mathcal{E}) = x_1 + x_2 + x_3, \quad c_2(\mathcal{E}) = x_1x_2 + x_1x_3 + x_2x_3, \qquad c_3(\mathcal{E}) = x_1x_2x_3.$$

Furthermore,

$$\mathrm{Sym}^d(\mathcal{E}) = \bigoplus_{v_1+v_2+v_3=d} L_1^{v_1} L_2^{v_2} L_3^{v_3}.$$

Letting

$$c_1(L_1^{v_1} L_2^{v_2} L_3^{v_3}) = v_1x_1 + v_2x_2 + v_3x_3 = \langle \mathbf{v}, \mathbf{x} \rangle$$

where $\mathbf{x} = (x_1, x_2, x_3)$ and $\mathbf{v} = (v_1, v_2, v_3)$ with $|\mathbf{v}| = v_1 + v_2 + v_3$ one obtains

$$c(\mathrm{Sym}^d(\mathcal{E})) = \prod_{|\mathbf{v}|=d} (1 + \langle \mathbf{v}, \mathbf{x} \rangle)$$

and, by (36),

$$c(\mathcal{E}_d) = \left(\prod_{|\mathbf{v}|=d} (1 + \langle \mathbf{v}, \mathbf{x} \rangle)\right) \cdot \left(\prod_{|\mathbf{v}|=d-2} (1 + \langle \mathbf{v}, \mathbf{x} \rangle)\right)^{-1} . \tag{37}$$

Now, the top Chern class $c_{3r-1}(\mathcal{E}_d)$ is the homogeneous component $\eta(x_1, x_2, x_3)$ of degree $3r - 1$ in the right hand side of (37) written as a formal power series in x_1, x_2, x_3. This is a symmetric form of degree $3r - 1$ in x_1, x_2, x_3. It can be expressed via the elementary symmetric polynomials in x_1, x_2, x_3, i.e., in terms of $c_1(\mathcal{E}), c_2(\mathcal{E}), c_3(\mathcal{E})$.

In order to compute $c_{3r-1}(\mathcal{E}_d)$ effectively, we prefer to use the Bott residue formula. The standard diagonal action of $T = (\mathbb{C}^*)^{r+1}$ on $\mathbb{P}^r$, see 2.3, induces an action of T on $\mathbb{G}(2, r)$ and on $\mathcal{H}_{c,r}$.

Lemma 7.5 *The action of T on $\mathcal{H}_{c,r}$ has exactly $r(r^2 - 1)$ isolated fixed points.*

Proof Let Γ be a fixed point for the T-action on $\mathcal{H}_{c,r}$. Then $\Pi = \langle \Gamma \rangle$ is fixed under the action of T on $\mathbb{G}(2, r)$. Hence Π is one of the coordinate planes in $\mathbb{P}^r$, and these are $\binom{r+1}{3}$ in number. We let x, y, z be the three coordinate axes in Π. Then the only conics on Π fixed by the T-action are the singular conics $x + y, x + z, y + z, 2x, 2y, 2z$. Thus, we get in total $6\binom{r+1}{3} = r(r^2 - 1)$ fixed points of T in $\mathcal{H}_{c,r}$.

□

We denote by $\mathcal{F}$ the set of fixed points for the T-action on $\mathcal{H}_{c,r}$. Bott's residue formula, applied in our setting, has the form

$$\deg(\Sigma_c(d, r)) = \int_{\mathcal{H}_{c,r}} c_{3r-1}(\mathcal{E}_d) = \sum_{\Gamma \in \mathcal{F}} \frac{c_\Gamma}{e_\Gamma},$$

where $\frac{c_\Gamma}{e_\Gamma}$ is the local contribution of a fixed point $\Gamma \in \mathcal{F}$. Recall that c_Γ results from the local contribution of $c_{3r-1}(\mathcal{E}_d)$ at Γ, and e_Γ is determined by the torus action on the tangent space to $\mathcal{H}_{c,r}$ at the point corresponding to Γ, see 2.3.

To compute e_Γ we have to compute the characters of the T-action on the tangent space

$$T_\Gamma(\mathcal{H}_{c,r}) \simeq H^0(\Gamma, N_{\Gamma|\mathbb{P}^r}) \simeq H^0(\Gamma, \mathcal{O}_\Gamma(1))^{\oplus(r-2)} \oplus H^0(\Gamma, \mathcal{O}_\Gamma(2)) \simeq \mathcal{E}_\Gamma^{\oplus(r-2)} \oplus \mathcal{E}_{2,\Gamma} .$$

Let $\Pi = \langle \Gamma \rangle$. Then Π is a coordinate plane which corresponds to a subset $I = \{i, j, k\} \subset \{0, \dots, r\}$ consisting of 3 distinct elements. Let $\mathcal{I}_3$ be the set of all the $\binom{r+1}{3}$ such subsets I. The characters of the T-action on $\mathcal{E}_\Gamma$ have weights $-t_\alpha$ with $\alpha \in I$. Let $I^{(2)}$ be the symmetric square of I; it consists of 6 unordered pairs $\{\alpha, \beta\}$, $\alpha, \beta \in I$. The characters of the T-action on $\mathcal{E}_{2,\Gamma}$ have weights $t_\alpha + t_\beta$ with $\{\alpha, \beta\} \in I^{(2)} \setminus \{a, b\}$, where $x_a x_b = 0$ is the equation of Γ in Π. Then

$$e_\Gamma = (-1)^{3(r-2)}(t_i t_j t_k)^{r-2} \prod_{\{\alpha,\beta\}\in I^{(2)}\setminus\{a,b\}} (t_\alpha + t_\beta)\,.$$

As for c_Γ, with the same notation as above we have

$$c_\Gamma = \eta(-t_i, -t_j, -t_k) = (-1)^{3r-1}\eta(t_i, t_j, t_k) \quad \text{where} \quad I = \{i, j, k\}.$$

In conclusion we find the formula

$$\deg(\Sigma_c(d,r)) = -\sum_{I=\{i,j,k\}\in\mathcal{I}_3} \sum_{\{a,b\}\in I^{(2)}} \frac{\eta(t_i, t_j, t_k)}{(t_i t_j t_k)^{r-2}\prod_{\{\alpha,\beta\}\in I^{(2)}\setminus\{a,b\}}(t_\alpha + t_\beta)}\,. \tag{38}$$

Again, the right hand side of this formula is, a priori, a rational function in the variables $t_0, \ldots, t_r$. In fact, this is a constant and a positive integer. Letting $t_i = 1$ for all $i = 0, \ldots, r$ we arrive at the following conclusion.

Theorem 7.6 *Assuming that* $\epsilon(d, r) = 2d + 2 - 3r > 0$ *and* $(d, r) \neq (4, 3)$ *one has*

$$\deg(\Sigma_c(d,r)) = -\frac{5}{32}\binom{r+1}{3}\eta(1, 1, 1)\,, \tag{39}$$

where η *is the homogeneous form of degree* $3r - 1$ *in the formal power series decomposition of the right hand side of* (37).

Remark 7.7 In the case of the surfaces in $\mathbb{P}^3$, one can find in [35, Proposition 7.1] a formula for the degree of $\Sigma_c(d, 3)$ expressed as a polynomial in d for $d \geq 5$. This formula was deduced by applying Bott's residue formula. After dividing by 2, this formula gives also the correct value $\deg(\Sigma_c(4, 3)) = 2508$.

Acknowledgements The first author acknowledges the MIUR Excellence Department Project awarded to the Department of Mathematics, University of Rome Tor Vergata, CUP E83C18000100006. He also thanks the GNSAGA of INdAM. This research was partially done during a visit of the second author at the Department of Mathematics, University of Rome Tor Vergata (supported by the project"Families of curves: their moduli and their related varieties", CUP E8118000100005, in the framework of Mission Sustainability). The work of the second author was also partially supported by the grant 346300 for IMPAN from the Simons Foundation and the matching 2015–2019 Polish MNiSW fund (code: BCSim-2018-s09). The authors thank all these Institutions and programs for the support and excellent working conditions.

Our special thanks are due to Laurent Manivel who kindly suggested to complete some references omitted in the first draft of this paper. We are grateful to the peer reviewer whose suggestions allowed to improve the style and to avoid some typos and arithmetical mistakes.

References

1. Altman, A.B., Kleiman, S.L.: Foundations of the theory of Fano schemes. Compos. Math. **34**, 3–47 (1977)

2. Araujo, C., Casagrande, C.: On the Fano variety of linear spaces contained in two odd-dimensional quadrics. Geom. Topol. **21**, 3009–3045 (2017)
3. Barth, W., Van de Ven, A.: Fano varieties of lines on hypersurfaces. Arch. Math. (Basel) **31**, 96–104 (1978)
4. Bastianelli, F., Ciliberto, C., Flamini, F., Supino, P.: On complete intersections containing a linear subspace. Geom. Dedicata 9 pp (2019). https://doi.org/10.1007/s10711-019-00452-2
5. Beauville, A.: Sous-variétés spéciales des variétés de Prym. Comp. Math. **45**, 357–383 (1982)
6. Beauville, A.: Quantum cohomology of complete intersections. Mat. Fiz. Anal. Geom. **2**, 384–398 (1995)
7. Behesti, R., Mohan Kumar, N.: Spaces of rational curves in complete intersections. Compos. Math. **149**, 1041–1060 (2013)
8. Bombieri, E., Swinnerton-Dyer, H.P.F.: On the local zeta function of a cubic threefold. Annali della Scuola Normale Superiore di Pisa Classe di Scienze, Ser. **3**(21), 1–29 (1967)
9. Bonavero, L., Höring, A.: Counting conics in complete intersections. Acta Math. Vietnam. **35**, 23–30 (2010)
10. Borcea, C.: Deforming varieties of k-planes of projective complete intersections. Pacific J. Math. **143**, 25–36 (1990)
11. Bott, R.: A residue formula for holomorphic vector-fields. J. Diff. Geom. **1**, 311–330 (1967)
12. Brion, M.: Equivariant cohomology and equivariant intersection theory 49 p (2008). arXiv:math/9802063
13. Casagrande, C.: Rank 2 quasiparabolic vector bundles on $\mathbb{P}^1$ and the variety of linear subspaces contained in two odd-dimensional quadrics. Math. Z. **280**, 981–988 (2015)
14. Ciliberto, C., Zaidenberg, M.: Lines, conics, and all that 37 p (2020). arXiv:math/1910.11423v2
15. Clemens, C.H., Griffiths, PhA: The intermediate Jacobian of the cubic threefold. Ann. Math. **95**, 281–356 (1972)
16. Collino, A.: The Abel-Jacobi isomorphism for the cubic fivefold. Pacific J. Math. **122**, 43–55 (1986)
17. Cordovez, J., Valenzano, M.: On Fano scheme of k-planes in a projective complete intersection. Universitá di Torino, Quaderni del Dipartimento di Matematica **21**, 12p (2005)
18. Debarre, O., Manivel, L.: Sur la variété des espaces linéaires contenus dans une intersection compléte. Math. Ann. **312**, 549–574 (1998)
19. Desale, U.V., Ramanan, S.: Classification of Vector Bundles of Rank 2 on Hyperelliptic Curves. Invent. Math. **38**, 161–185 (1976)
20. Donagi, R.: Group law on the intersection of two quadrics. Ann. Scuola Norm. Sup. Pisa Cl. Sci. **7**(4), 217–239 (1980)
21. Edidin, D., Graham, W.: Localization in equivariant intersection theory and the Bott residue formula. Am. J. Math. **120**, 619–636 (1998)
22. Eisenbud, D., Harris, J.: 3264 and All That. Intersection Theory in Algebraic Geometry. Cambridge University Press, Cambridge (2016)
23. Fano, G.: Sul sistema ∞^2 di rette contenuto in una varietà cubica generate dello spazio a quattro dimensioni. Atti R. Acc. Sc. Torino **XXXIX**, 778–792 (1904)
24. Fulton, W.: Introduction to Intersection Theory in Algebraic Geometry. American Mathematical Society, Providence, Rhode Island (1980)
25. Furukawa, K.: Rational curves on hypersurfaces. J. Reine Angew. Math. **665**, 157–188 (2012)
26. Gherardelli, F.: Un osservazione sullia varieta cubica di $\mathbb{P}^4$. Rend. sem. mat. e fisicodi Milano **37**, 157–160 (1967)
27. Harris, J., Roth, M., Starr, J.: Rational curves on hypersurfaces of low degree. J. Reine Angew. Math. **571**, 73–106 (2004)
28. Hiep, Dang Tuan: On the degree of the Fano schemes of linear subspaces on hypersurfaces. Kodai Math. J. **39**, 110–118 (2016)
29. Hiep, D.T.: Numerical invariants of Fano schemes of linear subspaces on complete intersections, 7 p (2017). arXiv:1602.03659
30. Iskovskikh, V.A.: Anticanonical models of three-dimensional algebraic varieties. J. Soviet Math. **13**, 745–814 (1980). Translated from: Itogi Nauki i Tekhniki. Ser. Sovrem. Probl. Mat. VINITI, Moscow **12**, 59–157 (1979)

31. Jiang, Zhi: A Noether-Lefschetz theorem for varieties of r-planes in complete intersections. Nagoya Math. J. **206**, 39–66 (2012)
32. Kollár, J.: Rational Curves on Algebraic Varieties. Ergebnisse der Mathematik und ihrer Grenzgebiete. 3. Folge, vol. 32. Springer, Berlin (1999)
33. Langer, A.: Fano schemes of linear spaces on hypersurfaces. Manuscripta Math. **93**, 21–28 (1997)
34. Libgober, A.: Numerical characteristics of systems of straight lines on complete intersections. Math. Notes **13**, 51–56 (1973). Translated from: Mat. Zametki **13**, 87–96 (1973)
35. Maia, J.A., Rodrigues, A., Xavier, F., Vainsencher, I.: Enumeration of surfaces containing a curve of low degree. preprint 19p (2011)
36. Manivel, L.: Sur les hypersurfaces contenant des espaces linéaires [On hypersurfaces containing linear spaces]. C. R. Acad. Sci. Paris Ser. I Math. **328**, 307–312 (1999)
37. Manivel, L.: Symmetric functions, Schubert polynomials and degeneracy loci. AMS Texts and Monographs, vol. 6. Providence, RI (2001)
38. Markushevich, D.G.: Numerical invariants of families of lines on some Fano varieties. Math. USSR-Sb. **44**, 239–260 (1983). Translated from: Matem. sb. **116**(158), 265–288 (1981)
39. Meireles Araújo, A.L., Vainsencher, I., Equivariant intersection theory and Bott's residue formula, 16th School of Algebra, Part I (Portuguese) (Brasilia, 2000). Mat. Contemp. **20**, 1–70 (2001)
40. Miyazaki, C.: Remarks on r-planes in complete intersections. Tokyo J. Math. **39**, 459–467 (2016)
41. Morin, U.: Sull'insieme degli spazi lineari contenuti in una ipersuperficie algebrica. Atti Accad. Naz. Lincei Rend. Cl. Sci. Fis. Mat. Nat. **24**, 188–190 (1936)
42. Predonzan, A.: Intorno agli S_k giacenti sulla varietà intersezione completa di più forme. Rend. Accad. Naz. dei Lincei **5**(8), 238–242 (1948)
43. Reid, M.: The complete intersection of two or more quadrics. PhD thesis, Trinity College, Cambridge (1972)
44. Roulleau, X.: Elliptic curve configurations on Fano surfaces. Manuscripta Math. **129**, 381–399 (2009)
45. Tennison, B.R.: On the quartic threefold. Proc. London Math. Soc. **29**, 714–734 (1974)
46. Tyurin, A.N.: On the Fano surface of a nonsingular cubic in $\mathbb{P}^4$. Math. USSR Izvestia **4**, 1207–1214 (1970)
47. Tyurin, A.N.: The geometry of the Fano surface of a nonsingular cubic $F \subset \mathbb{P}^4$ and Torelli Theorems for Fano surfaces and cubics. Math. USSR Izvestia **5**, 517–546 (1971)
48. Tyurin, A.N.: Five lectures on three-dimensional varieties. Russian Math. Surveys **27**, 1–53 (1972)
49. Tyurin, A.N.: On intersection of quadrics. Russian Math. Surveys **30**, 51–105 (1975). Translated from: Itogi Nauki i Tekhniki. Ser. Sovrem. Probl. Mat. VINITI, Moscow **12**, 5–57 (1979)
50. van der Waerden, B.L.: Zur algebraischen Geometrie 2. Die geraden Linien auf den Hyperflächen des $\mathbb{P}^n$. Math. Ann. **108**, 253–259 (1933)

Locally Nilpotent Sets of Derivations

Daniel Daigle

Abstract Let B be an algebra over a field $\mathbf{k}$. We define what it means for a subset of $\operatorname{Der}_{\mathbf{k}}(B)$ to be a *locally nilpotent set*. We prove some basic results about that notion and explore the following questions. Let L be a Lie subalgebra of $\operatorname{Der}_{\mathbf{k}}(B)$; if $L \subseteq \operatorname{LND}(B)$ then does it follow that L is a locally nilpotent set? Does it follow that L is a nilpotent Lie algebra?

Keywords Locally nilpotent derivation · Nilpotent Lie algebra

2010 Mathematics Subject Classification Primary: 14R20 · 13N15. Secondary: 17B30 · 17B65 · 17B66

In this article, an *algebra* over a field $\mathbf{k}$ is a pair $(B, \cdot)$ where B is a $\mathbf{k}$-vector space and "$\cdot$" is an arbitrary $\mathbf{k}$-bilinear map $B \times B \to B$, $(x, y) \mapsto x \cdot y$. Let B be an algebra over a field $\mathbf{k}$ and let $\operatorname{Der}_{\mathbf{k}}(B)$ be the set of all $\mathbf{k}$-derivations $D : B \to B$. A derivation $D \in \operatorname{Der}_{\mathbf{k}}(B)$ is said to be *locally nilpotent* if for each $x \in B$ there exists an $n > 0$ such that $D^n(x) = 0$. We write $\operatorname{LND}(B)$ for the set of locally nilpotent derivations of B. We say that a subset Δ of $\operatorname{Der}_{\mathbf{k}}(B)$ is *locally nilpotent* if for each $x \in B$ the following holds:

> for every infinite sequence $(D_1, D_2, \dots)$ of elements of Δ, there exists n such that $(D_n \circ \dots \circ D_1)(x) = 0$.

We say that Δ is *uniformly locally nilpotent* if for each $x \in B$ there exists n such that $(D_n \circ \dots \circ D_1)(x) = 0$ for all $(D_1, \dots, D_n) \in \Delta^n$. Then it is clear that the implications

$$\Delta \text{ is uniformly locally nilpotent} \Rightarrow \Delta \text{ is locally nilpotent} \Rightarrow \Delta \subseteq \operatorname{LND}(B)$$

Research supported by grant RGPIN/2015-04539 from NSERC Canada.

D. Daigle (✉)
Department of Mathematics and Statistics, University of Ottawa, Ottawa, ON K1N 6N5, Canada
e-mail: ddaigle@uottawa.ca

S. Kuroda et al. (eds.), *Polynomial Rings and Affine Algebraic Geometry*,
Springer Proceedings in Mathematics & Statistics 319,
https://doi.org/10.1007/978-3-030-42136-6_2

are true, and it is easy to see that both converses are false. Since $\Delta \subseteq \text{LND}(B)$ does not imply that Δ is locally nilpotent, it is natural to ask whether the stronger condition $\text{Span}_{\mathbf{k}}(\Delta) \subseteq \text{LND}(B)$ would imply that Δ is locally nilpotent; it turns out that the answer is negative:

Example Let $\mathbf{k}$ be a field and $B = \mathbf{k}[x, y, z]$ (polynomial ring in 3 variables). Let $\Delta = \{D, E\}$ where $D = x\frac{\partial}{\partial y} + y\frac{\partial}{\partial z}$ and $E = y\frac{\partial}{\partial x} - z\frac{\partial}{\partial y}$. Then $\text{Span}_{\mathbf{k}}(\Delta) \subseteq \text{LND}(B)$. However, we have $D(E(x)) = x$, so Δ is not a locally nilpotent set.

In the above example the derivation $[D, E] = D \circ E - E \circ D$ is not locally nilpotent (it sends x to itself), so the Lie algebra generated by Δ is not included in $\text{LND}(B)$. So it makes sense to ask: if we make the stronger assumption that the Lie algebra generated by Δ is included in $\text{LND}(B)$, then does it follow that Δ is locally nilpotent? This is equivalent to asking:

Question 1 Let L be a Lie subalgebra of $\text{Der}_{\mathbf{k}}(B)$. If $L \subseteq \text{LND}(B)$, does it follow that L is a locally nilpotent subset of $\text{Der}_{\mathbf{k}}(B)$?

This question leads naturally to another question: can we characterize the Lie algebras that can be embedded in $\text{LND}(B)$? This question may be too ambitious, so we shall restrict ourselves to the following version of it:

Question 2 Let L be a Lie subalgebra of $\text{Der}_{\mathbf{k}}(B)$. If L satisfies one of:

- $L \subseteq \text{LND}(B)$,
- L is a locally nilpotent (or uniformly locally nilpotent) subset of $\text{Der}_{\mathbf{k}}(B)$,
- L is a Lie-locally nilpotent (or uniformly so) subset of $\text{Der}_{\mathbf{k}}(B)$,

then does it follow that L is a nilpotent Lie algebra, or that L satisfies some other nilpotency condition that makes sense for abstract Lie algebras? (The notion of a Lie-locally nilpotent subset of $\text{Der}_{\mathbf{k}}(B)$ is defined in Sect. 3.)

The aim of this article is to develop the basic theory of locally nilpotent sets and to explore what can be said about Questions 1 and 2 (where in fact Question 2 contains several questions). The basic theory is developed in Sect. 2; the first steps are done in the more general context of linear endomorphisms of vector spaces, and the later part of the Section is devoted to derivations. Section 3 deals with a variant of the notion of locally nilpotent set that is relevant in the context of Lie algebras of derivations. Section 4 studies five notions of nilpotency for Lie or associative algebras, needed in order to address Question 2. Three of them are classical (an algebra can be *nilpotent*, *nil*, or *locally nilpotent*). The other two are those of a *sequentially nilpotent* algebra and of a *locally nil* algebra; as far as we know these two have not been studied previously. Sections 5 and 6 are mainly devoted to Questions 1 and 2. Section 5 shows that if B is the polynomial ring in $|\mathbb{N}|$ variables over a field $\mathbf{k}$ then the following hold:

(1) (Example 5.5) There exists a Lie subalgebra L of $\text{Der}_{\mathbf{k}}(B)$ such that

   (a) every finitely generated Lie subalgebra of L is a locally nilpotent subset of $\text{Der}_{\mathbf{k}}(B)$ (so $L \subseteq \text{LND}(B)$);

(b) L is not a locally nilpotent subset of $\mathrm{Der}_{\mathbf{k}}(B)$.

So Question 1 has a negative answer.

(2) (Example 5.6) For each integer $m \geq 2$ there exists an m-generated Lie subalgebra L of $\mathrm{Der}_{\mathbf{k}}(B)$ satisfying:

(a) every $(m-1)$-generated Lie subalgebra of L is a locally nilpotent subset of $\mathrm{Der}_{\mathbf{k}}(B)$ (so $L \subseteq \mathrm{LND}(B)$);
(b) L is not a locally nilpotent subset of $\mathrm{Der}_{\mathbf{k}}(B)$.

So Question 1 has a negative answer even when L is a finitely generated Lie algebra.

(3) (Example 5.4) There exists a Lie subalgebra L of $\mathrm{Der}_{\mathbf{k}}(B)$ such that

(a) L is a uniformly locally nilpotent subset of $\mathrm{Der}_{\mathbf{k}}(B)$;
(b) L is the free Lie algebra on a countably infinite set (so L is as non-nilpotent as a Lie algebra can be).

So all questions that are part of Question 2 have negative answers.

(4) (Corollary 5.7) There exists an infinite subset Δ of $\mathrm{Der}_{\mathbf{k}}(B)$ satisfying:

(a) Δ is not a locally nilpotent subset of $\mathrm{Der}_{\mathbf{k}}(B)$;
(b) every finite subset of Δ is a locally nilpotent subset of $\mathrm{Der}_{\mathbf{k}}(B)$.

(5) (Corollary 5.7) For each integer $m \geq 2$, there exists an m-subset Δ of $\mathrm{Der}_{\mathbf{k}}(B)$ satisfying:

(a) Δ is not a locally nilpotent subset of $\mathrm{Der}_{\mathbf{k}}(B)$;
(b) every proper subset of Δ is a locally nilpotent subset of $\mathrm{Der}_{\mathbf{k}}(B)$.

Facts (4) and (5) show just how capriciously the notion of locally nilpotent set can behave. Note that (2) and (5) are based on Golod's famous example of an associative algebra that is nil and finitely generated but not nilpotent.

The above facts suggest that if we do not assume that B satisfies some finiteness condition then we cannot expect our questions to have affirmative answers. Section 6 re-examines Question 2 under the additional assumption that B is "derivation-finite" (this is more general than B being finitely generated, see Definition 6.1). Corollary 6.4 and Proposition 6.5 give some positive answers to Question 2, but Question 1 is left open.

In future work, we hope to apply these ideas to the following questions. Let B be a commutative algebra over a field $\mathbf{k}$ of characteristic zero, let Δ be a subset of $\mathrm{LND}(B)$ and consider the subset $Y = \left\{ \exp(D) \mid D \in \Delta \right\}$ of $\mathrm{Aut}_{\mathbf{k}}(B)$. *How are the properties of Δ related to those of Y? When is Y a group? When is Y included in an algebraic group?*

1 Preliminaries

1.1 Directed Trees. Consider a pair (S, E) where S is a set and E is a subset of $S \times S$ such that every element (u, v) of E satisfies $u \neq v$. Then (S, E) can be viewed as a directed graph where S is the vertex-set and E is the edge-set (if $(u, v) \in E$ then (u, v) is a directed edge from u to v). We allow S and E to be infinite sets. By a *path* in the graph (S, E), we mean a sequence $\gamma = (x_0, x_1, x_2, \dots)$ (finite or infinite) of elements of S such that $(x_i, x_{i+1}) \in E$ for all $i \geq 0$. If $\gamma = (x_0, x_1, \dots, x_n)$ is a finite path, we say that γ is a path *from x_0 to x_n*. If $x \in S$ then we regard (x) as a path from x to x.

Now suppose that (S, E) is a pair of the above type. Then it is not hard to see that there exists at most one vertex $v_0 \in S$ satisfying: for every $v \in S$ there exists exactly one path from v_0 to v. If such a vertex v_0 exists then we say that *(S, E) is a directed tree with root v_0*. We shall use the following fact several times; its proof is left to the reader.

Theorem 1.2 *Let (S, E) be a directed tree with root v_0, and suppose:*

- *for each $v \in S$, there exist finitely many $v' \in S$ such that $(v, v') \in E$;*
- *there exists no infinite path $(x_0, x_1, x_2, \dots)$ in (S, E) satisfying $x_0 = v_0$.*

Then S is a finite set.

1.3 Algebras. Let $\mathbf{k}$ be a field.

(a) A $\mathbf{k}$-*algebra* is a pair $(A, \cdot)$ where A is a $\mathbf{k}$-vector space and "$\cdot$" is an arbitrary $\mathbf{k}$-bilinear map $A \times A \to A$, $(x, y) \mapsto x \cdot y$. We call $\cdot$ the multiplication of A. Let A be a $\mathbf{k}$-algebra.
A *subalgebra* of A is a $\mathbf{k}$-subspace A' of A closed under the multiplication of A. If Δ is a subset of A then the *subalgebra of A generated by Δ* is the intersection of all subalgebras A' of A satisfying $\Delta \subseteq A'$.
A $\mathbf{k}$-*derivation* of A is a $\mathbf{k}$-linear map $D : A \to A$ satisfying $D(x \cdot y) = D(x) \cdot y + x \cdot D(y)$ for all $x, y \in A$.
A *homomorphism* $\varphi : A \to B$ of $\mathbf{k}$-algebras is a $\mathbf{k}$-linear map satisfying $\varphi(x \cdot y) = \varphi(x) \cdot \varphi(y)$ for all $x, y \in A$.
(b) An *associative algebra* over $\mathbf{k}$ is a $\mathbf{k}$-algebra whose multiplication is associative.
(c) A *Lie algebra* over $\mathbf{k}$ is a $\mathbf{k}$-algebra $(A, \cdot)$ satisfying $x \cdot x = 0$ and the Jacobi identity $x \cdot (y \cdot z) + y \cdot (z \cdot x) + z \cdot (x \cdot y) = 0$ for all $x, y, z \in A$.

For associative algebras and Lie algebras, the notions of *subalgebra*, *subalgebra generated by a set*, *homomorphism* and $\mathbf{k}$-*derivation* are those defined in part (a) for general algebras.

Notation 1.4 (a) If $(A, \cdot)$ is an associative algebra and if we define $a_1 * a_2 = a_1 \cdot a_2 - a_2 \cdot a_1$ for all $a_1, a_2 \in A$ then $(A, *)$ is a Lie algebra that we denote A_{L}.

(b) If V is a vector space over $\mathbf{k}$ then $\mathcal{L}_{\mathbf{k}}(V)$ denotes the set of all $\mathbf{k}$-linear maps $V \to V$. We always regard $\mathcal{L}_{\mathbf{k}}(V)$ as an associative algebra, the multiplication being the composition of maps. We write $\mathcal{L}_{\mathbf{k}}(V)_{\mathrm{L}}$ for the corresponding Lie algebra, as explained in part (a).
(c) Let A be an algebra over a field $\mathbf{k}$. The set of all $\mathbf{k}$-derivations of A is denoted $\mathrm{Der}_{\mathbf{k}}(A)$, and is a $\mathbf{k}$-subspace of $\mathcal{L}_{\mathbf{k}}(A)$. If $D, E \in \mathrm{Der}_{\mathbf{k}}(A)$ then $D \circ E - E \circ D \in \mathrm{Der}_{\mathbf{k}}(A)$, so $\mathrm{Der}_{\mathbf{k}}(A)$ is a subalgebra of the Lie algebra $\mathcal{L}_{\mathbf{k}}(A)_{\mathrm{L}}$.

Notation 1.5 Let A be an associative algebra over a field $\mathbf{k}$. If Δ is any subset of A then $\bar{\Delta}$ denotes the subalgebra of A generated by Δ and $\tilde{\Delta}$ denotes the subalgebra of A_{L} generated by Δ. We have $\Delta \subseteq \tilde{\Delta} \subseteq \bar{\Delta}$.

Note that $\bar{\Delta}$ is the intersection of all $\mathbf{k}$-subspaces A' of A that are closed under the multiplication of A and satisfy $\Delta \subseteq A'$. We also point out that $\bar{\Delta} = \mathrm{Span}_{\mathbf{k}}(\Delta_{\circ})$ where $\Delta_{\circ}$ is the set of **nonempty** products of elements of Δ.

The notation $\bar{\Delta}$ has the above meaning even when A happens to have a unity 1. For instance, if A is the commutative polynomial ring $\mathbf{k}[x, y]$ and $\Delta = \{x, y\}$ then $\bar{\Delta}$ is the ideal generated by x, y.

1.6 We adopt the **right-associativity convention**, which stipulates that any unparenthesized product $a_n \cdots a_1$ (where $n > 2$ and $a_1, \dots, a_n$ are elements of some algebra) is to be interpreted as meaning $a_n \cdot (a_{n-1} \cdots (a_3 \cdot (a_2 \cdot a_1)) \dots)$. For instance, $a \cdot b \cdot c \cdot d = a \cdot (b \cdot (c \cdot d))$. This convention is in effect throughout the paper, in all types of algebras.

1.7 Let $(A, \cdot)$ be a Lie algebra. It is customary to use the bracket notation for multiplication, i.e., to define $[x, y] = x \cdot y$ for all $x, y \in A$. We shall use both notations. If $x_0, x_1, \dots, x_n \in A$, the element $[x_n, [x_{n-1}, \dots [x_2, [x_1, x_0]] \dots]]$ of A is simply denoted $[x_n, \dots, x_0]$. This agrees with the right-associativity convention; for instance, going back and forth between the two notations,

$$[x_3, x_2, x_1, x_0] = [x_3, [x_2, [x_1, x_0]]] = x_3 \cdot (x_2 \cdot (x_1 \cdot x_0)) = x_3 \cdot x_2 \cdot x_1 \cdot x_0 .$$

In general we have $[x_n, \cdots, x_0] = x_n \cdots x_0$, where we allow the case $n = 0$: $[x_0] = x_0$.

If $x \in A$ then the map $y \mapsto [x, y]$ is denoted $\mathrm{ad}(x) : A \to A$. The following properties of ad are well known, and easily verified:

(i) $\mathrm{ad} : A \to \mathcal{L}_{\mathbf{k}}(A)_{\mathrm{L}}$ is a homomorphism of Lie algebras;
(ii) $\mathrm{ad}(A) \subseteq \mathrm{Der}_{\mathbf{k}}(A)$.

Note that if $n > 0$ and $x_0, \dots, x_n \in A$ then

$$[x_n, \dots, x_0] = \big(\mathrm{ad}(x_n) \circ \cdots \circ \mathrm{ad}(x_1)\big)(x_0) .$$

We leave it to the reader to check the following fact (which is trivial in the associative case):

Lemma 1.8 *Let $(A, \cdot)$ be an associative algebra or a Lie algebra over a field* $\mathbf{k}$.

(a) Let $x_1, \dots, x_n \in A$ and let $x \in A$ be any parenthesization of the product $x_n \cdots x_1$. Then x is a finite sum $\sum_i r_i\, x_{i,n} \cdots x_{i,1}$ where, for each i, r_i belongs to the prime subring of $\mathbf{k}$ and $(x_{i,1}, \dots, x_{i,n})$ is a permutation of $(x_1, \dots, x_n)$.

(b) If H is a generating set for A then

$$A = \operatorname{Span}_{\mathbf{k}} \left\{ x_m \cdots x_0 \mid m \in \mathbb{N} \text{ and } (x_0, \dots, x_m) \in H^{m+1} \right\}.$$

2 Locally Nilpotent Sets of Linear Maps

Throughout this section, we consider a vector space V over a field $\mathbf{k}$ and we let $\mathfrak{L}_{\mathbf{k}}(V)$ denote the set of all $\mathbf{k}$-linear maps $V \to V$. An element F of $\mathfrak{L}_{\mathbf{k}}(V)$ is said to be *locally nilpotent* if for each $x \in V$ there exists $n > 0$ such that $F^n(x) = 0$. We write $\operatorname{LN}(\mathfrak{L}_{\mathbf{k}} V)$ for the set of all locally nilpotent elements of $\mathfrak{L}_{\mathbf{k}}(V)$.

If Δ is a set then $\Delta^{\mathbb{N}}$ denotes the set of all infinite sequences $(a_0, a_1, a_2, \dots)$ of elements of Δ.

Definition 2.1 Given a subset Δ of $\mathfrak{L}_{\mathbf{k}}(V)$, we define the subsets $\operatorname{Nil}(\Delta)$ and $\operatorname{UNil}(\Delta)$ of V as follows. If $\Delta = \varnothing$, we set $\operatorname{Nil}(\Delta) = V = \operatorname{UNil}(\Delta)$. If $\Delta \neq \varnothing$,

- $\operatorname{Nil}(\Delta)$ is the set of all $x \in V$ satisfying:

  for every sequence $(F_0, F_1, \dots) \in \Delta^{\mathbb{N}}$ there exists $n \in \mathbb{N}$ such that $(F_n \circ \cdots \circ F_0)(x) = 0$;

- $\operatorname{UNil}(\Delta)$ is the set of all $x \in V$ satisfying:

  there exists $n \in \mathbb{N}$ such that for every $(F_1, \dots, F_n) \in \Delta^n$ we have $(F_n \circ \cdots \circ F_1)(x) = 0$.

Note that $\operatorname{UNil}(\Delta) \subseteq \operatorname{Nil}(\Delta)$ are linear subspaces of V. If $\operatorname{Nil}(\Delta) = V$, we say that Δ is a *locally nilpotent subset of* $\mathfrak{L}_{\mathbf{k}}(V)$ (or simply, that Δ is *locally nilpotent*); if $\operatorname{UNil}(\Delta) = V$, we say that Δ is *uniformly locally nilpotent.*

If Δ is a locally nilpotent subset of $\mathfrak{L}_{\mathbf{k}}(V)$ then $\Delta \subseteq \operatorname{LN}(\mathfrak{L}_{\mathbf{k}} V)$; however, the converse is not true.

If Δ is a uniformly locally nilpotent subset of $\mathfrak{L}_{\mathbf{k}}(V)$ then Δ is locally nilpotent, but the converse is not true (see Example 6.2). However, note the following:

Lemma 2.2 *Let Δ be a finite subset of $\mathfrak{L}_{\mathbf{k}}(V)$. Then* $\operatorname{Nil}(\Delta) = \operatorname{UNil}(\Delta)$. *In particular, Δ is locally nilpotent if and only if it is uniformly locally nilpotent.*

Proof It suffices to show that $\operatorname{Nil}(\Delta) \subseteq \operatorname{UNil}(\Delta)$. Let $x \in \operatorname{Nil}(\Delta) \setminus \{0\}$. Consider the set S of finite sequences $(F_1, \dots, F_n)$ of elements of Δ satisfying $(F_n \circ \cdots \circ F_1)(x) \neq 0$ (where the empty sequence () belongs to S). Let $E \subseteq S \times S$ be the set of ordered pairs of the form $\big((F_1, \dots, F_n), (F_1, \dots, F_n, F_{n+1})\big)$. Then (S, E) is a directed tree with root () (see Sect. 1). For each vertex $v \in S$, the number of

$v' \in S$ satisfying $(v, v') \in E$ is at most $|\Delta|$, which is finite. The fact that $x \in \mathrm{Nil}(\Delta)$ implies that there is no infinite path $\big((), (F_1), (F_1, F_2), \dots\big)$ in this tree. It follows from Theorem 1.2 that S is a finite set, and this implies that $x \in \mathrm{UNil}(\Delta)$. □

Definition 2.3 Given a nonempty subset Δ of $\mathfrak{L}_{\mathbf{k}}(V)$ we define a map $\deg_\Delta : V \to \mathbb{N} \cup \{-\infty, \infty\}$ by declaring that $\deg_\Delta(0) = -\infty$, if $x \in V \setminus \mathrm{UNil}(\Delta)$ then $\deg_\Delta(x) = \infty$, and if $x \in \mathrm{UNil}(\Delta) \setminus \{0\}$ then $\deg_\Delta(x)$ is equal to:

$$\max\big\{ n \in \mathbb{N} \mid \text{ there exists } (F_1, \dots, F_n) \in \Delta^n \text{ satisfying } (F_n \circ \cdots \circ F_1)(x) \neq 0 \big\}$$

(to be clear, if $x \neq 0$ and $F(x) = 0$ for all $F \in \Delta$ then we define $\deg_\Delta(x) = 0$). If $\Delta = \varnothing$ then we define $\deg_\Delta(0) = -\infty$ and $\deg_\Delta(x) = 0$ for all $x \in V \setminus \{0\}$.

Lemma 2.4 *Let Δ be a subset of $\mathfrak{L}_{\mathbf{k}}(V)$. Then the map $\deg_\Delta : V \to \mathbb{N} \cup \{\pm\infty\}$ has the following properties, where x, y are arbitrary elements of V.*

(a) $\deg_\Delta(x) < 0$ *if and only if* $x = 0$.
(b) $\deg_\Delta(x) \leq 0$ *if and only if* $F(x) = 0$ *for all* $F \in \Delta$.
(c) $\deg_\Delta(x) < \infty$ *if and only if* $x \in \mathrm{UNil}(\Delta)$.
(d) *If* $x \in \mathrm{UNil}(\Delta) \setminus \{0\}$ *and* $F \in \Delta$ *then* $\deg_\Delta(F(x)) < \deg_\Delta(x)$.
(e) $\deg_\Delta(x + y) \leq \max(\deg_\Delta(x), \deg_\Delta(y))$

Verification of Lemma 2.4 is left to the reader.
The notations $\bar{\Delta}$ and $\tilde{\Delta}$ are defined in Notation 1.5.

Proposition 2.5 *Let Δ be a subset of $\mathfrak{L}_{\mathbf{k}}(V)$, let $\bar{\Delta}$ be the associative subalgebra of $\mathfrak{L}_{\mathbf{k}}(V)$ generated by Δ and let $\tilde{\Delta}$ be the Lie subalgebra of $\mathfrak{L}_{\mathbf{k}}(V)_L$ generated by Δ. Then $\Delta \subseteq \tilde{\Delta} \subseteq \bar{\Delta} \subseteq \mathfrak{L}_{\mathbf{k}}(V)$ and the following hold.*

(a) $\mathrm{Nil}(\bar{\Delta}) = \mathrm{Nil}(\tilde{\Delta}) = \mathrm{Nil}(\Delta)$
(b) $\deg_{\bar{\Delta}} = \deg_{\tilde{\Delta}} = \deg_\Delta$ *and* $\mathrm{UNil}(\bar{\Delta}) = \mathrm{UNil}(\tilde{\Delta}) = \mathrm{UNil}(\Delta)$
(c) *If Δ is locally nilpotent then so are $\bar{\Delta}$ and $\tilde{\Delta}$.*
(d) *If Δ is uniformly locally nilpotent then so are $\bar{\Delta}$ and $\tilde{\Delta}$.*

Proof We may assume that $\Delta \neq \varnothing$. Let $\Delta_\circ$ be as in Notation 1.5 and observe that $\Delta \subseteq \Delta_\circ \subseteq \bar{\Delta}$ and $\Delta \subseteq \tilde{\Delta} \subseteq \bar{\Delta}$. These inclusions have the following trivial consequences:

$$\mathrm{Nil}(\bar{\Delta}) \subseteq \mathrm{Nil}(\Delta_\circ) \subseteq \mathrm{Nil}(\Delta) \text{ and } \mathrm{Nil}(\bar{\Delta}) \subseteq \mathrm{Nil}(\tilde{\Delta}) \subseteq \mathrm{Nil}(\Delta), \quad (1)$$

$$\mathrm{UNil}(\bar{\Delta}) \subseteq \mathrm{UNil}(\Delta_\circ) \subseteq \mathrm{UNil}(\Delta) \text{ and } \mathrm{UNil}(\bar{\Delta}) \subseteq \mathrm{UNil}(\tilde{\Delta}) \subseteq \mathrm{UNil}(\Delta), \quad (2)$$

for all $x \in V$ we have
$$\deg_\Delta(x) \leq \deg_{\Delta_\circ}(x) \leq \deg_{\bar{\Delta}}(x) \text{ and } \deg_\Delta(x) \leq \deg_{\tilde{\Delta}}(x) \leq \deg_{\bar{\Delta}}(x). \quad (3)$$

Let $x \in \mathrm{Nil}(\Delta)$. If $(G_0, G_1, \dots) \in \Delta_\circ^{\mathbb{N}}$ then there exist $(F_0, F_1, \dots) \in \Delta^{\mathbb{N}}$ and $1 \leq n_0 < n_1 < \cdots$ in $\mathbb{N}$ such that $G_0 = F_{n_0 - 1} \circ \cdots \circ F_0$, $G_1 = F_{n_1 - 1} \circ \cdots \circ F_{n_0}$, $G_2 = F_{n_2 - 1} \circ \cdots \circ F_{n_1}$, and so on. Since $(F_k \circ \cdots \circ F_0)(x) = 0$ for some $k \in \mathbb{N}$, it follows that $(G_m \circ \cdots \circ G_0)(x) = 0$ for some $m \in \mathbb{N}$. This shows that $x \in \mathrm{Nil}(\Delta_\circ)$ and hence that $\mathrm{Nil}(\Delta) \subseteq \mathrm{Nil}(\Delta_\circ)$. By (1), we get $\mathrm{Nil}(\Delta_\circ) = \mathrm{Nil}(\Delta)$.

We continue to consider $x \in \mathrm{Nil}(\Delta) = \mathrm{Nil}(\Delta_\circ)$. Let $(H_0, H_1, \dots) \in \bar{\Delta}^{\mathbb{N}}$; let us prove that there exists n such that $(H_n \circ \cdots \circ H_0)(x) = 0$. Recall that $\bar{\Delta} = \mathrm{Span}_{\mathbf{k}}(\Delta_\circ)$; for each $i \in \mathbb{N}$, we choose $m_i \in \mathbb{N}$, $\lambda_{i,0}, \dots, \lambda_{i,m_i} \in \mathbf{k}$ and $G_{i,0}, \dots, G_{i,m_i} \in \Delta_\circ$ satisfying

$$H_i = \textstyle\sum_{j=0}^{m_i} \lambda_{i,j} G_{i,j}\,. \tag{4}$$

The element x, the sequence $(H_0, H_1, \dots)$ and the relations (4) determine a directed tree T that we now define. The elements of the vertex-set S are all finite sequences $(G_{0,j_0}, G_{1,j_1}, \dots, G_{n,j_n})$ such that $(G_{n,j_n} \circ \cdots \circ G_{0,j_0})(x) \neq 0$ (where $0 \le j_i \le m_i$ for all $i = 0, \dots, n$ and where the $G_{i,j}$ are the same as in (4)). The edge-set of T is the subset E of $S \times S$ whose elements are the pairs of the form $\big((G_{0,j_0}, \dots, G_{n,j_n}), (G_{0,j_0}, \dots, G_{n,j_n}, G_{n+1,j_{n+1}})\big)$. The root of T is the empty sequence $() \in S$. Note that if $\big((), (G_{0,j_0}), (G_{0,j_0}, G_{1,j_1}), (G_{0,j_0}, G_{1,j_1}, G_{2,j_2}), \dots\big)$ is an infinite path in T then the sequence $(G_{0,j_0}, G_{1,j_1}, G_{2,j_2}, \dots) \in \Delta_\circ^{\mathbb{N}}$ satisfies $(G_{n,j_n} \circ \cdots \circ G_{0,j_0})(x) \neq 0$ for all $n \in \mathbb{N}$, which contradicts the fact that $x \in \mathrm{Nil}(\Delta_\circ)$. So, *there is no infinite path in T starting at the root.* We also note that if $v = (G_{0,j_0}, \dots, G_{n,j_n})$ is any vertex then the number of vertices $v' \in S$ satisfying $(v, v') \in E$ is at most m_{n+1}, which is finite. It follows from Theorem 1.2 that T has finitely many vertices. So there exists $n \in \mathbb{N}$ satisfying:

$$(G_{n,j_n} \circ \cdots \circ G_{0,j_0})(x) = 0 \quad \text{for all } (j_0, \dots, j_n) \in \mathbb{I}_{m_0} \times \cdots \times \mathbb{I}_{m_n} \tag{5}$$

where we write $\mathbb{I}_m = \big\{ i \in \mathbb{N} \mid 0 \le i \le m \big\}$. Let n be as in (5) and note that $(H_n \circ \cdots \circ H_0)(x)$ is a linear combination

$$\sum_{(j_0,\dots,j_n)\in\mathbb{I}_{m_0}\times\cdots\times\mathbb{I}_{m_n}} \mu_{(j_0,\dots,j_n)} (G_{n,j_n} \circ \cdots \circ G_{0,j_0})(x) \tag{6}$$

where $\mu_{(j_0,\dots,j_n)} \in \mathbf{k}$ for all $(j_0, \dots, j_n) \in \mathbb{I}_{m_0} \times \cdots \times \mathbb{I}_{m_n}$. By (5), all terms of the sum (6) are zero, so $(H_n \circ \cdots \circ H_0)(x) = 0$. This proves that $x \in \mathrm{Nil}(\bar{\Delta})$ and hence that $\mathrm{Nil}(\Delta) \subseteq \mathrm{Nil}(\bar{\Delta})$. Then (1) gives $\mathrm{Nil}(\bar{\Delta}) = \mathrm{Nil}(\tilde{\Delta}) = \mathrm{Nil}(\Delta)$, so (a) is proved.

Let us prove (b). If $(G_1, \dots, G_n) \in \Delta_\circ^n$ then $G_n \circ \cdots \circ G_1 = F_N \circ \cdots \circ F_1$ for some $N \ge n$ and $(F_1, \dots, F_N) \in \Delta^N$. This implies that $\deg_{\Delta_\circ}(x) \le \deg_\Delta(x)$ for all $x \in V$, so we obtain $\deg_{\Delta_\circ} = \deg_\Delta$ by (3).

Consider $x \in V \setminus \{0\}$ and let us prove that $\deg_{\bar{\Delta}}(x) \le \deg_{\Delta_\circ}(x)$. We may assume that $\deg_{\Delta_\circ}(x) = n \in \mathbb{N}$. If $(H_1, \dots, H_{n+1}) \in \bar{\Delta}^{n+1}$ then $(H_{n+1} \circ \cdots \circ H_1)(x)$ is a finite sum $\sum_i \lambda_i t_i$ where $\lambda_i \in \mathbf{k}$ and $t_i = (G_{i,n+1} \circ \cdots \circ G_{i,1})(x)$ with $G_{i,1}, \dots, G_{i,n+1} \in \Delta_\circ$. Since $\deg_{\Delta_\circ}(x) = n$ we have $t_i = 0$ for all i, so $(H_{n+1} \circ \cdots \circ H_1)(x) = 0$. Consequently, $\deg_{\bar{\Delta}}(x) \le n = \deg_{\Delta_\circ}(x)$. We get $\deg_{\bar{\Delta}} = \deg_{\Delta_\circ}$ by (3), and since $\deg_{\Delta_\circ} = \deg_\Delta$ it follows that $\deg_{\bar{\Delta}} = \deg_\Delta$. Using (3) again gives $\deg_\Delta = \deg_{\tilde{\Delta}} = \deg_{\bar{\Delta}}$, and hence $\mathrm{UNil}(\Delta) = \mathrm{UNil}(\tilde{\Delta}) = \mathrm{UNil}(\bar{\Delta})$. So (b) is proved. Assertions (c) and (d) follow from (a) and (b). □

Corollary 2.6 *Let A be either an associative subalgebra of $\mathfrak{L}_{\mathbf{k}}(V)$ or a Lie subalgebra of $\mathfrak{L}_{\mathbf{k}}(V)_L$. If A is finitely generated then* $\text{Nil}(A) = \text{UNil}(A)$ *and, consequently, A is a locally nilpotent subset of $\mathfrak{L}_{\mathbf{k}}(V)$ if and only if it is a uniformly locally nilpotent subset of $\mathfrak{L}_{\mathbf{k}}(V)$.*

Proof Let Δ be a finite generating set for the algebra A. Then $\text{Nil}(A) = \text{Nil}(\Delta) = \text{UNil}(\Delta) = \text{UNil}(A)$ by Proposition 2.5 and Lemma 2.2. □

A Special Case: Sets of Derivations

Let $(A, \cdot)$ be an algebra over a field $\mathbf{k}$, in the sense of 1.3(a). Recall (from 1.4) that this determines a Lie subalgebra $\text{Der}_{\mathbf{k}}(A)$ of $\mathfrak{L}_{\mathbf{k}}(A)_{\text{L}}$. If Δ is a subset of $\text{Der}_{\mathbf{k}}(A)$ then $\Delta \subseteq \mathfrak{L}_{\mathbf{k}}(A)$, so it makes sense to consider the subsets $\text{Nil}(\Delta)$ and $\text{UNil}(\Delta)$ of A, and the map $\deg_\Delta : A \to \mathbb{N} \cup \{\pm\infty\}$.

The rest of the section is devoted to the proof of:

Theorem 2.7 *Let $(A, \cdot)$ be an algebra over a field* $\mathbf{k}$. *If Δ is a subset of* $\text{Der}_{\mathbf{k}}(A)$ *then the following hold.*

(a) The sets $\text{Nil}(\Delta)$ *and* $\text{UNil}(\Delta)$ *are subalgebras of* $(A, \cdot)$.
(b) The map $\deg_\Delta : A \to \mathbb{N} \cup \{\pm\infty\}$ *satisfies* $\deg_\Delta(x \cdot y) \leq \deg_\Delta(x) + \deg_\Delta(y)$ *for all $x, y \in$* $\text{UNil}(\Delta)$.

We already know that $\text{Nil}(\Delta)$ and $\text{UNil}(\Delta)$ are linear subspaces of A. So, to prove (a), it suffices to show that those two sets are closed under the multiplication of A.

Remark 2.8 If $\mathbf{k} \subseteq A$ then $\mathbf{k} \subseteq \text{UNil}(\Delta) \subseteq \text{Nil}(\Delta)$, because $\mathbf{k} \subseteq \ker D$ for all $D \in \text{Der}_{\mathbf{k}}(A)$.

Until the end of the section, we assume that $(A, \cdot)$ is a $\mathbf{k}$-algebra. The Theorem follows from Lemmas 2.13 and 2.14 and Corollary 2.15.

Notation 2.9 If $n \in \mathbb{N}$, we write $\mathbb{I}_n = \{\, i \in \mathbb{N} \mid 0 \leq i \leq n \,\}$. Let $n \in \mathbb{N}$ and $F_0, \dots, F_n \in \mathfrak{L}_{\mathbf{k}}(A)$. If $I = \{i_1, \dots, i_m\}$ is a subset of $\mathbb{I}_n$ and $i_1 < \cdots < i_m$, we write $F_I = F_{i_m} \circ \cdots \circ F_{i_1}$ (if $I = \varnothing$ then $F_I = \text{id}_A$). If S is a set then $\wp(S)$ denotes the set of all subsets of S.

Lemma 2.10 *Let $x, y \in A$, $n \in \mathbb{N}$ and $D_0, \dots, D_n \in$* $\text{Der}_{\mathbf{k}}(A)$. *Then*

$$(D_n \circ \cdots \circ D_0)(x \cdot y) = \sum_{I \in \wp(\mathbb{I}_n)} D_I(x) \cdot D_{\mathbb{I}_n \setminus I}(y).$$

Proof This can be proved by induction on n. The straightforward argument is left to the reader. □

Definition 2.11 We define a partial order $\preceq$ on the set $\wp_{\text{fin}}(\mathbb{N})$ of finite subsets of $\mathbb{N}$ by declaring that the condition $I \preceq J$ (where $I, J \in \wp_{\text{fin}}(\mathbb{N})$) is equivalent to:

$$I \subseteq J \text{ and for every } i \in I \text{ and } j \in J \setminus I \text{ we have } i < j.$$

Note in particular that $\varnothing \preceq J$ for all $J \in \wp_{\text{fin}}(\mathbb{N})$. Also observe that if $J \in \wp_{\text{fin}}(\mathbb{N})$ and $m \in \mathbb{N}$ then $\mathbb{I}_m \cap J \preceq J$.

Lemma 2.12 *Let X be a subset of $\wp_{\text{fin}}(\mathbb{N})$ such that*

- *for all $I, J \in \wp_{\text{fin}}(\mathbb{N})$ satisfying $I \preceq J$ and $J \in X$, we have $I \in X$;*
- *$(X, \preceq)$ satisfies the ascending chain condition (ACC).*

Then the set

$$Z = \{ n \in \mathbb{N} \mid \text{ there exists a subset } I \text{ of } \mathbb{I}_n \text{ such that } I \in X \text{ and } \mathbb{I}_n \setminus I \in X \}$$

is finite.

Proof By contradiction, assume that Z is an infinite set. For each $n \in Z$, choose $I_n \in \wp(\mathbb{I}_n)$ such that $I_n \in X$ and $\mathbb{I}_n \setminus I_n \in X$. We inductively define an infinite sequence of sets $Z_0, Z_1, \dots$. Choose an infinite subset Z_0 of Z such that $(\mathbb{I}_0 \cap I_n)_{n \in Z_0}$ is a constant sequence. Let $m \in \mathbb{N}$ and suppose that $Z_0, \dots, Z_m$ satisfy:

- $Z \supseteq Z_0 \supseteq Z_1 \supseteq \cdots \supseteq Z_m$ are infinite sets;
- for each $j \in \{0, \dots, m\}$, $\min Z_j \geq j$ and $(\mathbb{I}_j \cap I_n)_{n \in Z_j}$ is a constant sequence.

Since $(\mathbb{I}_{m+1} \cap I_n)_{n \in Z_m}$ is an infinite sequence of elements of the finite set $\wp(\mathbb{I}_{m+1})$, we may choose an infinite subset Z_{m+1} of Z_m such that $(\mathbb{I}_{m+1} \cap I_n)_{n \in Z_{m+1}}$ is a constant sequence and $\min Z_{m+1} \geq m+1$. By induction, it follows that there exists an infinite sequence $Z_0, Z_1, Z_2, \dots$ satisfying:

(i) $Z \supseteq Z_0 \supseteq Z_1 \supseteq Z_2 \supseteq \cdots$ are infinite sets;
(ii) for each $j \in \mathbb{N}$, $\min Z_j \geq j$ and $(\mathbb{I}_j \cap I_n)_{n \in Z_j}$ is a constant sequence.

For each $j \in \mathbb{N}$, define $B_j = \mathbb{I}_j \cap I_n$ for any $n \in Z_j$, and let $B_j^* = \mathbb{I}_j \setminus B_j$. We shall now prove that the two sequences $(B_j)_{j \in \mathbb{N}}$ and $(B_j^*)_{j \in \mathbb{N}}$ eventually stabilize; this is absurd, because $B_j \cup B_j^* = \mathbb{I}_j$.

Let $j \in \mathbb{N}$. Given any $n \in Z_j$ we have $B_j = \mathbb{I}_j \cap I_n \preceq I_n$ and $I_n \in X$, so $B_j \in X$. Also, $\mathbb{I}_j \cap (\mathbb{I}_n \setminus I_n) \preceq \mathbb{I}_n \setminus I_n \in X$ implies $\mathbb{I}_j \cap (\mathbb{I}_n \setminus I_n) \in X$; since $\mathbb{I}_j \cap (\mathbb{I}_n \setminus I_n) = (\mathbb{I}_j \cap \mathbb{I}_n) \setminus (\mathbb{I}_j \cap I_n) = \mathbb{I}_j \setminus B_j = B_j^*$ (where we used $n \in Z_j \Rightarrow n \geq j \Rightarrow \mathbb{I}_j \cap \mathbb{I}_n = \mathbb{I}_j$), we have $B_j^* \in X$. Thus $(B_j)_{j \in \mathbb{N}}$ and $(B_j^*)_{j \in \mathbb{N}}$ are sequences in X.

Again, let $j \in \mathbb{N}$. Pick $n \in Z_{j+1}$ (so $n \in Z_j$), then $\mathbb{I}_j \cap B_{j+1} = \mathbb{I}_j \cap (\mathbb{I}_{j+1} \cap I_n) = \mathbb{I}_j \cap I_n = B_j$, so $B_j \preceq B_{j+1}$. Thus $B_0 \preceq B_1 \preceq B_2 \preceq \cdots$, and since $(X, \preceq)$ satisfies ACC there exists $M \in \mathbb{N}$ such that $(B_j)_{j \geq M}$ is a constant sequence. For $j \geq M$ we have $B_j = B_{j+1} = B_M$, so $\mathbb{I}_j \cap B_{j+1}^* = \mathbb{I}_j \cap (\mathbb{I}_{j+1} \setminus B_{j+1}) = \mathbb{I}_j \cap (\mathbb{I}_{j+1} \setminus B_M) = \mathbb{I}_j \setminus B_M = \mathbb{I}_j \setminus B_j = B_j^*$, so $B_j^* \preceq B_{j+1}^*$. Thus $B_M^* \preceq B_{M+1}^* \preceq \cdots$, so the sequence $(B_j^*)_{j \geq M}$ stabilizes. As we already explained, this is absurd. □

Recall that A is an algebra over a field $\mathbf{k}$.

Lemma 2.13 *For every subset* Δ *of* $\mathrm{Der}_{\mathbf{k}}(A)$, $\mathrm{Nil}(\Delta)$ *is closed under the multiplication of* A.

Proof We may assume that $\Delta \neq \varnothing$. Let $x, y \in \mathrm{Nil}(\Delta)$ and $(D_0, D_1, \dots) \in \Delta^{\mathbb{N}}$. To prove the Lemma, it suffices to show that there exists $n \in \mathbb{N}$ such that $(D_n \circ \cdots \circ D_0)(x \cdot y) = 0$.

Define $X_x = \{ I \in \wp_{\text{fin}}(\mathbb{N}) \mid D_I(x) \neq 0 \}$. Let us prove:

(i) for all $I, J \in \wp_{\text{fin}}(\mathbb{N})$ satisfying $I \preceq J$ and $J \in X_x$, we have $I \in X_x$;
(ii) $(X_x, \preceq)$ satisfies the ACC.

Consider $I, J \in \wp_{\text{fin}}(\mathbb{N})$ satisfying $I \preceq J$ and $J \in X_x$. Then $0 \neq D_J(x) = (D_{J \setminus I} \circ D_I)(x)$, so $D_I(x) \neq 0$, so $I \in X_x$. This proves (i).

Consider an infinite sequence $I_0 \preceq I_1 \preceq \cdots$ in X_x. By contradiction, assume that $(I_n)_{n \in \mathbb{N}}$ does not stabilize. Then $I = \bigcup_{n \in \mathbb{N}} I_n$ is an infinite subset of $\mathbb{N}$. Let $i_0 < i_1 < i_2 < \cdots$ be the elements of I and consider $(D_{i_0}, D_{i_1}, \dots) \in \Delta^{\mathbb{N}}$. Given any $k \in \mathbb{N}$ there exists $n \in \mathbb{N}$ such that $\{i_0, i_1, \dots, i_k\} \subseteq I_n$; then $\{i_0, i_1, \dots, i_k\} \preceq I_n \in X_x$, so $\{i_0, i_1, \dots, i_k\} \in X_x$ by (i), so $(D_{i_k} \circ \cdots \circ D_{i_0})(x) \neq 0$. Since this holds for all $k \in \mathbb{N}$, we have $x \notin \mathrm{Nil}(\Delta)$, a contradiction. So (ii) is proved.

Let $X_y = \{ I \in \wp_{\text{fin}}(\mathbb{N}) \mid D_I(y) \neq 0 \}$; then it is clear that (i) and (ii) are true with "X_y" in place of "X_x". Consequently, if we define $X = X_x \cup X_y$ then:

(iii) for all $I, J \in \wp_{\text{fin}}(\mathbb{N})$ satisfying $I \preceq J$ and $J \in X$, we have $I \in X$;
(iv) $(X, \preceq)$ satisfies the ACC.

By (iii), (iv) and Lemma 2.12, the set

$$Z = \{ n \in \mathbb{N} \mid \text{ there exists a subset } I \text{ of } \mathbb{I}_n \text{ such that } I \in X \text{ and } \mathbb{I}_n \setminus I \in X \}$$

is finite. Choose $n \in \mathbb{N} \setminus Z$. Then for each subset I of $\mathbb{I}_n$ we have $I \notin X$ (i.e., $D_I(x) = 0 = D_I(y)$) or $\mathbb{I}_n \setminus I \notin X$ (i.e., $D_{\mathbb{I}_n \setminus I}(x) = 0 = D_{\mathbb{I}_n \setminus I}(y)$). So all terms are zero in the right-hand-side of

$$(D_n \circ \cdots \circ D_0)(x \cdot y) = \sum_{I \in \wp(\mathbb{I}_n)} D_I(x) \cdot D_{\mathbb{I}_n \setminus I}(y)$$

(the equality follows from Lemma 2.10). Thus $(D_n \circ \cdots \circ D_0)(x \cdot y) = 0$, as desired. □

Lemma 2.14 *Let* Δ *be a subset of* $\mathrm{Der}_{\mathbf{k}}(A)$. *Then the map* $\deg_\Delta : A \to \mathbb{N} \cup \{\pm\infty\}$ *satisfies* $\deg_\Delta(x \cdot y) \leq \deg_\Delta(x) + \deg_\Delta(y)$ *for all* $x, y \in \mathrm{UNil}(\Delta)$.

Proof The case $\Delta = \varnothing$ is trivial, so let us assume that $\Delta \neq \varnothing$. It suffices to show that $P(n)$ is true for all $n \in \mathbb{N}$, where we define

$$P(n): \quad \forall_{x,y \in A \setminus \{0\}} \big(\deg_\Delta(x) + \deg_\Delta(y) \leq n \implies \deg_\Delta(x \cdot y) \leq n \big).$$

Suppose that $x, y \in A \setminus \{0\}$ satisfy $\deg_\Delta(x) + \deg_\Delta(y) \leq 0$. Then $\deg_\Delta(x) = 0 = \deg_\Delta(y)$, so for all $D \in \Delta$ we have $D(x) = 0 = D(y)$ and hence $D(x \cdot y) = D(x) \cdot y + x \cdot D(y) = 0$, so $\deg_\Delta(x \cdot y) \leq 0$, showing that $P(0)$ is true.

Let $n \in \mathbb{N}$ and assume that $P(n)$ is true. To prove that $P(n+1)$ is true, consider $x, y \in A \setminus \{0\}$ satisfying $\deg_\Delta(x) + \deg_\Delta(y) \leq n + 1$. We have to show that $\deg_\Delta(x \cdot y) \leq n + 1$. To do this, it suffices to show that $(D_{n+2} \circ \cdots \circ D_1)(x \cdot y) = 0$ for all $(D_1, \dots, D_{n+2}) \in \Delta^{n+2}$. So let us consider $(D_1, \dots, D_{n+2}) \in \Delta^{n+2}$. As a first step, we claim that

$$\deg_\Delta(D_1(x) \cdot y) \leq n \quad \text{and} \quad \deg_\Delta(x \cdot D_1(y)) \leq n. \tag{7}$$

To see this, first note that if $D_1(x) = 0$ then $\deg_\Delta(D_1(x) \cdot y) = -\infty \leq n$; so, to prove the first part of (7), we may assume that $D_1(x) \neq 0$. Note that $\deg_\Delta(D_1(x)) < \deg_\Delta(x)$ by part (d) of Lemma 2.4, so $\deg_\Delta(D_1(x)) + \deg_\Delta(y) \leq n$; since both $D_1(x)$ and y belong to $A \setminus \{0\}$ and $P(n)$ is true, we get $\deg_\Delta(D_1(x) \cdot y) \leq n$. By the same argument we have $\deg_\Delta(x \cdot D_1(y)) \leq n$, so (7) is proved. Then

$$\begin{aligned} \deg_\Delta(D_1(x \cdot y)) &= \deg_\Delta(D_1(x) \cdot y + x \cdot D_1(y)) \\ &\leq \max\big(\deg_\Delta(D_1(x) \cdot y), \deg_\Delta(x \cdot D_1(y))\big) \leq n \end{aligned} \tag{8}$$

by part (e) of Lemma 2.4 and (7). Now (8) implies that

$$(D_{n+2} \circ \cdots \circ D_1)(x \cdot y) = (D_{n+2} \circ \cdots \circ D_2)(D_1(x \cdot y)) = 0,$$

which is the desired conclusion. So $\deg_\Delta(x \cdot y) \leq n + 1$ and hence $P(n+1)$ is true. □

Corollary 2.15 *Let Δ be a subset of* $\mathrm{Der}_{\mathbf{k}}(A)$. *Then* $\mathrm{UNil}(\Delta)$ *is closed under the multiplication of A.*

Proof This follows from Lemma 2.14 and Part (c) of Lemma 2.4. □

This completes the Proof of Theorem 2.7.

3 Lie-Locally Nilpotent Sets of Linear Maps

Let V be a vector space over a field $\mathbf{k}$.

Caution 3.1 Let $x \in V$ and $F_0, \dots, F_{n+1} \in \mathcal{L}_{\mathbf{k}}(V)$. It is clear that $(F_n \circ \cdots \circ F_0)(x) = 0$ implies $(F_{n+1} \circ F_n \circ \cdots \circ F_0)(x) = 0$ and that $[F_n, \dots, F_0] = 0$ implies $[F_{n+1}, F_n, \dots, F_0] = 0$, but it is not the case that $[F_n, \dots, F_0](x) = 0$ implies $[F_{n+1}, F_n, \dots, F_0](x) = 0$. (This should be kept in mind while reading this section.)

Definition 3.2 Given a subset Δ of $\mathcal{L}_{\mathbf{k}}(V)$, we define the subsets $\mathrm{Nil}^{\mathrm{L}}(\Delta)$ and $\mathrm{UNil}^{\mathrm{L}}(\Delta)$ of V as follows. If $\Delta = \varnothing$, we set $\mathrm{Nil}^{\mathrm{L}}(\Delta) = V = \mathrm{UNil}^{\mathrm{L}}(\Delta)$. If $\Delta \neq \varnothing$,

- $\text{Nil}^{\text{L}}(\Delta)$ is the set of all $x \in V$ satisfying:

  for each $(F_0, F_1, \dots) \in \Delta^{\mathbb{N}}$ there exists $N \in \mathbb{N}$ such that for all $m \geq 0$ and $(G_1, \dots, G_m) \in \Delta^m$ we have $[G_m, \dots, G_1, F_N, \dots, F_0](x) = 0$.[1]

- $\text{UNil}^{\text{L}}(\Delta)$ is the set of all $x \in V$ satisfying:

  there exists $N \in \mathbb{N}$ such that for every $n \geq N$ and $(F_1, \dots, F_n) \in \Delta^n$ we have $[F_n, \dots, F_1](x) = 0$.

Note that $\text{UNil}^{\text{L}}(\Delta) \subseteq \text{Nil}^{\text{L}}(\Delta)$ are linear subspaces of V. If $\text{Nil}^{\text{L}}(\Delta) = V$, we say that the set Δ is *Lie-locally nilpotent*; if $\text{UNil}^{\text{L}}(\Delta) = V$, we say that Δ is *uniformly Lie-locally nilpotent*.

Remark 3.3 The condition that Δ is a Lie-locally nilpotent (or a uniformly Lie-locally nilpotent) subset of $\mathcal{L}_{\mathbf{k}}(V)$ does not imply that $\Delta \subseteq \text{LN}(\mathcal{L}_{\mathbf{k}} V)$. For instance, if $F \in \mathcal{L}_{\mathbf{k}}(V) \setminus \text{LN}(\mathcal{L}_{\mathbf{k}} V)$ then $\Delta = \{F\}$ is Lie-locally nilpotent.

Example 3.4 In general there are no inclusion relations between $\text{Nil}(\Delta)$ and $\text{Nil}^{\text{L}}(\Delta)$. The Δ of Remark 3.3 satisfies $\text{Nil}(\Delta) \not\supseteq \text{Nil}^{\text{L}}(\Delta)$. We now give an example satisfying $\text{Nil}(\Delta) \not\subseteq \text{Nil}^{\text{L}}(\Delta)$.

Let $(e_i)_{i \in \mathbb{N}}$ be a basis of a vector space V over a field $\mathbf{k}$. Let $\Delta = \{F_1, F_2, F_3, \dots\}$ where for each $n \geq 1$ we define the $\mathbf{k}$-linear map $F_n : V \to V$ by

$$F_n(e_i) = 0 \text{ for all } i \in \{1, \dots, n\} \qquad \text{and} \qquad F_n(e_i) = e_n \text{ for all } i \in \mathbb{N} \setminus \{1, \dots, n\}.$$

Then we have:

$$F_n \circ F_m = 0 \quad \text{for all } m, n \text{ such that } m \leq n. \tag{9}$$

We claim that $\text{Nil}(\Delta) = V$. Indeed, consider a sequence $(F_{i_0}, F_{i_1}, F_{i_2}, \dots) \in \Delta^{\mathbb{N}}$. Then $(i_0, i_1, i_2, \dots)$ cannot be strictly decreasing, so there must exist $n \geq 1$ such that $i_{n-1} \leq i_n$. Then $F_{i_n} \circ F_{i_{n-1}} = 0$ by (9), so $F_{i_n} \circ \cdots \circ F_{i_0} = 0$. This implies in particular that $\text{Nil}(\Delta) = V$.

Now consider the sequence $(F_1, F_2, F_3, \dots) \in \Delta^{\mathbb{N}}$. Using (9), it is easy to show by induction that $[F_n, \dots, F_1] = (-1)^{n+1} F_1 \circ F_2 \circ \cdots F_n$ for all $n \geq 1$. Since $(F_1 \circ F_2 \circ \cdots F_n)(e_0) = e_1$ for all $n \geq 1$, we have $[F_n, \dots, F_1](e_0) \neq 0$ for all $n \geq 1$, so $e_0 \notin \text{Nil}^{\text{L}}(\Delta)$. So $\text{Nil}(\Delta) \not\subseteq \text{Nil}^{\text{L}}(\Delta)$.

In contrast with Example 3.4, we have:

Lemma 3.5 *For every subset Δ of $\mathcal{L}_{\mathbf{k}}(V)$, we have* $\text{UNil}(\Delta) \subseteq \text{UNil}^{L}(\Delta)$.

Proof Let $x \in \text{UNil}(\Delta)$. Then there exists $N \geq 1$ such that $(F_N \circ \cdots \circ F_1)(x) = 0$ for all $(F_1, \dots, F_N) \in \Delta^N$. So,

$$(F_n \circ \cdots \circ F_1)(x) = 0 \quad \text{for all } n \geq N \text{ and all } (F_1, \dots, F_n) \in \Delta^n. \tag{10}$$

[1] We compute $[G_m, \dots, G_1, F_N, \dots, F_0]$ in the Lie algebra $\mathcal{L}_{\mathbf{k}}(V)_{\text{L}}$ (see 1.4 and 1.7).

Now given any $n \geq N$ and $(G_1, \dots, G_n) \in \Delta^n$ we have

$$[G_n, \dots, G_1] \in \operatorname{Span}_{\mathbf{k}} \left\{ F_n \circ \cdots \circ F_1 \mid (F_1, \dots, F_n) \in \Delta^n \right\},$$

so $[G_n, \dots, G_1](x) = 0$ by (10). This shows that $x \in \mathrm{UNil}^{\mathrm{L}}(\Delta)$, as desired. □

Lemma 3.6 *Let Δ be a finite subset of $\mathfrak{L}_{\mathbf{k}}(V)$. Then* $\mathrm{Nil}^{L}(\Delta) = \mathrm{UNil}^{L}(\Delta)$. *In particular, Δ is Lie-locally nilpotent if and only if it is uniformly Lie-locally nilpotent.*

Proof It suffices to show that $\mathrm{Nil}^{\mathrm{L}}(\Delta) \subseteq \mathrm{UNil}^{\mathrm{L}}(\Delta)$. Let $x \in \mathrm{Nil}^{\mathrm{L}}(\Delta) \setminus \{0\}$. Consider the set S of finite sequences $(F_0, F_1 \dots, F_N)$ of elements of Δ for which there exist $m \geq 0$ and $(G_1, \dots, G_m) \in \Delta^m$ satisfying $[G_m, \dots, G_1, F_N, \dots, F_0](x) \neq 0$. Let $E \subseteq S \times S$ be the set of ordered pairs of the form

$$\big((F_0, \dots, F_n), (F_0, \dots, F_n, F_{n+1})\big) .$$

Then (S, E) is a directed tree and the empty sequence $() \in S$ is the root of this tree (see Sect. 1). For each vertex $v \in S$, the number of $v' \in S$ satisfying $(v, v') \in E$ is at most $|\Delta|$, which is finite. The fact that $x \in \mathrm{Nil}^{\mathrm{L}}(\Delta)$ implies that there is no infinite path $\big((), (F_0), (F_0, F_1), \dots\big)$ in this tree. It follows from Theorem 1.2 that S is a finite set, and this implies that $x \in \mathrm{UNil}^{\mathrm{L}}(\Delta)$. □

Proposition 3.7 *Let Δ be a subset of $\mathfrak{L}_{\mathbf{k}}(V)$ and let $\tilde{\Delta}$ be the Lie subalgebra of $\mathfrak{L}_{\mathbf{k}}(V)_L$ generated by Δ. Then the following hold.*

(a) Let $x \in V$. If N is a positive integer satisfying

$$[F_n, \dots, F_1](x) = 0 \quad \text{for all } n \geq N \text{ and all } (F_1, \dots, F_n) \in \Delta^n, \tag{11}$$

then the same N satisfies

$$[F_n, \dots, F_1](x) = 0 \quad \text{for all } n \geq N \text{ and all } (F_1, \dots, F_n) \in \tilde{\Delta}^n.$$

(b) $\mathrm{UNil}^{L}(\tilde{\Delta}) = \mathrm{UNil}^{L}(\Delta)$
(c) If Δ is uniformly Lie-locally nilpotent then so is $\tilde{\Delta}$.

Proof (a) Let $x \in V$ and let N be a positive integer satisfying (11). Let $n \geq N$. To prove (a), it suffices to show that

$$[H_n, \dots, H_1](x) = 0 \quad \text{for all } (H_1, \dots, H_n) \in \tilde{\Delta}^n. \tag{12}$$

Consider the set $\Delta_\Box = \big\{ [F_k, \dots, F_1] \mid k \geq 1 \text{ and } (F_1, \dots, F_k) \in \Delta^k \big\}$ and note that $\Delta \subseteq \Delta_\Box \subseteq \tilde{\Delta}$ and (by Lemma 1.8(b)) $\tilde{\Delta} = \operatorname{Span}_{\mathbf{k}}(\Delta_\Box)$. The first step in the proof of (12) consists in proving

$$[G_n, \dots, G_1](x) = 0 \quad \text{for all } (G_1, \dots, G_n) \in (\Delta_\Box)^n. \tag{13}$$

Let $(G_1, \dots, G_n) \in (\Delta_\square)^n$. For each $i \in \{1, \dots, n\}$, write

$$G_i = [F_{i,k_i}, \dots, F_{i,2}, F_{i,1}]$$

where $k_i \geq 1$ and $F_{i,j} \in \Delta$. Then Lemma 1.8(a) implies that $[G_n, \dots, G_1] = \sum_i \lambda_i [W_{i,k}, \dots, W_{i,1}]$ where $k = k_1 + \cdots + k_n \geq n$ and (for each i) $\lambda_i \in \mathbf{k}$ and $(W_{i,1}, \dots, W_{i,k})$ is a permutation of $(F_{1,1}, \dots, F_{1,k_1}, F_{2,1}, \dots, F_{n,k_n}) \in \Delta^k$. Since $k \geq n \geq N$, (11) implies that $[W_{i,k}, \dots, W_{i,1}](x) = 0$ for all i, so $[G_n, \dots, G_1](x) = 0$, proving (13).

Now let us prove (12). Let $(H_1, \dots, H_n) \in \tilde{\Delta}^n$. For each i we have $H_i = \sum_j \lambda_{i,j} G_{i,j}$ with $\lambda_{i,j} \in \mathbf{k}$ and $G_{i,j} \in \Delta_\square$. Thus $[H_n, \dots, H_1]$ is a linear combination of terms of the form $[G_{n,j_n}, \dots, G_{1,j_1}]$. By (13), $[G_{n,j_n}, \dots, G_{1,j_1}](x) = 0$ for every choice of $(j_1, \dots, j_n)$, so $[H_n, \dots, H_1](x) = 0$. This proves (12), so (a) is proved. Assertions (b) and (c) follow from (a). □

Theorem 3.8 *Let $(A, \cdot)$ be an algebra over a field* $\mathbf{k}$. *If Δ is a subset of* $\mathrm{Der}_{\mathbf{k}}(A)$ *then the sets* $\mathrm{Nil}^L(\Delta)$ *and* $\mathrm{UNil}^L(\Delta)$ *are subalgebras of* $(A, \cdot)$.

Proof Consider a sequence $\mathbf{D} = (D_0, D_1, \dots) \in \Delta^{\mathbb{N}}$. For each $N \in \mathbb{N}$, let $K_N(\mathbf{D})$ be the intersection of the kernels of the derivations $[G_m, \dots, G_1, D_N, \dots, D_0]$, for all $m \geq 0$ and $(G_1, \dots, G_m) \in \Delta^m$. Since $K_0(\mathbf{D}) \subseteq K_1(\mathbf{D}) \subseteq K_2(\mathbf{D}) \subseteq \cdots$ are subalgebras of A, so is $K(\mathbf{D}) = \bigcup_{N=0}^{\infty} K_N(\mathbf{D})$. Then $\mathrm{Nil}^L(\Delta) = \bigcap_{\mathbf{D} \in \Delta^{\mathbb{N}}} K(\mathbf{D})$ is a subalgebra of A.

For each $N \in \mathbb{N}$, let K_N be the intersection of the $\ker[D_n, \dots, D_1]$ for all $n \geq N$ and $(D_1, \dots, D_n) \in \Delta^n$. Then $K_0 \subseteq K_1 \subseteq K_2 \subseteq \cdots$ is a chain of subalgebras of A, so $\mathrm{UNil}^L(\Delta) = \bigcup_{N=0}^{\infty} K_N$ is a subalgebra of A. □

4 Nilpotency Conditions for Algebras

Throughout Sects. 4.1–4.4, $(A, \cdot)$ is an algebra over a field $\mathbf{k}$ in the sense of 1.3(a). Starting in 4.5, we restrict our attention to the special case where A is either an associative or a Lie algebra.

Definition 4.1 Given a subset H of A, let us define the subsets $\mathrm{Nil}'(H)$ and $\mathrm{UNil}'(H)$ of A as follows. If $H = \varnothing$, we set $\mathrm{Nil}'(H) = A = \mathrm{UNil}'(H)$. If $H \neq \varnothing$,

- $\mathrm{Nil}'(H)$ is the set of all $x \in A$ satisfying:

  for every sequence $(a_0, a_1, \dots) \in H^{\mathbb{N}}$ there exists $n \in \mathbb{N}$ such that $a_n \cdots a_0 \cdot x = 0$;[2]

- $\mathrm{UNil}'(H)$ is the set of all $x \in A$ satisfying:

  there exists $n \in \mathbb{N}$ such that for every $(a_1, \dots, a_n) \in H^n$ we have $a_n \cdots a_1 \cdot x = 0$.

---

[2] Recall that $a_n \cdots a_0 \cdot x = 0$ means $a_n \cdot (a_{n-1} \cdots (a_1 \cdot (a_0 \cdot x)) \dots) = 0$, by the right-associativity convention (1.4).

Define $\deg'_H : A \to \mathbb{N} \cup \{\pm\infty\}$ by declaring that $\deg'_H(0) = -\infty$, $\deg'_H(x) = \infty$ for all $x \in A \setminus \text{UNil}'(H)$, and if $x \in \text{UNil}'(H) \setminus \{0\}$ then $\deg'_H(x)$ is

$$\max\left\{ n \in \mathbb{N} \mid \right\} \text{there exists } (a_1, \dots, a_n) \in H^n \text{ satisfying } a_n \cdots a_1 \cdot x \neq 0 \,.$$

If $H = \varnothing$ then we set $\deg'_H(0) = -\infty$ and $\deg'_H(x) = 0$ for all $x \in A \setminus \{0\}$.

Let us now connect these notions to the formalism of Sect. 2.

4.2 Define the map $\varphi : A \to \mathcal{L}_{\mathbf{k}}(A)$ by stipulating that, given $a \in A$, $\varphi(a) : A \to A$ is the $\mathbf{k}$-linear map $x \mapsto a \cdot x$. Note that φ is a $\mathbf{k}$-linear map, but not necessarily a homomorphism of algebras. If H is a subset of A then $\varphi(H)$ is a subset of $\mathcal{L}_{\mathbf{k}}(A)$, so we may consider (as in Sect. 2) the subsets $\text{Nil}(\varphi(H))$ and $\text{UNil}(\varphi(H))$ of A and the map $\deg_{\varphi(H)} : A \to \mathbb{N} \cup \{\pm\infty\}$. Then we have

$$\text{Nil}'(H) = \text{Nil}(\varphi(H)),\ \text{UNil}'(H) = \text{UNil}(\varphi(H)) \text{ and } \deg'_H = \deg_{\varphi(H)} . \qquad (14)$$

Indeed, the right-associativity convention implies that $a_n \cdots a_0 \cdot x = \big(\varphi(a_n) \circ \cdots \circ \varphi(a_0)\big)(x)$ for all $x, a_0, \dots, a_n \in A$, and (14) immediately follows.

Corollary 4.3 *If H is a finite subset of A then* $\text{Nil}'(H) = \text{UNil}'(H)$.

Proof Apply Lemma 2.2 to the finite set $\varphi(H)$ and use (14). □

Lemma 4.4 *The sets* $\text{Nil}'(A)$ *and* $\text{UNil}'(A)$ *are left ideals of A. Moreover, for each $n \in \mathbb{N}$ the set $Z_n = \left\{ x \in A \mid \deg'_A(x) < n \right\}$ is a left ideal of A.*

Proof We have $\{0\} = Z_0 \subseteq Z_1 \subseteq Z_2 \subseteq \cdots$ and $\text{UNil}'(A) = \bigcup_{n=0}^{\infty} Z_n$; so, to prove the Lemma, it suffices to show that $\text{Nil}'(A)$ and all Z_n are left ideals of A. It is clear that $\text{Nil}'(A)$ and Z_n are $\mathbf{k}$-subspaces of A.

Let $x \in A$ and $y \in \text{Nil}'(A)$. To prove that $x \cdot y \in \text{Nil}'(A)$, we have to show that for every $(x_0, x_1, \dots) \in A^{\mathbb{N}}$ there exist an n such that $x_n \cdots x_0 \cdot (x \cdot y) = 0$. Note that $x_n \cdots x_0 \cdot (x \cdot y) = x_n \cdots x_0 \cdot x \cdot y$ by the right-associativity convention. So the claim is true, because $(x, x_0, x_1, \dots) \in A^{\mathbb{N}}$ and $y \in \text{Nil}'(A)$. So $x \cdot y \in \text{Nil}'(A)$, showing that $\text{Nil}'(A)$ is a left ideal of A.

Note that $Z_0 = \{0\}$ is indeed a left ideal of A. Let $n > 0$, $x \in A$ and $y \in Z_n$. Then $x_n \cdots x_1 \cdot y = 0$ for all $(x_1, \dots, x_n) \in A^n$. In particular, $x_n \cdots x_2 \cdot x \cdot y = 0$ for all $(x_2, \dots, x_n) \in A^{n-1}$. Since $x_n \cdots x_2 \cdot (x \cdot y) = x_n \cdots x_2 \cdot x \cdot y = 0$, this shows that $x \cdot y \in Z_{n-1} \subseteq Z_n$. So Z_n is a left ideal of A. □

Corollary 4.3 and Lemma 4.4 are valid for all algebras A. We can say more if we assume that A is either associative or Lie.

Observation 4.5 *Let A be a* $\mathbf{k}$*-algebra and consider the map φ of 4.2.*

(a) If A is an associative algebra then $\varphi : A \to \mathcal{L}_{\mathbf{k}}(A)$ is a homomorphism of associative algebras.

(b) *If A is a Lie algebra then* $\varphi : A \to \mathfrak{L}_{\mathbf{k}}(A)_L$ *is a homomorphism of Lie algebras and* $\varphi(A) \subseteq \mathrm{Der}_{\mathbf{k}}(A)$.

Proof Part (a) is clear and (b) is the same as 1.7(i,ii) (we have $\varphi = \mathrm{ad}$). □

Corollary 4.6 *Let* $(A, \cdot)$ *be either an associative algebra or a Lie algebra, let* H *be a subset of* A *and let* $\hat{H}$ *be the subalgebra of* $(A, \cdot)$ *generated by* H.

(a) $\mathrm{Nil}'(H) = \mathrm{Nil}'(\hat{H})$, $\mathrm{UNil}'(H) = \mathrm{UNil}'(\hat{H})$ *and* $\deg'_H = \deg'_{\hat{H}}$.
(b) *If* $(A, \cdot)$ *is a Lie algebra then* $\mathrm{Nil}'(H)$ *and* $\mathrm{UNil}'(H)$ *are Lie subalgebras of* A *and* $\deg'_H(x \cdot y) \le \deg'_H(x) + \deg'_H(y)$ *for all* $x, y \in \mathrm{UNil}'(H)$.

Proof (a) Let $\mathbf{k}$ be the field over which A is an algebra. Consider the algebra $(\mathfrak{L}_{\mathbf{k}}(A), \star)$, where for $F, G \in \mathfrak{L}_{\mathbf{k}}(A)$ we define $F \star G = F \circ G$ (resp. $F \star G = F \circ G - G \circ F$) if we are proving the case where A is associative (resp. A is Lie). Let $\widehat{\varphi(H)}$ be the subalgebra of $(\mathfrak{L}_{\mathbf{k}}(A), \star)$ generated by $\varphi(H)$. Since $\varphi : (A, \cdot) \to (\mathfrak{L}_{\mathbf{k}}(A), \star)$ is a homomorphism of algebras by Observation 4.5, we have $\widehat{\varphi(H)} = \varphi(\hat{H})$, so $\mathrm{Nil}'(H) = \mathrm{Nil}(\varphi(H)) = \mathrm{Nil}(\widehat{\varphi(H)}) = \mathrm{Nil}(\varphi(\hat{H})) = \mathrm{Nil}'(\hat{H})$, where the second equality is Proposition 2.5 and the first and last equalities are (14). It is clear that $\mathrm{UNil}'(H) = \mathrm{UNil}'(\hat{H})$ and $\deg'_H = \deg'_{\hat{H}}$ follow by the same argument, so (a) is proved.

(b) If $(A, \cdot)$ is a Lie algebra then $\varphi(H) \subseteq \mathrm{Der}_{\mathbf{k}}(A, \cdot)$ by Observation 4.5, so Theorem 2.7 implies that $\mathrm{Nil}(\varphi(H))$ and $\mathrm{UNil}(\varphi(H))$ are Lie subalgebras of $(A, \cdot)$ and that $\deg_{\varphi(H)}(x \cdot y) \le \deg_{\varphi(H)}(x) + \deg_{\varphi(H)}(y)$ for all $x, y \in \mathrm{UNil}(\varphi(H))$. In view of (14), this is the desired conclusion. □

Notation 4.7 Let A be either an associative algebra or a Lie algebra. Given a subset H of A, define $s(H) = \sup\big\{ \deg'_H(x) \mid x \in H \big\} \in \mathbb{N} \cup \{\pm\infty\}$, where we agree that $s(\varnothing) = -\infty$. Note that if $H \subseteq \{0\}$ then $s(H) = -\infty$, and if $H \nsubseteq \{0\}$ then $s(H) = \sup\big\{ n \in \mathbb{N} \mid x_n \cdots x_0 \neq 0 \text{ for some } (x_0, \dots, x_n) \in H^{n+1} \big\}$.

Lemma 4.8 *Let* A *be either an associative algebra or a Lie algebra, let* H *be a subset of* A *and let* $\hat{H}$ *be the subalgebra of* A *generated by* H.

(a) *If* $H \subseteq \mathrm{Nil}'(H)$ *then* $\hat{H} \subseteq \mathrm{Nil}'(\hat{H})$.
(b) *If* $H \subseteq \mathrm{UNil}'(H)$ *then* $\hat{H} \subseteq \mathrm{UNil}'(\hat{H})$.
(c) $s(H) = s(\hat{H})$

Proof (a) Assume that $H \subseteq \mathrm{Nil}'(H)$. Corollary 4.6 gives $\mathrm{Nil}'(H) = \mathrm{Nil}'(\hat{H})$, so $H \subseteq \mathrm{Nil}'(\hat{H})$ and hence $H \subseteq \hat{H} \cap \mathrm{Nil}'(\hat{H})$. Applying Lemma 4.4 *to the algebra* $\hat{H}$ shows that $\hat{H} \cap \mathrm{Nil}'(\hat{H})$ is a left ideal of $\hat{H}$ and hence a subalgebra of $\hat{H}$; since that subalgebra contains H, it must then contain $\hat{H}$, so $\hat{H} \subseteq \mathrm{Nil}'(\hat{H})$. This proves (a). We obtain a proof of (b) by replacing Nil' by UNil' everywhere.

(c) It is clear that $s(H) \le s(\hat{H})$ and that equality holds whenever $s(H) \notin \mathbb{N}$. Assume that $s(H) = n \in \mathbb{N}$. Since $\deg'_H = \deg'_{\hat{H}}$ by Corollary 4.6, we have $\deg'_{\hat{H}}(x) \le n$ for all $x \in H$. Thus $H \subseteq Z_{n+1}$, where we define $Z_j = \big\{ x \in \hat{H} \mid \deg'_{\hat{H}}(x) < j \big\}$. Lemma 4.4 implies that Z_{n+1} is a left ideal (and hence a subalgebra) of $\hat{H}$, so $\hat{H} \subseteq Z_{n+1}$. It follows that $s(\hat{H}) \le n$, so $s(\hat{H}) = s(H)$. □

Corollary 4.9 *Let A be either an associative algebra or a Lie algebra and let H be a generating set for A. Then the following hold.*

(a) If for every sequence $(a_0, a_1, \dots) \in H^{\mathbb{N}}$ there exists an n such that $a_n \cdots a_0 = 0$, then for every sequence $(a_0, a_1, \dots) \in A^{\mathbb{N}}$ there exists an n such that $a_n \cdots a_0 = 0$.

(b) Suppose that n is a positive integer such that for every $(a_1, \dots, a_n) \in H^n$ we have $a_n \cdots a_1 = 0$. Then we have $a_n \cdots a_1 = 0$ for all $(a_1, \dots, a_n) \in A^n$.

Proof We have $\hat{H} = A$ by assumption, where $\hat{H}$ denotes the subalgebra of A generated by H. Part (a) asserts that $H \subseteq \operatorname{Nil}'(H) \Rightarrow \operatorname{Nil}'(A) = A$; since $\hat{H} = A$, the claim follows from Lemma 4.8(a). Part (b) asserts that if n satisfies $s(H) < n - 1$, then $s(A) < n - 1$; Lemma 4.8(c) implies that $s(H) = s(\hat{H}) = s(A)$, so the claim is true. □

Definition 4.10 Let $(A, \cdot)$ be either an associative algebra or a Lie algebra. Consider the map $\varphi : A \to \mathcal{L}_{\mathbf{k}}(A)$ of 4.2. We say that

- A is *nilpotent* (N) if there exists an $n \geq 1$ such that for all $(x_1, \dots, x_n) \in A^n$ we have $x_n \cdots x_1 = 0$;
- A is *sequentially nilpotent* (SN) if for every infinite sequence $(x_0, x_1, \dots)$ of elements of A, there exists an $n \geq 0$ such that $x_n \cdots x_0 = 0$;
- A is *locally nilpotent* (LN) if every finitely generated subalgebra of A is nilpotent;
- A is *nil* (nil) if for each $x \in A$ the map $\varphi(x) : A \to A$ is nilpotent;
- A is *locally nil* (Lnil) if for each $x \in A$ the map $\varphi(x) : A \to A$ is locally nilpotent.

Remark 4.11 We found nothing in the literature about (SN) or (Lnil); as far as we know, these two conditions have not been considered previously. The definitions of (N), (LN) and (nil) given in Definition 4.10 are compatible with standard usage of terminology. Regarding (nil), note the following.

(1) Proposition 4.13(d) shows that an associative algebra A satisfies (nil) if and only if every element of A is nilpotent (which is the standard meaning of "nil" for associative algebras).

(2) Our definition of "nil" for Lie algebras is the standard one. Note that a Lie algebra that is nil is also said to be *Engel*. However, an associative algebra A is said to be *Engel* if the Lie algebra A_{L} is Engel. So "Engel" and "nil" have the same meaning for Lie algebras but not for associative algebras.

Remark 4.12 It follows from Lemma 1.8 that if an associative or Lie algebra is both finitely generated and nilpotent, then it is finite dimensional (as a vector space over $\mathbf{k}$). Consequently, if A is (LN) then every finitely generated subalgebra of A is nilpotent *and finite dimensional*.

Proposition 4.13 *Let A be either an associative algebra or a Lie algebra.*

(a) We have the following implications for A: $\begin{array}{ccc} & & \text{(SN)} \Rightarrow \text{(LN)} \\ \text{(N)} & \begin{array}{c} \nearrow \\ \searrow \end{array} & \Downarrow \\ & & \text{(nil)} \Rightarrow \text{(Lnil)}. \end{array}$

(b) If A is finitely generated then (N) $\Leftrightarrow$ (SN) $\Leftrightarrow$ (LN).
(c) If A is finite dimensional then (LN) $\Leftrightarrow$ (SN) $\Leftrightarrow$ (N) $\Leftrightarrow$ (nil) $\Leftrightarrow$ (Lnil).
(d) If A is associative then (SN) $\Rightarrow$ (nil) and moreover

$$(Lnil) \Leftrightarrow (nil) \Leftrightarrow \textit{every element of } A \textit{ is nilpotent.}$$

Proof (a), (SN) $\Rightarrow$ (LN). Assume that A satisfies (SN) and let H be a finite subset of A; we have to show that $\hat{H}$ is nilpotent, where $\hat{H}$ denotes the subalgebra of A generated by H. The assumption that A satisfies (SN) implies that $H \subseteq \text{Nil}'(H)$. Since H is finite we have $\text{UNil}'(H) = \text{Nil}'(H)$ by Corollary 4.3, so $H \subseteq \text{UNil}'(H)$ and consequently $\deg'_H(x) < \infty$ for each $x \in H$. Since H is finite, it follows that $s(H) < \infty$. Then Lemma 4.8 implies that $s(H) = s(\hat{H})$, so $s(\hat{H}) < \infty$, so $\hat{H}$ is nilpotent. So A satisfies (LN).

(LN) $\Rightarrow$ (Lnil). Suppose that A is (LN). To show that A is (Lnil), it suffices to show that for all $x, y \in A$ there exists an $n > 0$ such that $\varphi(x)^n(y) = 0$. Consider $x, y \in A$. As A is (LN), the subalgebra A_0 of A generated by $\{x, y\}$ is nilpotent, so there exists $n > 0$ such that $x_n \cdots x_0 = 0$ for all $(x_0, \dots, x_n) \in A_0^{n+1}$, i.e., $\big(\varphi(x_n) \circ \cdots \circ \varphi(x_1)\big)(x_0) = 0$ for all $(x_0, \dots, x_n) \in A_0^{n+1}$. As $x, y \in A_0$ it follows that $\varphi(x)^n(y) = 0$. So A is (Lnil).

The implications (Lnil) $\Leftarrow$ (nil) $\Leftarrow$ (N) $\Rightarrow$ (SN) are clear, so (a) is proved.

If A is finitely generated then (LN) $\Rightarrow$ (N) is clear, by definition of (LN); so (b) is clear.

Let us prove (d) before (c). Suppose that A is associative.

If A satisfies (Lnil) then, for every choice of $x, y \in A$, there exists n such that $\varphi(x)^n(y) = 0$; so for every $x \in A$ there exists n such that $x^{n+1} = \varphi(x)^n(x) = 0$. This shows that (Lnil) implies that every element of A is nilpotent.

Since $\varphi : A \to \mathcal{L}_{\mathbf{k}}(A)$ is a homomorphism of associative algebras, it is clear that if every element of A is nilpotent then every element of $\varphi(A)$ is nilpotent, i.e., A satisfies (nil).

This proves the equivalences in (d). Now suppose that A satisfies (SN). Given any $x \in A$, applying the property (SN) to the sequence $(x, x, x, \dots) \in A^{\mathbb{N}}$ implies that there exists n such that $x^n = 0$. So (SN) implies that every element of A is nilpotent, and the proof of (d) is complete.

(c) By (a), it suffices to show that (Lnil) $\Rightarrow$ (nil) $\Rightarrow$ (N) when A is finite dimensional.

If A is finite dimensional then every locally nilpotent linear map $A \to A$ is in fact nilpotent, so (Lnil) implies (nil).

For finite dimensional Lie algebras, the fact that (nil) implies (N) is *Engel's theorem on abstract Lie algebras*, see p. 36 of [5].

Any finite dimensional associative $\mathbf{k}$-algebra A is (isomorphic to) a subalgebra of a matrix algebra $M_n(\mathbf{k})$, for some n. If A satisfies (nil), then part (d) implies that every element of A is a nilpotent matrix. It is then well known that A is a nilpotent algebra (see for instance Theorem 1, p. 33 of [5]). So (nil) implies (N) in the associative case as well, and this proves (c). □

4.14 By Example 6.2, (SN) does not imply (nil) for Lie algebras.

4.15 It is known that (nil) does not imply (LN). Indeed, the following statements are valid over an arbitrary field and are consequences of more precise results due to Golod ([2, 3], cited in [6, p. 6] and [7, Theorem 6.2.9]):

(a) *For each integer $m \geq 2$, there exists an infinite dimensional associative algebra A generated by m elements such that every $(m-1)$-subset of A is nilpotent.*[3] *Moreover,* $\bigcap_{n=1}^{\infty} A^n = \{0\}$.
(b) *There exists a* 3*-generated Lie algebra that satisfies* (nil) *but not* (N).

In (a), the fact that every $(m-1)$-subset of A is nilpotent implies that A is nil and the fact that A is finitely generated and infinite dimensional implies that A is not nilpotent (Remark 4.12).

4.16 If (P)$\in$ { (N), (SN), (LN), (nil), (Lnil) } and A is an associative algebra satisfying (P), does it follow that A_L satisfies (P)? Example 4.18 shows that there exists an associative algebra A such that A satisfies (SN) but A_L doesn't. The following Lemma gives positive results.

Lemma 4.17 *Let A be an associative algebra.*

(a) Let (P)$\in$ {(N), (nil), (LN), (Lnil)}. *If A satisfies* (P), *then A_L satisfies* (P).
(b) If A satisfies (SN), *then A_L satisfies* (LN).

Proof (a) (i) Define $A_n = \operatorname{Span}_{\mathbf{k}} \left\{ a_n \cdots a_1 \mid (a_1, \dots, a_n) \in A^n \right\}$. If A satisfies (N) then $A_n = 0$ for n large enough; since $[a_n, \dots, a_1] \in A_n$ for all $(a_1, \dots, a_n) \in A^n$, it follows that A_L satisfies (N).

(ii) Suppose that A satisfies (Lnil). Then Proposition 4.13(d) implies that each element of A is nilpotent. We claim that A_L satisfies (nil). Indeed, let $x \in A_L$; to prove the assertion, we have to show that $\operatorname{ad}(x) : A_L \to A_L$ is nilpotent. There exists m such that $x^m = 0$ in A. For any $y \in A_L$,

$$\left(\operatorname{ad}(x)^{2m-1}\right)(y) = [\underbrace{x, \dots, x}_{2m-1}, y] \in \operatorname{Span}_{\mathbf{k}} \left\{ x^i y x^j \mid i + j = 2m - 1 \right\}.$$

As $x^i y x^j = 0$ for all i, j satisfying $i + j = 2m - 1$, we have $\left(\operatorname{ad}(x)^{2m-1}\right)(y) = 0$ for all $y \in A_L$. So $\operatorname{ad}(x)^{2m-1} = 0$ and consequently A_L satisfies (nil).

(iii) Suppose that A satisfies (LN). Consider a finite subset H of A_L; to show that A_L is (LN), it suffices to show that $\tilde{H}$ is nilpotent, where $\tilde{H}$ denotes the subalgebra of A_L generated by H. Note that $H \subseteq A$ (since $A = A_L$ as sets) and consider the subalgebra $\bar{H}$ of A generated by H. Then $\bar{H}$ satisfies (N), so part (i) of the proof shows that $(\bar{H})_L$ satisfies (N). Since $\tilde{H} \subseteq \bar{H}$, $\tilde{H}$ is a subalgebra of the nilpotent Lie algebra $(\bar{H})_L$; so $\tilde{H}$ is nilpotent, which shows that A_L satisfies (LN). This proves (a).

---

[3]A subset H of an associative algebra A is *nilpotent* if there exists n such that $h_n \cdots h_1 = 0$ for all $(h_1, \dots, h_n) \in H^n$. By Corollary 4.9, H is nilpotent if and only if $\bar{H}$ is nilpotent, where $\bar{H}$ is the subalgebra of A generated by H.

(b) Since A satisfies (SN), it satisfies (LN) by Proposition 4.13, so A_L satisfies (LN) by part (a). □

Example 4.18 Let V be a vector space of dimension $|\mathbb{N}|$ over a field $\mathbf{k}$. Then there exists an associative subalgebra A of $\mathfrak{L}_{\mathbf{k}}(V)$ satisfying:

(a) A is (SN) but the Lie algebra A_L is not (SN).
(b) A is a uniformly locally nilpotent subset of $\mathfrak{L}_{\mathbf{k}}(V)$.

Proof Let $(e_i)_{i\in\mathbb{N}}$ be a basis of V, let $I = \{ (i, j) \in \mathbb{N}^2 \mid i \geq j \}$ and, for each $(i, j) \in I$, define the linear map $T_{i,j} : V \to V$ by

$$T_{i,j}(e_k) = \begin{cases} 0, & \text{if } k \leq i, \\ e_j, & \text{if } k > i. \end{cases}$$

Let $A = \operatorname{Span}_{\mathbf{k}} \{ T_{i,j} \mid (i, j) \in I \} \subseteq \mathfrak{L}_{\mathbf{k}}(V)$. Note that if $(i_1, j_1), (i_2, j_2) \in I$ then

(α) if $j_1 \leq i_2$ then $T_{i_2,j_2} \circ T_{i_1,j_1} = 0$;
(β) if $j_1 > i_2$ then $(i_1, j_2) \in I$ and $T_{i_2,j_2} \circ T_{i_1,j_1} = T_{i_1,j_2} \in A$.

So A is a subalgebra of $\mathfrak{L}_{\mathbf{k}}(V)$.

Proof that A is (SN). Consider the subset $\Delta = \{ T_{i,j} \mid (i, j) \in I \}$ of A. By Corollary 4.9(a), it suffices to show that for every sequence $(S_0, S_1, \dots) \in \Delta^{\mathbb{N}}$ there exists n such that $S_n S_{n-1} \cdots S_0 = 0$. In other words, it suffices to show that for every sequence $(T_{u_0,v_0}, T_{u_1,v_1}, T_{u_2,v_2}, \dots) \in \Delta^{\mathbb{N}}$ (where $(u_\nu, v_\nu) \in I$ for all $\nu \in \mathbb{N}$) there exists n such that $T_{u_n,v_n} T_{u_{n-1},v_{n-1}} \cdots T_{u_0,v_0} = 0$. Note that the above statement (α) implies that if $(i_1, j_1), (i_2, j_2) \in I$ are such that $T_{i_2,j_2} \circ T_{i_1,j_1} \neq 0$, then $i_2 < i_1$ and $j_2 < j_1$. Consequently, if $T_{u_n,v_n} T_{u_{n-1},v_{n-1}} \cdots T_{u_0,v_0} \neq 0$ then in particular $u_n < u_{n-1} < \cdots < u_0$. So there must exist an n such that $T_{u_n,v_n} T_{u_{n-1},v_{n-1}} \cdots T_{u_0,v_0} = 0$.

Proof that A_L is not (SN). Consider the sequence $(S_0, S_1, S_2, \dots) \in A^{\mathbb{N}}$, where we define $S_i = T_{i,i}$ for all i. Note that $S_i S_j = 0$ whenever $i \geq j$. This implies that $[S_1, S_0] = -S_0 S_1$, $[S_2, S_1, S_0] = [S_2, -S_0 S_1] = S_0 S_1 S_2$, and by induction we find $[S_n, \dots, S_0] = (-1)^n S_0 \cdots S_n$ for all $n \geq 0$. As $(S_0 \cdots S_n)(e_{n+1}) = e_0$, we have $S_0 S_1 \cdots S_n \neq 0$ and hence $[S_n, \dots, S_0] \neq 0$ (for all $n \geq 0$). So A_L is not (SN) and (a) is proved.

(b) Let $k \in \mathbb{N}$; let us show that $e_k \in \operatorname{UNil}(\Delta)$. Indeed, if $T_{i,j}$ is any element of Δ such that $T_{i,j}(e_k) \neq 0$, then $i < k$. It follows that if $(T_{u_1,v_1}, \dots, T_{u_n,v_n}) \in \Delta^n$ satisfies $(T_{u_n,v_n} \circ \cdots \circ T_{u_1,v_1})(e_k) \neq 0$ then $k > u_1 > \cdots > u_n \geq 0$ ($T_{u_n,v_n} \circ \cdots \circ T_{u_1,v_1} \neq 0$ implies $u_1 > \cdots > u_n$ by the preceding paragraph), so $n \leq k$, proving that $e_k \in \operatorname{UNil}(\Delta)$. It follows that $\operatorname{UNil}(\Delta) = V$. As $\bar{\Delta} = A$, Proposition 2.5 implies that $\operatorname{UNil}(\Delta) = \operatorname{UNil}(A)$, so $\operatorname{UNil}(A) = V$, so (b) is proved. □

5 Locally Nilpotent Sets of Derivations

In this section we assume that B is an algebra over a field $\mathbf{k}$ (in the sense of 1.3(a)) and we consider the Lie subalgebra $\mathrm{Der}_{\mathbf{k}}(B)$ of $\mathcal{L}_{\mathbf{k}}(B)_{\mathrm{L}}$ (see 1.4).

A derivation $D \in \mathrm{Der}_{\mathbf{k}}(B)$ is said to be *locally nilpotent* if it is locally nilpotent as a linear map $B \to B$. We write $\mathrm{LND}(B)$ for the set of locally nilpotent derivations of B, i.e., $\mathrm{LND}(B) = \mathrm{LN}(\mathcal{L}_{\mathbf{k}}B) \cap \mathrm{Der}_{\mathbf{k}}(B)$. For any subset Δ of $\mathrm{Der}_{\mathbf{k}}(B)$, the sets $\mathrm{UNil}(\Delta) \subseteq \mathrm{Nil}(\Delta)$ are subalgebras of B (by Theorem 2.7). If $\mathrm{Nil}(\Delta) = B$, we say that Δ is a locally nilpotent subset of $\mathrm{Der}_{\mathbf{k}}(B)$. If $\mathrm{UNil}(\Delta) = B$, we say that Δ is a uniformly locally nilpotent subset of $\mathrm{Der}_{\mathbf{k}}(B)$.

This section explores Questions 1 and 2, which are stated in the Introduction. In particular, we give a series of examples showing that both questions have negative answers when B is the commutative polynomial ring in $|\mathbb{N}|$ variables over an arbitrary field $\mathbf{k}$. The next section studies the case where B is assumed to satisfy some finiteness condition.

Lemma 5.1 *Let V be a vector space over a field $\mathbf{k}$ and B a commutative polynomial ring over $\mathbf{k}$ satisfying $\mathrm{trdeg}_{\mathbf{k}}(B) = \dim_{\mathbf{k}}(V)$. Then there exist an injective $\mathbf{k}$-linear map $\nu : V \to B$ and an injective homomorphism of Lie algebras $\psi : \mathcal{L}_{\mathbf{k}}(V)_L \to \mathrm{Der}_{\mathbf{k}}(B)$ such that $\mathbf{k}[\operatorname{im} \nu] = B$ and*

$$\text{for every } F \in \mathcal{L}_{\mathbf{k}}(V), \text{ the diagram } \quad \begin{array}{ccc} B & \xrightarrow{\psi(F)} & B \\ {\scriptstyle \nu}\uparrow & & \uparrow{\scriptstyle \nu} \\ V & \xrightarrow[F]{} & V \end{array} \quad \text{commutes.} \tag{15}$$

Moreover, the following statements are true for every subset Δ of $\mathcal{L}_{\mathbf{k}}(V)$:

(a) $\Delta \subseteq \mathrm{LN}(\mathcal{L}_{\mathbf{k}}V)$ *if and only if* $\psi(\Delta) \subseteq \mathrm{LND}(B)$.
(b) Δ *is a locally nilpotent subset of* $\mathcal{L}_{\mathbf{k}}(V)$ *if and only if* $\psi(\Delta)$ *is a locally nilpotent subset of* $\mathrm{Der}_{\mathbf{k}}(B)$.
(c) Δ *is a uniformly locally nilpotent subset of* $\mathcal{L}_{\mathbf{k}}(V)$ *if and only if* $\psi(\Delta)$ *is a uniformly locally nilpotent subset of* $\mathrm{Der}_{\mathbf{k}}(B)$.

Proof Let $(x_i)_{i \in I}$ be a family of indeterminates over $\mathbf{k}$ such that $B = \mathbf{k}[(x_i)_{i \in I}]$. Then there exists a basis $(e_i)_{i \in I}$ of V indexed by the same set I. Consider the $\mathbf{k}$-linear map $\nu : V \to B$ given by $\nu(e_i) = x_i$ for all $i \in I$. Then ν is injective and $\nu(V) = B_1$, where we define $B_1 = \mathrm{Span}_{\mathbf{k}}\big\{ x_i \mid i \in I \big\}$. We might as well assume that $V = B_1$ and that ν is the inclusion map $B_1 \hookrightarrow B$. For each $F \in \mathcal{L}_{\mathbf{k}}(B_1)$, there exists a unique $D_F \in \mathrm{Der}_{\mathbf{k}}(B)$ satisfying $D_F(v) = F(v)$ for all $v \in B_1$. Consider the map $\psi : \mathcal{L}_{\mathbf{k}}(B_1) \to \mathrm{Der}_{\mathbf{k}}(B)$, $F \mapsto D_F$. If $a, b \in \mathbf{k}$ and $F, G \in \mathcal{L}_{\mathbf{k}}(B_1)$ then the derivations D_{aF+bG} and $aD_F + bD_G$ have the same restriction to B_1, and hence must be equal; so ψ is a $\mathbf{k}$-linear map (and is clearly injective). Similarly, the derivations $D_{F \circ G - G \circ F}$ and $D_F \circ D_G - D_G \circ D_F$ have the same restriction to B_1, and hence must be equal. Thus ψ is a homomorphism of Lie algebras $\mathcal{L}_{\mathbf{k}}(B_1)_{\mathrm{L}} \to \mathrm{Der}_{\mathbf{k}}(B)$

and (15) is true. To prove (b) and (c), consider a subset Δ of $\mathfrak{L}_{\mathbf{k}}(V)$. Since $\mathrm{Nil}(\Delta)$ and $\mathrm{UNil}(\Delta)$ are linear subspaces of V and $(x_i)_{i\in I}$ spans V, we have:

$$\mathrm{Nil}(\Delta) = V \Leftrightarrow \forall_i\ x_i \in \mathrm{Nil}(\Delta) \quad \text{and} \quad \mathrm{UNil}(\Delta) = V \Leftrightarrow \forall_i\ x_i \in \mathrm{UNil}(\Delta).$$

Since $\mathrm{Nil}(\psi(\Delta))$ and $\mathrm{UNil}(\psi(\Delta))$ are subalgebras of $B = \mathbf{k}[(x_i)_{i\in I}]$ by Theorem 2.7, and contain $\mathbf{k}$ by Remark 2.8, we have:

$$\mathrm{Nil}(\psi(\Delta)) = B \Leftrightarrow \forall_i\ x_i \in \mathrm{Nil}(\psi(\Delta))$$
$$\mathrm{UNil}(\psi(\Delta)) = B \Leftrightarrow \forall_i\ x_i \in \mathrm{UNil}(\psi(\Delta)).$$

It is easily verified that for each $i \in I$ we have

$$x_i \in \mathrm{Nil}(\Delta) \Leftrightarrow x_i \in \mathrm{Nil}(\psi(\Delta)) \quad \text{and} \quad x_i \in \mathrm{UNil}(\Delta) \Leftrightarrow x_i \in \mathrm{UNil}(\psi(\Delta)).$$

Consequently, we have $\mathrm{Nil}(\Delta) = V \Leftrightarrow \mathrm{Nil}(\psi(\Delta)) = B$ and $\mathrm{UNil}(\Delta) = V \Leftrightarrow \mathrm{UNil}(\psi(\Delta)) = B$, i.e., assertions (b) and (c) are true. Assertion (a) follows from (b). Indeed, the condition $\Delta \subseteq \mathrm{LN}(\mathfrak{L}_{\mathbf{k}} V)$ is equivalent to "for each $F \in \Delta$, $\{F\}$ is a locally nilpotent subset of $\mathfrak{L}_{\mathbf{k}}(V)$;" by (b), this is equivalent to "for each $F \in \Delta$, $\{\psi(F)\}$ is a locally nilpotent subset of $\mathrm{Der}_{\mathbf{k}}(B)$," which is itself equivalent to $\psi(\Delta) \subseteq \mathrm{LND}(B)$. □

In view of Question 2, it is natural to ask whether an arbitrary Lie algebra L can be embedded as a Lie subalgebra of $\mathrm{Der}_{\mathbf{k}}(B)$ (for some B) in such a way that $L \subseteq \mathrm{LND}(B)$. A preliminary question is whether L can be embedded in $\mathrm{Der}_{\mathbf{k}}(B)$ at all (for some B). The answer is affirmative:

Proposition 5.2 *Let L be a Lie algebra over a field* $\mathbf{k}$*. Then there exists a commutative polynomial ring B over* $\mathbf{k}$ *such that L is isomorphic to a Lie-subalgebra of* $\mathrm{Der}_{\mathbf{k}}(B)$*. Moreover, if L is finite dimensional then we can choose B to be of finite transcendence degree over* $\mathbf{k}$*.*

Proof It is known[4] that there exists a vector space V over $\mathbf{k}$ such that L is isomorphic to a subalgebra of $\mathfrak{L}_{\mathbf{k}}(V)_{\mathrm{L}}$. It is also known that if $\dim L < \infty$ then V can be chosen to be finite dimensional (if $\mathrm{char}\,\mathbf{k} = 0$ this is called *Ado's Theorem* [1]; the general case is due to Iwasawa [4]). Consider a polynomial ring B over $\mathbf{k}$ such that $\mathrm{trdeg}_{\mathbf{k}}(B) = \dim_{\mathbf{k}}(V)$. By Lemma 5.1, $\mathfrak{L}_{\mathbf{k}}(V)_{\mathrm{L}}$ is isomorphic to a Lie subalgebra of $\mathrm{Der}_{\mathbf{k}}(B)$. As L is isomorphic to a Lie subalgebra of $\mathfrak{L}_{\mathbf{k}}(V)_{\mathrm{L}}$, we are done. □

The following is a preliminary step to Example 5.4.

Example 5.3 If $\mathbf{k}$ is a field and V is a $\mathbf{k}$-vector space of dimension $|\mathbb{N}|$, then there exists an associative subalgebra A of $\mathfrak{L}_{\mathbf{k}}(V)$ satisfying:

(a) A is the free associative algebra on a countably infinite set;

[4]See for instance p. 6 of [5].

(b) A is a uniformly locally nilpotent subset of $\mathcal{L}_{\mathbf{k}}(V)$.

Proof Let $\mathbf{k}$ be a field, $\mathcal{V} = \{x_1, x_2, x_3, \dots\}$ a countably infinite set of indeterminates and $\mathbf{k}\langle\mathcal{V}\rangle$ the polynomial ring over $\mathbf{k}$ in the **noncommutative** variables $x_1, x_2, x_3, \dots$. By a *nonempty monomial*, we mean a nonempty finite product of elements of $\mathcal{V}$, i.e., a product $x_{i_1} \cdots x_{i_n}$ with $n \geq 1$. Let $W \subset \mathbf{k}\langle\mathcal{V}\rangle$ be the set of nonempty monomials and consider the associative algebra $\mathscr{A} = \operatorname{Span}_{\mathbf{k}} W$. Note that $\mathscr{A}$ is the associative subalgebra $\bar{\mathcal{V}}$ of $\mathbf{k}\langle\mathcal{V}\rangle$ generated by $\mathcal{V}$; so $\mathscr{A}$ is the free associative algebra on $\mathcal{V}$.

We say that a nonempty monomial $x_{i_1} \cdots x_{i_n} \in W$ is *admissible* if it satisfies $n > i_n$. Let W_0 be the set of all admissible monomials and consider the $\mathbf{k}$-subspace $\mathscr{A}_0 = \operatorname{Span}_{\mathbf{k}}(W_0)$ of $\mathscr{A}$. Then $V = \mathscr{A}/\mathscr{A}_0$ is a vector space over $\mathbf{k}$ of dimension $|\mathbb{N}|$. In fact $\mathscr{A}_0$ is a left ideal of $\mathscr{A}$, so V is also a left $\mathscr{A}$-module; so if $a \in \mathscr{A}$ and $x \in V$ then $ax \in V$ is defined, and if also $b \in \mathscr{A}$ then $(ab)x = a(bx)$. If $a \in \mathscr{A}$, let $\mu(a) : V \to V$ be the $\mathbf{k}$-linear map $x \mapsto ax$ (for $x \in V$). Then $\mu : \mathscr{A} \to \mathcal{L}_{\mathbf{k}}(V)$ is a homomorphism of associative $\mathbf{k}$-algebras.

We claim that μ is injective. To see this, consider $a \in \mathscr{A} \setminus \{0\}$. Write $a = \lambda_1 w_1 + \cdots + \lambda_n w_n$ where $w_1, \dots, w_n$ are distinct elements of W, $n \geq 1$ and $\lambda_1, \dots, \lambda_n \in \mathbf{k}^*$. We have $w_1 = x_{i_1} \cdots x_{i_m}$; then the monomial $w_1 x_{m+1} \in W$ is not admissible. Since $a x_{m+1} = \lambda_1 w_1 x_{m+1} + \cdots + \lambda_n w_n x_{m+1}$ where $w_1 x_{m+1}, \dots, w_n x_{m+1}$ are distinct elements of W and $w_1 x_{m+1} \notin W_0$, $a x_{m+1} \notin \mathscr{A}_0$, so $\big(\mu(a)\big)(x) = ax \neq 0$ where $x = x_{m+1} + \mathscr{A}_0 \in \mathscr{A}/\mathscr{A}_0$, showing that $\mu(a) \neq 0$. So μ is injective. Define $A = \mu(\mathscr{A})$, then A is an associative subalgebra of $\mathcal{L}_{\mathbf{k}}(V)$ and satisfies (a).

We claim that A is a uniformly locally nilpotent subset of $\mathcal{L}_{\mathbf{k}}(V)$ (meaning $\operatorname{UNil}(A) = V$). Since $\big\{\, w + \mathscr{A}_0 \mid w \in W \,\big\}$ is a spanning set for the vector space V, in order to prove the claim it suffices to show that $w + \mathscr{A}_0 \in \operatorname{UNil}(A)$ for each $w \in W$. So consider $w \in W$ and write $w = x_{i_1} \cdots x_{i_m}$. Then for every $(w_1, w_2, \dots, w_{i_m}) \in W^{i_m}$ we have $w_{i_m} \cdots w_2 \cdot w_1 \cdot w \in W_0$ (because $w_1, \dots, w_{i_m}$ are *nonempty* monomials); it follows that $a_{i_m} \cdots a_1 \cdot w \in \mathscr{A}_0$ for all $(a_1, a_2, \dots, a_{i_m}) \in \mathscr{A}^{i_m}$, which implies that $(F_{i_m} \circ \cdots \circ F_1)(w + \mathscr{A}_0) = 0$ for all $(F_1, \dots, F_{i_m}) \in A^{i_m}$. Thus $w + \mathscr{A}_0 \in \operatorname{UNil}(A)$ and consequently $\operatorname{UNil}(A) = V$. □

The next example is interesting because (a) says that L is as non-nilpotent as a Lie algebra can be, while (b) says that it is as locally nilpotent as a subset of $\operatorname{Der}_{\mathbf{k}}(B)$ can be. So this answers Question 2 in the negative.

Example 5.4 If $\mathbf{k}$ is a field and B is the commutative polynomial algebra in $|\mathbb{N}|$ variables over $\mathbf{k}$, then there exists a Lie subalgebra L of $\operatorname{Der}_{\mathbf{k}}(B)$ satisfying:

(a) L is the free Lie algebra on a countably infinite set;
(b) L is a uniformly locally nilpotent subset of $\operatorname{Der}_{\mathbf{k}}(B)$.

Proof Write $B = \mathbf{k}[x_1, x_2, x_3, \dots]$ and let $V = \operatorname{Span}_{\mathbf{k}}\{x_1, x_2, x_3, \dots\} \subseteq B$. By Example 5.3, there exists an associative subalgebra A of $\mathcal{L}_{\mathbf{k}}(V)$ satisfying:

- A is the free associative algebra on a countably infinite set;
- A is a uniformly locally nilpotent subset of $\mathcal{L}_{\mathbf{k}}(V)$.

Consider a countably infinite subset S of A such that A is the free associative algebra on S. Let $\bar{S}$ (resp. $\tilde{S}$) be the associative subalgebra of $\mathfrak{L}_{\mathbf{k}}(V)$ (resp. the Lie subalgebra of $\mathfrak{L}_{\mathbf{k}}(V)_{\mathrm{L}}$) generated by S, then $\tilde{S} \subseteq \bar{S} = A$ and $\tilde{S}$ is the free Lie algebra on S. Consider the injective homomorphism of Lie algebras $\psi : \mathfrak{L}_{\mathbf{k}}(V)_{\mathrm{L}} \to \mathrm{Der}_{\mathbf{k}}(B)$ of Lemma 5.1 and define $L = \psi(\tilde{S})$. Then L is the free Lie algebra on a countably infinite set, and is a Lie subalgebra of $\mathrm{Der}_{\mathbf{k}}(B)$. We have $\mathrm{UNil}(\tilde{S}) = \mathrm{UNil}(\bar{S}) = \mathrm{UNil}(A) = V$, so $\tilde{S}$ is a uniformly locally nilpotent subset of $\mathfrak{L}_{\mathbf{k}}(V)$; then part (c) of Lemma 5.1 implies that L is a uniformly locally nilpotent subset of $\mathrm{Der}_{\mathbf{k}}(B)$. □

The following gives a negative answer to Question 1.

Example 5.5 If B is the commutative polynomial algebra in $|\mathbb{N}|$ variables over a field $\mathbf{k}$ then there exists a Lie subalgebra L of $\mathrm{Der}_{\mathbf{k}}(B)$ satisfying:

(a) $\mathrm{Nil}(L) \neq B$ and if $\mathrm{char}\,\mathbf{k} = 0$ then $\mathrm{Nil}(L) = \mathbf{k}$;
(b) every finitely generated subalgebra L_0 of L satisfies $\mathrm{UNil}(L_0) = B$;
(c) the Lie algebra L satisfies (LN).

Note that (b) implies that $L \subseteq \mathrm{LND}(B)$.

Proof We use the notation $B = \mathbf{k}[x_0, x_1, x_2, \dots]$. For each $n \in \mathbb{N}$, define $D_n \in \mathrm{Der}_{\mathbf{k}}(B)$ by

$$D_n(x_i) = \begin{cases} x_{i+1} & \text{if } i \leq n, \\ 0 & \text{if } i > n. \end{cases}$$

Let L be the Lie subalgebra of $\mathrm{Der}_{\mathbf{k}}(B)$ generated by $\{D_0, D_1, D_2, \dots\}$. Let $n \in \mathbb{N}$; then $x_n \notin \mathrm{Nil}(L)$ because the sequence $(D_n, D_{n+1}, \dots) \in L^{\mathbb{N}}$ is such that $(D_N \circ \cdots \circ D_{n+1} \circ D_n)(x_n) = x_{N+1} \neq 0$ for all $N \geq n$. In particular, $\mathrm{Nil}(L) \neq B$.

Assume that $\mathrm{char}\,\mathbf{k} = 0$. To prove $\mathrm{Nil}(L) = \mathbf{k}$, it suffices to show that

$$\text{for each } f \in B \setminus \mathbf{k} \text{ there exists } n \in \mathbb{N} \text{ such that } D_n(f) \in B \setminus \mathbf{k}. \tag{16}$$

Let $f \in B \setminus \mathbf{k}$. There is a unique $n \in \mathbb{N}$ such that $f \in \mathbf{k}[x_n, x_{n+1}, \dots] \setminus \mathbf{k}[x_{n+1}, x_{n+2}, \dots]$. Let $R = \mathbf{k}[x_{n+1}, x_{n+2}, \dots]$; then $f = P(x_n)$ for some $P(T) \in R[T] \setminus R$. Since $\mathrm{char}\, R = 0$ and $R \subseteq \ker(D_n)$, we have $D_n(f) = P'(x_n)D_n(x_n) = x_{n+1}P'(x_n) \in B \setminus \mathbf{k}$. This proves (16), so $\mathrm{Nil}(L) = \mathbf{k}$ and (a) is proved.

We now drop the assumption on $\mathrm{char}\,\mathbf{k}$ (so $\mathbf{k}$ is an arbitrary field until the end of the proof). Let L_0 be a finitely generated Lie subalgebra of L. If we choose n large enough and define $\Delta = \{D_0, D_1, \dots, D_n\}$ then $L_0 \subseteq \tilde{\Delta}$, where $\tilde{\Delta}$ is the Lie subalgebra of $\mathrm{Der}_{\mathbf{k}}(B)$ generated by Δ. Let us fix n with this property. Note that $\mathrm{UNil}(L_0) \supseteq \mathrm{UNil}(\tilde{\Delta})$. So, to prove (b) and (c), it suffices to show that

(b′) $\mathrm{UNil}(\tilde{\Delta}) = B$ (c′) $\tilde{\Delta}$ is a nilpotent Lie algebra.

Given $\delta \in \Delta = \{D_0, D_1, \dots, D_n\}$, we have $\delta(x_i) \in \{0, x_{i+1}\}$ for all $i \in \mathbb{N}$ and $\delta(x_i) = 0$ for all $i > n$. It follows that $(\delta_{n+2} \circ \cdots \circ \delta_1)(x_j) = 0$ for all $(\delta_1, \delta_2, \dots, \delta_{n+2}) \in \Delta^{n+2}$ and all $j \in \mathbb{N}$, i.e., $\delta_{n+2} \circ \cdots \circ \delta_1 = 0$ for all $(\delta_1, \delta_2, \dots, \delta_{n+2}) \in$

Δ^{n+2}. This certainly implies that $\text{UNil}(\Delta) = B$ (hence $\text{UNil}(\tilde{\Delta}) = B$ by Proposition 2.5), and it also implies (by Corollary 4.9) that $\bar{\Delta}$ satisfies (N), where $\bar{\Delta}$ is the associative subalgebra of $\mathfrak{L}_{\mathbf{k}}(B)$ generated by Δ. By Lemma 4.17, we obtain that $(\bar{\Delta})_{\mathrm{L}}$ satisfies (N), and since $\tilde{\Delta}$ is a subalgebra of $(\bar{\Delta})_{\mathrm{L}}$ it follows that $\tilde{\Delta}$ satisfies (N). So (b′) and (c′) are proved and we are done. □

In view of the above Example, one may ask whether Question 1 always has an affirmative answer when L is finitely generated. The answer is negative:

Example 5.6 Let B be the commutative polynomial algebra in $|\mathbb{N}|$ variables over a field $\mathbf{k}$. Then for each integer $m \geq 2$ there exists an m-generated Lie subalgebra L of $\text{Der}_{\mathbf{k}}(B)$ satisfying:

(a) L is not a locally nilpotent subset of $\text{Der}_{\mathbf{k}}(B)$.
(b) Every $(m-1)$-generated Lie subalgebra of L is a locally nilpotent subset of $\text{Der}_{\mathbf{k}}(B)$ and a nilpotent Lie algebra.
(c) $L \subseteq \text{LND}(B)$ and L is a nil Lie algebra.

Proof Let $m \geq 2$. By the result of Golod stated in 4.15(a), there exists an m-generated associative $\mathbf{k}$-algebra A such that A is not nilpotent but every $(m-1)$-subset of A is nilpotent (so A is nil). Consider the map $\varphi : A \to \mathfrak{L}_{\mathbf{k}}(A)$ of 4.2. We noted in 4.5 that φ is a homomorphism of associative algebras. Since the $\mathbf{k}$-vector space A has dimension $|\mathbb{N}|$, we have $\dim_{\mathbf{k}}(A) = \text{trdeg}_{\mathbf{k}}(B)$; so we may consider an injective $\mathbf{k}$-linear map $\nu : A \to B$ and an injective homomorphism of Lie algebras $\psi : \mathfrak{L}_{\mathbf{k}}(A)_{\mathrm{L}} \to \text{Der}_{\mathbf{k}}(B)$ satisfying the conditions of Lemma 5.1 (with $V = A$); in particular, $B = \mathbf{k}[\text{im}\,\nu]$ and the diagram in part (i) of (17) commutes for every $F \in \mathfrak{L}_{\mathbf{k}}(A)$. Let H be a generating set of A with $|H| = m$, let $\Delta = \varphi(H) \subset \mathfrak{L}_{\mathbf{k}}(A)$, let $\Lambda = \psi(\Delta) \subset \text{Der}_{\mathbf{k}}(B)$ and let L be the Lie subalgebra of $\text{Der}_{\mathbf{k}}(B)$ generated by Λ. Let us prove that L satisfies the desired conditions.

$$\text{(i)}\quad \begin{array}{ccc} B & \xrightarrow{\psi(F)} & B \\ \uparrow{\scriptstyle\nu} & & \uparrow{\scriptstyle\nu} \\ A & \xrightarrow{F} & A \end{array} \qquad \text{(ii)}\quad \begin{array}{ccccc} A & \xrightarrow{\varphi} & \mathfrak{L}_{\mathbf{k}}(A) = \mathfrak{L}_{\mathbf{k}}(A)_{\mathrm{L}} & \xrightarrow{\psi} & \text{Der}_{\mathbf{k}}(B) \\ \| & & \uparrow & & \uparrow \\ A & \xrightarrow{\text{surjective}} & \bar{\Delta} \supseteq \tilde{\Delta} & \xrightarrow{\cong} & L \end{array} \qquad (17)$$

Let $\bar{\Delta}$ (resp. $\tilde{\Delta}$) be the subalgebra of $\mathfrak{L}_{\mathbf{k}}(A)$ (resp. of $\mathfrak{L}_{\mathbf{k}}(A)_{\mathrm{L}}$) generated by Δ. Since A is generated by H and φ is a homomorphism of associative algebras, $\varphi(A) = \bar{\Delta}$. Since ψ is an injective homomorphism of Lie algebras, it restricts to an isomorphism $\tilde{\Delta} \to L$. See part (ii) of (17).

Since A is finitely generated but not nilpotent, Proposition 4.13(b) implies that A is not (SN). So there exists an infinite sequence $(a, a_0, a_1, a_2, \dots) \in A^{\mathbb{N}}$ such that $a_n \cdots a_0 \cdot a \neq 0$ for all $n \in \mathbb{N}$. If we define $F_i = \varphi(a_i)$ for all $i \in \mathbb{N}$ then $(F_0, F_1, \dots) \in (\bar{\Delta})^{\mathbb{N}}$ satisfies $(F_n \circ \cdots \circ F_0)(a) \neq 0$ for all n, so $a \notin \text{Nil}(\bar{\Delta}) = \text{Nil}(\Delta)$ (cf. Proposition 2.5) and hence Δ is not a locally nilpotent subset of $\mathfrak{L}_{\mathbf{k}}(A)$. By Lemma 5.1(a), $\Lambda = \psi(\Delta)$ is not a locally nilpotent subset of $\text{Der}_{\mathbf{k}}(B)$. In particular, (a) is true.

To prove (b), consider an $(m-1)$-subset Λ' of L and let L' be the subalgebra of L generated by Λ'. There exists an $(m-1)$-subset H' of A such that $\psi(\varphi(H'))=\Lambda'$. By our choice of A, H' is a nilpotent subset of A; it follows that $\Delta'=\varphi(H')$ is a nilpotent subset of $\bar{\Delta}$, i.e., there exists $n>0$ such that

$$F_n \circ \cdots \circ F_1 = 0 \text{ for all } (F_1,\ldots,F_n)\in(\Delta')^n. \tag{18}$$

In particular, Δ' is a locally nilpotent subset of $\mathcal{L}_{\mathbf{k}}(A)$; by Lemma 5.1(b), $\Lambda'=\psi(\Delta')$ is a locally nilpotent subset of $\mathrm{Der}_{\mathbf{k}}(B)$, so Proposition 2.5 implies that L' is a locally nilpotent subset of $\mathrm{Der}_{\mathbf{k}}(B)$. On the other hand, (18) and Corollary 4.9 imply that the associative algebra $\overline{\Delta'}$ is nilpotent. By Lemma 4.17, it follows that the Lie algebra $(\overline{\Delta'})_{\mathrm{L}}$ is nilpotent. Since $\widetilde{\Delta'}$ is a subalgebra of $(\overline{\Delta'})_{\mathrm{L}}$, $\widetilde{\Delta'}$ is nilpotent. Since ψ maps $\widetilde{\Delta'}$ onto L', we obtain that L' is nilpotent. This proves (b).

We have $L\subseteq \mathrm{LND}(B)$ by (b). Since $\bar{\Delta}$ is a homomorphic image of the nil algebra A, $\bar{\Delta}$ is nil (as an associative algebra). So the Lie algebra $(\bar{\Delta})_{\mathrm{L}}$ is nil, by Lemma 4.17. As $\tilde{\Delta}$ is a subalgebra of $(\bar{\Delta})_{\mathrm{L}}$, it is a nil Lie algebra. Since $L\cong\tilde{\Delta}$, L is nil. So (c) is proved. □

Corollary 5.7 *Let B be the commutative polynomial algebra in $|\mathbb{N}|$ variables over a field* $\mathbf{k}$.

(a) There exists an infinite subset Δ of $\mathrm{Der}_{\mathbf{k}}(B)$ *satisfying:*

- *Δ is not a locally nilpotent subset of* $\mathrm{Der}_{\mathbf{k}}(B)$*;*
- *every finite subset of Δ is a locally nilpotent subset of* $\mathrm{Der}_{\mathbf{k}}(B)$.

(b) For each integer $m\geq 2$, there exists an m-subset Δ of $\mathrm{Der}_{\mathbf{k}}(B)$ *satisfying:*

- *Δ is not a locally nilpotent subset of* $\mathrm{Der}_{\mathbf{k}}(B)$*;*
- *every proper subset of Δ is a locally nilpotent subset of* $\mathrm{Der}_{\mathbf{k}}(B)$.

Proof The following is an obvious consequence of Proposition 2.5:

(∗) For any subset Δ of $\mathrm{Der}_{\mathbf{k}}(B)$, Δ is a locally nilpotent subset of $\mathrm{Der}_{\mathbf{k}}(B)$ if and only if the Lie subalgebra of $\mathrm{Der}_{\mathbf{k}}(B)$ generated by Δ is a locally nilpotent subset of $\mathrm{Der}_{\mathbf{k}}(B)$.

Proof of (a) Consider the Lie subalgebra L of $\mathrm{Der}_{\mathbf{k}}(B)$ given in Example 5.5 and let Δ be any generating set of L. Then, by (∗), Δ satisfies (a).

Proof of (b) Let $m\geq 2$. Consider the Lie subalgebra L of $\mathrm{Der}_{\mathbf{k}}(B)$ given by Example 5.6. Let Δ be a generating set of L with $|\Delta|=m$. Then, by (∗), Δ satisfies (b). □

6 The Case of Derivation-Finite Algebras

Definition 6.1 Let B be an algebra over a field $\mathbf{k}$. We say that B is *derivation-finite* if there exists a finite subset X of B satisfying:

if $D \in \mathrm{Der}_{\mathbf{k}}(B)$ satisfies $D(x) = 0$ for all $x \in X$, then $D = 0$.

For instance, if B is a finitely generated $\mathbf{k}$-algebra then it is derivation-finite. If $\operatorname{char}\mathbf{k} = 0$ and B is a commutative integral domain of finite transcendence degree over $\mathbf{k}$ then B is derivation-finite.

This section re-examines Question 2 under the assumption that B is a derivation-finite algebra over a field $\mathbf{k}$. We begin with an example.

Example 6.2 Consider the polynomial ring $B = \mathbf{k}[X_1, \dots, X_n]$ where $n \geq 2$ and $\mathbf{k}$ is a field of characteristic zero. Let L be the set of all $D \in \mathrm{Der}_{\mathbf{k}}(B)$ satisfying $D(X_1) \in \mathbf{k}$ and $D(X_i) \in \mathbf{k}[X_1, \dots, X_{i-1}]$ for all $i = 2, \dots, n$. Note that L is a Lie subalgebra of $\mathrm{Der}_{\mathbf{k}}(B)$. We claim:

(a) L is a locally nilpotent subset of $\mathrm{Der}_{\mathbf{k}}(B)$.
(b) L is not a uniformly locally nilpotent subset of $\mathrm{Der}_{\mathbf{k}}(B)$.
(c) The Lie algebra L satisfies (SN) but not (nil).

Proof The proof of (a) uses the following trivial observation: *if $b \in B$ is such that $D(b) \in \mathrm{Nil}(L)$ for all $D \in L$, then $b \in \mathrm{Nil}(L)$*. We have $D(X_1) \in \mathbf{k} \subseteq \mathrm{Nil}(L)$ for all $D \in L$, so $X_1 \in \mathrm{Nil}(L)$, so $\mathbf{k}[X_1] \subseteq \mathrm{Nil}(L)$ by Theorem 2.7. Let $i \in \{2, \dots, n\}$ be such that $\mathbf{k}[X_1, \dots, X_{i-1}] \subseteq \mathrm{Nil}(L)$. Since $D(X_i) \in \mathbf{k}[X_1, \dots, X_{i-1}] \subseteq \mathrm{Nil}(L)$ for all $D \in L$, $X_i \in \mathrm{Nil}(L)$ and hence $\mathbf{k}[X_1, \dots, X_i] \subseteq \mathrm{Nil}(L)$ by Theorem 2.7. It follows by induction that $\mathrm{Nil}(L) = B$, so (a) is true.

(b) Given any $m > 0$ let $D_m = X_1^m \frac{\partial}{\partial X_2}$, and let $E = \frac{\partial}{\partial X_1}$; then $\{D_m, E\} \subset L$ and (since $\operatorname{char}\mathbf{k} = 0$) $(E^m \circ D_m)(X_2) \neq 0$, so $X_2 \notin \mathrm{UNil}(L)$.

(c) Let us check that L is not a nil Lie algebra. Let $E = \frac{\partial}{\partial X_1} \in L$. We claim that $\mathrm{ad}(E) : L \to L$ is not nilpotent. Indeed, let $m > 0$ and let us prove that $\mathrm{ad}(E)^m \neq 0$. It suffices to show that $\mathrm{ad}(E)^m(D_m) = [E, \dots, E, D_m]$ is not zero (where $D_m = X_1^m \frac{\partial}{\partial X_2}$ and where there are m "E" in the bracket); so it suffices to show that $[E, \dots, E, D_m](X_2) \neq 0$. Since $E(X_2) = 0$, we have $[E, \dots, E, D_m](X_2) = (E^m \circ D_m)(X_2) \neq 0$. So $\mathrm{ad}(E)$ is not nilpotent and consequently the Lie algebra L is not nil.

To show that the Lie algebra L is (SN), we use the following notation. Let $B_0 = \mathbf{k}$ and $B_j = \mathbf{k}[X_1, \dots, X_j]$ for all $j = 1, \dots, n$. If $j \in \{1, \dots, n\}$ and $f \in B_{j-1}$, define $D^j_f = f \frac{\partial}{\partial X_j} \in L$. Let H be the set of D^j_f, for all pairs (j, f) such that $j \in \{1, \dots, n\}$ and $f \in B_{j-1}$ (thus $H \subseteq L$). Consider $D^j_f, D^k_g \in H$ (where $f \in B_{j-1}$ and $g \in B_{k-1}$). A straightforward calculation gives $[D^j_f, D^k_g] = 0$ when $j = k$, and $[D^j_f, D^k_g] = D^k_{D^j_f(g)}$ if $j < k$. Since $D^j_f(g) = 0$ when $j = k$, we get

$$[D^j_f, D^k_g] = D^k_{D^j_f(g)} \quad \text{whenever } j \leq k. \tag{19}$$

We have $L = \mathrm{Span}_{\mathbf{k}}(H)$, so in particular H is a generating set for the Lie algebra L. By Corollary 4.9(a), to show that L is (SN) it suffices to show that for every sequence $(d_0, d_1, d_2, \dots) \in H^{\mathbb{N}}$ there exists an m such that $[d_m, \dots, d_0] = 0$. So consider $(d_0, d_1, d_2, \dots) \in H^{\mathbb{N}}$. Then for each i we have $d_i = D_{f_i}^{j_i}$, where $j_i \in \{1, \dots, n\}$ and $f_i \in \mathbf{k}[X_1, \dots, X_{j_i-1}]$. Choose $\nu \in \mathbb{N}$ such that $j_\nu = \max\{\, j_i \mid i \in \mathbb{N} \,\}$. Then (19) implies that

$$\text{for each } m \geq \nu,\ [d_m, \dots, d_0] = D_{g_m}^{j_\nu} \text{ for some } g_m \in B_{j_\nu - 1}.$$

We have $[d_{m+1}, d_m, \dots, d_0] = [d_{m+1}, [d_m, \dots, d_0]] = [d_{m+1}, D_{g_m}^{j_\nu}] = [D_{f_{m+1}}^{j_{m+1}}, D_{g_m}^{j_\nu}]$ $= D^{j_\nu}_{D_{f_{m+1}}^{j_{m+1}}(g_m)}$, so $g_{m+1} = D_{f_{m+1}}^{j_{m+1}}(g_m) = d_{m+1}(g_m)$ (for all $m \geq \nu$). Thus $g_m = (d_m \circ \dots \circ d_{\nu+1})(g_\nu)$ for all $m > \nu$. Since $(d_i)_{i=\nu+1}^{\infty}$ is an infinite sequence in H, and H is a locally nilpotent subset of $\mathrm{Der}_{\mathbf{k}}(B)$, it follows that for m large enough we have $g_m = 0$ and hence $[d_m, \dots, d_0] = 0$. This shows that the Lie algebra L is (SN). □

Until the end of the section, we assume:

B is an algebra over a field k (cf. 1.3(a)) and is derivation-finite.

For each element D of the Lie algebra $\mathrm{Der}_{\mathbf{k}}(B)$, we may consider the map $\mathrm{ad}(D) : \mathrm{Der}_{\mathbf{k}}(B) \to \mathrm{Der}_{\mathbf{k}}(B)$, $E \mapsto [D, E]$ (this is the map defined in 1.7). Then we have the following.

Lemma 6.3 *If $D \in \mathrm{LND}(B)$ then the map $\mathrm{ad}(D) : \mathrm{Der}_{\mathbf{k}}(B) \to \mathrm{Der}_{\mathbf{k}}(B)$ is locally nilpotent.*

Proof We have to show that for each $E \in \mathrm{Der}_{\mathbf{k}}(B)$ there exists $N > 0$ such that $\mathrm{ad}(D)^N(E) = 0$. Let $E \in \mathrm{Der}_{\mathbf{k}}(B)$. Choose a nonempty finite subset X of B such that

$$\text{if } F \in \mathrm{Der}_{\mathbf{k}}(B) \text{ satisfies } F(x) = 0 \text{ for all } x \in X, \text{ then } F = 0.$$

Choose $n > 0$ such that $D^n(x) = 0$ for all $x \in X$. Since

$$Y = \{\, (E \circ D^i)(x) \mid (i, x) \in \{0, \dots, n-1\} \times X \,\}$$

is a finite subset of B, there exists $m > 0$ such that $D^m(y) = 0$ for all $y \in Y$; then $(D^m \circ E \circ D^i)(x) = 0$ for all $(i, x) \in \{0, \dots, n-1\} \times X$. Let $F = \mathrm{ad}(D)^{m+n-1}(E) = [D, D, \dots, D, E]$. Since $F \in \mathrm{Der}_{\mathbf{k}}(B)$, in order to show that $F = 0$ it suffices to show that $F(x) = 0$ for all $x \in X$. Let $x \in X$; then $F(x)$ is a linear combination (over $\mathbf{k}$) of terms of the form $(D^j \circ E \circ D^i)(x)$ where $i, j \geq 0$ and $i + j = m + n - 1$. In each one of these terms we have either $i \geq n$ or ($i < n$ and $j \geq m$), so $(D^j \circ E \circ D^i)(x) = 0$. It follows that $F(x) = 0$ and hence that $\mathrm{ad}(D)^{m+n-1}(E) = 0$. □

The next facts give some positive answers to Question 2. Observe that the conclusion of Corollary 6.4(a) cannot be strengthened from (Lnil) to (nil), by Example 6.2.

Corollary 6.4 *Let L be a Lie subalgebra of $\mathrm{Der}_{\mathbf{k}}(B)$ such that $L \subseteq \mathrm{LND}(B)$.*

(a) *L is locally nil* (Lnil).
(b) *If L is finite dimensional then it is a nilpotent Lie algebra.*

Proof Assertion (a) follows immediately from Lemma 6.3. Part (b) follows from the fact (Proposition 4.13) that (Lnil) $\Leftrightarrow$ (N) for finite dimensional Lie algebras. □

Proposition 6.5 *For a finitely generated Lie subalgebra L of* $\text{Der}_{\mathbf{k}}(B)$*, the following are equivalent:*

(a) *L is a nilpotent Lie algebra;*
(b) *L is a Lie-locally nilpotent subset of* $\text{Der}_{\mathbf{k}}(B)$*;*
(c) *L is a uniformly Lie-locally nilpotent subset of* $\text{Der}_{\mathbf{k}}(B)$.

Moreover, if L is a locally nilpotent subset of $\text{Der}_{\mathbf{k}}(B)$ *then (a–c) are satisfied.*

Proof Choose a finite subset Δ of L such that $\tilde{\Delta} = L$. Choose a nonempty finite subset X of B such that

$$\text{if } F \in \text{Der}_{\mathbf{k}}(B) \text{ satisfies } F(x) = 0 \text{ for all } x \in X, \text{ then } F = 0.$$

It is obvious that (a) implies (b). If (b) is true then Δ is a finite Lie-locally nilpotent subset of $\text{Der}_{\mathbf{k}}(B)$; by Lemma 3.6, Δ is uniformly Lie-locally nilpotent; so $L = \tilde{\Delta}$ is uniformly Lie-locally nilpotent by Proposition 3.7, and this shows that (b) implies (c).

Assume that (c) is true. Since X is a finite set and L is uniformly Lie-locally nilpotent, there exists $N \in \mathbb{N}$ satisfying

$$[D_n, \ldots, D_1](x) = 0 \quad \text{for all } n \geq N, (D_1, \ldots, D_n) \in L^n \text{ and } x \in X. \tag{20}$$

In (20), the fact that $[D_n, \ldots, D_1]$ is a derivation that annihilates each element of X implies that $[D_n, \ldots, D_1] = 0$. So we have $[D_N, \ldots, D_1] = 0$ for all $(D_1, \ldots, D_N) \in L^N$, i.e., L is nilpotent. So (c) implies (a) and consequently the three conditions are equivalent.

Assume that L is a locally nilpotent subset of $\text{Der}_{\mathbf{k}}(B)$. Then Δ is a finite locally nilpotent subset of $\text{Der}_{\mathbf{k}}(B)$, so Δ is uniformly locally nilpotent by Lemma 2.2, so $L = \tilde{\Delta}$ is uniformly locally nilpotent by Proposition 2.5, so L is uniformly Lie-locally nilpotent by Lemma 3.5, so (a–c) are satisfied. □

References

1. Ado, I.D.: The representation of Lie algebras by matrices. Am. Math. Soc. Transl. **1949**(2), 21 (1949)
2. Golod, E.S.: On nil-algebras and finitely approximable p-groups. Izv. Akad. Nauk SSSR Ser. Math. **28**, 273–276 (1964). (Russian)

3. Golod, E.S.: Some problems of Burnside type. In: Proceedings of the International Congress of Mathematicians (Moscow, 1966), pp. 284–289. Izdat. "Mir", Moscow (1968) (Russian)
4. Iwasawa, K.: On the representation of Lie algebras. Jpn. J. Math. **19**, 405–426 (1948)
5. Jacobson, N.: Lie Algebras. Dover Publications, Inc., New York (1979). Republication of the 1962 original
6. Kostrikin, A.I.: Around Burnside. Ergebnisse der Mathematik und ihrer Grenzgebiete (3) [Results in Mathematics and Related Areas (3)], vol. 20. Springer, Berlin (1990). Translated from the Russian and with a preface by James Wiegold
7. Rowen, L.H.: Ring Theory, vol. II. Pure and Applied Mathematics, vol. 128. Academic Press, Inc., Boston (1988)

On the Theory of Gordan-Noether on Homogeneous Forms with Zero Hessian (Improved Version)

Junzo Watanabe and Michiel de Bondt

Abstract We give a detailed proof for Gordan-Noether's results in "Ueber die algebraischen Formen, deren Hesse'sche Determinante identisch verschwindet." C. Lossen has written a paper in a similar direction as the present paper, but did not provide a proof for every result. In our paper, every result is proved. Furthermore, our paper is independent of Lossen's paper and includes a considerable number of new observations.

Keywords Hessian form · Homogeneous polynomial

An earlier version of this paper, namely [16], has been printed in Proceedings of the School of Science of Tokai University, Vol. 49, Mar. 2014. In this version, a serious error has been corrected, namely [16, Lemma 5.2]. Lemma 5.2 has been replaced by a weaker statement, and Proposition 6.2 has been weakened along with that. Lemma 5.2 no longer suffices for the proof of Proposition 7.2. For that reason, a new section (Sect. 8) has been added to complete the proof of Proposition 7.2. Furthermore, several trivial errors have been corrected, and a section with new results has been added (Sect. 9).

1 Introduction

In 1851 and 1859, O. Hesse wrote two papers [10, 11] in Crelle's Journal Bd. 42 and Bd. 56 in which he claimed that if the Hessian determinant of a homogeneous polynomial identically vanishes, then a variable can be eliminated by a linear trans-

J. Watanabe (✉)
Department of Mathematics, Tokai University, Hiratsuka, Kanagawa 259-1292, Japan
e-mail: watanabe.junzo@tokai-u.jp

M. de Bondt
Institute for Mathematics, Astrophysics and Particle Physics,
Radboud University Nijmegen, Nijmegen, The Netherlands
e-mail: M.deBondt@math.ru.nl

S. Kuroda et al. (eds.), *Polynomial Rings and Affine Algebraic Geometry*,
Springer Proceedings in Mathematics & Statistics 319,
https://doi.org/10.1007/978-3-030-42136-6_3

formation of the variables. Unfortunately his claim is not true in general. In fact Hesse's proof was unconventional and the validity of the proof was questioned from the beginning [8]. Nonetheless it should have been easy to see that Hesse's claim is true for binary forms as well as quadrics. In 1875, M. Pasch proved that Hesse's claim is true for ternary cubics and quaternary cubics [14].

In 1876, Gordan and Noether [8] finally established the correct statement which says that if a form has zero Hessian, then one variable can be eliminated from the form itself and its partial derivatives simultaneously by way of a birational transformation of the variables. Moreover they proved in the same paper that Hesse's claim is true, if the number of variables is at most four, and furthermore they determined all homogeneous polynomials in five variables for which the Hessian determinant identically vanishes.

The present paper goes beyond the necessity and desire to understand their proof. The Hessian of a homogeneous polynomial is essential to the theory of Artinian Gorenstein rings because it is used with higher Hessians to determine the set of the strong Lefschetz elements in a zero-dimensional Gorenstein algebra [13]. In particular, if the Hessian of a homogeneous polynomial is identically zero, we get a Gorenstein algebra which lacks the strong Lefschetz property. To explain this further, let $R = K[x_1, \ldots, x_n]$ be the polynomial ring over a field K of characteristic zero and let $G \in R$ be a homogeneous polynomial. In addition let $I \subset R$ be the ideal:

$$I = \left\{ f(x_1, \ldots, x_n) \in R \,\middle|\, f\left(\left(\frac{\partial}{\partial x_1}, \ldots, \frac{\partial}{\partial x_n} \right) G \right) = 0 \right\},$$

and let $A = R/I$. Then A is a zero-dimensional Gorenstein graded algebra $A = \bigoplus_{i=0}^{d} A_i$, where $d = \deg G$. It is easy to see that if the polynomial G contains properly n variables, then the partial derivatives $G_1, \ldots, G_n$ of G are linearly independent. Moreover if $L = \xi_1 x_1 + \cdots + \xi_n x_n \in R_1$ is a linear form, it defines a linear map

$$\times L^{d-2} : A_1 \longrightarrow A_{d-1}$$

by $A_1 \ni a \mapsto aL^{d-2} \in A_{d-1}$. Let M be the matrix for this linear map with respect to the bases $\langle x_1, \ldots, x_n \rangle$ and $\langle G_1, \ldots, G_n \rangle$ for A_1 and A_{d-1} respectively. Then $\det M$ is the Hessian of G evaluated at $(\xi_1, \ldots, \xi_n)$ (up to a constant multiple). If $\dim A_1 \leq 4$, and if we take the results of [8] for granted, this is to say that there exists a linear form L such that L^{d-2} is bijective. If $n = 5$, Gordan-Noether's paper enables us to determine all homogeneous polynomials G such that L^{d-2} is not bijective for any choice of $\xi_1, \ldots, \xi_n$.

Gordan-Noether's paper [8] has been cited by several authors [6, 9, 13, 15]; however each time it had to be accompanied with a proviso that the result is yet to be confirmed. Gordan-Noether's paper is difficult to understand. Not only their results but also the methods have been completely forgotten. Thus, it seems necessary to consider the paper from the viewpoint of contemporary algebra. Yamada [17] devoted considerable efforts to constructing a modernized translation of [8]; however, it was not completely successful and therefore it was unpublished.

The purpose of this paper is to give detailed proofs for most of the results that were obtained by Gordan-Noether in [8]. The foundation of their theory lies in the fact that if the Hessian determinant of a homogeneous polynomial is identically zero, then the polynomial satisfies a certain linear partial differential equation. For simplicity we assume that the Hessian matrix $M := (\partial^2 f/\partial x_i \partial x_j)$ has corank one. Then the left null space of the matrix M has dimension one over the function field $K(x)$. Let $(h_1, \dots, h_n)$ be a vector with $h_j \in K(x)$ such that $(h_1, \dots, h_n)M = 0$. Then it is easy to see that f and its partial derivatives satisfy the partial differential equation

$$h_1 \frac{\partial F}{\partial x_1} + h_2 \frac{\partial F}{\partial x_2} + \cdots + h_n \frac{\partial F}{\partial x_n} = 0. \tag{1}$$

By clearing the denominator we may think h_j are homogeneous polynomials of the same degree. (Cf. Remark 4.6.) Gordan and Noether discovered that each coefficient h_j *itself* of (1) satisfies the partial differential Eq. (1). This readily proves that a variable can be eliminated from f and its partial derivatives simultaneously by a birational transformation. (See Theorem 4.7.) The Gordan-Noether called the functions satisfying (1) "die Functionen Φ". The coefficients of this partial linear differential equation were termed as a "self-vanishing system" by Yamada [17]. The solution of this type of differential equation behaves as if the coefficients were constants. According to [8], Jacobi considered this type of differential equation and it is the key to understanding the Gordan and Noether theory.

To prove that Hesse's claim is true for homogeneous polynomials for $n \le 4$, it is necessary to consider the fundamental locus and the image of the rational map $\mathbb{P}^{n-1} \to \mathbb{P}^{n-1}$ defined by $(x_1, \dots, x_n) \mapsto (h_1, \dots, h_n)$. From the fact that $h_1, \dots, h_n$ is a "self-vanishing system," it follows that the dimension of the image is at most $n - 3$ if $n \ge 3$. This enables us to determine the forms with zero Hessian for $n \le 4$. For $n = 5$, the dimension of the image of $h_1, \dots, h_n$ may be exactly $n - 3 = 2$, which is too large to determine the forms with zero Hessian in dimension 5. But this cannot occur in the context of forms with zero Hessian. So the dimension is at most 1 and we can determine the forms with zero Hessian for $n = 5$ as well.

Gordan and Noether's idea and proof techniques appear quite new and they are interesting in their own right, and they give us a series of new problems, some of which will be investigated in our subsequent papers.

Historical Notes

In 1990, Yamada wrote a paper [17], under Grant-in-Aid no. 20022551 entitled "On the hypersurface defined by a form whose Hessian identically vanishes." It was unpublished because it was incomplete; however, he defined and systematically studied "the self-vanishing system of polynomials," as he named them. This paper was written to finish Yamada's paper [17].

An earlier version of this paper was written by the first author only without a prior knowledge of Lossen's paper [12]. The first author gave a 3 hour lecture on this subject at the workshop "Aspects of SLP and WLP" held in Hawaii Tokai International College in Honolulu in September 2012, where he learned that Lossen had written a paper [12] in the same direction and that there were other related papers [1, 2]. Clearly the objectives of Lossen's paper and this paper are identical. However the methods employed are different, although both are based on the same source, i.e., on the paper of Gordan and Noether [8].

The main differences with Lossen's paper are listed below.

1. Theorem 4.7 is not explicitly written in Lossen [12], which tells us that Hesse was in some sense correct in his intuition, when he said that if the Hessian determinant vanishes, then a variable can be eliminated by a linear transformation.
2. In this paper, a so-called "self-vanishing system" is defined and studied systematically.
3. Our observation Proposition 6.2 considerably simplifies the entire argument.
4. In Lossen's paper [12], the connection to the Lefschetz properties of Artinian rings is not indicated.
5. In this paper it is proved that a cubic form in five variables is essentially unique, while in [12], the provided proof is incomplete.

In 2014, the first author published a version of this paper in Proceedings of the School of Science of Tokai University. But Lemma 5.2 of that paper contains a serious error. The error can however be fixed, and the second author has been added to the paper to do this. In the current version, Lemma 5.2 has been replaced by a weaker statement. Proposition 6.2 has been replaced along with that, and Sect. 8 has been added to complete the proof of Proposition 7.2 (the new version of Lemma 5.2 does not suffice for this). Section 9 is due to the second author as well, and was inspired by the final remark (Remark 7.6) in the 2014 version of the paper [16].

The second author has written some related papers on his own. In [4], the forms with zero Hessian are determined for $n \leq 4$. In [5], using results of [6], the forms with zero Hessian are determined for $n = 5$. But the proofs diverge from the techniques in Gordan-Noether's paper [8] on some points (see also Remark 8.8). Nonhomogeneous polynomials with zero Hessians are considered in [4, 6] as well. In [3], all these results are generalized to arbitrary dimension, with the zero Hessian condition replaced by that the Hessian matrix has fixed small rank (one less than the original dimension).

The first author would like to thank H. Nasu and T. Tsukioka for insightful discussions for Remark 7.5.

2 Notation and Preliminaries

Throughout the paper we denote by K an algebraically closed field of characteristic zero. We denote by $K[x_1, x_2, \dots, x_n]$ the polynomial ring in the variables $x_1, \dots, x_n$ and by $K(x_1, \dots, x_n)$ the function field. Let $R = K[x_1, \dots, x_n]$. An element of R is sometimes abbreviated as $f(x)$ or simply as f.

A **system** of homogeneous polynomials (or forms) of R is a vector $(f_1, f_2, \dots, f_n)$ consisting of homogeneous polynomials $f_i \in R$ of the same degree. A system of forms is denoted by a bold face letter as $\boldsymbol{f} = (f_1, \ldots, f_n)$. To avoid triviality we will always assume that $\boldsymbol{f} \neq 0$. To indicate that a system is a vector of polynomials depending on the argument vector $x = (x_1, \ldots, x_n)$, we write $\boldsymbol{f}(x)$ as well as $\boldsymbol{f}$. Although a coordinate system like $x = (x_1, \ldots, x_n)$ is a "system of forms," we do not apply the rule to use a bold face letter to denote it. If $y = (y_1, y_2, \ldots, y_n)$ is a new set of variables and if $\boldsymbol{f}(x)$ is a system of forms in R, then $\boldsymbol{f}(y)$ obtained from $\boldsymbol{f}(x)$ by the substitution $x_j \mapsto y_j$ is a system of forms in $K[y_1, \ldots, y_n]$.

We treat vectors both as row vectors and as column vectors, so if $A = (a_{ij})$ is an $n \times n$ matrix with $a_{ij} \in K$, and if $x = (x_1, \ldots, x_n)$ is a vector, then $y = Ax$ means that $y = (y_1, \ldots, y_n)$ is a vector defined by

$$y_i = \sum_{j=1}^{n} a_{ij} x_j.$$

Likewise $y = xA$ means that

$$y_j = \sum_{i=1}^{n} a_{ij} x_i.$$

The same rule applies to systems of polynomials as well as coordinate systems. Thus, if $\boldsymbol{f} = (f_1, \ldots, f_n)$ is a system of forms, then $\boldsymbol{f}' = A\boldsymbol{f}$ is a system of forms for any $n \times n$ invertible matrix A over K.

Lemma 2.1 *Let $K[x] = K[x_1, \ldots, x_n]$ be the polynomial ring and let $f(x) = f(x_1, \ldots, x_n) \in K[x]$. Let $A = (a_{ij}) \in \mathrm{GL}(n, K)$, and put*

$$(x'_1, x'_2, \ldots, x'_n) = A(x_1, x_2, \ldots, x_n)$$

Consider $(x'_1. \ldots, x'_n)$ as a new coordinate system and let f' be the polynomial in $x'_1, \ldots, x'_n$ defined by

$$f'(x') := f(A^{-1}x') = f(x).$$

Then

$$(f'_1, f'_2, \ldots, f'_n) = {}^tA^{-1}(f_1, f_2, \ldots, f_n),$$

where $f_j = \frac{\partial f}{\partial x_j}$, and $f'_j = \frac{\partial f'}{\partial x'_j}$.

Proof is left to the reader.

Lemma 2.2 *Let $f(x) = f(x_1, \ldots, x_n)$ be a homogeneous polynomial of positive degree. Put $f_i = \frac{\partial f}{\partial x_i}$ and let $\boldsymbol{f} = (f_1, \ldots, f_n)$ be a system of polynomials. If $\dim_K \sum_{i=1}^{n} K f_i = s$, then $n - s$ variables can be eliminated from f by means of a linear transformation of the variables. In other words, there exists an invertible matrix $A = (a_{ij}) \in \mathrm{GL}(n, K)$ such that, if we let $x' = Ax$, then the polynomial $f'(x'_1, \ldots, x'_n) = f(A^{-1}x')$ does not depend on $x'_{s+1}, x'_{s+2}, \ldots, x'_n$.*

Proof By assumption there exists an invertible matrix B such that $B(f_1, f_2, \ldots, f_n) = (f'_1, f'_2, \ldots, f'_s, 0, 0, \ldots, 0)$. Let A be the matrix such that ${}^tA^{-1} = B$, and put $x' = Ax$ and $f'(x') = f(A^{-1}x')$. Then

$$\frac{\partial f'}{\partial x'_j} = 0 \text{ for } j \geq s+1,$$

by Lemma 2.1. Thus $f'(x')$ does not depend on the variables $x'_{s+1}, x'_{s+2}, \ldots, x'_n$.

Proposition 2.3 *Let $f = (f_1, \ldots, f_n)$ be a system of forms in $K[x_1, \ldots, x_n]$, where $f_1, \ldots, f_n$ are algebraically dependent. Let $y = (y_1, \ldots, y_n)$ be a new set of variables independent of x. Let*

$$\phi : K[y_1, y_2, \ldots, y_n] \to K[x_1, x_2, \ldots, x_n]$$

be the homomorphism defined by $y_i \mapsto f_i$, and let $g = g(y_1, \ldots, y_n)$ be a nonzero element in $\ker \phi$ of the least degree. Put

$$h_j = h_j(x) := \frac{\partial g}{\partial y_j}(f_1, \ldots, f_n).$$

In words, $h_j(x)$ is the polynomial obtained from $\frac{\partial g}{\partial y_j}$ by substituting $(f_1, \ldots, f_n)$ for $(y_1, \ldots, y_n)$. Let $W = \sum_{j=1}^n K h_j$ be the vector space over K spanned by the elements $h_1, \ldots, h_n$. Let $s = \dim_K W$. Then $n - s$ variables can be eliminated from $g(y)$ by means of a linear transformation of the variables $y_1, \ldots, y_n$.

Proof Put $V = \sum_{i=1}^n K \frac{\partial g}{\partial y_i}$. We claim that $\dim_K V = s$. It is clear that $\dim_K V \geq s$. Consider the restriction

$$\phi|_V : V \to K[x_1, x_2, \ldots, x_n].$$

We see that $(\ker \phi) \cap V = 0$ by minimality of the degree of g. Thus we have $\dim_K V = s$, since $\phi|_V$ is injective and $\operatorname{im} \phi|_V$ is W. By the previous lemma, proof is complete.

Remark 2.4 In the above Proposition, it is possible that $h_j = 0$ for some j. It means that $\frac{\partial g}{\partial y_j} = 0$. Hence $\boldsymbol{h} := (h_1, \ldots, h_n) \neq 0$. If we drop the condition that $\deg g$ is minimal in $\mathcal{I}$, we may define $\boldsymbol{h}$ as well, but $\boldsymbol{h}$ can be 0.

3 Self-vanishing Systems of Polynomials

Let $x = (x_1, \ldots, x_n)$, $y = (y_1, \ldots, y_n)$ be two sets of indeterminates and let $K(x, y) = K(x_1, \ldots, x_n, y_1, \ldots, y_n)$ denote the rational function field. We introduce the differential operator

$$\mathcal{D}_x(y) : K(x, y) \to K(x, y),$$

which is defined by

$$\mathcal{D}_x(y) f(x, y) := \sum_{j=1}^{n} y_j \frac{\partial f(x, y)}{\partial x_j}$$

for $f(x, y) \in K(x, y)$. For a homogeneous polynomial $f(x) = f(x_1, \dots, x_n) \in K[x_1, \dots, x_n]$ and for $j \geq 0$, define $f^{(j)}(x, y)$ to be the polynomial in $K[y_1, \dots, y_n, x_1, \dots, x_n]$ given by

$$f^{(j)}(x, y) = \frac{1}{j!} \mathcal{D}_x(y)^j (f(x)).$$

It is easy to see that

$$f^{(j)}(x, x) = \binom{d}{j} f(x), \quad j = 0, 1, 2, \cdots, d$$

and

$$f^{(d)}(x, y) = f(y)$$

where $d = \deg f$.

Proposition 3.1 *Let $A = (a_{ij})$ be an invertible matrix with $a_{ij} \in K$ and let*

$$x' = (x'_1, \dots, x'_n) = A(x_1, \dots, x_n),$$

$$y' = (y'_1, \dots, y'_n) = A(y_1, \dots, y_n).$$

Let $f'(x')$ be the polynomial in the coordinate x' defined by $f'(x') = f(x) \in K[x_1, \dots, x_n]$. Then we have

$$\mathcal{D}_{x'}(y') f'(x') = \mathcal{D}_x(y) f(x).$$

Proof Let $\boldsymbol{f}$ be the system of forms:

$$\boldsymbol{f} = \frac{\partial f(x)}{\partial x} := \left(\frac{\partial f(x)}{\partial x_1}, \frac{\partial f(x)}{\partial x_2}, \dots, \frac{\partial f(x)}{\partial x_n} \right).$$

Likewise, let $\boldsymbol{f}'$ be the system of forms:

$$\boldsymbol{f}' = \frac{\partial f'(x')}{\partial x'} = \left(\frac{\partial f'(x')}{\partial x'_1}, \frac{\partial f'(x')}{\partial x'_2}, \dots, \frac{\partial f'(x')}{\partial x'_n} \right).$$

Then, by definition,

$$\mathcal{D}_x(y)f(x) = y \cdot \boldsymbol{f} = y_1 \frac{\partial f}{\partial x_1} + \cdots + y_n \frac{\partial f}{\partial x_n}.$$

By Lemma 2.1 we have

$$\begin{aligned}\mathcal{D}_x(y)f(x) = y \cdot \boldsymbol{f} &= (A^{-1}y') \cdot ({}^tA\boldsymbol{f}') \\ &= (y'({}^tA^{-1})) \cdot ({}^tA\boldsymbol{f}') \\ &= y' \cdot \boldsymbol{f}' = \mathcal{D}_{x'}(y')f'(x').\end{aligned}$$

Proposition 3.2 *For $f(x) \in K[x]$, we have*

$$f(x + ty) = \sum_{j=0}^{\infty} t^j f^{(j)}(x, y),$$

for an indeterminate t.

Proof This is a direct consequence of the Taylor expansion.

Notation 3.3 Let $\boldsymbol{h} = (h_1(x), \ldots, h_n(x))$ be a system of homogeneous polynomials $h_i \in K[x_1, \ldots, x_n]$. We define the differential operator $\mathcal{D}_x(\boldsymbol{h}) : K(x) \to K(x)$ associated to $\boldsymbol{h}$ by

$$\mathcal{D}_x(\boldsymbol{h})f(x) = \sum_{j=1}^{n} h_j(x) \frac{\partial f(x)}{\partial x_j} = f^{(1)}(x, h(x)).$$

Furthermore we denote by

$$\mathrm{Sol}(\boldsymbol{h}; R)$$

the set of solutions in $R \subseteq K[x]$ of the differential equation

$$\mathcal{D}_x(\boldsymbol{h})f(x) = 0.$$

Namely,

$$\mathrm{Sol}(\boldsymbol{h}; R) = \left\{ f(x) \in R \;\middle|\; \sum_{j=1}^{n} h_j(x) \frac{\partial f(x)}{\partial x_j} = 0 \right\}.$$

Note that $\mathrm{Sol}(\boldsymbol{h}; K[x])$ is a graded subalgebra of $K[x]$.

Definition 3.4 A system $\boldsymbol{h} = (h_1(x), \ldots, h_n(x))$ of polynomials is called self-vanishing, if $h_j(x) \in \mathrm{Sol}(\boldsymbol{h}; K[x])$ for all $j = 1, 2, \ldots, n$. In addition to it, if $\mathrm{GCD}(h_1, \ldots, h_n) = 1$, we will say that $\boldsymbol{h}$ is a reduced self-vanishing system.

Example 3.5 A constant vector $\boldsymbol{h} = (c_1, c_2, \ldots, c_n) \in K^n$ is obviously a self-vanishing system.

Example 3.6 Let $h_j \in K[x]$ be homogeneous polynomials (of the same degree). Suppose that $\boldsymbol{h} = (h_1, \ldots, h_n)$ satisfy the following conditions.

1. $h_1 = \cdots = h_r = 0$, for some integer r; $1 \leq r < n$.
2. The polynomials $h_{r+1}, \ldots, h_n$ do not involve the variables $x_{r+1}, \ldots, x_n$.

Then $\boldsymbol{h}$ is a self-vanishing system of forms.

Definition 3.7 Let $\boldsymbol{f} = (f_1, \ldots, f_n)$ be a system of forms in $K[x]$. We denote by $\partial \boldsymbol{f}/\partial x_j$ the system of forms:

$$\frac{\partial \boldsymbol{f}}{\partial x_j} = \frac{\partial}{\partial x_j} \boldsymbol{f} = \left(\frac{\partial f_1}{\partial x_j}, \frac{\partial f_2}{\partial x_j}, \ldots, \frac{\partial f_n}{\partial x_j} \right).$$

This should not be confused with the notation already used:

$$\frac{\partial f(x)}{\partial x} := \left(\frac{\partial f}{\partial x_1}, \frac{\partial f}{\partial x_2}, \ldots, \frac{\partial f}{\partial x_n} \right).$$

Proposition 3.8 *Let $\boldsymbol{f} := (f_1, f_2, \ldots, f_n)$ be a system of forms in $K[x]$, in which the components are algebraically dependent. Let $y = (y_1, \ldots, y_n)$ be a coordinate system algebraically independent of x. Let $\phi : K[y] \to K[x]$ be the homomorphism defined by*

$$y_j \mapsto f_j, (j = 1, 2, \ldots, n).$$

Let $g = g(y) \in \ker \phi$ be a non-zero homogeneous polynomial of the least degree in $\ker \phi$. As in Proposition 2.3, define $h_j \in K[x]$ by

$$h_j = \frac{\partial g}{\partial y_j}(f_1, \ldots, f_n).$$

Let $\boldsymbol{h} = (h_1, \ldots, h_n)$. Then

(a) $\boldsymbol{h}$ is a syzygy of $\boldsymbol{f}$.
(b) $\boldsymbol{h}$ is a syzygy of $\partial \boldsymbol{f}/\partial x_j$ for every $j = 1, 2, \ldots, n$.

Proof Since $g = g(y)$ is homogeneous, we have

$$\frac{\partial g}{\partial y_1} y_1 + \frac{\partial g}{\partial y_2} y_2 + \cdots + \frac{\partial g}{\partial y_n} y_n = (\deg g) g.$$

Make the substitution $y_i \mapsto f_i$. Then we have

$$h_1 f_1 + h_2 f_2 + \cdots + h_n f_n = 0.$$

This shows the first assertion. By definition of $g = g(y)$, we have $g(f_1, f_2, \ldots, f_n) = 0$. Apply the operator $\frac{\partial}{\partial x_j}$ to the this equality. Then we have

$$0 = \frac{\partial g(f_1, f_2, \ldots, f_n)}{\partial x_j} = \sum_{k=1}^{n} \frac{\partial g}{\partial y_k}(f_1, \ldots, f_n)\frac{\partial f_k}{\partial x_j} = \sum_{k=1}^{n} h_k \frac{\partial f_k}{\partial x_j}.$$

This shows the second assertion.

Theorem 3.9 *Let $f = f(x) \in K[x]$ be a homogeneous polynomial and put $f_j = \frac{\partial f}{\partial x_j}$. Assume that $f_1, \ldots, f_n$ are algebraically dependent. Let $\boldsymbol{f} = (f_1, \ldots, f_n)$ and let $\boldsymbol{h} = (h_1, \ldots, h_n)$ be a system of forms as defined in Proposition 3.8 for $\boldsymbol{f}$. Then*

(a) $f(x) \in \mathrm{Sol}(\boldsymbol{h}; K[x])$.
(b) $f_j(x) \in \mathrm{Sol}(\boldsymbol{h}; K[x])$ for $j = 1, 2, \ldots, n$.
(c) $f(x) \in \mathrm{Sol}(\partial\boldsymbol{h}/\partial x_j; K[x])$ for $j = 1, 2, \ldots, n$.
(d) $\boldsymbol{h}$ is a self-vanishing system of forms.

Proof The assertion (a) follows immediately from Proposition 3.8(a). Proposition 3.8(b) says that $\boldsymbol{h}$ is a syzygy of $\partial \boldsymbol{f}/\partial x_j$. This means that

$$\sum_{k=1}^{n} h_k \frac{\partial f_k}{\partial x_j} = 0.$$

But $f_k = \frac{\partial f}{\partial x_k}$. Hence we have

$$\sum_{k=1}^{n} h_k \frac{\partial f_k}{\partial x_j} = \sum_{k=1}^{n} h_k \frac{\partial f_j}{\partial x_k} = 0.$$

This shows assertion (b).

Again by Proposition 3.8(a) we have $\boldsymbol{h} \cdot \boldsymbol{f} = 0$. For each j, we have

$$\frac{\partial}{\partial x_j}(\boldsymbol{h} \cdot \boldsymbol{f}) = \frac{\partial \boldsymbol{h}}{\partial x_j} \cdot \boldsymbol{f} + \boldsymbol{h} \cdot \frac{\partial \boldsymbol{f}}{\partial x_j} = 0.$$

Again by Proposition 3.8(b) we have $\boldsymbol{h} \cdot \partial \boldsymbol{f}/\partial x_j = 0$. Hence $\partial \boldsymbol{h}/\partial x_j \cdot \boldsymbol{f} = 0$. This shows that $f(x) \in \mathrm{Sol}(\frac{\partial}{\partial x_j}\boldsymbol{h}; K[x])$. Thus (c) is proved.

Since $\mathrm{Sol}(\boldsymbol{h}; K[x])$ is a commutative ring, we have

$$K[f_1, \ldots, f_n] \subset \mathrm{Sol}(\boldsymbol{h}; K[x]).$$

Since $h_j(x)$ are polynomials in $f_1, \ldots, f_n$, this shows that $h_j \in \mathrm{Sol}(\boldsymbol{h}; K[x])$. Thus (d) is proved.

Proposition 3.10 *Let $K[x] = K[x_1, \ldots, x_n]$ be the polynomial ring and let $\boldsymbol{h}$ be a self-vanishing system of forms in $K[x]$. Then, for any $f(x) \in K[x]$, we have*

$$\mathcal{D}_x(\boldsymbol{h})\left(f^{(i)}(x, \boldsymbol{h})\right) = (i+1) f^{(i+1)}(x, \boldsymbol{h}).$$

Proof Let $y = (y_1, y_2, \dots y_n)$ be a coordinate system independent of x. Note that

$$\frac{\partial f^{(i)}(x,\boldsymbol{h})}{\partial x_j} = \left\{\frac{\partial f^{(i)}(x,y)}{\partial x_j}\right\}\Bigg|_{y\mapsto h} + \left\{\sum_{k=1}^{n}\frac{\partial f^{(i)}(x,y)}{\partial y_k}\frac{\partial h_k}{\partial x_j}\right\}\Bigg|_{y\mapsto h},$$

so $\mathcal{D}_x(\boldsymbol{h})f^{(i)}(x,\boldsymbol{h})$ is equal to

$$\sum_{j=1}^{n} h_j(x)\left\{\frac{\partial f^{(i)}(x,y)}{\partial x_j}\right\}\Bigg|_{y\mapsto \boldsymbol{h}} + \sum_{j=1}^{n} h_j(x)\left\{\sum_{k=1}^{n}\frac{\partial f^{(i)}(x,y)}{\partial y_k}\frac{\partial h_k(x)}{\partial x_j}\right\}\Bigg|_{y\mapsto \boldsymbol{h}}, \quad (2)$$

where $y \mapsto \boldsymbol{h}$ means substitution. Since $\mathcal{D}_x(\boldsymbol{h})h_k(x) = \sum_{j=1}^{n} h_j(x)\frac{\partial h_k(x)}{\partial x_j} = 0$ for every $k = 1, 2, \dots, n$, the second summand of formula (2) vanishes. The first summand of formula (2) is equal to

$$\left\{\sum_{j=1}^{n} y_j\frac{\partial f^{(i)}(x,y)}{\partial x_j}\right\}\Bigg|_{y\mapsto \boldsymbol{h}} = \{\mathcal{D}_x(y)f^{(i)}(x,y)\}|_{y\mapsto \boldsymbol{h}},$$

so

$$\mathcal{D}_x(\boldsymbol{h})f^{(i)}(x,\boldsymbol{h}) = \left\{\mathcal{D}_x(y)f^{(i)}(x,y)\right\}\big|_{y\mapsto \boldsymbol{h}} = (i+1)f^{(i+1)}(x,\boldsymbol{h})$$

by definition of $f^{(j)}(x,y)$.

Theorem 3.11 *Suppose that $\boldsymbol{h} = (h_1, \dots, h_n)$ is a self-vanishing system of forms in $K[x]$. Then, for a homogeneous polynomial $f(x) \in K[x]$, the following conditions are equivalent:*

(a) $f(x) \in \mathrm{Sol}(\boldsymbol{h}; K[x])$.
(b) $f^{(j)}(x,\boldsymbol{h}) = 0$, *for* $j = 1, 2, \dots$
(c) $f(x + t\boldsymbol{h}(x)) = f(x)$ *for any* $t \in K'$, *where* K' *is any extension field of* K.

Proof $(a) \Leftrightarrow (b)$ By Proposition 3.10, it is enough to show the case for $j = 1$.

$$\begin{aligned} f^{(1)}(x,\boldsymbol{h}) &= \{\mathcal{D}_x(y)f(x)\}|_{y\mapsto \boldsymbol{h}} \\ &= \mathcal{D}_x(\boldsymbol{h})f(x) \\ &= h_1\frac{\partial f}{\partial x_1} + h_2\frac{\partial f}{\partial x_2} + \cdots + h_n\frac{\partial f}{\partial x_n} = 0. \end{aligned}$$

The equivalence of (b) and (c) follows immediately from Proposition 3.2.

Corollary 3.12 *Let $\boldsymbol{h}$ be a self-vanishing system in $K[x]$. If $f(x), g(x) \in K[x] \setminus \{0\}$, and if $f(x)g(x) \in \mathrm{Sol}(\boldsymbol{h}; K[x])$, then both $f(x), g(x) \in \mathrm{Sol}(\boldsymbol{h}; K[x])$.*

Proof Using Taylor expansion (Proposition 3.2), we have

$$f(x+t\boldsymbol{h})g(x+t\boldsymbol{h}) = \sum_{j=0}^{\infty}\sum_{k+l=j} t^{k+l} f^{(k)}(x,\boldsymbol{h})g^{(l)}(x,\boldsymbol{h}).$$

Let k_0 be the highest degree for which $f^{(k_0)} \neq 0$, and similarly l_0 for $g^{(l_0)}(x,\boldsymbol{h})$. Then by Theorem 3.11, $f(x)g(x) \in \mathrm{Sol}(\boldsymbol{h}; R)$ implies $k_0 + l_0 = 0$. Hence $k_0 = l_0 = 0$. Again by Theorem 3.11, proof is complete.

Corollary 3.13 *Suppose that $\boldsymbol{h} = (h_1, \dots, h_n)$ is a self-vanishing system in $R = K[x]$ such that $h_n(x) \neq 0$. Put*

$$s_i(x) = x_i - \frac{h_i(x)}{h_n(x)}x_n, \ (1 \leq i \leq n).$$

Let $f(x) \in \mathrm{Sol}(\boldsymbol{h}; R)$. Then

$$f(x) = f(s_1, s_2, \dots, s_{n-1}, 0),$$

and

$$\mathrm{Sol}(\boldsymbol{h}; K[x]) = K[s_1, s_2, \dots, s_{n-1}] \cap K[x].$$

Proof It is easy to check by direct computation that

$$\mathcal{D}_x(\boldsymbol{h})\left(\frac{1}{h_n(x)}\det\begin{pmatrix} x_i & x_n \\ h_i(x) & h_n(x)\end{pmatrix}\right) = 0.$$

Hence $\mathcal{D}_x(\boldsymbol{h})s_j(x) = 0$. This shows that

$$K[s_1, s_2, \dots, s_{n-1}] \cap K[x] \subset \mathrm{Sol}(\boldsymbol{h}; K[x]).$$

To show the converse, let $f(x) \in \mathrm{Sol}(\boldsymbol{h}; K[x])$. Then $f(x) = f(x + t\boldsymbol{h})$ for an indeterminate t by Theorem 3.11. Replace t for $t = -\frac{x_n}{h_n(x)}$. Then we get

$$f(x) = f(s_1, s_2, \dots, s_{n-1}, 0).$$

Corollary 3.14 *Let $\boldsymbol{h} = (h_1, \dots, h_n)$ be a self-vanishing system of polynomials. Let $f(x) \in \mathrm{Sol}(\boldsymbol{h}; K[x])$. Suppose that $f(x)$ is homogeneous of positive degree d. Then we have*

$$f(h_1, h_2, \dots, h_n) = 0.$$

In particular, if each h_j has positive degree, then

$$h_j(h_1, h_2, \dots, h_n) = 0 \textit{ for all } j = 1, 2, \dots, n.$$

Proof Let $y = (y_1, \dots, y_n)$ be a coordinate system independent of x, and put $\mathcal{D}_x(y)^j f(x) = \frac{1}{j!} f^{(j)}(x, y)$. We have shown that $f^{(d)}(x, y) = f(y)$. In this equation substitute y for $\boldsymbol{h}$. Then we have $f^{(d)}(x, \boldsymbol{h}) = f(\boldsymbol{h})$. In Theorem 3.11, we showed that $f^{(j)}(x, \boldsymbol{h}) = 0$ for $j > 0$. Thus we have $f(\boldsymbol{h}) = 0$.

4 Forms with Zero Hessian and Reduced Self-vanishing Systems

Proposition 4.1 *Let $R = K[x_1, \dots, x_n]$. Let $\boldsymbol{f} = (f_1, f_2, \dots, f_n)$ be a system of forms in R. Then* rank $\left(\frac{\partial f_i}{\partial x_j}\right) =$ tr. $\deg_K$ $K(f_1, f_2, \dots, f_n)$. *In particular the following conditions are equivalent.*

(1) $f_{i_1}, f_{i_2}, \dots, f_{i_r}$ are algebraically dependent for every $i_1, i_2, \dots, i_r \in \{1, 2, \dots, n\}$.
(2) The rank of Jacobian matrix $(\frac{\partial f_i}{\partial x_j})$ is $< r$.
(3) tr. $\deg_K$ $K(f_{i_1}, f_{i_2}, \dots, f_{i_r}) < r$ *for every $i_1, i_2, \dots, i_r \in \{1, 2, \dots, n\}$.*

Proof Left to the reader.

We say that a homogeneous polynomial $f \in R = K[x_1, \dots, x_n]$ has zero Hessian if det $\left(\frac{\partial^2 f}{\partial x_i \partial x_j}\right) = 0$. (Whenever we discuss forms f with zero Hessian, we assume $\deg f \geq 2$.) Let f be a homogeneous form in $K[x_1, \dots, x_n]$. Since the Hessian determinant of f is the Jacobian determinant of the partial derivatives of f, the following proposition follows immediately from Proposition 4.1.

Proposition 4.2 *A homogeneous polynomial $f \in R$ is a form with zero Hessian if and only if the partial derivatives*

$$\frac{\partial f}{\partial x_1}, \frac{\partial f}{\partial x_2}, \dots, \frac{\partial f}{\partial x_n}$$

are algebraically dependent.

Definition 4.3 Let $y = (y_1, \dots, y_n)$ be a coordinate system independent of x. Let $f \in K[x]$ be a form with zero Hessian, and let

$$\phi : K[y_1, \dots, y_n] \to K[x_1, \dots, x_n]$$

be the homomorphism defined by $\phi(y_j) = \frac{\partial f}{\partial x_j}$. We denote by $\mathcal{I}(f)$ the kernel of ϕ. (By Proposition 4.2, $\mathcal{I}(f) \neq 0$.)

Definition 4.4 Let $f \in R = K[x]$ be a form with zero Hessian. Let $\mathcal{I}(f) \subset K[y]$ be as in Definition 4.3. Put $f_j = \frac{\partial f}{\partial x_j}$. Let $g(y) = g(y_1, \dots, y_n) \in \mathcal{I}(f)$ be a homogeneous form of the least degree in $\mathcal{I}(f)$. Let

$$h'_i(x_1, x_2, \dots, x_n) = \frac{\partial g}{\partial y_i}(f_1, \dots, f_n),$$

$$h_i(x_1, x_2, \dots, x_n) = \frac{1}{\mathrm{GCD}(h'_1, h'_2, \dots, h'_n)} h'_i(x_1, \dots, x_n).$$

We call the vector $\boldsymbol{h}' := (h'_1, h'_2, \dots, h'_n)$ a system of polynomials arising from $f(x)$, and $\boldsymbol{h} := (h_1, \dots, h_n)$ a reduced system of polynomials arising from $f(x)$.

Remark 4.5 By Remark 2.4, $\boldsymbol{h} \neq 0$ as well as $\boldsymbol{h}' \neq 0$.

Remark 4.6 Assume that rank $\left(\frac{\partial^2 f}{\partial x_i \partial x_j}\right) = n - 1$. Then the ideal $\mathcal{I}(f)$ is a principal ideal of $K[y]$. In this case $g \in \mathcal{I}(f) \setminus \{0\}$ with the smallest degree is uniquely determined (up to a constant multiple). Hence $\boldsymbol{h}$, in Definition 4.4, is uniquely determined by f (up to a nonzero element of K). On the other hand by Proposition 3.8(b), we see that $\boldsymbol{h}$ is a null vector of the matrix $\left(\frac{\partial^2 f}{\partial x_i \partial x_j}\right)$. Such a polynomial vector is unique up to a multiple of a polynomial. Hence, for any system $\boldsymbol{h}$ of forms, we can show that $\boldsymbol{h}$ is the self-vanishing system as defined in Definition 4.4 if and only if the following two conditions are satisfied.

(1) $(h_1, h_2, \dots, h_n)\left(\frac{\partial^2 f}{\partial x_i \partial x_j}\right) = 0,$
(2) $\mathrm{GCD}(h_1, h_2 \dots, h_n) = 1.$

Theorem 4.7 (Gordan-Noether) *Suppose that $f(x) \in K[x_1, \dots, x_n]$ is a form with zero Hessian. Then a variable can be eliminated from f and its partial derivatives simultaneously by means of a birational transformation of the variables.*

Proof Let $\boldsymbol{h}$ be a reduced self-vanishing system arising from f. Then we have $f(x) \in \mathrm{Sol}(\boldsymbol{h}; K[x])$ and $\partial f(x)/\partial x_j \in \mathrm{Sol}(\boldsymbol{h}; K[x])$ for $j = 1, 2, \dots, n$ by Theorem 3.9(a) and (b) respectively. By Lemma 2.1 and Remark 2.4, we may assume that $h_n \neq 0$. Put $s_j = x_j - \frac{h_j}{h_n} x_n$, for $j = 1, \dots, n-1$. Set $s_n = 0$. Then by Corollary 3.13, f is a polynomial in $s_1, \dots, s_{n-1}$, and so are $\partial f(x)/\partial x_j$ for $j = 1, 2, \dots, n$. We claim that

$$K(s_1, \dots, s_{n-1}, x_n) = K(x_1, \dots, x_n).$$

In fact we have

$$x_j = s_j + \frac{h_j(x)}{h_n(x)} x_n, \quad j = 1, 2, \dots, n-1.$$

Since $h_j(x) \in \mathrm{Sol}(\boldsymbol{h}; K[x])$, we have $h_j(x + t\boldsymbol{h}(x)) = h_j(x)$ for any t in any extension field of K by Theorem 3.11(c). Now let $t = -\frac{x_n}{h_n}$. Then $h_j(s) = h_j(x)$. This shows that $x_j \in K(s_1, \dots, s_{n-1}, x_n)$ for all j, as desired.

Proposition 4.8 *Suppose that $f(x) \in K[x]$ is a form with zero Hessian. Let $\mathcal{I}(f(x))$ be the ideal of $K[y]$ as defined in Definition 4.3. Then the following conditions are equivalent.*

(a) *The ideal* $\mathcal{I}(f(x))$ *contains a linear form.*
(b) *The partial derivatives of* $f(x)$ *are linearly dependent.*
(c) *A variable can be eliminated from* $f(x)$ *by means of a linear transformation of the variables.*

Proof The equivalence of (a) and (b) is clear. Suppose that there exists a non-trivial relation

$$a_1 f_1 + a_2 f_2 + \cdots + a_n f_n = 0,$$

where $f_j = \frac{\partial f}{\partial x_j}$ and $a_j \in K$. It is possible to choose a set of linearly independent linear forms $y_1, \ldots, y_n$ in $x_1, \ldots, x_n$ such that $\frac{\partial x_j}{\partial y_1} = a_j$. Then

$$\frac{\partial f}{\partial y_1} = \frac{\partial f}{\partial x_1}\frac{\partial x_1}{\partial y_1} + \frac{\partial f}{\partial x_2}\frac{\partial x_2}{\partial y_1} + \cdots + \frac{\partial f}{\partial x_n}\frac{\partial x_n}{\partial y_1} = 0.$$

This shows that if f is expressed in terms of y_j, then f does not contain y_1. Thus (b) $\Rightarrow$ (c). The same argument shows (b) $\Leftarrow$ (c) as well.

Theorem 4.9 *Let* $f(x) \in K[x]$ *be a form with zero Hessian and let* $\boldsymbol{h}' = (h'_1, \ldots, h'_n)$ *be a system of polynomials associated to an element* $g = g(y_1, \ldots, y_n) \in \mathcal{I}(f(x))$ *of the least degree. Similarly let* $\boldsymbol{h} = (h_1, \ldots, h_n)$ *be the reduced system of polynomials defined by*

$$\boldsymbol{h} = \frac{1}{\mathrm{GCD}(\boldsymbol{h}')}\boldsymbol{h}'.$$

(See Definition 4.4.) Then we have:

(a) $\boldsymbol{h}$ *and* $\boldsymbol{h}'$ *are self-vanishing systems of polynomials.*
(b) $f(x) \in \mathrm{Sol}(\boldsymbol{h}; K[x])$.
(c) $\frac{\partial}{\partial x_j} f(x) \in \mathrm{Sol}(\boldsymbol{h}; K[x])$, *for all* $j = 1, 2, \ldots, n$.

Proof (a) By Theorem 3.9, $\boldsymbol{h}'(x)$ is a self-vanishing system. Note that $\mathrm{Sol}(\boldsymbol{h}'; K[x]) = \mathrm{Sol}(\boldsymbol{h}; K[x])$. By Corollary 3.12, $\boldsymbol{h}(x)$ is a self-vanishing system. (b) and (c) are proved in Theorem 3.9.

Example 4.10 If $f = (x_1^2x_3 + 2x_1x_2x_4 + x_2^2x_5)(x_1^2x_4 + 2x_1x_2x_5 + x_2^2x_6)$, then $\mathcal{I}(f)$ is principal, and

$$\boldsymbol{h} = (0, 0, x_2^3, -x_2^2x_1, x_1^2x_2, -x_1^3),$$

is the unique reduced self-vanishing system arising from f, which is as in Example 3.6.

But if $f = (x_3x_1 + x_4x_2)(x_5x_1 + x_6x_2)$, then $\mathcal{I}(f)$ is principal as well, and

$$h = \big(0, 0, (x_1x_3 + x_2x_4)x_2, -(x_1x_3 + x_2x_4)x_1, -x_2(x_1x_5 + x_2x_6), x_1(x_1x_5 + x_2x_6)\big),$$

is the unique reduced self-vanishing system arising from f, which is *not* as in Example 3.6.

5 Binary and Ternary Forms with Zero Hessian

Theorem 5.1 *Assume that $n = 2$ and let $f \in K[x_1, x_2]$ be a form of degree d with zero Hessian. Then $f = (a_1x_1 + a_2x_2)^d$ for some $a_1, a_2 \in K$.*

Proof Let $\mathcal{I}(f) \subset K[y_1, y_2]$ be the ideal defined in Definition 4.3. Since $\mathcal{I}(f)$ is a prime ideal, it is a principal ideal generated by a linear form. Hence the assertion follows from Proposition 4.8.

Lemma 5.2 *Let $\boldsymbol{h}$ be a homogeneous self-vanishing system in dimension n. Then*

$$\operatorname{rank}\left(\frac{\partial h_i}{\partial x_j}\right) = \operatorname{tr.deg}_K K(\boldsymbol{h}) = \operatorname{Krull\,dim} K[\boldsymbol{h}] \le n - 1$$

and if $n \ge 3$ in addition, then

$$\operatorname{rank}\left(\frac{\partial h_i}{\partial x_j}\right) = \operatorname{tr.deg}_K K(\boldsymbol{h}) = \operatorname{Krull\,dim} K[\boldsymbol{h}] \le n - 2$$

Proof The case where $\boldsymbol{h} = 0$ is trivial, so let us assume without loss of generality that $h_1 \neq 0$. From Theorem 3.11(c), it follows that $h_1(x + t\boldsymbol{h}(x)) = h_1(x)$. If we look at the leading coefficient with respect to t, we see that $h_1(\boldsymbol{h}) = 0$. So

$$\operatorname{rank}\left(\frac{\partial h_i}{\partial x_j}\right) = \operatorname{tr.deg}_K K(\boldsymbol{h}) \le n - 1.$$

This gives the case $n \le 2$, so assume from now on that $n \ge 3$.

Let f be an irreducible factor of h_1. If each h_i is a K-multiple of a power of f, then the components of $\boldsymbol{h}$ are linearly dependent in pairs, and

$$\operatorname{rank}\left(\frac{\partial h_i}{\partial x_j}\right) = \operatorname{tr.deg}_K K(\boldsymbol{h}) \le 1 \le n - 2.$$

Otherwise, there exists an h_i with an irreducible factor f' which is not a K-multiple of f. By Corollary 3.12, we have $f'(x) \in \operatorname{Sol}(\boldsymbol{h}; K[x])$ as well as $f(x) \in \operatorname{Sol}(\boldsymbol{h}; K[x])$. From Theorem 3.11 it follows that $f(x + t\boldsymbol{h}(x)) = f(x)$ and $f'(x + t\boldsymbol{h}(x)) = f'(x)$. If we look at the leading coefficient with respect to t, we see that

$$f(\boldsymbol{h}(x)) = f'(\boldsymbol{h}(x)) = 0, \tag{3}$$

which gives the second claim of Lemma 5.2.

Theorem 5.3 *Suppose that $f = f(x) \in K[x_1, x_2, x_3]$ is a form with zero Hessian. Then a variable can be eliminated from f by means of a linear transformation of variables.*

Proof Let $\boldsymbol{h} = (h_1, h_2, h_3)$ be a reduced system of polynomials arising from $f(x)$ (See Definition 4.4). We claim that $\boldsymbol{h}$ is a constant vector. Suppose that it is not. Then $\text{Krull dim } K[h_1, h_2, h_3] = 1$ by Lemma 5.2. Thus any two of the elements h_1, h_2, h_3 has a homogeneous algebraic relation. Since $K[h_1, h_2, h_3]$ is an integral domain, and K is algebraically closed, they should be linear relations. Thus we have $\dim_K Kh_1 + Kh_2 + Kh_3 = 1$. Since $\text{GCD}(h_1, h_2, h_3) = 1$, this is impossible unless they are constants. Recall that we have

$$h_1 \frac{\partial f}{\partial x_1} + h_2 \frac{\partial f}{\partial x_2} + h_3 \frac{\partial f}{\partial x_3} = 0.$$

By Proposition 4.8, proof is complete.

6 The Rational Map Defined by $\boldsymbol{h}$

In this section, we assume that $n \geq 4$. Let $R = K[x_1, \dots, x_n]$ and let $\boldsymbol{h} = (h_1, \dots, h_n)$ be any homogeneous reduced self-vanishing system. So $\text{GCD}(h_1, \dots, h_n) = 1$.

Let

$$Z : \mathbb{P}^{n-1}(x) \to \mathbb{P}^{n-1}(y)$$

be the rational map defined by the correspondence $x = (x_1 : \cdots : x_n) \mapsto (h_1 : \cdots : h_n)$. Let W be the image of Z and T the fundamental locus of Z in $\mathbb{P}^{n-1}(x)$ defined by the equations $h_1(x) = h_2(x) = \cdots = h_n(x) = 0$. The algebraic set $W \subset \mathbb{P}^{n-1}(y)$ is defined by the kernel of the homomorphism defined by

$$y_j \mapsto h_j, (j = 1, 2, \dots, n).$$

which corresponds to Z.

Proposition 6.1 *The following conditions are equivalent.*

(a) deg $(h_j) = 0$, *i.e.,* $\boldsymbol{h}$ *is a constant vector.*
(b) dim $W = 0$, *i.e.,* W *is a one-point set.*
(c) T *is empty, i.e.,* Z *is a morphism.*

Proof Suppose that $\boldsymbol{h}$ is not a constant. Then we have

$$h_j(h_1, \ldots, h_n) = 0 \text{ for every } j = 1, \ldots, n.$$

by Corollary 3.14. Thus any specialization of $(h_1, \ldots, h_n)$ is a point of T. This shows that if T is empty, then $\boldsymbol{h}$ is a constant vector. All other implications are trivial.

Proposition 6.2 *If* dim $W \geq 1$, *then*

$$2 \leq \text{Krulldim } K[x]/(h_1, \ldots, h_n) \leq n - 2,$$

or equivalently,

$$1 \leq \dim T \leq n - 3.$$

Proof Since T is not empty, the ideal $(h_1, h_2, \ldots, h_n) \subset K[x]$ is not the unit ideal by the previous proposition. On the other hand, since $\text{GCD}(h_1, \ldots, h_n) = 1$, it is not a principal ideal. Hence $\text{ht}(h_1, \ldots, h_n) \geq 2$. This shows that $\dim T =$ Krull dim $K[x]/(h_1, \ldots, h_n) - 1 \leq n - 3$. On the other hand there exists a surjective homomorphism of rings:

$$K[x_1, \ldots, x_n]/(h_1, \ldots, h_n) \to K[h_1, \ldots, h_n],$$

$$x_j \mapsto h_j,$$

provided that $\deg \boldsymbol{h} > 0$. Note that $K[x_1, \ldots, x_n]/(h_1, \ldots, h_n)$ is the fiber at the origin of the inclusion map $K[h_1, \ldots, h_n] \to K[x_1, \ldots, x_n]$ and Krull dim $K[h_1, \ldots, h_n] \leq n - 2$ by Lemma 5.2. Hence we have Krull dim $K[x_1, \ldots, x_n]/(h_1, \ldots, h_n) \geq 2$.

For the rest of this section we assume that dim $W = 1$. In this case the fiber of $Z : \mathbb{P}^{n-1}(x) \to \mathbb{P}^{n-1}(y)$ is a hypersurface of $\mathbb{P}^{n-1}(x)$. Thus, for any $\omega \in \mathbb{P}^{n-1}(y)$, $Z^{-1}(\omega)$ is defined by one homogeneous polynomial in $x = (x_1, \ldots, x_n)$.

Definition 6.3 Let $Z : \mathbb{P}^{n-1}(x) \to \mathbb{P}^{n-1}(y)$ be as above. Let $\omega = (\omega_1 : \omega_2 : \cdots : \omega_n) \in \mathbb{P}^{n-1}(y)$. We denote by $g^{(\omega)}(x)$ the square-free polynomial in $K[x]$ that defines the hypersurface of the fiber of Z at $\omega \in W \subset \mathbb{P}^{n-1}(y)$.

For each point $\omega = (\omega_1 : \omega_2 : \cdots : \omega_n) \in \mathbb{P}^{n-1}$, we define the differential operator $\mathcal{D}_x(\omega)$ on $R = K[x_1, \ldots, x_n]$ by

$$\mathcal{D}_x(\omega) f(x) = \sum_{j=1}^{n} \omega_j \frac{\partial f(x)}{\partial x_j}$$

whose value is determined up to a non-zero constant factor. Hence we may speak of the set of solutions of the equation $\mathcal{D}_x(\omega) f(x) = 0$. We denote the space of solutions of $\mathcal{D}_x(\omega) f(x) = 0$ in $R = K[x_1, \ldots, x_n]$ by $\text{Sol}(\omega; R)$. It is a subring of R. For any subset U of $\mathbb{P}^{n-1}$ we denote by $\mathrm{Sol}(U; R)$ the space of solutions of the system of linear differential equations

$$\mathcal{D}_x(\omega)(f(x)) = 0, \ \omega \in U.$$

If we denote by $L(U)$ the linear closure of U in $\mathbb{P}^{n-1}$ it is easy to see that

$$\mathrm{Sol}(U; R) = \mathrm{Sol}(L(U); R).$$

If $U = \{\omega^{(1)}, \omega^{(2)}, \ldots, \omega^{(s)}\}$ is a finite set, then $\mathrm{Sol}(U; R)$ is also denoted as

$$\mathrm{Sol}(\omega^{(1)}, \omega^{(2)}, \ldots, \omega^{(s)}; R).$$

The same notation is used if we replace $\mathbb{P}^{n-1}(y)$ for the vector space K^n in the obvious sense. Namely for a linear subspace L of K^n, we denote by $\mathrm{Sol}(L; R)$, the set of solutions of the differential equations

$$\mathcal{D}_x(\boldsymbol{a})F(x) = 0 \text{ for all } \boldsymbol{a} \in L,$$

where $\boldsymbol{a}$ denotes a row vector in L regarded as a system of constants. In a set theoretic notation,

$$\mathrm{Sol}(L; R) = \left\{ F(x) \in R \,\middle|\, \left(a_1 \frac{\partial}{\partial x_1} + a_2 \frac{\partial}{\partial x_2} + \cdots + a_n \frac{\partial}{\partial x_n} \right) F(x) = 0, (a_1, \ldots, a_n) \in L \right\}.$$

Note that $\mathrm{Sol}(L; R)$ is a subring of R generated by homogeneous linear forms.

The relation between the subring $\mathrm{Sol}(L; R)$ of R and the ideal I for the linear subspace in $\mathbb{P}^{n-1}$ is very important for us. In the next theorem and corollary we describe a set of generators of the subring $\mathrm{Sol}(L; R)$ and a set of generators that defines the linear space L as a subspace of $\mathbb{P}^{n-1}$.

Theorem 6.4 *Let $R = K[x_1, \ldots, x_n]$ be the polynomial ring over K. Let $A = (a_i^{(j)})$ be a $k \times n$ matrix, $a_i^{(j)} \in K$, with $\boldsymbol{a}^{(j)} = (a_1^{(j)}, a_2^{(j)}, \ldots, a_n^{(j)})$ as the j-th row. Suppose that the rows are linearly independent. Assume that $k < n$. Then the set of solutions as a subring of $K[x]$*

$$\mathrm{Sol}(\boldsymbol{a}^{(1)}, \ldots, \boldsymbol{a}^{(k)}; R) := \bigcap_{j=1}^{k} \mathrm{Sol}(\boldsymbol{a}^{(j)}; R)$$

is isomorphic to the polynomial ring in $n - k$ variables. It is generated by the $(k+1) \times (k+1)$ minors of the matrix

$$A' := \begin{pmatrix} a_1^{(1)} & a_2^{(1)} & \cdots & a_{n-1}^{(1)} & a_n^{(1)} \\ a_1^{(2)} & a_2^{(2)} & \cdots & a_{n-1}^{(2)} & a_n^{(2)} \\ & & & & \\ a_1^{(k)} & a_2^{(k)} & \cdots & a_{n-1}^{(k)} & a_n^{(k)} \\ \hline x_1 & x_2 & \cdots & x_{n-1} & x_n \end{pmatrix}$$

Proof Let V be the vector space of common syzygies of $\boldsymbol{a}^{(1)}, \ldots, \boldsymbol{a}^{(k)}$ over K. Take $\boldsymbol{v} = (v_1, v_2, \ldots, v_n) \in V$ and define a linear form

$$l_{\boldsymbol{v}} := v_1 x_1 + v_2 x_2 + \cdots + v_n x_n \tag{4}$$

Let $k = n - \dim V$. Take a basis $\boldsymbol{v}_1, \boldsymbol{v}_2, \ldots, \boldsymbol{v}_{n-k}$ of V, and extend it to a basis $\boldsymbol{v}_1, \boldsymbol{v}_2, \ldots, \boldsymbol{v}_n$ of K^n. Let $l_1, l_2, \ldots, l_n$ be the corresponding linear forms as defined in (4). Then

$$R = K[l_1, l_2, \ldots, l_n].$$

Now take $f \in R$. Then we can write $f = g(l_1, l_2, \ldots, l_n)$ where $g \in K[y_1, y_2, \ldots, y_n]$. Write $f_i = \partial f / \partial x_i$ and $g_i = \partial g / \partial y_i$ for $i = 1, 2, \ldots, n$. Then

$$(f_1, f_2, \ldots, f_n) = g_1(l_1, l_2, \ldots, l_n)\boldsymbol{v}_1 + g_2(l_1, l_2, \ldots, l_n)\boldsymbol{v}_2 + \cdots + g_n(l_1, l_2, \ldots, l_n)\boldsymbol{v}_n$$

Hence $f \in \mathrm{Sol}(\boldsymbol{a}^{(1)}, \ldots, \boldsymbol{a}^{(k)}; R)$ if and only if $g_{n-k+1} = g_{n-k+2} = \cdots = g_n = 0$, i.e., $f \in K[l_1, l_2, \ldots, l_{n-k}]$. Indeed, $K[l_1, l_2, \ldots, l_{n-k}]$ is isomorphic to the polynomial ring in $n - k$ variables.

If we take for f a $(k+1) \times (k+1)$ minor of A', then $f(\boldsymbol{a}^{(j)}) = 0$ because $f(\boldsymbol{a}^{(j)})$ is the determinant of a matrix of which the last row coincides with row j. From this, we infer that the $(k+1) \times (k+1)$ minor of A' are linear combinations of $l_1, l_2, \ldots, l_{n-k}$. So it remains to show the converse, i.e., that $l_{\boldsymbol{v}}$ is a linear combination of the $(k+1) \times (k+1)$ minors of A' for every $\boldsymbol{v} \in V$.

Suppose first that $k = n - 1$. Then $V = Kw_1$ has dimension 1, and $l_{\boldsymbol{v}}$, l_1 and $\det A'$ are the same up to a nonzero constant for every nonzero $\boldsymbol{v} \in V$. So $l_{\boldsymbol{v}}$ is a linear combination of the $(k+1) \times (k+1)$ minors of A' for every $\boldsymbol{v} \in V$.

Suppose next that $k < n - 1$. Then we can extend $\boldsymbol{a}^{(1)}, \ldots, \boldsymbol{a}^{(k)}$ to a basis $\boldsymbol{a}^{(1)}, \ldots, \boldsymbol{a}^{(n-1)}$ of the syzygies over K of $\boldsymbol{v}$. Now the case $k = n - 1$ yields $l_{\boldsymbol{v}}$ as a determinant of an $n \times n$ matrix up to a nonzero constant. If we expand this matrix along rows $k+1, k+2, \ldots, n-1$, then we get a linear combination of the $(k+1) \times (k+1)$ minors of A'.

Corollaries 6.5 and 6.6 that follow are other ways to describe the set of linear forms as generators for $\mathrm{Sol}(L; R)$.

Corollary 6.5 *Let R and A be the same as Theorem 6.4, and furthermore let L be the vector subspace in K^n generated by the rows of A. Then the set of solutions* $\mathrm{Sol}(L; R)$, *as a subring of $R = K[x_1, \ldots, x_n]$, is generated by the linear forms $l = l(x)$ such that*

$$l(a_1^{(j)}, a_2^{(j)}, \dots, a_n^{(j)}) = 0,$$

for all $j = 1, 2, \dots, k$.

Proof This follows from the proof of Theorem 6.4.

Corollary 6.6 *Suppose that* $U \subset \mathbb{P}^{n-1}(y)$ *is a subset and let* $L(U)$ *be the linear closure of* U. *Then* $\mathrm{Sol}(U; R) = \mathrm{Sol}(L(U); R)$ *is generated by the linear forms of*

$$\{ l(x) \in K[x_1, x_2, \dots, x_n] | l(\omega_1, \omega_2, \dots, \omega_n) = 0 \text{ for all } (\omega_1 : \omega_2 : \cdots : \omega_n) \in L(U) \},$$

as a subring of $K[x]$.

The following two theorems are very important to determine the forms in four and five variables with zero Hessian.

Theorem 6.7 *Let* $\boldsymbol{h} = (h_1, \dots, h_n)$ *be a self-vanishing system of forms in* $R = K[x_1, \dots, x_n]$. *Let* $Z : \mathbb{P}^{n-1}(x) \to \mathbb{P}^{n-1}(y)$ *be the rational map defined by* $y_j = h_j(x)$. *Let* $T \subset \mathbb{P}^{n-1}(x)$ *be the fundamental locus and* W *the image of* Z. *Assume that* $\dim W = 1$. *Let*

$$i : \mathbb{P}^{n-1}(y) \to \mathbb{P}^{n-1}(x)$$

be the natural map $y_j \to x_j$. *Then* $i(L(W)) \subset T$, *where* $L(W)$ *is the linear closure of* W *in* $\mathbb{P}^{n-1}(y)$.

Proof Recall that T is defined by the polynomials $h_1(x), \dots, h_n(x)$. On the other hand $i(L(W))$ is defined by linear forms which vanish on the set $i(L(W))$. Hence, in view of Corollary 6.6, this follows immediately from Theorem 6.8 below.

Theorem 6.8 *With the same notations and assumptions as in Theorem 6.7, we have*

$$h_j(x) \in \mathrm{Sol}(L(W); R) \text{ for all } j = 1, 2, \dots, n.$$

We prove it after some propositions. For the rest of this section we fix notation and assumption of Theorem 6.7. In particular it is assumed that $\dim W = 1$.

Proposition 6.9 *For any* $\omega \in W$, *we have* $g^{(\omega)}(x) \in \mathrm{Sol}(\boldsymbol{h}; R)$.

Proof Choose a hyperplane $H \subset \mathbb{P}^{n-1}(y)$ such that $\omega \in H$, and suppose that H is defined by the linear equation

$$a_1 y_1 + \cdots + a_n y_n = 0.$$

Put $f(x) = \sum_{j=1}^{n} a_j h_j(x)$. Since K is infinite, it is possible to choose H such that $f(x) \neq 0$. Then for any $\alpha \in Z^{-1}(\omega)$, if $\alpha \in T$, then $h_j(\alpha) = 0$ for all j by definition of T. Hence $f(\alpha) = 0$. If $\alpha \notin T$, then

$$f(\alpha) = \sum_{j=1}^{n} a_j h_j(\alpha) = c \sum_{j=1}^{n} a_j \omega_j = 0.$$

This shows that any point on $Z^{-1}(\omega)$ is a zero of $f(x)$. Hence $f(x)$ is a multiple of $g^{(\omega)}(x)$. By Corollary 3.12, this proves that $g^{(\omega)}(x) \in \mathrm{Sol}(\boldsymbol{h}; R)$ for any $\omega \in W$.

Proposition 6.10 *For each point $\omega \in W$, we have $g^{(\omega)}(x) \in \mathrm{Sol}(L(W); R)$.*

Proof is preceded by two lemmas.

Lemma 6.11 *For $\omega, \omega' \in W$, the polynomial $\mathcal{D}_x(\omega')g^{(\omega)}(x)$ is divisible by $g^{(\omega')}(x)$.*

Proof In Proposition 6.9, we proved that $g^{(\omega)}(x) \in \mathrm{Sol}(\boldsymbol{h}; R)$. So if $\mathcal{D}_x(\boldsymbol{h})$ is applied to $g^{(\omega)}(x)$, it becomes 0. Namely,

$$h_1(x)\frac{\partial g^{(\omega)}(x)}{\partial x_1} + \cdots + h_n(x)\frac{\partial g^{(\omega)}(x)}{\partial x_n} = 0.$$

Choose $\alpha \in Z^{-1}(\omega')$ and make the substitution $x \mapsto \alpha$ in the above equality. Then the left hand side is the same as

$$\mathcal{D}_x(\omega')g^{(\omega)}(x)$$

evaluated at $x = \alpha$. We have shown that the zero locus of $g^{(\omega')}(x)$ is contained in that of $\mathcal{D}_x(\omega')g^{(\omega)}(x)$. Since $g^{(\omega')}(x)$ is square-free, this proves the assertion.

Lemma 6.12 *For any point $\omega \in W$, we have $\mathcal{D}_x(\omega)g^{(\omega)}(x) = 0$.*

Proof In the previous lemma let $\omega' = \omega$. Then we have proved that $\mathcal{D}_x(\omega)g^{(\omega)}(x) = r(x)g^{(\omega)}(x)$ for some $r(x) \in R$. For the degree reason, we get the assertion.

Proof *of Proposition* 6.10 By Lemma 6.12, we have

$$\mathcal{D}_x(\omega)\left(\mathcal{D}_x(\omega')g^{(\omega)}(x)\right) = \mathcal{D}_x(\omega')\left(\mathcal{D}_x(\omega)g^{(\omega)}(x)\right) = 0.$$

By Lemma 6.11, we may write $\mathcal{D}_x(\omega')g^{(\omega)}(x) = r(x)g^{(\omega')}(x)$ for some $r(x)$. Since

$$\mathcal{D}_x(\omega)\left(r(x)g^{(\omega')}(x)\right) = 0,$$

$r(x)g^{(\omega')}(x) \in \mathrm{Sol}(\omega; R)$. Therefore, since a constant vector is a self-vanishing system, we have $g^{(\omega')}(x) \in \mathrm{Sol}(\omega; R)$ by Corollary 3.12.

Proof *of Theorem* 6.8 We have to show that $h_j(x) \in \mathrm{Sol}(L(W); R)$. We may assume that $h_j \neq 0$. Let W_j be the intersection of W and the hyperplane defined by $y_j = 0$ in the space $\mathbb{P}^{n-1}(y)$, and let $H_j \subset \mathbb{P}^{n-1}(x)$ be the hypersurface defined by $h_j(x) = 0$.

Then W_j is a finite set, since dim $W = 1$ and W is not contained in the above hyperplane. Thus, if we write $W_j = \{\omega^{(1)}, \omega^{(2)}, \ldots, \omega^{(s)}\}$, then

$$H_j \supset Z^{-1}(\omega^{(1)}) \cup Z^{-1}(\omega^{(2)}) \cup \cdots \cup Z^{-1}(\omega^{(s)}).$$

Conversely if $\alpha \in H_j \setminus T$, then $Z(\alpha) \in W_j$. Thus

$$H_j \setminus T \subset Z^{-1}(\omega^{(1)}) \cup Z^{-1}(\omega^{(2)}) \cup \cdots \cup Z^{-1}(\omega^{(s)}).$$

Since H_j is purely of codimension 1 and T has codimension at least 2 (Proposition 6.2), it in fact shows that

$$H_j \subset Z^{-1}(\omega^{(1)}) \cup Z^{-1}(\omega^{(2)}) \cup \cdots \cup Z^{-1}(\omega^{(s)}).$$

Thus we have proved that up to a nonzero constant factor, the square-free part of $h_j(x)$ is equal to

$$g^{(\omega^{(1)})}(x) \times g^{(\omega^{(2)})}(x) \times \cdots \times g^{(\omega^{(s)})}(x).$$

By Proposition 6.10, this completes the proof of Theorem 6.8.

7 Quaternary and Quinary Forms with Zero Hessian

Theorem 7.1 *Let $f = f(x_1, x_2, x_3, x_4) \in R = K[x_1, x_2, x_3, x_4]$ be a form with zero Hessian. Then f can be transformed into a form with three variables via a linear transformation of the variables x_1, x_2, x_3, x_4.*

Proof Let $\mathbf{h} = (h_1, h_2, h_3, h_4)$ be a reduced self-vanishing system of forms arising from f. Then by Lemma 5.2, Krull dim $K[h_1, h_2, h_3, h_4] \leq n - 2 = 2$. This shows dim $W \leq 1$. Put $s =$ dim $L(W)$ (as a linear variety in $\mathbb{P}^3$). Note that $s + 1 = \dim_K(Kh_1 + \cdots + Kh_4)$. Since $i(L(W)) \subset T$ by Theorem 6.7 and dim $T \leq 1$ by Proposition 6.2, we have $s \leq 1$. If $s = 0$, we get the result by Proposition 6.1. If $s = 1$, we may assume that $h_1(x) = h_2(x) = 0$ by Lemmas 2.1 and 2.2. Let $g(y) \in K[y_1, \ldots, y_4]$ be the polynomial through which $h_1, \ldots, h_4$ are defined (cf. Definition 4.4). Then $h_1 = h_2 = 0$ implies that $\frac{\partial g}{\partial y_1} = \frac{\partial g}{\partial y_2} = 0$ by the minimality of the degree. (cf. Proposition 2.3 and Remark 2.4.) Hence g is a form only in two variables. Since g should be irreducible, g is a linear form. This implies that there exists a linear relation among the partial derivatives of f. Hence the assertion follows from Proposition 4.8.

Proposition 7.2 *Let $f = f(x_1, x_2, x_3, x_4, x_5) \in R = K[x_1, x_2, x_3, x_4, x_5]$ be a form with zero Hessian. Let $\mathbf{h} = (h_1, h_2, h_3, h_4, h_5)$ be a self-vanishing system of polynomials associated to f as defined in Definition 4.4. Assume that there exists no linear relations among the partial derivatives of f. Then by a linear change of variables, f*

can be transformed into a form so that $h_1 = h_2 = 0$, *and* h_3, h_4, h_5 *are polynomials only in* x_1, x_2.

Proof Let $Z : \mathbb{P}^4(x) \to \mathbb{P}^4(y)$ be the rational map defined by $y_j = h_j(x)$ and let T be the fundamental locus and W the image of Z. *We will prove that* $\dim W \leq 1$ *in the next section.* If $\dim W = 0$, then a variable can be eliminated from f, since $\boldsymbol{h}$ is a constant vector by Proposition 6.1. Assume that $\dim W = 1$. Then we have $i(L(W)) \subset T$ by Theorem 6.7. On the other hand we have $\dim T \leq 2$ by Proposition 6.2. This shows $\dim L(W) \leq 2$ as a linear subspace in $\mathbb{P}^4$. Let $s = \dim L(W)$ or equivalently $s + 1 = \dim_K \sum_{j=1}^{5} Kh_j$. Since $\dim W = 1$, $s \geq 1$. If $s = 1$, we may assume $h_1 = h_2 = h_3 = 0$. As in the proof of Theorem 7.1 this would imply that there exists a linear relation among the partial derivatives of f. Since we have excluded this case, we are left with the case $s = 2$. Then we may assume that $h_1 = h_2 = 0$ and $h_j \in \mathrm{Sol}(i(L(W)); R)$ for $j = 3, 4, 5$. The linear subspace $i(L(W))$ consists of vectors $(0, 0, *, *, *)$. Hence h_3, h_4, h_5 should be polynomials only in x_1 and x_2.

Theorem 7.3 *Let* $R = K[x_1, x_2, x_3, x_4, x_5]$, *and let* Δ *be a homogeneous polynomial of the form*

$$\Delta = p_3(x_1, x_2)x_3 + p_4(x_1, x_2)x_4 + p_5(x_1, x_2)x_5$$

Then any element in the algebra $K[x_1, x_2][\Delta]$ *is a polynomial with zero Hessian.*

Conversely, let f *be a homogeneous form in five variables with zero Hessian and assume that* f *properly involves five variables. Then we can choose* Δ *such that* f *can be transformed into a homogeneous polynomial in the algebra* $K[x_1, x_2][\Delta]$ *by means of a linear change of variables.*

Proof Put $\boldsymbol{f} = (f_1, \dots, f_5)$, where $f_j = \frac{\partial f}{\partial x_j}$. Suppose first that $f \in K[x_1, x_2][\Delta]$. Assume without loss of generality that $f_5 \neq 0$. Then $f_3/f_5 = p_3/p_5 \in K(x_1/x_2)$ and $f_4/f_5 = p_4/p_5 \in K(x_1/x_2)$, so $f_3, f_4, f_5 \in K(x_1/x_2, f_5)$. Hence $\mathrm{tr.\,deg}_K\, K(f_3, f_4, f_5) \leq 2$ and $\mathrm{tr.\,deg}_K\, K(f_1, f_2, f_3, f_4, f_5) \leq 4$. On account of Proposition 4.1, f has zero Hessian.

Suppose next that f has zero Hessian. On account of Proposition 4.2, f_1, f_2, f_3, f_4, f_5 are algebraically dependent over K. Put $\boldsymbol{f} = (f_1, \dots, f_5)$, where $f_j = \frac{\partial f}{\partial x_j}$. Let $\boldsymbol{h} = (h_1, \dots, h_5)$ be a reduced self-vanishing system arising from $f(x)$. We have proved that $\boldsymbol{h} \cdot \frac{\partial}{\partial x_j}\boldsymbol{f} = 0$, $j = 1, \dots, 5$ in Theorem 3.9. This shows that $f(x) \in \mathrm{Sol}(\frac{\partial}{\partial x_1}\boldsymbol{h}, \dots, \frac{\partial}{\partial x_5}\boldsymbol{h}; R)$. By Proposition 7.2, we may assume that $h_1 = h_2 = 0$ and h_3, h_4, h_5 involve only x_1, x_2. Let $\tilde{K}$ be the algebraic closure of $K(x_1, x_2)$. It follows from Theorem 6.4 that f is a polynomial over $\tilde{K}$ in

$$\tilde{A}' := \begin{vmatrix} \frac{\partial h_3}{\partial x_1} & \frac{\partial h_4}{\partial x_1} & \frac{\partial h_5}{\partial x_1} \\ \frac{\partial h_3}{\partial x_2} & \frac{\partial h_4}{\partial x_2} & \frac{\partial h_5}{\partial x_2} \\ \hline x_3 & x_4 & x_5 \end{vmatrix}.$$

Notice that $\tilde{A}' \in R$ has only one irreducible factor which is *not* contained in $K[x_1, x_2]$. Furthermore, we can choose Δ to be this factor. Then $f \in \tilde{K}[\Delta]$, say

$$f = b_0 + b_1\Delta + b_2\Delta^2 + \cdots .$$

Then the coefficient of x_j^i of f as a polynomial in $K[x_1, x_2][x_3, x_4, x_5]$ equals $b_i p_j^i$, for $j = 3, 4, 5$. So $b_i \in K(x_1, x_2)$. Since $\mathrm{GCD}(p_3^i, p_4^i, p_5^i) = 1$, we infer that $b_i \in K[x_1, x_2]$. This holds for all i, so $f \in K[x_1, x_2][\Delta]$.

If Δ is cubic, then we can take $\Delta = A'$. But if Δ has larger degree, then this is not always possible. Take e.g. Δ irreducible of even degree, then we cannot take $\Delta = A'$, because A' has always odd degree. It is however possible to write Δ in the same way as A', but with the $h_i^{(j)}$ replaced by other polynomials in $K[x_1, x_2]$.

Theorem 7.4 *Let $n \geq 4$. Let $\Delta \in K[x_1, x_2, \ldots, x_n]$ be a homogeneous polynomial of degree at least 2 of the form*

$$\Delta = p_3(x_1, x_2)x_3 + p_4(x_1, x_2)x_4 + \cdots + p_n(x_1, x_2)x_n$$

Then there are polynomials $a_i^{(j)}$ in $K[x_1, x_2]$, such that $\mathbf{a}^{(j)} = (a_3^{(j)}, a_4^{(j)}, \ldots, a_n^{(j)})$ is homogeneous for all $3 \leq j \leq n-1$, and

$$\Delta = \left| \begin{array}{cccc} a_3^{(3)} & a_4^{(3)} & \cdots & a_n^{(3)} \\ a_3^{(4)} & a_4^{(4)} & \cdots & a_n^{(4)} \\ \vdots & \vdots & & \vdots \\ a_3^{(n-1)} & a_4^{(n-1)} & \cdots & a_n^{(n-1)} \\ \hline x_3 & x_4 & \cdots & x_n \end{array} \right| .$$

Proof If Theorem 7.4 holds for $\Delta\big(x_1, x_2, A(x_3, x_4, \ldots, x_n)\big)$ for some $A \in \mathrm{GL}(n-2, K)$ instead of Δ itself, then we can substitute $(x_3, x_4, \ldots, x_n)$ by $A^{-1}(x_3, x_4, \ldots, x_n)$ in the matrix with determinant $\Delta\big(x_1, x_2, A(x_3, x_4, \ldots, x_n)\big)$, to obtain a matrix with determinant Δ. Without affecting its determinant, we can get the last matrix of the form of Theorem 7.4 by way of column operations and multiplying the first row with a nonzero constant. This allows us to replace $\big(p_3(x_1, x_2), p_4(x_1, x_2), \ldots, p_n(x_1, x_2)\big)$ by $\big(p_3(x_1, x_2), p_4(x_1, x_2), \ldots, p_n(x_1, x_2)\big)\, A$ for any $A \in \mathrm{GL}(n-2, K)$.

Suppose first that $p_3(x_1, x_2), p_4(x_1, x_2), \ldots, p_n(x_1, x_2)$ are linearly dependent over K. Then we may assume that $p_n(x_1, x_2) = 0$. If $n = 4$, then we take $a_3^{(3)} = 0$ and $a_4^{(3)} = -p_3(x_1, x_2)$. So assume that $n \geq 5$. Then we take $a_i^{(n-1)} = 0$ for all $3 \leq i < n$ and $a_n^{(n-1)} = -1$. Furthermore, we take $a_n^{(j)} = 0$ for all $3 \leq j < n-1$. By induction on n, there exist $a_i^{(j)}$ as claimed.

Suppose next that $p_3(x_1, x_2), p_4(x_1, x_2), \ldots, p_n(x_1, x_2)$ are linearly independent over K. Take $\alpha_3 \in K$, such that $p_3(\alpha_3, 1) \neq 0$, and assume without loss of generality that $p_i(\alpha_3, 1) = 0$ for all $i \neq 3$. Take $\alpha_4 \in K$, such that $p_4(\alpha_4, 1) \neq 0$, and assume without loss of generality that $p_i(\alpha_4, 1) = 0$ for all $i \neq 4$. Do the same with $\alpha_5, \alpha_6, \ldots, \alpha_n$, and let

$$\sigma := (x_1 - \alpha_3 x_2)(x_1 - \alpha_4 x_2) \cdots (x_1 - \alpha_n x_2).$$

Then

$$\Delta' := \frac{(x_1 - \alpha_3 x_2)\, p_3}{\sigma} x_3 + \frac{(x_1 - \alpha_4 x_2)\, p_4}{\sigma} x_4 + \cdots + \frac{(x_1 - \alpha_n x_2)\, p_n}{\sigma} x_n$$

is a polynomial, and by induction on the degree, we have

$$\Delta' = \begin{vmatrix} a_3^{(3)} & a_4^{(3)} & \cdots & a_n^{(3)} \\ a_3^{(4)} & a_4^{(4)} & \cdots & a_n^{(4)} \\ \vdots & \vdots & & \vdots \\ a_3^{(n-1)} & a_4^{(n-1)} & \cdots & a_n^{(n-1)} \\ \hline x_3 & x_4 & \cdots & x_n \end{vmatrix},$$

with the above properties on the $a_i^{(j)}$. Consequently,

$$\Delta = \begin{vmatrix} (x_1 - \alpha_3 x_2)\, a_3^{(3)} & (x_1 - \alpha_4 x_2)\, a_4^{(3)} & \cdots & (x_1 - \alpha_n x_2)\, a_n^{(3)} \\ (x_1 - \alpha_3 x_2)\, a_3^{(4)} & (x_1 - \alpha_4 x_2)\, a_4^{(4)} & \cdots & (x_1 - \alpha_n x_2)\, a_n^{(4)} \\ \vdots & \vdots & & \vdots \\ (x_1 - \alpha_3 x_2)\, a_3^{(n-1)} & (x_1 - \alpha_4 x_2)\, a_4^{(n-1)} & \cdots & (x_1 - \alpha_n x_2)\, a_n^{(n-1)} \\ \hline x_3 & x_4 & \cdots & x_n \end{vmatrix}.$$

This is essentially equivalent to Hilbert-Burch Theorem. The interested reader may wish to see [7] Theorem 20.15.

Remark 7.5 Hirokazu Nasu showed that the variety $X = (\Delta = 0) \subset \mathbb{P}^4$ is isomorphic to the projection of the Segre variety $S \subset \mathbb{P}^5$ of degree three from a general point outside of S. Here $S \subset \mathbb{P}^5$ is defined as the image of the Segre embedding

$$\mathbb{P}^1 \times \mathbb{P}^2 \hookrightarrow \mathbb{P}^5.$$

This is the locus of the maximal minors of a generic 2×3 matrix.

It is easy to see that any degree three homogeneous polynomial $F \in K[x_1, x_2, \Delta]$ in Theorem 7.3 which properly involves five variables can be transformed into the canonical form $x_1^2 x_3 + x_1 x_2 x_4 + x_2^2 x_5$ by means of a linear transformation of the variables. This form is also known as the Macaulay dual of the trivial extension of the algebra $K[x_1, x_2]/(x_1, x_2)^3$ by the canonical module.

8 Proof of dim $W \leq 1$ in the Proof of Proposition 7.2

We first formulate a result about homogeneous self-vanishing systems in dimension 5 in general.

Theorem 8.1 *Let $\boldsymbol{h} = (h_1, h_2, h_3, h_4, h_5)$ be a homogeneous self-vanishing system, for which*

$$\operatorname{rank}\left(\frac{\partial h_i}{\partial x_j}\right) = \operatorname{tr.deg}_K K(\boldsymbol{h}) = \operatorname{Krull\,dim} K[\boldsymbol{h}] \geq 3.$$

Let $Z : \mathbb{P}^4 \dashrightarrow \mathbb{P}^4$ be the rational map, defined by

$$x = (x_1 : \cdots : x_5) \mapsto (h_1 : \cdots : h_5),$$

and let W be the closure of the image of Z.

If $\dim L(W) > 1$, *then one of the following holds.*

(1) $\dim L(W) = 3$ *(where $L(W)$ is the K-linear span of W as a variety in $\mathbb{P}^4$).*
(2) W has a vertex, i.e., a point $p \in \mathbb{P}^4$ such that W is a union of lines through p.

Using Theorem 8.1, we can prove that $\dim W \leq 1$. Cases (1) and (2) are covered by Lemmas 8.2 and 8.3 respectively. Lemma 8.2 and its proof are essentially from [8, p. 567]. Lemma 8.3 is proved on [8, p. 568], and we think the proof is correct, but it could use some justification. For that reason, we formulated an alternative proof. Both Lemmas 8.2 and 8.3 come from the second author's paper [5].

Lemma 8.2 *Suppose that $f \in K[x_1, x_2, x_3, x_4, x_5]$ is homogeneous with zero Hessian, and let $\boldsymbol{h}$ be a reduced self-vanishing system arising from f. Then $\boldsymbol{h}$ is not as in* (1) *of Theorem 8.1.*

Proof Suppose that $\boldsymbol{h}$ is as in (1) of Theorem 8.1. Then $\dim L(W) = 3$, so we may assume that the last coordinate of every point of W is zero. Develop f into powers of x_5, and write

$$f = x_5^{\mu} A + x_5^{\mu+1} B,$$

where $A \in K[x_1, x_2, x_3, x_4]$ and $B \in K[x_1, x_2, x_3, x_4, x_5]$. Take g as in Definition 4.4. Then

$$g\Big(\frac{\partial f}{\partial x_1}, \frac{\partial f}{\partial x_2}, \frac{\partial f}{\partial x_3}, \frac{\partial f}{\partial x_4}, \frac{\partial f}{\partial x_5}\Big) = 0.$$

Since g has minimum degree as such, $\frac{\partial g}{\partial y_5} = 0$. Hence $g \in K[y_1, y_2, y_3, y_4]$. If we look at the trailing coefficient with respect to x_5, we see that

$$g\Big(\frac{\partial A}{\partial x_1}, \frac{\partial A}{\partial x_2}, \frac{\partial A}{\partial x_3}, \frac{\partial A}{\partial x_4}\Big), \tag{5}$$

i.e., $A \in K[x_1, x_2, x_3, x_4]$ has Hessian determinant zero (in dimension 4). From Theorem 7.1, it follows that we may assume that $A \in K[x_1, x_2, x_3]$.

If $y_4 \mid g$, then $g = y_4$ because g has minimum degree, so $\deg \boldsymbol{h} = 0$. Consequently, $y_4 \nmid g$. It follows from $A \in K[x_1, x_2, x_3]$ and (5) that $A \in K[x_1, x_2, x_3]$ has Hessian determinant zero (in dimension 3). From Theorem 5.3, it follows that we may assume that $A \in K[x_1, x_2]$.

Using Theorems 3.9(a) and 3.11, we deduce that $f\big(x + t\,\boldsymbol{h}(x)\big) = f(x)$. As $h_5 = 0$, we have

$$A\big(x + t\,\boldsymbol{h}(x)\big)x_5^{\mu} + B\big(x + t\,\boldsymbol{h}(x)\big)x_5^{\mu+1} = x_5^{\mu} A(x) + x_5^{\mu+1} B(x).$$

If we look at the leading coefficient with respect to t, we see that $A(\boldsymbol{h}(x)) = 0$. Since $A \in K[x_1, x_2]$, we see that h_1 and h_2 are algebraically dependent over K. Consequently, h_1 and h_2 are linearly dependent over K, say that $h_1 = 0$. Then $h_1 = h_5 = 0$, so $\dim L(W) \le 2$. Contradiction.

Lemma 8.3 *Suppose that $f \in K[x] = K[x_1, x_2, \ldots, x_n]$ is homogeneous with zero Hessian, and let $\boldsymbol{h}$ be a reduced self-vanishing system arising from f. Suppose that p is a vertex of W. Then $f \in \mathrm{Sol}(p; K[x])$. In particular, a variable can be eliminated from f by way of a linear transformation of the variables.*

Proof Write $p = (p_1 : p_2 : \cdots : p_n)$ and take $d := \deg \boldsymbol{h}$. Take $\tilde{\boldsymbol{h}} = \boldsymbol{h} + x_{n+1}^d p$. Since p is a vertex of W, we infer that the image of

$$\tilde{Z} : \mathbb{P}^n(x) \to \mathbb{P}^{n-1}(y),$$

defined by $(x_1 : x_2 : \cdots : x_n : x_{n+1}) \mapsto (\tilde{h}_1 : \tilde{h}_2 : \cdots : \tilde{h}_n)$, is equal to W. Hence it follows from Lemma 5.2 that

$$\mathrm{rank}\,(\partial \tilde{h}_i / \partial x_j) = \text{Krull dim } K[\tilde{\boldsymbol{h}}] = \dim W + 1 = \text{Krull dim } K[\boldsymbol{h}] = \mathrm{rank}\,(\partial h_i / \partial x_j).$$

Since $(\partial \tilde{h}_i / \partial x_j)$ can be obtained from $(\partial h_i / \partial x_j)$ by adding $\partial \tilde{\boldsymbol{h}} / \partial x_{n+1}$ as a column to the right hand side, we infer from $\mathrm{rank}\,(\partial \tilde{h}_i / \partial x_j) = \mathrm{rank}\,(\partial h_i / \partial x_j)$ that $\partial \tilde{\boldsymbol{h}} / \partial x_{n+1} = d\, x_{n+1}^{d-1}\, p$ is contained in the column space of $(\partial h_i / \partial x_j)$.

Hence p is dependent over $K(x)$ on $\partial \boldsymbol{h} / \partial x_1, \partial \boldsymbol{h} / \partial x_2, \ldots, \partial \boldsymbol{h} / \partial x_n$. From Theorem 3.9(c), we infer that $f \in \mathrm{Sol}(p; K[x])$.

The rest of this section will be devoted to the proof of Theorem 8.1.

Suppose that $\dim W > 1$, and let T be the fundamental locus of Z in $\mathbb{P}^4(x)$, defined by the equations $h_1(x) = h_2(x) = \cdots = h_5(x) = 0$. From Theorem 3.11(c), it follows that $\boldsymbol{h}(x + t\boldsymbol{h}(x)) = \boldsymbol{h}(x)$. If we look at the leading coefficient with respect to t, we see that

$$\boldsymbol{h}(\boldsymbol{h}(x)) = 0, \qquad \text{i.e.,} \qquad W \subseteq T. \tag{6}$$

From Proposition 6.2, it follows that $\dim T \le 2$. Since $\dim W \ge 2$, (6) tells us that

$$\dim W = \dim T = 2.$$

As W is the closure of the image of Z, we see that W is a component of T.

On [8, p. 565], the authors claim that for every $c \in T$, there exists a $p \in W$ such that the line through c and p is contained in T. But if $c = p$, then 'the line through c and p' shrinks to a single point, so it must be shown that p can be taken different from c. The following lemma can be used for that.

Lemma 8.4 *Let $\boldsymbol{h} = (h_1, h_2, \ldots, h_n)$ be a homogeneous self-vanishing system. Let S be a hyperplane in $\mathbb{P}^{n-1}$. Then every irreducible component of $Z^{-1}(S)$ contains W and has dimension $n-2$.*

Proof We can take $\boldsymbol{\alpha} = (\alpha_1, \alpha_2, \ldots, \alpha_n)$, such that

$$S = \{(\sigma_1 : \sigma_2 : \cdots : \sigma_n) \mid \alpha_1\sigma_1 + \alpha_2\sigma_2 + \cdots + \alpha_n\sigma_n = 0\}.$$

Now $Z^{-1}(S)$ is contained in

$$\{(\tau_1 : \tau_2 : \cdots : \tau_n) \mid \alpha_1 h_1(\boldsymbol{\tau}) + \alpha_2 h_2(\boldsymbol{\tau}) + \cdots + \alpha_n h_n(\boldsymbol{\tau}) = 0\},$$

where $\boldsymbol{\tau} = (\tau_1, \tau_2, \ldots, \tau_n)$, and any irreducible component X of $Z^{-1}(S)$ is of the form

$$\{(\tau_1 : \tau_2 : \cdots : \tau_n) \mid f(\boldsymbol{\tau}) = 0\},$$

where f is an irreducible factor of $\alpha_1 h_1 + \alpha_2 h_2 + \cdots + \alpha_n h_n$. So $\dim X = n-2$.

It suffices to show that $f(p) = 0$ for every p in the image of Z (i.e., skip the points of W which were added by taking closure). So let p be an image point of W. From Lemma 5.2, it follows that

$$\operatorname{rank}\left(\frac{\partial h_i}{\partial x_j}\right) = \operatorname{Krull\,dim} K[\boldsymbol{h}] \le n-1.$$

On account of the fiber dimension theorem, the closure of $Z^{-1}(p)$ has dimension at least 1. As $\dim X = n-2$, the intersection of X and the closure of $Z^{-1}(p)$ is nonempty, say that θ is contained in this intersection.

If $\theta \in Z^{-1}(p)$, then $p = \boldsymbol{h}(\theta)$, and from Theorem 3.11 we infer that $\boldsymbol{h}(\boldsymbol{\theta} + t\boldsymbol{p}) = \boldsymbol{h}(\boldsymbol{\theta})$, where $\boldsymbol{\theta}$ and $\boldsymbol{p}$ are vectors over K which correspond to θ and p respectively. The last equality holds in general as well because θ is contained in the closure of $Z^{-1}(p)$. Hence

$$\deg_t f(\boldsymbol{\theta} + t\boldsymbol{p}) \le \deg_t\big(\alpha_1 h_1(\boldsymbol{\theta} + t\boldsymbol{p}) + \alpha_2 h_2(\boldsymbol{\theta} + t\boldsymbol{p}) + \cdots + \alpha_n h_n(\boldsymbol{\theta} + t\boldsymbol{p})\big) = 0.$$

So $f(\boldsymbol{\theta} + t\boldsymbol{p}) = f(\boldsymbol{\theta})$. From $\theta \in X$, it follows that $f(\boldsymbol{\theta} + t\boldsymbol{p}) = f(\boldsymbol{\theta}) = 0$. If we look at the leading coefficient with respect to t, we see that $f(\boldsymbol{p}) = 0$, which completes the proof.

Corollary 8.5 *Let $n \geq 3$ and $\boldsymbol{h} = (h_1, h_2, \ldots, h_n)$ be a homogeneous self-vanishing system. Let S be a hyperplane in $\mathbb{P}^{n-1}$, and Y be a component of $S \cap W$. Assume that Y contains an image point of Z. Then for every $c \in W$, there exists a $p \in Y$ such that the line through c and p is contained in T.*

Proof Let X be any component of the closure of $Z^{-1}(Y)$. For every $c \in X$ for which $\boldsymbol{h}(c) \neq 0$, there exists a $p \in Y$ such that $\boldsymbol{h}(\boldsymbol{c} + t\boldsymbol{p}) = \boldsymbol{h}(\boldsymbol{c})$, where $\boldsymbol{c}$ and $\boldsymbol{p}$ are vectors over K which correspond to c and p respectively. To see this just set $p = \boldsymbol{h}(c)$ and use Theorem 3.11.

Let

$$U = \{(c, p) \in X \times Y | \boldsymbol{h}(\boldsymbol{c} + t\boldsymbol{p}) = \boldsymbol{h}(\boldsymbol{c})\}.$$

Since Y is a complete variety, the projection of U on X is a closed morphism. Hence

$$\tilde{X} = \{c \in X | \text{ there is a } p \in Y \text{ such that } \boldsymbol{h}(\boldsymbol{c} + t\boldsymbol{p}) = \boldsymbol{h}(\boldsymbol{c})\}$$

is closed. The $c \in X$ for which $\boldsymbol{h}(c) \neq 0$ form a dense subset of X and are contained in $\tilde{X}$, so $\tilde{X} = X$.

From Lemma 8.4, it follows that X contains W, so $W \subseteq \tilde{X}$. So for every $c \in W$, there exists a $p \in Y$ such that $\boldsymbol{h}(\boldsymbol{c} + t\boldsymbol{p}) = \boldsymbol{h}(\boldsymbol{c}) = 0$, i.e., the line through c and p is contained in T.

So let us take $c \in W$. It is possible to take S such that $c \notin S$ and S contains an image point of Z. From Corollary 8.5, it follows that there exist a $c' \in S$ such that the line through c and c' is contained in T. As $c \notin S$ and $c' \in S$, we see that 'the line through c and c'' does not shrink to a single point.

Since W is a component of T, the interior W° of W as a subspace of T is nonempty. Now take $p \in W^\circ$ in the image of Z. There exists a $p' \in W$ such that $p' \neq p$ and such that the line L_p through p and p' is contained in T. But since W is the only component of T which contains p, it follows that $L_p \subseteq W$.

Taking $Y = L_p$ in Corollary 8.5, we deduce that for every $q \in W$, there exists a $q' \in L_p$ such that the line L_q through q and q' is contained in T. This is also claimed on [8, p. 565]. If we take $q \in W^\circ$, then L_q is even contained in W.

Now let us fix L_p, and range q over W°. There are infinitely many points $q \in W^\circ$, but this does not mean automatically that there are infinitely many lines L_q. This is because for a line L_q, there may be infinitely many candidates for the point q.

However, since $\dim W = 2$, there are infinitely many lines L_q indeed, just as claimed on [8, p. 565]. On [8, pp. 565, 566], the following two cases (a) and (b) are distinguished:

(a) The set of points $q' \in L_p$ (which we get by ranging q over W°) is infinite.
(b) There exists a fixed point $q' \in L_p$, which is contained in infinitely many lines L_q.

We will treat these cases in essentially the same way as on [8, pp. 566, 568].

Assume first that case (b) above applies. Then one can show that the closure of the union of lines L_q through q' has dimension 2. As W is irreducible of dimension

2, W is just the closure of this union, so W is a union of lines through q'. Thus case (b) corresponds to case (2) of Theorem 8.1, as claimed on [8, p. 568].

Assume next that case (a) above applies. Take $r \in W^\circ$. Then there exists an $r' \in L_p$ such that the line L_r through r and r' is contained in W.

Now let us fix L_q as well, by choosing q in W° in the image of Z outside L_p. Range r over W°. On [8, p. 566], two subcases of case (a) are distinguished, which are essentially as follows:

(a1) There does not exist a line L_r as above, such that $L_r \cap L_q = \varnothing$.
(a2) There does exist a line L_r as above, such that $L_r \cap L_q = \varnothing$.

Both cases are treated essentially as follows on [8, p. 566].

Assume first that case (a1) above applies. Then one can show that the linear span of p, p' and q contains infinitely many lines L_r, and that the closure of these lines has dimension 2. As W is irreducible of dimension 2, W is just the closure of these lines, which corresponds to the linear span of p, p' and q.

So $\dim L(W) \le 2$. Hence $\dim L(W) = \dim W$, so $L(W) = W$ and every point of W is a vertex of W. So W is as in (2) of Theorem 8.1 (and case (b) above applies as well).

Assume next that case (a2) above applies. Take $r \in W^\circ$ and $r' \in L_p$, such that $L_r \cap L_q = \varnothing$. Take $s \in L_r \cap W^\circ$. Just as with $r' \in L_p$, there exists an $s' \in L_q$, such that the line L_s through s and s' is contained in W.

There are infinitely many lines L_s, if we range s over $L_r \cap W^\circ$. Let U be the closure of the lines L_s. Only finitely many lines can be a component of U, so there is a line which is not. Being irreducible, the line is fully contained in a component of U, and this component has larger dimension than the line. So U has dimension at least 2. Consequently, the intersection of W and the linear span of q, q', r and r' has dimension at least 2. As W is irreducible of dimension 2, W is just this intersection. So W is as in (2) of Theorem 8.1.

Remark 8.6 Self-vanishing systems corresponding to cases (a1), (a2), (b) indeed exist:

(a1) $H = (x_4^2, x_4x_5, x_1x_5 - x_2x_4, 0, 0)$,
(a2) $H = (x_5^2(ax_1 - x_5^2x_2), a(ax_1 - x_5^2x_2), x_5^2(ax_3 - x_5^2x_4), a(ax_3 - x_5^2x_4), 0)$ with $a = x_1x_4 - x_2x_3$,
(b) $H = (x_5^5, bx_5^3, b^2x_5, -b^2x_1 + 2bx_2x_5^2 - x_3x_5^4, 0)$ with $b = x_1x_3 - x_2^2 + x_4x_5$.

All these examples are counterexamples to the original version of Lemma 5.2. They were taken from the introduction of the second author's paper [5].

Remark 8.7 Gordan and Noether only prove that $\dim L(W) \le 3$ in Theorem 8.1 (1). This makes that the conclusion that $\dim L(W) \le 2$ at the end of of the proof of Lemma 8.2 is not sufficient to prove Lemma 8.2. Gordan and Noether advance as follows.

On account of $\dim L(W) \le 2$, we may assume that $h_1 = h_2 = 0$. If $f \in K[x_1, x_2]$, then $h_i \in K[x_1, x_2]$ for all i as well, so $\dim W = \text{Krull}\dim K[\boldsymbol{h}] - 1 = \text{tr.}\deg_K$

$K(\boldsymbol{h}) - 1 \le 1$. If $f \notin K[x_1, x_2]$, then it follows from Theorem 3.9(c) that the last three rows of the Jacobian matrix $(\partial h_i/\partial x_j)$ of $\boldsymbol{h}$ are dependent. The first two rows are zero, so $\operatorname{rank}(\partial h_i/\partial x_j) \le 2$ and $\dim W = \operatorname{Krull\,dim} K[\boldsymbol{h}] - 1 = \operatorname{rank}(\partial h_i/\partial x_j) - 1 \le 1$.

Remark 8.8 Theorem 8.1 is proved in [5] by the second author as well, but in a different way, because the proof by Gordan and Noether was not fully understood. The second author only proved that $\dim L(W) \le 2$ in [5], because that was the missing link in [6] to obtain Theorem 7.3.

The focus is on $\dim L(W)$ rather than $\dim W$ in other papers of the second author as well. In [4], only $\dim L(W) \le 1$ is used in the classification of all homogeneous polynomials with zero Hessian in dimension 4. In [3], all homogeneous polynomials with zero Hessian in dimension 6 are classified, under the assumption that $\dim L(W) \le 3$. Non-homogeneous polynomials are classified under similar assumptions in [3, 6].

9 Section 6 of Gordan and Noether [8]

We start with formulating a result which, as opposed to Theorem 3.9, applies to $\boldsymbol{h}$ which are not constructed as in Proposition 3.8. The result is inspired by Lemma 3.1 of [12].

Proposition 9.1 *Let $\boldsymbol{h} = (h_1, h_2, \ldots, h_n)$ be a system of forms of the same degree, and let $f \in K[x]$ be a homogeneous polynomial. Put $f_j = \frac{\partial f}{\partial x_j}$ for $j = 1, 2, \ldots, n$, and let $\boldsymbol{f} = (f_1, \ldots, f_n)$ Then the following statements are equivalent:*

(a) $\partial \boldsymbol{h}/\partial x_j$ is a syzygy of $\boldsymbol{f}$ for $j = 1, 2, \ldots, n$.
(b) $\boldsymbol{h}$ is a syzygy of $\partial \boldsymbol{f}/\partial x_j$ for $j = 1, 2, \ldots, n$.

Furthermore, both (a) *and* (b) *imply that $\boldsymbol{h}$ is a syzygy of $\boldsymbol{f}$.*

Proof Using
$$\frac{\partial \boldsymbol{h}}{\partial x_1}x_1 + \frac{\partial \boldsymbol{h}}{\partial x_2}x_2 + \cdots + \frac{\partial \boldsymbol{h}}{\partial x_n}x_n = (\deg \boldsymbol{h})\boldsymbol{h}$$

we can obtain the last claim from (a). Similarly, the last claim can be obtained from (b). Having these results, the equivalence of (a) and (b) follows from the formula

$$\frac{\partial}{\partial x_j}(\boldsymbol{h} \cdot \boldsymbol{f}) = \frac{\partial \boldsymbol{h}}{\partial x_j} \cdot \boldsymbol{f} + \boldsymbol{h} \cdot \frac{\partial \boldsymbol{f}}{\partial x_j} = 0.$$

for $j = 1, 2, \ldots, n$. This formula appears in the proof of Theorem 3.9.

Let $f \in K[x]$ be a homogeneous polynomial, and put $f_j = \frac{\partial f}{\partial x_j}$. Assume that $f_1, f_2, \ldots, f_n$ are algebraically dependent over K. Then by Proposition 4.2, there exists a system $\boldsymbol{h} = (h_1, h_2, \ldots, h_n)$ of forms of the same degree, such that Proposition 9.1(b) is satisfied.

Assume that

$$h_j(x) \in \mathrm{Sol}(i(L(W)); R) \text{ for all } j. \tag{7}$$

Then by a linear change of variables, $\boldsymbol{h}$ coincides with a self-vanishing system described in Example 3.6. We assume this from now on, so

- $h_1(x) = \cdots = h_r(x) = 0$, for some $1 \le r < n$, and
- $h_j(x)$ is a function only in $x_1, \ldots, x_r$ for $j > r$.

In Sect. 6 of [8] and Sect. 3 of [12], all f which satisfy Proposition 9.1(b) are classified for this particular $\boldsymbol{h}$.

Although $\boldsymbol{h}$ is a self-vanishing system which has many properties of reduced self-vanishing systems arising from f, especially if $\mathrm{GCD}(h_1, h_2, \ldots, h_n) = 1$, we will show in Example 9.5 below that it is possible for $\boldsymbol{h}$ to satisfy $\mathrm{GCD}(h_1, h_2, \ldots, h_n) = 1$ and not to be of this form for some f which satisfies Proposition 9.1(b).

But if $\mathrm{GCD}(h_1, h_2, \ldots, h_n) = 1$ and $\mathcal{I}(f)$ is a principal ideal, then $\boldsymbol{h}$ is indeed a reduced self-vanishing systems of f. This is because there is only one $\boldsymbol{h}$ with $\mathrm{GCD}(h_1, h_2, \ldots, h_n) = 1$ which satisfy Proposition 9.1(a).

Theorem 9.2 *Suppose that the Jacobian* $M := (\partial h_i / \partial x_j)$ *of* $\boldsymbol{h} = (h_1, h_2, \ldots, h_n)$ *has rank* k. *Then we can choose* k *columns of* M *which generate its column space, say with indices* $i_1, i_2, \ldots, i_k$. *Then* $i_1, i_2, \ldots, i_k \in \{1, 2, \ldots, r\}$.

Suppose that $f \in K(x_1, x_2, \ldots, x_r)[x_{r+1}, x_{r+2}, \ldots, x_n]$. *Then* f *is a polynomial over* $K(x_1, x_2, \ldots, x_r)$ *in the* $(k+1) \times (k+1)$ *minors of*

$$\begin{pmatrix} \frac{\partial h_{r+1}}{\partial x_{i_1}} & \frac{\partial h_{r+2}}{\partial x_{i_1}} & \cdots & \frac{\partial h_n}{\partial x_{i_1}} \\ \frac{\partial h_{r+1}}{\partial x_{i_2}} & \frac{\partial h_{r+2}}{\partial x_{i_2}} & \cdots & \frac{\partial h_n}{\partial x_{i_2}} \\ \vdots & \vdots & & \vdots \\ \frac{\partial h_{r+1}}{\partial x_{i_k}} & \frac{\partial h_{r+2}}{\partial x_{i_k}} & \cdots & \frac{\partial h_n}{\partial x_{i_k}} \\ \hline x_{r+1} & x_{x+2} & \cdots & x_n \end{pmatrix}$$

if and only if f *satisfies Proposition 9.1(b).*

Proof Let $\tilde{K}$ be the algebraic closure of $K(x_1, x_2, \ldots, x_r)$. From Theorem 6.4, it follows that f is a polynomial over $\tilde{K}$ in the above-described minors, if and only if f satisfies Proposition 9.1(a).

Suppose that f is a polynomial over $\tilde{K}$ in the above-described minors. Let $\mathcal{B}$ be a $K(x_1, x_2, \ldots, x_r)$-basis of $\tilde{K}$, such that $1 \in \mathcal{B}$. Then $\mathcal{B}$ is also a basis of $\tilde{K}[x_{r+1}, x_{r+2}, \ldots, x_n]$ as a free module over $K(x_1, x_2, \ldots, x_r)[x_{r+1}, x_{r+2}, \ldots, x_n]$. Taking coefficients of $1 \in \mathcal{B}$ yields f as a polynomial over $K(x_1, x_2, \ldots, x_r)$ in the above-described minors.

Remark 9.3 The proof of Theorem 6.4 tells us that the $(k+1) \times (k+1)$-minors can be replaced by $n-r-k$ homogeneous polynomials $l_1, l_2, \dots, l_{n-r-k}$ which are linear in $x_{r+1}, x_{r+2}, \dots, x_n$.

In Sect. 2.3 of [2], a similar construction is given as above, but some conditions on that construction makes it incomplete for classification. This is however not a real problem, because there is no classification result in [2]. There is just a definition of so-called GN-polynomials of type (r, s, μ, n).

Let us call a polynomial f as in Theorem 9.2 a *GN-polynomial of type*$(n-1, n-1-r, k-1)$. Then GN-polynomials of type $(n-1, n-1-r, k-1, n')$, as defined in [2], are GN-polynomials of type $(n-1, n-1-r, k-1)$ with additional conditions.

Remark 9.4 Since $\operatorname{tr.deg}_K K(h_{r+1}, h_{r+2}, \dots, h_r) = k$, there exists a transcendence basis $A_1, A_2, \dots, A_k$ over K of $K(h_{r+1}, h_{r+2}, \dots, h_r)$. In Sect. 6 of [8], Sect. 3 of [12], and Sect. 2.3 of [2], derivatives are taken with respect to $A_1, A_2, \dots, A_k$ instead of $x_{i_1}, x_{i_2}, \dots, x_{i_k}$ in the matrix of Theorem 9.2.

For other choices of $A_1, A_2, \dots, A_k$, there are problems with the meaning of differentiating h_j with respect to A_i. To obtain meaning, we choose $A_{k+1}, A_{k+2}, \dots, A_r$, such that $A_1, A_2, \dots, A_r$ becomes a transcendence basis of $K(x_1, x_2, \dots, x_r)$. Now the only condition is that the first k rows of the matrix of Theorem 9.2 are independent. With this condition, it is possible to take for $A_1, A_2, \dots, A_r$ a permutation of $x_1, x_2, \dots, x_r$, which is exactly what we do in Theorem 9.2.

Example 9.5 If $\mathcal{I}(f)$ is not a principal ideal, then is possible for $\boldsymbol{h}$ to satisfy Proposition 9.1(b) and $\mathrm{GCD}(h_1, h_2, \dots, h_n) = 1$ without being a reduced self-vanishing systems arising from f. Take for instance

$$f = x_1^2 x_3 + x_1 x_2 x_4 + x_2^2 x_5 + z_1^2 z_3 + z_1 z_2 z_4 + z_2^2 z_5,$$

where $z_1, z_2, z_3, z_4, z_5 = x_6, x_7, x_8, x_9, x_{10}$. Then for a reduced self-vanishing system arising from f, the degree which respect to x_1, x_2, x_3, x_4, x_5 has the same parity as the degree with respect to z_1, z_2, z_3, z_4, z_5. This is however not the case for

$$\boldsymbol{h} = \big(0, 0, (x_1 \cdot x_2^2), (x_1 \cdot -2x_1 x_2), (x_1 \cdot x_1^2), 0, 0, (x_2 \cdot z_2^2), (x_2 \cdot -2z_1 z_2), (x_2 \cdot z_1^2)\big).$$

References

1. Alice, A., Repetto, F.: A geometrical approach to Gordan-Noether's and Franchetta's contributions to a question posed by Hesse. Collect. Math. **60**(1), 27–41 (2009)
2. Ciliberto, C., Russo, F., Simis, A.: Homaloidal hypersurfaces and hypersurfaces with vanishing Hessian. Adv. Math. **218**(6), 1759–1805 (2008)
3. de Bondt, M.C.: Polynomial Hessians with small rank. arxiv:1609.03904
4. de Bondt, M.C.: Quasi-translations and singular Hessians. Colloq. Math. **152**(2), 175–198 (2018)

5. de Bondt, M.C.: Homogeneous quasi-translations in dimension 5. Beitr. Algebra Geom. **59**(2), 295–326 (2018)
6. de Bondt, M.C., van den Essen, A.R.P.: Singular Hessians. J. Algebra **282**(1), 195–204 (2004)
7. Eisenbud, D.: Commutative Algebra with a View Toward Algebraic Geometry. Graduate Texts in Mathematics, vol. 150. Springer, New York (1995)
8. Gordan, P., Noether, M.: Ueber die algebraischen Formen, deren Hesse'sche Determinante identisch verschwindet. Math. Ann. **10**, 547–568 (1876)
9. Harima, T., Migliore, J., Nagel, U., Watanabe, J.: The weak and strong Lefschetz properties for Artinian K-algebras. J. Algebra **262**, 99–126 (2003)
10. Hesse, O.: Über die Bedingung, unter welcher eine homogene ganze Function von n unabhängigen Variabeln durch lineäre Substitutionen von n andern unabhängigen Variabeln auf eine homogene Function sich zurükfüren läfst, die eine Variable weniger enthält. Journal für die reine und angewandte Mathematik **42**, 117–124 (1851)
11. Hesse, O.: Zur Theorie der Ganzen homogenen Functionen. Journal für die reine und angewandte Mathematik **56**, 263–269 (1859)
12. Lossen, C.: When does the Hessian determinant vanish identically? Bull. Braz. Math. Soc. **35**(1), 71–82 (2004)
13. Maeno, T., Watanabe, J.: Lefschetz elements of Artinian Gorenstein algebras and Hessians of Homogeneous polynomials. Illinois J. Mathematics **53**(2), 591–603 (2009)
14. Pasch, M.: Zur Theorie der Hesseschen Determinante. Journal für die reine und angewandte Mathematik **80**, 169–176 (1875)
15. Watanabe, J.: A remark on the Hessian of homogeneous polynomials. In: The Curves Seminar at Queen's Volume XIII, Queen's Papers in Pure and Applied Mathematics, vol. 119, pp. 171–178 (2000)
16. Watanabe, J.: On the theory of Gordan-Noether on homogeneous forms with zero Hessian. Proc. School Sci. Tokai Univ. **49**, 1–21 (2014)
17. Yamada, H.: On a theorem of Hesse—P. Gordan and M. Noether's Theory—, Unpublished

Rational Real Algebraic Models of Compact Differential Surfaces with Circle Actions

Adrien Dubouloz and Charlie Petitjean

Abstract We give an algebro-geometric classification of smooth real affine algebraic surfaces endowed with an effective action of the real algebraic circle group $\mathbb{S}^1$ up to equivariant isomorphisms. As an application, we show that every compact differentiable surface endowed with an action of the circle S^1 admits a unique smooth rational real quasi-projective model up to $\mathbb{S}^1$-equivariant birational diffeomorphism.

Keywords Affine surface · Real form · Circle group action

1 Introduction

A description of normal complex affine surfaces admitting an effective action of the complex multiplicative group $\mathbb{G}_{m,\mathbb{C}} = (\mathbb{C}^*, \times)$ was given by Flenner and Zaidenberg [9] in terms of their graded coordinate rings. Generalizing earlier constructions due to Dolgachev-Pinkham-Demazure [6, 7, 13], they described these graded rings as rings of sections of divisors with rational coefficients on suitable one-dimensional rational quotients of the given action. This type of presentation, which is nowadays called the DPD-*presentation* of normal affine surfaces with $\mathbb{G}_{m,\mathbb{C}}$-actions, was generalized vastly by Altmann and Hausen [1] to give presentations of normal complex affine varieties of any dimension endowed with effective actions of tori $\mathbb{G}^r_{m,\mathbb{C}}$ in terms of so-called polyhedral Weil divisors on suitable rational quotients obtained as limits of GIT quotients.

For normal affine varieties over arbitrary base fields k endowed with effective actions of non necessarily split tori, that is, commutative k-groups schemes G whose base extension to a separable closure k^s of k are isomorphic to split tori $\mathbb{G}^r_{m,k^s}$, only

A. Dubouloz (✉)
IMB UMR5584 CNRS, Université Bourgogne Franche-Comté, 21000 Dijon, France
e-mail: adrien.dubouloz@u-bourgogne.fr

C. Petitjean
I.U.T. Dijon-Département GMP, Boulevard Dr. Petitjean, 21078 Dijon, France
e-mail: charlie.petitjean@u-bourgogne.fr

S. Kuroda et al. (eds.), *Polynomial Rings and Affine Algebraic Geometry*,
Springer Proceedings in Mathematics & Statistics 319,
https://doi.org/10.1007/978-3-030-42136-6_4

partial extensions of the Altmann-Hausen formalism are known so far. Besides the toric case considered by several authors, a first step was made by Langlois [10] who obtained a generalization to affine varieties endowed with effective actions of complexity one. In another direction, Liendo and the first author [8] recently extended the Altmann-Hausen framework to describe normal real affine varieties endowed with an effective action of the 1-dimensional non-split real torus, the circle

$$\mathbb{S}^1 = \mathrm{Spec}(\mathbb{R}[u, v]/(u^2 + v^2 - 1)) \cong \mathrm{SO}_2(\mathbb{R}).$$

The common approach in these generalizations is based on the understanding of the interplay between the algebro-combinatorial data in an Altmann-Hausen presentation of the variety with split torus action obtained by base extension to a separable closure k^s of the base field k and Galois descent with respect to the Galois group $\mathrm{Gal}(k^s/k)$. In the real case, this amounts to describe normal real affine varieties as normal complex affine varieties endowed with an anti-regular involution σ, called a *real structure*. The results in [8] then essentially consist of a description of $\mathbb{S}^1$-actions on normal real affine varieties in the language of [1] extended to complex affine varieties with real structures.

The goal of this article is to give a survey of this description and some applications in the special case of normal real affine surfaces, formulated in the language of DPD-presentations of Flenner and Zaidenberg. The first main result, Theorem 14, describes the one-to-one correspondence between normal real affine surfaces with effective $\mathbb{S}^1$-actions and certain pairs (D, h) called *real* DPD-*pairs* on smooth real affine curves C, consisting of a Weil $\mathbb{Q}$-divisor D on the complexification of C and a rational function on C. We also characterize which such pairs correspond to smooth real affine surfaces. A second main result, Theorem 20, consists of a complete classification of real $\mathbb{S}^1$-orbits on a smooth real affine surface in relation to the structure of the real fibers of the real quotient morphism of the $\mathbb{S}^1$-action.

To give an illustration of the flavor of these results, consider the smooth complex affine surface

$$S = \{xy^2 = 1 - z^2\} \subset \mathrm{Spec}(\mathbb{C}[x^{\pm 1}, y, z]).$$

The group $\mathbb{G}_{m,\mathbb{C}}$ acts effectively on S by $t \cdot (x, y, z) = (t^2 x, t^{-1} y, z)$ and the categorical quotient for this action is the projection $\pi = \mathrm{pr}_z : S \to \mathbb{A}^1_{\mathbb{C}} = \mathrm{Spec}(\mathbb{C}[z])$. All fibers of π consist of a unique $\mathbb{G}_{m,\mathbb{C}}$-orbit, isomorphic to $\mathbb{G}_{m,\mathbb{C}}$ acting on itself by translations except for $\pi^{-1}(-1)$ and $\pi^{-1}(1)$ which are isomorphic to $\mathbb{G}_{m,\mathbb{C}}$ on which $\mathbb{G}_{m,\mathbb{C}}$ acts with stabilizer equal to the group μ_2 of complex square roots of unity. The composition of the involution $t \mapsto t^{-1}$ of $\mathbb{G}_{m,\mathbb{C}}$ with the complex conjugation defines a real structure σ_0 on $\mathbb{G}_{m,\mathbb{C}}$ for which the pair $(\mathbb{G}_{m,\mathbb{C}}, \sigma_0)$ describes a real algebraic group isomorphic to $\mathbb{S}^1$. The composition of the involution $(x, y, z) \mapsto (x^{-1}, xy, z)$ of S with the complex conjugation defines a real structure σ on S, making the pair (S, σ) into a smooth real affine surface. The $\mathbb{G}_{m,\mathbb{C}}$-action on S is compatible with these two real structures and defines a real action of $\mathbb{S}^1$ on the real affine surface (S, σ). The quotient morphism π can in turn be interpreted as a real morphism $\pi : (S, \sigma) \to \mathbb{A}^1_{\mathbb{R}} = \mathrm{Spec}(\mathbb{R}[z])$ which is the categorical quotient of (S, σ)

for the $\mathbb{S}^1$-action in the category of real algebraic varieties. A real DPD-pair (D, h) describing the $\mathbb{S}^1$-action on (S, σ) is then given for instance by the Weil $\mathbb{Q}$-divisor $D = \frac{1}{2}\{-1\} + \frac{1}{2}\{1\}$ on $\mathbb{A}^1_{\mathbb{C}}$ and the rational function $h = 1 - z^2$ on $\mathbb{A}^1_{\mathbb{R}}$. The fibers of $\pi : (S, \sigma) \to \mathbb{A}^1_{\mathbb{R}}$ over real points c of $\mathbb{A}^1_{\mathbb{R}}$ all consist of a single $\mathbb{S}^1$-orbit which is isomorphic to $\mathbb{S}^1$-acting on itself by translations if $c \in]-1, 1[$, to $\mathbb{S}^1$ acting on itself with stabilizer μ_2 if $c \in \{-1, 1\}$, and to the real affine curve without real point $\{u^2 + v^2 = -1\}$ otherwise.

It can be checked (see Sect. 4.1.4) that the set of real points of the above surface (S, σ) endowed with the induced euclidean topology is diffeomorphic to the Klein bottle K. Furthermore, the $\mathbb{S}^1$-action on (S, σ) induces a differentiable action of the real circle S^1 on K which coincides with the standard S^1-action on K viewed as the S^1-equivariant connected sum $\mathbb{RP}^2 \sharp_{S^1} \mathbb{RP}^2$ of two copies of the projective plane $\mathbb{RP}^2$. In other words, (S, σ) endowed with its $\mathbb{S}^1$-action is an equivariant *real affine algebraic model* of the Klein bottle K endowed with its differentiable S^1-action. By a result of Comessatti [5], a compact connected differential manifold of dimension 2 without boundary admits a projective rational real algebraic model if and only if it is non-orientable or diffeomorphic to the sphere S^2 or the torus $T = S^1 \times S^1$. It was established later on by Biswas and Huisman [3] that such a projective model is unique up to so-called *birational diffeomorphisms*, that is, diffeomorphisms induced by birational maps defined at every real point and admitting an inverse of the same type. As an application of the real DPD-presentation formalism, we establish the following uniqueness property of rational models of compact differentiable surfaces with S^1 -actions among all smooth rational quasi-projective real algebraic surfaces with $\mathbb{S}^1$-actions.

Theorem 1 *A connected compact real differential manifold of dimension* 2 *without boundary endowed with an effective differentiable S^1-action admits a smooth rational quasi-projective real algebraic model with $\mathbb{S}^1$-action, unique up to $\mathbb{S}^1$-equivariant birational diffeomorphism.*

The article is organized as follows. In the first section we review the classical equivalence of categories between quasi-projective real varieties and quasi-projective complex varieties equipped with real structures and recall the interpretation of $\mathbb{S}^1$-actions on such varieties as forms of $\mathbb{G}_{m,\mathbb{C}}$-actions on complex varieties with real structures. We also describe a correspondence between $\mathbb{S}^1$-torsors and certain pairs consisting of an invertible sheaf and a rational function on their base space which to our knowledge did not appear before in the literature. The second section is devoted to the presentation of smooth real affine surfaces with $\mathbb{S}^1$-action in terms of real DPD-pairs and to the study of their real $\mathbb{S}^1$-orbits. In the third section, we first present different constructions of rational projective and affine real algebraic models of compact differential surfaces with S^1-actions and then proceed to the proof of Theorem 1.

2 Preliminaries

In what follows the term algebraic variety over field k refers to a geometrically integral scheme of finite type over k.

2.1 *Real and Complex Quasi-projective Algebraic Varieties*

Every complex algebraic variety V can be viewed as a scheme over the field $\mathbb{R}$ of real numbers via the composition of its structure morphism $p : V \to \mathrm{Spec}(\mathbb{C})$ with theétale double cover $\mathrm{Spec}(\mathbb{C}) \to \mathrm{Spec}(\mathbb{R})$ induced by the inclusion $\mathbb{R} \hookrightarrow \mathbb{C} = \mathbb{R}[i]/(i^2 + 1)$. The Galois group $\mathrm{Gal}(\mathbb{C}/\mathbb{R}) = \mu_2$ acts on $\mathrm{Spec}(\mathbb{C})$ by the usual complex conjugation $z \mapsto \overline{z}$.

Definition 2 A *real structure* on a complex algebraic variety V consists of an involution of $\mathbb{R}$-scheme σ of V which lifts the complex conjugation, so that we have a commutative diagram

$$\begin{array}{ccc} V & \xrightarrow{\sigma} & V \\ {\scriptstyle p}\downarrow & & \downarrow{\scriptstyle p} \\ \mathrm{Spec}(\mathbb{C}) & \xrightarrow{z \mapsto \overline{z}} & \mathrm{Spec}(\mathbb{C}) \\ & \searrow \quad \swarrow & \\ & \mathrm{Spec}(\mathbb{R}). & \end{array}$$

When $V = \mathrm{Spec}(A)$ is affine, a real structure σ is equivalently determined by its comorphism $\sigma^* : A \to A$, which is an involution of A viewed as an $\mathbb{R}$-algebra.

A *real morphism (resp. real rational map)* between complex algebraic varieties with real structures (V', σ') and (V, σ) is a morphism (resp. a rational map) of complex algebraic varieties $f : V' \to V$ such that $\sigma \circ f = f \circ \sigma'$ as morphisms of $\mathbb{R}$-schemes.

For every real algebraic variety X, the *complexification*

$$X_{\mathbb{C}} = X \times_{\mathbb{R}} \mathbb{C} := X \times_{\mathrm{Spec}(\mathbb{R})} \mathrm{Spec}(\mathbb{C})$$

of X comes equipped with a canonical real structure σ_X given by the action of $\mathrm{Gal}(\mathbb{C}/\mathbb{R})$ by complex conjugation on the second factor. Conversely, if a complex variety $p : V \to \mathrm{Spec}(\mathbb{C})$ is equipped with a real structure σ and covered by σ-invariant affine open subsets—so for instance if V is quasi-projective—, then the quotient $\pi : V \to V/\langle\sigma\rangle$ exists in the category of schemes and the structure morphism $p : V \to \mathrm{Spec}(\mathbb{C})$ descends to a morphism $V/\langle\sigma\rangle \to \mathrm{Spec}(\mathbb{R}) = \mathrm{Spec}(\mathbb{C})/\langle z \mapsto \overline{z}\rangle$

making $V/\langle\sigma\rangle$ into a real algebraic variety X such that $V \cong X_{\mathbb{C}}$. In the case where $V = \mathrm{Spec}(A)$ is affine, the algebraic variety $X = V/\langle\sigma\rangle$ is affine, equal to the spectrum of the ring A^{σ^*} of σ^*-invariant elements of A. This correspondence extends to a well-known equivalence of categories:

Lemma 3 [4] *The functor $X \mapsto (X_{\mathbb{C}}, \sigma_X)$ is an equivalence between the category of quasi-projective real algebraic varieties and the category of pairs (V, σ) consisting of a complex quasi-projective variety V endowed with a real structure σ.*

In what follows, we will switch freely from one point of view to the other, by viewing a quasi-projective real algebraic variety X either as a geometrically integral $\mathbb{R}$-scheme of finite type or as a pair (V, σ) consisting of a quasi-projective complex algebraic variety V endowed with a real structure σ. A *real form* of a given real algebraic variety (V, σ) is a real algebraic variety (V', σ') such that the complex varieties V and V' are isomorphic. A *real closed subscheme* Z of real algebraic variety (V, σ) is a σ-invariant closed subscheme of V, endowed with the induced real structure $\sigma|_Z$.

The set $V(\mathbb{C})$ of complex points of a smooth complex algebraic variety V can be endowed with a natural structure of real smooth manifold locally inherited from that on $\mathbb{A}^n_{\mathbb{C}}(\mathbb{C}) \simeq \mathbb{C}^n \simeq \mathbb{R}^{2n}$ [14, Lemme 1 and Proposition 2]. Every morphism of smooth complex algebraic varieties $f : V' \to V$ induces a differentiable map $f(\mathbb{C}) : V'(\mathbb{C}) \to V(\mathbb{C})$ which is a diffeomorphism when f is an isomorphism. Similarly, a real structure σ on V induces a differentiable involution of $V(\mathbb{C})$, whose set of fixed points $V(\mathbb{C})^{\sigma}$, called the *real locus* of (V, σ), is a smooth differential real manifold M. The real algebraic variety (V, σ) is then said to be an *algebraic model* of this differential manifold.

Definition 4 A *birational diffeomorphism* $\varphi : (V', \sigma') \dashrightarrow (V, \sigma)$ between smooth real algebraic varieties with non empty real loci is a real birational map whose restriction to the real locus $V'(\mathbb{C})^{\sigma'}$ of (V', σ') is a diffeomorphism onto the real locus $V(\mathbb{C})^{\sigma}$ of (V, σ), and which admits a rational inverse of the same type.

Example 5 Let $(Q_{1,\mathbb{C}}, \sigma_{Q_1})$ be the complexification of the smooth affine curve Q_1 in $\mathbb{A}^2_{\mathbb{R}} = \mathrm{Spec}(\mathbb{R}[u, v])$ defined by the equation $u^2 + v^2 = 1$. The stereographic projection from the real point $N = (0, 1)$ of $(Q_{1,\mathbb{C}}, \sigma_{Q_1})$ induces an everywhere defined birational diffeomorphism

$$\pi_N : (Q_{1,\mathbb{C}} \setminus \{N\}, \sigma_{Q_1}|_{Q_{1,\mathbb{C}}\setminus\{N\}}) \to (\mathbb{A}^1_{\mathbb{C}} = \mathrm{Spec}(\mathbb{C}[z]), \sigma_{\mathbb{A}^1_{\mathbb{R}}}), \quad (u, v) \mapsto \frac{u}{1 - v}$$

with image equal to $\mathbb{A}^1_{\mathbb{C}} \setminus \{\pm i\}$. Its inverse is given by

$$z \mapsto (u, v) = \left(\frac{2z}{z^2 + 1}, \frac{z^2 - 1}{z^2 + 1}\right).$$

2.2 Circle Actions as Real Forms of Hyperbolic $\mathbb{G}_m$-Actions

Definition 6 The *circle* $\mathbb{S}^1$ is the nontrivial real form $(\mathbb{G}_{m,\mathbb{C}}, \sigma_0)$ of $(\mathbb{G}_{m,\mathbb{C}}, \sigma_{\mathbb{G}_{m,\mathbb{R}}})$ whose real structure is the composition of the involution $t \mapsto t^{-1}$ of $\mathbb{G}_{m,\mathbb{C}}$ with the complex conjugation. It is a real algebraic group isomorphic to the group

$$SO_2(\mathbb{R}) = \mathrm{Spec}(\mathbb{C}[t^{\pm 1}]^{\sigma_0^*}) \cong \mathrm{Spec}(\mathbb{R}[u, v]/(u^2 + v^2 - 1)),$$

with group law given by $(u, v) \cdot (u', v') = (uu' - vv', uv' + u'v)$.

An action of $\mathbb{S}^1$ on a real algebraic variety (V, σ) is a real action of $(\mathbb{G}_{m,\mathbb{C}}, \sigma_0)$ on (V, σ), that is, an action $\mu : \mathbb{G}_{m,\mathbb{C}} \times V \to V$ of $\mathbb{G}_{m,\mathbb{C}}$ on V for which the following diagram commutes

$$\begin{array}{ccc} \mathbb{G}_{m,\mathbb{C}} \times V & \xrightarrow{\mu} & V \\ \downarrow{\scriptstyle \sigma_0 \times \sigma} & & \downarrow{\scriptstyle \sigma} \\ \mathbb{G}_{m,\mathbb{C}} \times V & \xrightarrow{\mu} & V. \end{array}$$

Let $\pi : (V, \sigma) \to (C, \tau)$ be an affine morphism between real algebraic varieties and let $\mu : \mathbb{G}_{m,\mathbb{C}} \times V \to V$ be an $\mathbb{S}^1$-action on (V, σ) by real (C, τ)-automorphisms. Putting $\mathcal{A} = \pi_* \mathcal{O}_V$, μ is uniquely determined by its associated $\mathcal{O}_C$-algebra co-action homomorphism

$$\mu^* : \mathcal{A} \to \mathcal{A} \otimes_{\mathcal{O}_C} \mathcal{O}_C[t^{\pm 1}].$$

The latter determines a $\mathbb{Z}$-grading of $\mathcal{A}$ by its $\mathcal{O}_C$-submodules

$$\mathcal{A}_m = \left\{ f \in \mathcal{A},\ \mu^* f = f \otimes t^m \right\},\ m \in \mathbb{Z},$$

of semi-invariants germs of sections of weight m. The action μ is said to be *effective* if the set $\{m \in \mathbb{Z},\ \mathcal{A}_m \neq \{0\}\}$ is not contained in a proper subgroup of $\mathbb{Z}$, and *hyperbolic* if there exists $m < 0$ and $m' > 0$ such that $\mathcal{A}_m$ and $\mathcal{A}_{m'}$ are non-zero. The following lemma is an extension in the relative affine setting of [8, Lemma 1.7].

Lemma 7 *Let $\pi : (V, \sigma) \to (C, \tau)$ be an affine morphism between real algebraic varieties and let $\mu \colon \mathbb{G}_{m,\mathbb{C}} \times V \to V$ be an effective $\mathbb{S}^1$-action on (V, σ) by (C, τ)-automorphisms. Let $\mathcal{A} = \bigoplus_{m \in \mathbb{Z}} \mathcal{A}_m$ be the corresponding decomposition of the quasi-coherent $\mathcal{O}_C$-algebra $\mathcal{A} = \pi_* \mathcal{O}_V$ into semi-invariants $\mathcal{O}_C$-submodules. Then the following hold:*

(1) The action μ is hyperbolic and $\sigma^* \mathcal{A}_m = \tau_* \mathcal{A}_{-m}$ for every $m \in \mathbb{Z}$.

(2) The $\mathcal{O}_C$-module $\mathcal{A}_0$ is a quasi-coherent $\mathcal{O}_C$-subalgebra of finite type of $\mathcal{A}$. Furthermore, the restriction $\sigma^* : \mathcal{A}_0 \to \tau_* \mathcal{A}_0$ of $\pi_* \sigma^*$ is the comorphism of a real structure τ_0 on $\mathrm{Spec}_C(\mathcal{A}_0)$ for which the induced morphism $\pi_0 : (\mathrm{Spec}_C(\mathcal{A}_0), \tau_0) \to (C, \tau)$ is a real morphism.

Proof Since $\pi : (V, \sigma) \to (C, \tau)$ is an affine morphism, σ is equivalently determined by its comorphism $\sigma^* : \mathcal{A} \to \tau_* \mathcal{A}$. Since μ is a non trivial action, there exists a nonzero element $m \in \mathbb{Z}$ such that $\mathcal{A}_m \neq \{0\}$. The commutativity of the diagram in Definition 6 implies that for a local section f of $\mathcal{A}_m$,

$$\mu^*(\sigma^*(f)) = (\sigma^* \otimes \sigma_0^*)(f \otimes t^m) = \sigma^*(f) \otimes t^{-m}$$

hence that $\sigma^*(f) \in \tau_*\mathcal{A}_{-m}$. Thus $\sigma^*\mathcal{A}_m \subseteq \tau_*\mathcal{A}_{-m}$, and since $(\tau_*\sigma^*) \circ \sigma^* = \mathrm{id}_{\mathcal{A}}$, it follows that the equality $\sigma^*\mathcal{A}_m = \tau_*\mathcal{A}_{-m}$ holds. This shows that the action μ is hyperbolic and that $\sigma^*\mathcal{A}_0 = \tau_*\mathcal{A}_0$. The fact that $\mathcal{A}_0$ is a quasi-coherent $\mathcal{O}_C$-algebra of finite type is well-known [11, Theorem 1.1], and the fact that $\sigma^*|_{\mathcal{A}_0}$ is the comorphism of a real structure τ_0 on $\mathrm{Spec}_C(\mathcal{A}_0)$ making $\pi_0 : (\mathrm{Spec}_C(\mathcal{A}_0), \tau_0) \to (C, \tau)$ into a real morphism is a straightforward consequence of the definitions. □

Definition 8 In the setting of Lemma 7, the real affine morphism $(V, \sigma) \to (\mathrm{Spec}_C(\mathcal{A}_0), \tau_0)$ is called the real (categorical) quotient morphism of the $\mathbb{S}^1$-action on (V, σ).

2.3 Principal Homogeneous $\mathbb{S}^1$-Bundles

Recal that a $\mathbb{G}_{m,\mathbb{C}}$-*torsor* over a complex algebraic variety C is a C-scheme $\rho : P \to C$ endowed with an action $\mu : \mathbb{G}_{m,\mathbb{C}} \times P \to P$ of $\mathbb{G}_{m,\mathbb{C}}$ by C-scheme automorphisms, such that P is Zariski locally isomorphic over C to $C \times \mathbb{G}_{m,\mathbb{C}}$ on which $\mathbb{G}_{m,\mathbb{C}}$ acts by translations on the second factor.

Definition 9 An $\mathbb{S}^1$-*torsor* (also called a principal homogeneous $\mathbb{S}^1$-bundle) over a real algebraic variety (C, τ) is a real algebraic variety $\rho : (P, \sigma) \to (C, \tau)$ endowed with an $\mathbb{S}^1$-action $\mu : \mathbb{G}_{m,\mathbb{C}} \times P \to P$ for which $\rho : P \to C$ is a $\mathbb{G}_{m,\mathbb{C}}$-torsor.

Recall that isomorphism classes of $\mathbb{G}_{m,\mathbb{C}}$-torsors $\rho : P \to C$ over C are in one-to-one correspondence with elements of the Picard group $\mathrm{Pic}(C) \cong H^1(C, \mathcal{O}_C^*)$ of C. More explicitly, for every such P, there exists an invertible $\mathcal{O}_C$-submodule $\mathcal{L}$ of the sheaf of rational functions $\mathcal{K}_C$ of C and an isomorphism of $\mathbb{Z}$-graded algebras

$$\rho_*\mathcal{O}_P \cong \bigoplus_{m\in\mathbb{Z}} \mathcal{L}^{\otimes m},$$

where for $m < 0$, $\mathcal{L}^{\otimes m}$ denotes the $-m$-th tensor power of the dual $\mathcal{L}^\vee$ of $\mathcal{L}$. Furthermore, two invertible $\mathcal{O}_C$-submodules of $\mathcal{K}_C$ determine isomorphic $\mathbb{G}_{m,\mathbb{C}}$-torsors if and only if they are isomorphic as abstract invertible $\mathcal{O}_C$-modules. For $\mathbb{S}^1$-torsors, we have the following counterpart:

Lemma 10 *For every $\mathbb{S}^1$-torsor $\rho : (P, \sigma) \to (C, \tau)$ there exists a pair $(\mathcal{L}, h)$ consisting of an invertible $\mathcal{O}_C$-submodule $\mathcal{L} \subset \mathcal{K}_C$ and a nonzero real rational function h on (C, τ) such that $\rho_*\mathcal{O}_P = \bigoplus_{m\in\mathbb{Z}} \mathcal{L}^{\otimes m}$ and $\mathcal{L} \otimes \tau^*\mathcal{L} = h^{-1}\mathcal{O}_C$ as $\mathcal{O}_C$-submodules of $\mathcal{K}_C$.*

Furthermore, two such pairs $(\mathcal{L}_1, h_1)$ and $(\mathcal{L}_2, h_2)$ determine $\mathbb{S}^1$-equivariantly isomorphic $\mathbb{S}^1$-torsors if and only if there exists a rational function $f \in \Gamma(C, \mathcal{K}_C^*)$ such that $\mathcal{L}_1^\vee \otimes \mathcal{L}_2 = f^{-1}\mathcal{O}_C$ and $h_2 = (f\tau^* f)h_1$.

Proof Let $\mathcal{A} = \bigoplus_{m\in\mathbb{Z}} \mathcal{A}_m$ be the decomposition of $\mathcal{A} = \rho_*\mathcal{O}_P$ into $\mathcal{O}_C$-submodules of semi-invariants with respect to the action μ and let $\mathcal{L}$ be an invertible $\mathcal{O}_C$-submodule of $\mathcal{K}_C$ for which we have an isomorphism of graded $\mathcal{O}_C$-algebras

$$\Psi : \mathcal{A} = \bigoplus_{m\in\mathbb{Z}} \mathcal{A}_m \xrightarrow{\cong} \bigoplus_{m\in\mathbb{Z}} \mathcal{L}^{\otimes m}.$$

By Lemma 7(1), we have $\sigma^*\mathcal{A}_m = \tau_*\mathcal{A}_{-m}$ for every $m \in \mathbb{Z}$. It follows that for every $m \in \mathbb{Z}$, the composition

$$\varphi_m : \tau_*\Psi \circ \sigma^* \circ \Psi^{-1} : \mathcal{L}^{\otimes m} \to \tau_*\mathcal{L}^{\otimes -m}$$

is an isomorphism of $\mathcal{O}_C$-modules such that $\varphi_0 = \tau^* : \mathcal{O}_C = \mathcal{L}^{\otimes 0} \to \tau_*\mathcal{L}^{\otimes -0} = \tau_*\mathcal{O}_C$ and

$$\varphi_{m+m'} = \varphi_m \otimes \varphi_{m'} : \mathcal{L}^{\otimes(m+m')} = \mathcal{L}^{\otimes m} \otimes \mathcal{L}^{\otimes m'} \to \tau_*\mathcal{L}^{\otimes(-m+-m')} = \tau_*\mathcal{L}^{\otimes(-m)} \otimes \tau_*\mathcal{L}^{\otimes(-m')}$$

for every $m, m' \in \mathbb{Z}$. Furthermore, since $\tau_*\sigma^* \circ \sigma^* = \mathrm{id}_{\mathcal{A}}$ and $\tau_*^2 = \mathrm{id}_{\mathcal{O}_C}$, we have

$$\tau_*\varphi_{-m} \circ \varphi_m = \mathrm{id}_{\mathcal{L}^{\otimes m}} \quad \text{and} \quad \tau_*\varphi_m \circ \varphi_{-m} = \mathrm{id}_{\mathcal{L}^{\otimes(-m)}}$$

for every $m \in \mathbb{Z}$. For $m = 1$ and $m' = -1$, the commutativity of the diagram

$$\begin{array}{ccc} \mathcal{L} \otimes \mathcal{L}^\vee & \xrightarrow{\varphi_1 \otimes \varphi_{-1}} & \tau_*\mathcal{L}^\vee \otimes \tau_*\mathcal{L} \\ \Big\downarrow{\scriptstyle \mathrm{ev}} & & \Big\downarrow{\scriptstyle \tau_*\mathrm{ev}} \\ \mathcal{O}_C & \xrightarrow{\tau^*} & \tau_*\mathcal{O}_C, \end{array}$$

where $\mathrm{ev} : \mathcal{L} \otimes \mathcal{L}^\vee \xrightarrow{\cong} \mathcal{O}_C$ is the canonical homomorphism $f \otimes f' \mapsto f'(f)$, implies that

$$\varphi_{-1} = ({}^t\varphi_1)^{-1} : \mathcal{L}^\vee \to \tau_*\mathcal{L}.$$

It follows that $\varphi_m = \varphi_1^{\otimes m}$ for $m \geq 1$ and that $\varphi_m = ({}^t\varphi_1)^{\otimes(-m)}$ for $m \leq -1$. The collection $(\varphi_m)_{m\in\mathbb{Z}}$ is thus uniquely determined by $\varphi_0 = \tau^*$ and an isomorphism $\varphi_1 : \mathcal{L} \to \tau_*\mathcal{L}^\vee$ satisfying the identity $\tau_*({}^t\varphi_1)^{-1} \circ \varphi_1 = \mathrm{id}_{\mathcal{L}}$, hence equivalently by an isomorphism $\psi = \tau^*\varphi_1 : \tau^*\mathcal{L} \to \mathcal{L}^\vee$ such that $({}^t\psi^{-1}) \circ \tau^*\psi = \mathrm{id}_{\mathcal{L}}$. An isomorphism $\psi : \tau^*\mathcal{L} \xrightarrow{\cong} \mathcal{L}^\vee$ is in turn equivalently determined by an isomorphism $\mathcal{O}_C \xrightarrow{\cong} \mathcal{L} \otimes \tau^*\mathcal{L}$, that is, by a rational function $h \in \Gamma(C, \mathcal{K}_C^*)$ such that $\mathcal{L} \otimes \tau^*\mathcal{L} = h^{-1}\mathcal{O}_C$ as $\mathcal{O}_C$-submodules of $\mathcal{K}_C$. The condition $({}^t\psi^{-1}) \circ \tau^*\psi = \mathrm{id}_{\mathcal{L}}$ then amounts to the property that $h^{-1}(\tau^*h) = 1$, i.e. that h is a real rational function on (C, τ).

Two invertible $\mathcal{O}_C$-submodules $\mathcal{L}_1, \mathcal{L}_2 \subset \mathcal{K}_C$ define equivariantly isomorphic $\mathbb{G}_{m,\mathbb{C}}$-torsors $\rho_1 : P_1 \to C$ and $\rho_2 : P_2 \to C$ if and only if there exists an isomorphism $\alpha : \mathcal{L}_1 \xrightarrow{\cong} \mathcal{L}_2$. When $\mathcal{L}_1$ and $\mathcal{L}_2$ come with respective isomorphisms $\psi_1 : \tau^*\mathcal{L}_1 \to \mathcal{L}_1^\vee$ and $\psi_2 : \tau^*\mathcal{L}_2 \to \mathcal{L}_2^\vee$ corresponding to $\mathbb{S}^1$-actions on (P_1, σ_1) and (P_2, σ_2), the condition that a given isomorphism $\alpha : \mathcal{L}_1 \xrightarrow{\cong} \mathcal{L}_2$ induces an $\mathbb{S}^1$-equivariant isomorphism between (P_1, σ_1) and (P_2, σ_2) is equivalent to the commutativity of the diagram

$$\begin{array}{ccc} \tau^*\mathcal{L}_1 & \xrightarrow{\psi_1} & \mathcal{L}_1^\vee \\ {\scriptstyle \tau^*\alpha}\downarrow & & \downarrow{\scriptstyle {}^t\alpha^{-1}} \\ \tau^*\mathcal{L}_2 & \xrightarrow{\psi_2} & \mathcal{L}_2^\vee . \end{array}$$

The isomorphism α is uniquely determined by a rational function $f \in \Gamma(C, \mathcal{K}_C^*)$ such that $\mathcal{L}_1^\vee \otimes \mathcal{L}_2 = f^{-1}\mathcal{O}_C$ as $\mathcal{O}_C$-submodules of $\mathcal{K}_C$. By definition of h_1 and h_2 as the unique nonzero real rational functions on (C, τ) such that $\mathcal{L}_i \otimes \tau^*\mathcal{L}_i = h_i^{-1}\mathcal{O}_C, i = 1, 2$, the commutativity of the above diagram is equivalent to the equality $h_2 = (f\tau^* f)h_1$. □

Example 11 (See also [8, Proposition 3.1]) By Hilbert's Theorem 90, every $\mathbb{G}_{m,\mathbb{C}}$-torsor over Spec($\mathbb{C}$) is isomorphic to the trivial one, that is, to $\mathbb{G}_{m,\mathbb{C}}$ acting on itself by translations. In contrast, there exists precisely two non-isomorphic $\mathbb{S}^1$-torsors over $\mathrm{Spec}(\mathbb{R}) = (\mathrm{Spec}(\mathbb{C}), \sigma_{\mathrm{Spec}(\mathbb{R})})$:

(1) The trivial one given by $\mathbb{S}^1 = (\mathbb{G}_{m,\mathbb{C}} = \mathrm{Spec}(\mathbb{C}[t^{\pm 1}]), \sigma_0)$ acting on itself by translations. A corresponding pair is $(\mathcal{L}, h) = (\mathbb{C}, 1)$,

(2) A nontrivial one $\hat{\mathbb{S}}^1 = (\mathrm{Spec}(\mathbb{C}[u^{\pm 1}]), \hat{\sigma}_0)$ whose real structure $\hat{\sigma}_0$ is the composition of the involution $u \mapsto -u^{-1}$ with the complex conjugation, endowed with the $\mathbb{S}^1$-action given by $t \cdot u = tu$. A corresponding pair is $(\mathcal{L}, h) = (\mathbb{C}, -1)$.

Note that the real locus of $\mathbb{S}^1$ is isomorphic to the real circle $S^1 = \{x^2 + y^2 = 1\} \subset \mathbb{R}^2$ whereas the real locus of $\hat{\mathbb{S}}^1$ is empty.

3 Circle Actions on Smooth Real Affine Surfaces

In this section, we first review the correspondence between normal real affine surfaces (S, σ) with effective $\mathbb{S}^1$-actions and suitable pairs consisting of a Weil $\mathbb{Q}$-divisor and a rational function on a smooth real affine curves (C, τ), which we call real DPD-pairs. We characterize smooth affine surfaces in terms of properties of their corresponding pairs. We then describe the structure of exceptional orbits of $\mathbb{S}^1$-actions on smooth surfaces (S, σ) in relation to degenerate fibers of their quotient morphisms.

3.1 Real DPD-Presentations of Smooth Real Affine Surfaces with $\mathbb{S}^1$-Actions

Recall that a Weil $\mathbb{Q}$-divisor on a smooth real affine curve $(C = \mathrm{Spec}(A_0), \tau)$ is an element of the abelian group consisting of formal sums

$$D = \sum_{c \in C} D(c)\{c\} \in \mathbb{Q} \otimes_{\mathbb{Z}} \mathrm{Div}(C)$$

such that $D(c) \in \mathbb{Q}$ is equal to zero for all but finitely many points $c \in C$. The *support* of D is the finite set of points $c \in C$ such that $D(c) \neq 0$. The group of Weil $\mathbb{Q}$-divisors is partially ordered by the relation

$$(D \leq D' \Leftrightarrow D(c) \leq D'(c) \quad \forall c \in C).$$

Every nonzero rational function f on C determines a Weil $\mathbb{Q}$-divisor $\mathrm{div}(f) = \sum_{c \in C}(\mathrm{ord}_c f)\{c\}$ with integral coefficients. For every Weil $\mathbb{Q}$-divisor D on C, we denote by $\Gamma(C, \mathcal{O}_C(D))$ the A_0-submodule of the field of fractions $\mathrm{Frac}(A_0)$ of A_0 generated by nonzero rational functions f on C such that $\mathrm{div}(f) + D \geq 0$.

Given an automorphism α of C as a scheme over $\mathbb{R}$ or $\mathbb{C}$, the pull-back of $D = \sum_{c \in C} D(c)\{c\}$ by α is the Weil $\mathbb{Q}$-divisor

$$\alpha^* D = \sum_{c \in C} D(c)\{\alpha^{-1}(c)\} = \sum_{c \in C} D(\alpha(c))\{c\}.$$

Definition 12 A *real DPD-pair* on a smooth real affine curve (C, τ) is a pair (D, h) consisting of a Weil $\mathbb{Q}$-divisor D on C and a nonzero real rational function h on (C, τ) satisfying $D + \tau^* D \leq \mathrm{div}(h)$.

We say that two rational numbers $r_i = p_i/q_i$, $i = 1, 2$, where $\gcd(p_i, q_i) = 1$, form a *regular pair* if $|p_1 q_2 - p_2 q_1| = 1$.

Definition 13 A real DPD-pair (D, h) on a smooth real affine curve (C, τ) is said to be *regular* if for every $c \in C$ such that $D(c) + D(\tau(c)) < \mathrm{ord}_c(h)$ the rational numbers $D(c)$ and $D(\tau(c)) - \mathrm{ord}_c(h)$ form a regular pair.

Given a smooth real affine curve (C, τ), a pair $(\mathcal{L}, h)$ consisting of an invertible $\mathcal{O}_C$-submodule $\mathcal{L} \subset \mathcal{K}_C$ and a real rational function h on (C, τ) such that $\mathcal{L} \otimes \tau^* \mathcal{L} = h^{-1} \mathcal{O}_C$ as $\mathcal{O}_C$-submodules of $\mathcal{K}_C$ determines a Cartier divisor D on C such that $D + \tau^* D = \mathrm{div}(h)$, hence a regular real DPD-pair (D, h) on (C, τ). By Lemma 10, every smooth real affine surface (S, σ) endowed with the structure of an $\mathbb{S}^1$-torsor $\rho : (S, \sigma) \to (C, \tau)$ over (C, τ) is determined by such a regular real DPD-pair (D, h). More generally, we have the following:

Theorem 14 *Every normal real affine surface (S, σ) with an effective $\mathbb{S}^1$-action $\mu : \mathbb{G}_{m,\mathbb{C}} \times S \to S$ is determined by a smooth real affine curve (C, τ) and a real DPD-pair (D, h) on it. Furthermore, the following hold:*

(1) Two DPD-pairs (D_1, h_1) and (D_2, h_2) on the same curve (C, τ) determine $\mathbb{S}^1$-equivariantly isomorphic real affine surfaces if and only if there exists a real automorphism ψ of (C, τ) and a rational function f on C such that

$$\psi^* D_2 = D_1 + \mathrm{div}(f) \quad \textit{and} \quad \psi^* h_2 = (f \tau^* f) h_1.$$

(2) The normal real affine surface (S, σ) determined by a DPD-pair (D, h) is smooth if and only if the pair is regular.

The correspondence between normal real affine surface (S, σ) with effective $\mathbb{S}^1$-actions and real DPD-pairs on smooth real affine curves (C, τ) was established in [8, Proposition 3.2] as a particular case of a general correspondence between $\mathbb{S}^1$-actions on normal real affine varieties and suitable pairs (D, h) on certain normal real semi-projective varieties [8, Corollary 2.16], whose proof uses the formalism of polyhedral divisors due to Altmann and Hausen [1]. Since this correspondence provides an explicit method to determine the data (S, σ) and (C, τ), (D, h) from each others, we will review it in detail using the DPD-formalism of [9] in the next subsections.

Proof of Theorem 14 Assertion (1) follows from Corollary 2.16 in [8]. Note that if (D_2, h_2) is a regular real DPD-pair then for every real automorphism ψ of (C, τ) and every rational function f on C, the real DPD-pair

$$(D_1, h_1) = (\psi^* D_2 - \mathrm{div}(f), (f \tau^* f)^{-1} \psi^* h_2)$$

is regular due to the fact that $\mathrm{div}(f)$ and $\mathrm{div}((f \tau^* f)^{-1} \psi^* h_2)$ are integral Weil divisors on C.

To prove (2), let (S, σ) be the normal real affine surface with $\mathbb{S}^1$-action determined by a real DPD-pair (D, h) on a smooth real affine curve (C, τ) as in Sect. 3.1.1 below. Let $\pi : (S, \sigma) \to (C, \tau)$ be its real quotient morphism and let $D_+ = D$ and $D_- = \tau^* D - \mathrm{div}(h)$. By [9, Theorem 4.15], the singular locus of S is contained in the fibers of the quotient morphism $\pi : S \to C$ over the points $c \in C$ such that

$$D_+(c) + D_-(c) = D_+(c) + D_+(\tau(c)) - \mathrm{ord}_c(h) < 0.$$

Furthermore, for such a point c, S is smooth at every point of $\pi^{-1}(c)$ if and only if the rational numbers $D_+(c)$ and $D_-(c)$ form a regular pair. □

3.1.1 From Real DPD-Pairs to Normal Real Affine Surfaces with Effective $\mathbb{S}^1$-Actions

Given a real DPD-pair (D, h) on a smooth real affine curve $(C = \mathrm{Spec}(A_0), \tau)$, we set $D_+ = D$ and $D_- = \tau^* D_+ - \mathrm{div}(h)$. The condition $D + \tau^* D \leq \mathrm{div}(h)$ implies that $D_+ + D_- \leq 0$, so that for every $m' \leq 0 \leq m$, the product

$$\Gamma(C, \mathcal{O}_C(-m'D_-)) \cdot \Gamma(C, \mathcal{O}_C(mD_+))$$

in $\mathrm{Frac}(A_0)$ is contained either in $\Gamma(C, \mathcal{O}_C(-(m'+m)D_-))$ if $m'+m \leq 0$ or in $\Gamma(C, \mathcal{O}_C((m'+m)D_+))$ if $m'+m \geq 0$. It follows that the graded A_0-module

$$A_0[D_-, D_+] = \bigoplus_{m<0} \Gamma(C, \mathcal{O}_C(-mD_-)) \oplus \Gamma(C, \mathcal{O}_C) \oplus \bigoplus_{m>0} \Gamma(C, \mathcal{O}_C(mD_+))$$

is a graded A_0-algebra for the multiplication law given by component wise multiplication in $\mathrm{Frac}(A_0)$. By [9, Sect. 4.2], $A_0[D_+, D_-]$ is finitely generated over $\mathbb{C}$ and normal. The grading then corresponds to an effective hyperbolic $\mathbb{G}_{m,\mathbb{C}}$-action $\mu : \mathbb{G}_{m,\mathbb{C}} \times S \to S$ on the normal complex affine surface $S = \mathrm{Spec}(A_0[D_-, D_+])$. The ring of invariants for this action is equal to A_0 and the morphism $\pi : S \to C = \mathrm{Spec}(A_0)$ induced by the inclusion $A_0 \subset A_0[D_-, D_+]$ is the categorical quotient morphism. Since $D_- = \tau^* D_+ - \mathrm{div}(h)$, for every $m \geq 1$, the homomorphism

$$\tau_m^* : \Gamma(C, \mathcal{O}_C(mD_+)) \to \Gamma(C, \mathcal{O}_C(mD_-)), \ f \mapsto h^m \tau^* f$$

is an isomorphism with inverse

$$\tau_{-m}^* : \Gamma(C, \mathcal{O}_C(-mD_-)) \to \Gamma(C, \mathcal{O}_C(-mD_+)), \ f \mapsto h^m \tau^* f.$$

Letting $\tau_0 = \tau$, these isomorphisms collect into an automorphism $\sigma^* = \bigoplus_{m\in\mathbb{Z}} \tau_m^*$ of $A_0[D_-, D_+]$ which is the comorphism of a real structure σ on S for which we have, by construction, $\sigma \circ \mu = \mu \circ (\sigma_0 \times \sigma)$. It follows that (S, σ) is a normal real affine surface and that $\mu : \mathbb{G}_{m,\mathbb{C}} \times S \to S$ is an effective $\mathbb{S}^1$-action on (S, σ) in the sense of Definition 6.

Example 15 Let $(C = \mathrm{Spec}(A_0), \tau)$ be a smooth real affine curve with a real point c whose defining ideal is a principal ideal generated by a real regular function h on (C, τ). Let D be the trivial divisor 0. Then (D, h) is a real DPD-pair on (C, τ) for which we have $D_+ = D = 0$ and $D_- = \tau^* D_- - \mathrm{div}(h) = -\{c\}$. It follows that $\Gamma(C, \mathcal{O}_C(mD_+)) = A_0$ and that $\Gamma(C, \mathcal{O}_C(mD_-)) = h^m A_0$ for every $m \geq 0$. The corresponding homomorphism

$$\tau_m^* : \Gamma(C, \mathcal{O}_C(mD_+)) = A_0 \to h^m A_0 = \Gamma(C, \mathcal{O}_C(mD_-))$$

is the multiplication by h^m. The algebra $A_0[D_+, D_-]$ is generated by the homogeneous elements

$$x = 1 \in \Gamma(C, \mathcal{O}_C(D_+)) = \Gamma(C, \mathcal{O}_C) \quad \text{and} \quad y = h \in \Gamma(C, \mathcal{O}_C(D_-)) = \Gamma(C, \mathcal{O}_C(-\{c\}))$$

of degree 1 and -1 respectively. These satisfy the obvious homogeneous relation $xy = h$, and we have

$$A_0[D_+, D_-] \cong A_0[x, y]/(xy - h).$$

The corresponding $\mathbb{G}_{m,\mathbb{C}}$-action μ on $S = \mathrm{Spec}(A_0[D_+, D_-])$ is given by $t \cdot (x, y) = (tx, t^{-1}y)$ and the real structure σ for which μ becomes an $\mathbb{S}^1$-action on (S, σ) is given by $\sigma^* x = y$ and $\sigma^* y = x$.

3.1.2 From Normal Real Affine Surfaces with Effective $\mathbb{S}^1$-Actions to Real DPD Pairs

Given a normal real affine surface (S, σ) with an effective $\mathbb{S}^1$-action $\mu : \mathbb{G}_{m,\mathbb{C}} \times S \to S$, it follows from Lemma 7 that the coordinate ring A of S decomposes as the direct sum $A = \bigoplus_{m\in\mathbb{Z}} A_m$ of semi-invariants sub-spaces such that $\sigma^*(A_m) = A_{-m}$ for every $m \in \mathbb{Z}$. The curve $C = \mathrm{Spec}(A_0)$ is the categorical quotient of the $\mathbb{G}_{m,\mathbb{C}}$-action on S. The restriction σ^* to A_0 induces a real structure τ on C. Let $s \in \mathrm{Frac}(A)$ be any semi-invariant rational function of weight 1 and let $h = s\sigma^* s \in \mathrm{Frac}(A)$. Since $\sigma^* s$ is a semi-invariant rational function of weight -1, h is a σ^*-invariant rational function of weight 0, hence a τ^*-invariant element of $\mathrm{Frac}(A_0)$. For every $m \in \mathbb{Z}$, $s^{-m} A_m$ is a locally free A_0-submodule of $\mathrm{Frac}(A_0)$. By [9, Sect. 4.2], there exists Weil $\mathbb{Q}$-divisors D_+ and D_- on C satisfying $D_+ + D_- \leq 0$ such that for every $m \geq 0$ we have

$$s^{-m} A_m = \Gamma(C, \mathcal{O}_C(mD_+)) \quad \text{and} \quad s^m A_{-m} = \Gamma(C, \mathcal{O}_C(mD_-))$$

as A_0-submodules of $\mathrm{Frac}(A_0)$. Since by Lemma 7 and the definition of h, we have

$$\tau^*(s^{-m} \cdot A_m) = h^{-m}(s^m \cdot A_{-m}) \quad \forall m \in \mathbb{Z},$$

it follows that $D_- = \tau^* D_+ - \mathrm{div}(h)$. So setting $D = D_+$, the pair (D, h) is a real DPD-pair on the smooth real affine curve (C, τ). By construction, $S \cong \mathrm{Spec}(A_0[D_-, D_+])$ and the real structure σ on S coincides with that constructed form (D, h) in the previous subsection.

Example 16 Let $(C = \mathrm{Spec}(R), \tau)$ be a smooth real affine curve with a real point c whose defining ideal is a principal ideal generated by some real regular function f on (C, τ). Let $A = R[x^{\pm 1}, y]/(xy^2 - f)$ and let $S = \mathrm{Spec}(A)$. The morphism $\mu : \mathbb{G}_{m,\mathbb{C}} \times S \to S$, $(t, (x, y)) \mapsto (t^2 x, t^{-1} y)$ defines an $\mathbb{G}_{m,\mathbb{C}}$-action on S by C-automorphisms, which becomes an $\mathbb{S}^1$-action by (C, τ)-automorphisms when S is endowed with the unique real structure σ lifting τ such $\sigma^* x = x^{-1}$ and $\sigma^* y = xy$. The ring of $\mathbb{G}_{m,\mathbb{C}}$-invariant A_0 is equal to $R[xy^2]/(xy^2 - f) \cong R$. Choosing $s = y^{-1}$ as semi-invariant rational function of weight 1, we have $h = y^{-1}\sigma^*(y^{-1}) = x^{-1}y^{-2} = f^{-1} \in \mathrm{Frac}(R)$. The decomposition of A into sub-spaces of semi-invariants functions is then given for every $m \geq 0$ by

$$\begin{aligned}
s^{-m} A_m &= s^{-m} R \cdot (xy)^m = R \cdot (xy^2)^m = f^m R = \Gamma(C, \mathcal{O}_C(mD_+)),\\
s^{2m+1} A_{-2m-1} &= s^{2m+1} R \cdot (x^{-m} y) = R \cdot (xy^2)^{-m} = f^{-m} R = \Gamma(C, \mathcal{O}_C((2m+1)D_-)),\\
s^{2m} A_{-2m} &= s^{2m} R \cdot x^{-m} = R \cdot (x^{-m} y^{-2m}) = f^{-m} R = \Gamma(C, \mathcal{O}_C(2mD_-)).
\end{aligned}$$

It follows that $D_+ = -rc$ for some rational number $r \in]0, 1]$ and that

$$D_- = \tau^*(D_+) - \mathrm{div}(h) = r\{c\} - \mathrm{div}(f^{-1}) = (1-r)\{c\}.$$

Since $f^{-m}R = \Gamma(C, \mathcal{O}_C((2m+1)D_-)) = \Gamma(C, \mathcal{O}_C(2mD_-))$ for every $m \geq 0$, it follows that

$$\frac{m}{2m+1} \leq (1-r) < \frac{m+1}{2m+1} \quad \text{and} \quad \frac{1}{2} \leq (1-r) < \frac{m+1}{2m}$$

for every $m \geq 0$. Thus $(1-r) = \frac{1}{2}$ and a real DPD-pair on (C, τ) corresponding to (S, σ) endowed with the $\mathbb{S}^1$-action μ is $(D, h) = (-\frac{1}{2}\{c\}, f^{-1})$.

3.2 Real Fibers of the Quotient Morphism: Principal and Exceptional Orbits

Let (S, σ) be a smooth real affine surface with an effective $\mathbb{S}^1$-action $\mu : \mathbb{G}_{m,\mathbb{C}} \times S \to S$, and let $\pi : (S, \sigma) \to (C, \tau)$ be its real quotient morphism. Recall that $\pi : S \to C = \mathrm{Spec}(\Gamma(S, \mathcal{O}_S)^{\mathbb{G}_{m,\mathbb{C}}})$ is a good quotient: it is surjective, and each fiber $\pi^{-1}(c)$ over a point $c \in C$ contains a unique closed $\mathbb{G}_{m,\mathbb{C}}$-orbit Z and is the union of all $\mathbb{G}_{m,\mathbb{C}}$-orbits in S containing Z in their closure. In the complex case, [9, Theorem 18] provides a description of the structure of the fiber $\pi^{-1}(c)$ in terms of a pair of Weil $\mathbb{Q}$-divisors D_+ and D_- on C for which $\Gamma(S, \mathcal{O}_S) = A_0[D_-, D_+]$ (see Sect. 3.1.2). In this subsection, we specialize this description to fibers of π over points in the real locus of (C, τ). We begin with the following example which illustrates different possibilities for such fibers.

Example 17 Let $S_\varepsilon \subset \mathbb{A}^3_{\mathbb{C}} = \mathrm{Spec}(\mathbb{C}[x, y, z])$ be the smooth complex affine surface with equation

$$xy = z^2 + \varepsilon, \qquad \text{where } \varepsilon = \pm 1,$$

endowed with the real structure σ defined as the composition of the involution $(x, y, z) \mapsto (y, x, z)$ with the complex conjugation. The effective $\mathbb{G}_{m,\mathbb{C}}$-action μ on S_ε given by $t \cdot (x, y, z) = (tx, t^{-1}y, z)$ defines an $\mathbb{S}^1$-action on (S_ε, σ) whose real quotient morphism coincides with the projection

$$\pi = \mathrm{pr}_z : (S_\varepsilon, \sigma) \to (C, \tau) = (\mathrm{Spec}(\mathbb{C}[z]), \sigma_{\mathbb{A}^1_{\mathbb{R}}}).$$

A corresponding real DPD-pair is for instance $(D, h) = (0, z^2 + \varepsilon)$ where 0 denotes the trivial Weil divisor.

The morphism $(x, y, z) \mapsto (-x, -y, -z)$ defines a fixed point free real action of $\mathbb{Z}_2$ on (S_ε, σ) commuting with the $\mathbb{S}^1$-action. The quotient surface $\overline{S}_\varepsilon = S_\varepsilon/\mathbb{Z}_2$ is smooth and σ descends to a real structure $\overline{\sigma}$ on it. The morphism π descends to a real morphism $\overline{\pi} : (\overline{S}_\varepsilon, \overline{\sigma}) \to (\overline{C}, \overline{\tau}) = (\mathrm{Spec}(\mathbb{C}[z^2]), \sigma_{\mathbb{A}^1_\mathbb{R}})$ which coincides with the real quotient morphism of the induced $\mathbb{S}^1$-action on $\overline{S}_\varepsilon$.

(1) If $\varepsilon = 1$, then since $z^2 + 1 = (z - i)(z + i) = f\tau^* f$, it follows from Theorem 14(2) and Lemma 10 that $\pi : (S_1, \sigma) \to (C, \tau)$ restricts to a trivial $\mathbb{S}^1$-torsor over the principal real affine open subset $(C_h = \mathrm{Spec}(\mathbb{C}[z]_{z^2+1}), \tau|_{C_h})$ of C. In particular, for every real point c of (C, τ), $(\pi^{-1}(c), \sigma|_{\pi^{-1}(c)})$ is isomorphic to $\mathbb{S}^1$ on which $\mathbb{S}^1$-acts by translations.

Since the real point $0 \in (C, \tau)$ is a fixed point of the $\mathbb{Z}_2$-action on C, the fiber of $\overline{\pi} : (\overline{S}_1, \overline{\sigma}) \to (\overline{C}, \overline{\tau})$ over the real point $0 \in (\overline{C}, \overline{\tau})$ has multiplicity two. When endowed with its reduced structure, it is isomorphic to the quotient of $\mathrm{Spec}(\mathbb{C}[x, y]/(xy - 1))$ by the involution $(x, y) \mapsto (-x, -y)$. It is thus isomorphic to $\mathbb{A}^1_* = \mathrm{Spec}(\mathbb{C}[w^{\pm 1}])$, where $w = x^2$, endowed with the real structure given by the composition of the involution $w \mapsto w^{-1}$ with the complex conjugation. The group $\mathbb{G}_{m,\mathbb{C}}$ acts on it by $t \cdot w = t^2 w$ and so, $(\overline{\pi}^{-1}(0)_{\mathrm{red}}, \overline{\sigma}|_{\overline{\pi}^{-1}(0)_{\mathrm{red}}})$ is isomorphic to $\mathbb{S}^1$ on which $\mathbb{S}^1$ acts with stabilizer μ_2.

(2) If $\varepsilon = -1$, then, in contrast with the previous case, there is no rational function $f \in \mathbb{C}(z)$ such that $z^2 - 1 = f\tau^* f$. Consequently, there is no real open subset of (C, τ) over which $\pi : (S_{-1}, \sigma) \to (C, \tau)$ restricts to the trivial $\mathbb{S}^1$-torsor. For a real point c of (C, τ), $h = z^2 - 1$ takes negative value at c if $c \in]-1, 1[$ and positive value if $c \in]-\infty, -1[\cup]1, +\infty[$. The fiber $(\pi^{-1}(c), \sigma|_{\pi^{-1}(c)})$ is thus isomorphic to the nontrivial $\mathbb{S}^1$-torsor $\hat{\mathbb{S}}^1$ of Example 11 in the first case, and to the trivial $\mathbb{S}^1$-torsor $\mathbb{S}^1$ in the second case.

The fiber of π over the point ± 1 is isomorphic to $\mathrm{Spec}(\mathbb{C}[x, y]/(xy))$ and thus consists of two affine lines $\overline{O}^+ = \mathrm{Spec}(\mathbb{C}[x])$ and $\overline{O}^- = \mathrm{Spec}(\mathbb{C}[y])$ exchanged by the real structure σ, intersecting at the real point $p = (0, 0, 0)$ of (S_{-1}, σ). The curves $O^\pm = \overline{O}^\pm \setminus \{p\} \cong \mathbb{A}^1_\mathbb{C} \setminus \{0\}$ endowed with the induced $\mathbb{G}_{m,\mathbb{C}}$-actions are trivial $\mathbb{G}_{m,\mathbb{C}}$-torsors and p is an $\mathbb{S}^1$-fixed point.

As in the previous case, since the real point $0 \in (C, \tau)$ is a fixed point of the $\mathbb{Z}_2$-action on C, the fiber of $\overline{\pi} : (\overline{S}_{-1}, \overline{\sigma}) \to (\overline{C}, \overline{\tau})$ over the real point $0 \in (\overline{C}, \overline{\tau})$ has multiplicity two. When endowed with its reduced structure, it is isomorphic to the quotient of $\mathrm{Spec}(\mathbb{C}[x, y]/(xy + 1))$ by the involution $(x, y) \mapsto (-x, -y)$ hence to $\mathbb{A}^1_* = \mathrm{Spec}(\mathbb{C}[w^{\pm 1}])$, where $w = x^2$. The induced real structure is the composition of the involution $w \mapsto w^{-1}$ with the complex conjugation. The induced $\mathbb{G}_{m,\mathbb{C}}$-action is given by $t \cdot w = t^2 w$. Thus $(\overline{\pi}^{-1}(0)_{\mathrm{red}}, \overline{\sigma}|_{\overline{\pi}^{-1}(0)_{\mathrm{red}}})$ is isomorphic to $\mathbb{S}^1$ on which $\mathbb{S}^1$-acts with stabilizer μ_2.

Lemma 18 *Let (S, σ) be a smooth real affine surface with an effective $\mathbb{S}^1$-action $\mu : \mathbb{G}_{m,\mathbb{C}} \times S \to S$ determined by a regular real DPD-pair (D, h) on a smooth real affine curve (C, τ), and let $\pi : (S, \sigma) \to (C, \tau)$ be the corresponding real quotient morphism. Then for every real point c of (C, τ) there exists a principal real affine open neighborhood $(U, \tau|_U)$ of c and a regular real DPD-pair (D', h') on $(U, \tau|_U)$ with the following properties:*

(1) $D'|_{U \setminus \{c\}}$ is the trivial divisor, $D'(c) \in [0, 1[$ and $h' \in \Gamma(U, \mathcal{O}_U) \cap \Gamma(U \setminus \{c\}, \mathcal{O}^*_{U \setminus \{c\}})$.

(2) The surface $(\pi^{-1}(U), \sigma|_{\pi^{-1}(U)})$ is $\mathbb{S}^1$-equivariantly isomorphic to that determined by the DPD-pair (D', h') on $(U, \tau|_U)$.

In particular, $\pi|_{\pi^{-1}(U\setminus\{c\})} : (\pi^{-1}(U \setminus \{c\}), \sigma|_{\pi^{-1}(U\setminus\{c\})}) \to (U \setminus \{c\}, \tau|_{U\setminus\{c\}})$ is an $\mathbb{S}^1$-torsor.

Proof Recall that by the construction described in Sect. 3.1.1, we have

$$\Gamma(S, \mathcal{O}_S) = A_0[D_+, D] = \bigoplus_{m<0} \Gamma(C, \mathcal{O}_C(-mD_-)) \oplus \Gamma(C, \mathcal{O}_C) \oplus \bigoplus_{m>0} \Gamma(C, \mathcal{O}_C(mD_+))$$

where $A_0 = \Gamma(C, \mathcal{O}_C)$, $D_+ = D$ and $D_- = \tau^* D - \mathrm{div}(h)$. Let $U = C_g$ be a real principal affine open neighborhood of c for some real regular function g on (C, τ), and let $(S|_U, \sigma|_U) = (\pi^{-1}(U), \sigma|_{\pi^{-1}(U)})$ be endowed with the induced $\mathbb{S}^1$-action. The graded coordinate ring of $S|_U$ is isomorphic to the homogeneous localization

$$\Gamma(S, \mathcal{O}_S)_{(g)} \cong \bigoplus_{m<0} \Gamma(U, \mathcal{O}_C(-mD_-)) \oplus \Gamma(U, \mathcal{O}_U) \oplus \bigoplus_{m>0} \Gamma(U, \mathcal{O}_C(mD_+))$$

of $\Gamma(S, \mathcal{O}_S)$ with respect to $g \in \Gamma(C, \mathcal{O}_C)$. It follows that $(S|_U, \sigma|_U)$ is $\mathbb{S}^1$ -equivariantly isomorphic to the real affine surface determined by the real DPD-pair $(D|_U, h|_U)$ on the smooth real affine curve $(U, \tau|_U)$. For a small enough such real affine neighborhood U of c, we have $D(c') = 0$ for every $c' \in U \setminus \{c\}$ and $h \in \Gamma(U \setminus \{c\}, \mathcal{O}^*_{U\setminus\{c\}})$. In particular, $D|_{U\setminus\{c\}}$ is a principal Cartier divisor such that $D|_{U\setminus\{c\}} + \tau|^*_{U\setminus\{c\}} D|_{U\setminus\{c\}} = \mathrm{div}(h|_{U\setminus\{c\}})$, which implies by Lemma 10 that $\pi : (S|_{U\setminus\{c\}}, \sigma|_{U\setminus\{c\}}) \to (U \setminus \{c\}, \tau|_{U\setminus\{c\}})$ is an $\mathbb{S}^1$-torsor. Shrinking U further if necessary, we can ensure in addition that $c = \mathrm{div}(f)$ for some real regular function on $(U, \tau|_U)$. Letting $\delta = \lfloor D(c) \rfloor$ be the round-down of the rational number $D(c)$, it follows from Theorem 14(2) that $(S|_U, \sigma|_U)$ is $\mathbb{S}^1$-equivariantly isomorphic to the surface determined by the regular real DPD-pair

$$(D', h') = (D|_U - \mathrm{div}(f^\delta), (f^{-\delta}\tau^* f^{-\delta}) \cdot h),$$

on $(U, \tau|_U)$. By construction, we have $D' = (D(c) - \delta)\{c\}$ where $D(c) - \delta \in [0, 1[$ and $h' \in \Gamma(U \setminus \{c\}, \mathcal{O}^*_{U\setminus\{c\}})$. Since (D', h') is a real DPD-pair,

$$\mathrm{ord}_c(h') \geq D'(c) + \tau^*(D')(c) = 2D'(c) \geq 0,$$

which implies that $h' \in \Gamma(U, \mathcal{O}_U) \cap \Gamma(U \setminus \{c\}, \mathcal{O}^*_{U\setminus\{c\}})$.

Definition 19 Let (S, σ) be a smooth real affine surface with an effective $\mathbb{S}^1$-action $\mu : \mathbb{G}_{m,\mathbb{C}} \times S \to S$. A $\mathbb{G}_{m,\mathbb{C}}$-orbit Z is called *principal* if Z endowed with the $\mathbb{G}_{m,\mathbb{C}}$-action induced by μ is the trivial $\mathbb{G}_{m,\mathbb{C}}$-torsor. It is called *exceptional* otherwise. If Z is in addition irreducible and σ-invariant, we say that $(Z, \sigma|_Z)$ is a principal $\mathbb{S}^1$-orbit if Z is a principal $\mathbb{G}_{m,\mathbb{C}}$-orbit, and an exceptional $\mathbb{S}^1$-orbit otherwise.

Theorem 20 *Let (S, σ) be a smooth real affine surface with an effective $\mathbb{S}^1$-action $\mu : \mathbb{G}_{m,\mathbb{C}} \times S \to S$ determined by a regular real DPD-pair (D, h) on a smooth real affine curve (C, τ). Let $\pi : (S, \sigma) \to (C, \tau)$ be the corresponding real quotient morphism and let c be a real point of (C, τ). Then exactly one of the following three possibilities occurs:*

(a) $D(c) \in \mathbb{Z}$ and $\mathrm{ord}_c(h) = 2D(c)$. In this case, there exists a real affine open neighborhood $(U, \tau|_U)$ of c such that $\pi|_{\pi^{-1}(U)} : (\pi^{-1}(U), \sigma|_{\pi^{-1}(U)}) \to (U, \tau|_U)$ is an $\mathbb{S}^1$-torsor. The fiber $(\pi^{-1}(c), \sigma|_{\pi^{-1}(c)})$ is an $\mathbb{S}^1$-torsor over $(c, \sigma_{\mathrm{Spec}(\mathbb{R})})$ which is either isomorphic to $\mathbb{S}^1$ if $\pi^{-1}(c)$ contains a real point of (S, σ), or to the nontrivial $\mathbb{S}^1$-torsor $\hat{\mathbb{S}}^1$ of Example 11 otherwise.
(b) $D(c) \in \frac{1}{2}\mathbb{Z} \setminus \mathbb{Z}$ and $\mathrm{ord}_c(h) = 2D(c)$. In this case, $(\pi^{-1}(c), \sigma|_{\pi^{-1}(c)})$ is a multiple fiber of multiplicity 2, whose reduction is an exceptional $\mathbb{S}^1$-orbit, isomorphic to $\mathbb{S}^1$ on which $\mathbb{S}^1$ acts with stabilizer μ_2.
(c) $D(c) \in \mathbb{Z}$ and $\mathrm{ord}_c(h) = 2D(c) + 1$. In this case, the fiber $\pi^{-1}(c)$ is reduced, consisting of the closures of two principal $\mathbb{G}_{m,\mathbb{C}}$ orbits O^+ and O^- exchanged by the real structure σ, whose closures $\overline{O}^{\pm}$ in S are affine lines intersecting transversally at an $\mathbb{S}^1$-fixed real point p of (S, σ).

Furthermore, in cases (b) and (c), for every real open neighborhood $(V, \tau|_V)$ of c, there exists a real affine open neighborhood $(U, \tau|_U)$ of $c \in C$ contained in V such that the restriction

$$\pi|_{\pi^{-1}(U\setminus\{c\})} : (\pi^{-1}(U \setminus \{c\}), \sigma|_{\pi^{-1}(U\setminus\{c\})}) \to (U \setminus \{c\}, \tau|_{U\setminus\{c\}})$$

is a nontrivial $\mathbb{S}^1$-torsor.

Proof By Lemma 18, there exists a real affine open neighborhood $(U, \tau|_U)$ of c such that $c = \mathrm{div}(f)$ for some real regular function f on $(U, \tau|_U)$ and such that $(S|_U, \sigma|_U)$ is $\mathbb{S}^1$-equivariantly isomorphic over $(U, \tau|_U)$ to the real surface determined by the regular real DPD-pair (D', h') such that $D' = D'(c)\{c\}$ where $D'(c) \in [0, 1[$ and $h' \in \Gamma(U, \mathcal{O}_U) \cap \Gamma(U \setminus \{c\}, \mathcal{O}^*_{U\setminus\{c\}})$. Since $D'(c) + \tau^* D'(c) \leq \mathrm{ord}_c(h')$ by definition of a real DPD-pair, this leads to the following dichotomy:

(I) If $2D'(c) = \mathrm{ord}_c(h')$ then since $D'(c) \in [0, 1[$ and $\mathrm{ord}_c(h')$ is an integer, we have either $D'(c) = 0$ and $\mathrm{ord}_c(h') = 0$ or $D'(c) = \frac{1}{2}$ and $\mathrm{ord}_c(h') = 1$. In the first case, D' is the trivial divisor, and so is $\tau^* D'$. Furthermore, since h' does not vanish on U, $\pi|_{\pi^{-1}(U)} : (S|_U, \sigma|_U) = (\pi^{-1}(U), \sigma|_{\pi^{-1}(U)}) \to (U, \tau|_U)$ is an $\mathbb{S}^1$-torsor by Lemma 10. By Theorem 14(2) and Example 11, $(\pi^{-1}(c), \sigma|_{\pi^{-1}(c)})$ is isomorphic either to $\mathbb{S}^1$ if $h_c(c) \in \mathbb{R}_{>0}$, or to $\hat{\mathbb{S}}^1$ otherwise. This yields case (a).

In the second case, we have $D' = \frac{1}{2}\{c\}$ and it follows from [9, Theorem 18(a)] that $\pi^{-1}(c) = 2Z$ where Z is an exceptional $\mathbb{G}_{m,\mathbb{C}}$-orbit isomorphic to a punctured affine line on which $\mathbb{G}_{m,\mathbb{C}}$ acts with stabilizer μ_2. The real curve $(Z, \sigma|_Z)$ endowed with the restriction of μ is thus isomorphic either to $\mathbb{S}^1$ if it contains a real point or to $\hat{\mathbb{S}}^1$ otherwise, on which $\mathbb{S}^1$ acts with stabilizer μ_2. We claim that the case where $(Z, \sigma|_Z)$ is isomorphic to $\hat{\mathbb{S}}^1$ cannot occur. Indeed, let $B = \Gamma(U, \mathcal{O}_U)$ and let $\tilde{B} = B[X]/(X^2 - f)$. The real structure $\tau|_U$ on U lifts in a unique way to a real structure $\tilde{\tau}$ on $\tilde{U} = \mathrm{Spec}(\tilde{B})$ such that $\tilde{\tau}^* X = X$, for which the morphism $\psi : (\tilde{U}, \tilde{\tau}) \to (U, \tau|_U)$ is a real double cover totally ramified over c and étale elsewhere. The normalization of the fiber product $S \times_U \tilde{U}$ is a smooth real affine surface $(\tilde{S}, \tilde{\sigma})$ and the action of the Galois group $\mathbb{Z}_2$ of the cover $\psi : (\tilde{U}, \tilde{\tau}) \to (U, \tau|_U)$ lifts to a free real $\mathbb{Z}_2$-action on $(\tilde{S}, \tilde{\sigma})$ for which we have $(S|_U, \sigma|_U) \cong (\tilde{S}, \tilde{\sigma})/\mathbb{Z}_2$, the quotient morphism $\Psi : (\tilde{S}, \tilde{\sigma}) \to (S|_U, \sigma|_U) \cong (\tilde{S}, \tilde{\sigma})/\mathbb{Z}_2$ being étale. The $\mathbb{S}^1$-action μ on S lifts to an effective $\mathbb{S}^1$-action $\tilde{\mu}$ on $(\tilde{S}, \tilde{\sigma})$, whose real quotient morphism $\tilde{\pi} : (\tilde{S}, \tilde{\sigma}) \to (\tilde{U}, \tilde{\tau})$ is equal to the composition of the normalization morphism $\nu : \tilde{S} \to S \times_U \tilde{U}$ with the projection $\mathrm{pr}_{\tilde{U}}$. Letting $\tilde{c} = \psi^{-1}(c)$, it follows from the construction that $(\tilde{\pi}^{-1}(c), \tilde{\sigma}|_{\tilde{\pi}^{-1}(c)})$ is an $\mathbb{S}^1$-torsor over $(\tilde{c}, \sigma_{\mathrm{Spec}(\mathbb{R})})$. By Example 11, it is isomorphic to $\mathbb{A}^1_{\mathbb{C}} \setminus \{0\} = \mathrm{Spec}(\mathbb{C}[u^{\pm 1}])$ endowed with a real structure given as the composition of the complex conjugation either with the involution $u \mapsto u^{-1}$ or with the involution $u \mapsto -u^{-1}$. Furthermore, the $\mathbb{Z}_2$-action on $\tilde{S}$ restricts to a $\mathbb{G}_{m,\mathbb{C}}$-equivariant free $\mathbb{Z}_2$-action on $\tilde{\pi}^{-1}(\tilde{c}) \cong \mathrm{Spec}(\mathbb{C}[u^{\pm 1}])$ compatible with the real structure $\tilde{\sigma}|_{\tilde{\pi}^{-1}(c)}$. The latter is thus necessarily given by $u \mapsto -u$, and the quotient morphism Ψ restricts on $(\tilde{\pi}^{-1}(\tilde{c}), \tilde{\sigma}|_{\tilde{\pi}^{-1}(c)})$ to an étale double cover $(\tilde{\pi}^{-1}(c), \tilde{\sigma}|_{\tilde{\pi}^{-1}(c)}) \to (Z, \sigma|_Z)$. We conclude that $Z \cong \mathrm{Spec}(\mathbb{C}[w^{\pm 1}])$, where $w = u^2$ and that $\sigma|_Z$ is the real structure given as the composition of the involution $w \mapsto w^{-1}$ with the complex conjugation, which shows that $(Z, \sigma|_Z)$ is isomorphic to $\mathbb{S}^1$. Finally, since $\mathrm{ord}_c(h') = 1$, there cannot exist any rational function on U such that $h' = g\tau^* g$. It follows that for every real affine open neighborhood $(V, \tau|_V)$ of c contained in $(U, \tau|_U)$, the restriction of π over $V \setminus \{c\}$ is a nontrivial $\mathbb{S}^1$-torsor. This yields case (b).

(II) Otherwise, if $2D'(c) - \mathrm{ord}_c(h) < 0$, then since by hypothesis (D, h) whence (D', h') is a regular real DPD-pair, the rational numbers $D'(c)$ and $D'(\tau(c)) - \mathrm{ord}_c(h') = D'(c) - \mathrm{ord}_c(h')$ form a regular pair. Since $D'(c) \in [0, 1[$ and $\mathrm{ord}_c(h') > 2D'(c)$ is an integer, the only possibility is that $D'(c) = 0$ and $\mathrm{ord}_c(h') = 1$, and hence that $\mathrm{ord}_c(h) = 2D(c) + 1$. By [9, Theorem 18(b)], the fiber $\pi^{-1}(c)$ is then reduced consisting of the closure of two principal $\mathbb{G}_{m,\mathbb{C}}$-orbits O^+ and O^- whose closures $\overline{O}^{\pm}$ in S are affine lines intersecting transversally at a $\mathbb{G}_{m,\mathbb{C}}$-fixed point $p \in \pi^{-1}(c)$. The defining ideals of $\overline{O}^+$ and $\overline{O}^-$ in the graded coordinate ring $\Gamma(S|_U, \mathcal{O}_S) \otimes_{\Gamma(U, \mathcal{O}_U)} (\Gamma(U, \mathcal{O}_U)/f)$ of the scheme theoretic fiber $\pi^{-1}(c)$ are the positive and negative part respectively. Since by Lemma 7, σ^* exchanges the positive and negative parts of the grading of $\Gamma(S|_U, \mathcal{O}_S)$, it follows that σ exchanges $\overline{O}^+$ and $\overline{O}^-$, hence that p is a σ-invariant point. As in the previous case, the fact that $\mathrm{ord}_c(h') = 1$ implies that for every real affine open neighborhood $(V, \tau|_V)$ of c contained in $(U, \tau|_U)$, the restriction of π over $V \setminus \{c\}$ is a nontrivial $\mathbb{S}^1$-torsor. This yields case (c). □

Since the only proper algebraic subgroups of $\mathbb{G}_{m,\mathbb{C}}$ are cyclic groups, the exceptional orbits of a $\mathbb{G}_{m,\mathbb{C}}$-action are either $\mathbb{G}_{m,\mathbb{C}}$-fixed points, or closed curves isomorphic to the punctured affine line $\mathbb{A}^1_{\mathbb{C}} \setminus \{0\}$ on which $\mathbb{G}_{m,\mathbb{C}}$ acts with stabilizer isomorphic to a cyclic group μ_m of order $m \geq 2$. While there exists smooth complex affine surfaces S endowed with hyperbolic $\mathbb{G}_{m,\mathbb{C}}$-actions admitting 1-dimensional exceptional orbits with stabilizers μ_m for every $m \geq 2$, for instance $(\mathbb{A}^1_{\mathbb{C}} \setminus \{0\}) \times \mathbb{A}^1_{\mathbb{C}} =$

$\mathrm{Spec}(\mathbb{C}[x^{\pm 1}, y])$ endowed with the $\mathbb{G}_{m,\mathbb{C}}$-action $t \cdot (x, y) = (t^m x, t^{-1} y)$, the possible types of exceptional $\mathbb{S}^1$-orbits on smooth real affine surfaces with effective $\mathbb{S}^1$-actions are much more restricted:

Corollary 21 *The exceptional $\mathbb{S}^1$-orbits on a smooth real affine surface (S, σ) with an effective $\mathbb{S}^1$-action are either real $\mathbb{S}^1$-fixed points or closed curves isomorphic to $\mathbb{S}^1$ on which $\mathbb{S}^1$ acts with stabilizer μ_2.*

Proof Let $\pi : (S, \sigma) \to (C, \tau)$ be the real quotient morphism for the given $\mathbb{S}^1$-action. The image by π of a real exceptional $\mathbb{S}^1$-orbit $(Z, \sigma|_Z)$ is a τ-invariant proper closed subset of C, which is irreducible since Z is irreducible. So $\pi(Z)$ is a real point c of (C, τ). Since Z is an exceptional $\mathbb{G}_{m,\mathbb{C}}$-orbit, the assertion follows from Theorem 20, cases (b) and (c). □

The following proposition records the possible structures of fibers of the real quotient morphism $\pi : (S, \sigma) \to (C, \tau)$ over pairs of non-real complex points q and $\tau(q)$ of C. Its proof, which is similar to that of Theorem 20, is left to the reader.

Proposition 22 *Let (S, σ) be the smooth real affine surface with effective $\mathbb{S}^1$-action $\mu : \mathbb{G}_{m,\mathbb{C}} \times S \to S$ determined by a regular real DPD-pair (D, h) on a smooth real affine curve (C, τ). Let $\pi : (S, \sigma) \to (C, \tau)$ be its real quotient morphism, and let q and $\tau(q)$ be a pair of non-real complex points of C exchanged by the real structure τ. Then exactly one of the following possibilities occurs:*

(1) $D(q) + D(\tau(q)) = \mathrm{ord}_q(h) = \mathrm{ord}_{\tau(q)}(h)$ *and:*
(a) Either $D(q)$ and $D(\tau(q))$ both belong to $\mathbb{Z}$ and then $\pi^{-1}(q)$ and $\pi^{-1}(\tau(q))$ are principal $\mathbb{G}_{m,\mathbb{C}}$-orbits. Furthermore, there exists a real affine open neighborhood $(U, \tau|_U)$ of $q \cup \tau(q)$ such that $\pi|_{\pi^{-1}(U)} : (\pi^{-1}(U), \sigma|_{\pi^{-1}(U)}) \to (U, \tau|_U)$ is an $\mathbb{S}^1$-torsor.
(b) Or $D(q)$ and $D(\tau(q))$ both belong to $\mathbb{Q} \setminus \mathbb{Z}$ and then $\pi^{-1}(q)$ and $\pi^{-1}(\tau(q)) = \sigma(\pi^{-1}(q))$ are 1-dimensional exceptional $\mathbb{G}_{m,\mathbb{C}}$-orbits of multiplicity $m \geq 2$ on which $\mathbb{G}_{m,\mathbb{C}}$ acts with stabilizer μ_m.

(2) $D(q) + D(\tau(q)) < \mathrm{ord}_q(h) = \mathrm{ord}_{\tau(q)}(h)$. *Then $\pi^{-1}(q)_{\mathrm{red}} = \overline{O_q^+} \cup \overline{O_q^-}$, where O_q^+ and O_q^- are 1-dimensional $\mathbb{G}_{m,\mathbb{C}}$-orbits whose closures $\overline{O_q^{\pm}}$ in S are affine lines intersecting transversally at a $\mathbb{G}_{m,\mathbb{C}}$-fixed point p. Furthermore, the fiber $\pi^{-1}(\tau(q))_{\mathrm{red}} = \sigma(\pi^{-1}(q)_{\mathrm{red}})$ is equal to*

$$\pi^{-1}(\tau(q))_{\mathrm{red}} = \overline{O_{\tau(q)}^+} \cup \overline{O_{\tau(q)}^-} = \sigma(\overline{O_q^-}) \cup \sigma(\overline{O_q^+}).$$

Example 23 Let (S_ε, σ), where $\varepsilon = \pm 1$, be the smooth real affine surface with equation $xy = \varepsilon(z^2 + 1)$ in $\mathbb{A}^3_{\mathbb{C}} = \mathrm{Spec}(\mathbb{C}[x, y, z])$ endowed with the real structure given by the composition of the involution $(x, y, z) \mapsto (y, x, z)$ with the complex conjugation. The $\mathbb{G}_{m,\mathbb{C}}$-action μ on S given by $t \cdot (x, y, z) = (tx, t^{-1}y, z)$ defines a real action of $\mathbb{S}^1$ on (S_ε, σ). The categorical quotient for the $\mathbb{G}_{m,\mathbb{C}}$-action is the affine line $C = \mathrm{Spec}(A_0)$, where $A_0 = \mathbb{C}[xy, z]/(xy - \varepsilon(z^2 + 1)) \cong \mathbb{C}[z]$, and the quotient morphism $\pi : S_\varepsilon \to C$ is a real morphism for the real structures σ and

$\tau = \sigma_{\mathbb{A}^1_{\mathbb{R}}}$ on S_ε and C respectively. The decomposition of the coordinate ring A_ε of S_ε into semi-invariant subspaces if given by

$$\begin{aligned} A_\varepsilon = \bigoplus_{m\in\mathbb{Z}} A_{\varepsilon,m} &= \bigoplus_{m<0} A_0 \cdot y^{-m} \oplus A_0 \oplus \bigoplus_{m>0} A_0 \cdot x^m \\ &= \bigoplus_{m<0} A_0 \cdot (xy)^{-m} x^m \oplus A_0 \oplus \bigoplus_{m>0} A_0 \cdot x^m \\ &= \bigoplus_{m<0} A_0 \cdot (\varepsilon(z^2+1))^{-m} s^m \oplus A_0 \oplus \bigoplus_{m>0} A_0 \cdot s^m \end{aligned}$$

where $s = x$. A corresponding real DPD-pair on (C, τ) is thus given by $(D, h) = (0, x\sigma^* x) = (0, \varepsilon(z^2+1))$. Noting that $1 + z^2 = (1+iz)(1-iz) = f\tau^* f$, we deduce from Theorem 14(2) that (S_ε, σ) is also given by the real DPD-pair $(D', h') = (D - \mathrm{div}(f), \varepsilon) = (1 \cdot \{i\}, \varepsilon)$.

It then follows from Lemma 10 and Example 11 that the restriction of $\pi : (S_\varepsilon, \sigma) \to (C, \tau)$ over the real affine open subset $U = C \setminus \{\pm i\}$ is either the trivial $\mathbb{S}^1$-torsor $(U, \tau|_U) \times \mathbb{S}^1$ if $\varepsilon = 1$, or the $\mathbb{S}^1$-torsor $(U, \tau|_U) \times \hat{\mathbb{S}}^1$ if $\varepsilon = -1$. On the other hand, $\pi^{-1}(\{\pm i\}) \cong \mathrm{Spec}(\mathbb{C}[x, y]/(xy))$ is the union of two copies $\overline{O^+_{\pm i}} = \{x = z \mp i = 0\}$ and $\overline{O^-_{\pm i}} = \{y = z \mp i = 0\}$ of the complex affine line intersecting at the point $\{x = y = z \mp i = 0\}$, and since $\sigma^* x = y$, we have $\sigma(\overline{O^\pm_{\pm i}}) = \overline{O^\mp_{\mp i}}$.

4 Rational Real Algebraic Models of Compact Differential Surfaces with Circle Actions

This section is devoted to the proof of Theorem 1. We first construct some explicit rational affine and projective real algebraic models of compact real manifolds of dimension 2 without boundary endowed with effective differentiable S^1-actions. Then we show that each rational quasi-projective real algebraic model of such a manifold is $\mathbb{S}^1$-equivariantly birationally diffeomorphic to one of these explicit affine models.

4.1 Rational Affine Models with Compact Real Loci

It is a classical result (see e.g. [2, I.3.a]) that a compact connected real manifold of dimension 2 without boundary endowed with an effective differentiable S^1-action is equivariantly diffeomorphic to one of the following manifolds: the torus $T = S^1 \times S^1$, the sphere S^2, the projective plane $\mathbb{RP}^2$ and the Klein bottle K. We now describe smooth rational projective and affine algebraic models with $\mathbb{S}^1$-actions of these compact differential surfaces.

4.1.1 Equivariant Rational Models of the Torus

The group S^1 acts on the torus $T = S^1 \times S^1$ by translations on the second factor. All the orbits are principal, and the orbit space is equal to S^1.

A rational projective model of T is the complexification $(\mathbb{P}^1_{\mathbb{C}} \times \mathbb{P}^1_{\mathbb{C}}, \sigma_{\mathbb{P}^1_{\mathbb{R}} \times \mathbb{P}^1_{\mathbb{R}}})$ of $\mathbb{P}^1_{\mathbb{R}} \times \mathbb{P}^1_{\mathbb{R}}$ on which $\mathbb{S}^1$ acts on the second factor via the projective representation induced by the representation $\mathbb{S}^1 \to SO_2(\mathbb{R})$ in Definition 6. The action of $\mathbb{S}^1$ on $(\mathbb{P}^1_{\mathbb{C}} = \mathrm{Proj}(\mathbb{C}[u, v]), \sigma_{\mathbb{P}^1_{\mathbb{R}}})$ induced by the representation $\mathbb{S}^1 \to SO_2(\mathbb{R})$ has a pair of non-real fixed points $[1 : i]$ and $[1 : -i]$ exchanged by the real structure $\sigma_{\mathbb{P}^1_{\mathbb{R}}}$. Their complement is isomorphic to the trivial $\mathbb{S}^1$-torsor. The affine open subset

$$S_1 = (\mathbb{P}^1_{\mathbb{C}} \times \mathbb{P}^1_{\mathbb{C}}, \sigma_{\mathbb{P}^1_{\mathbb{R}} \times \mathbb{P}^1_{\mathbb{R}}}) \setminus \{[1 : \pm i] \times \mathbb{P}^1_{\mathbb{C}} \cup \mathbb{P}^1_{\mathbb{C}} \times [1 : \pm i]\}$$

is $\sigma_{\mathbb{P}^1_{\mathbb{R}} \times \mathbb{P}^1_{\mathbb{R}}}$-invariant and $\mathbb{S}^1$-invariant. Furthermore, letting σ_1 be the restriction of $\sigma_{\mathbb{P}^1_{\mathbb{R}} \times \mathbb{P}^1_{\mathbb{R}}}$ to S_1, the inclusion $(S_1, \sigma_1) \hookrightarrow (\mathbb{P}^1_{\mathbb{C}} \times \mathbb{P}^1_{\mathbb{C}}, \sigma_{\mathbb{P}^1_{\mathbb{R}} \times \mathbb{P}^1_{\mathbb{R}}})$ is an $\mathbb{S}^1$-equivariant birational diffeomorphism. It follows that (S_1, σ_1) is a rational affine model of T, equivariantly isomorphic the product $(Q_{1,\mathbb{C}}, \sigma_{Q_1}) \times \mathbb{S}^1$ of the complexification of the smooth affine quadric curve $Q_1 \subset \mathrm{Spec}(\mathbb{R}[u, v])$ with equation $u^2 + v^2 = 1$ with $\mathbb{S}^1$, on which $\mathbb{S}^1$ acts by translations on the second factor.

The projection

$$\pi_1 = \mathrm{pr}_{Q_{1,\mathbb{C}}} : (S_1, \sigma_1) = (Q_{1,\mathbb{C}}, \sigma_{Q_1}) \times \mathbb{S}^1 \to (C_1, \tau_1) = (Q_{1,\mathbb{C}}, \sigma_{Q_1})$$

is the trivial $\mathbb{S}^1$-torsor. A corresponding real DPD-pair on (C_1, τ_1) is $(D_1, h_1) = (0, 1)$, where 0 denotes the trivial Weil divisor. The image by π_1 of the real locus of (S_1, σ_1) is equal to the real locus S^1 of (C_1, τ_1).

4.1.2 Equivariant Rational Models of the Sphere

The group S^1 acts on the unit sphere S^2 in $\mathbb{R}^3$ by rotations around a fixed axis. All the orbits are principal, except for the two fixed points where the axis meets the sphere, and the orbit space is a closed interval, each of its ends corresponding to a non-principal orbit. A rational projective model is given by the complexification of the smooth quadric

$$Q = \{u^2 + v^2 + z^2 - w^2 = 0\} \subset \mathbb{P}^3_{\mathbb{R}} = \mathrm{Proj}_{\mathbb{R}}(\mathbb{R}[u, v, z, w])$$

endowed with the restriction of the $\mathbb{S}^1$-action on $(\mathbb{P}^3_{\mathbb{C}}, \sigma_{\mathbb{P}^3_{\mathbb{R}}})$ defined by the projective representation induced by the direct sum of the representation $\mathbb{S}^1 \to SO_2(\mathbb{R})$ with the trivial 2-dimensional representation.

The $\mathbb{S}^1$-invariant real hyperplane section $H = \{w = 0\}$ of $(Q_{\mathbb{C}}, \sigma_Q)$ has empty real locus. Its complement is $\mathbb{S}^1$-equivariantly isomorphic to the complexification $(S_2, \sigma_2) = (\mathbb{S}^2_{\mathbb{C}}, \sigma_{\mathbb{S}^2})$ of the smooth affine quadric $\mathbb{S}^2$ in $\mathrm{Spec}(\mathbb{R}[u, v, z])$ defined by

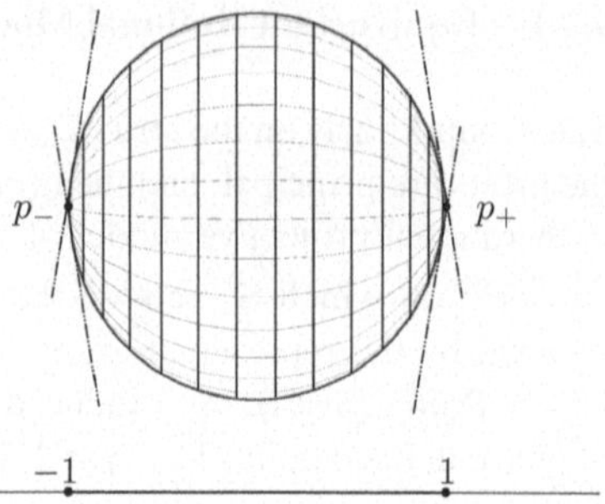

Fig. 1 Projection of the real locus S^2 of (S_2, σ_2) onto the interval $[-1, 1]$. The dashed lines represent the closures $\overline{O}^{\pm}$ of the two principal $\mathbb{G}_{m,\mathbb{C}}$-orbits in $\pi_2^{-1}(\pm 1)$ exchanged by σ_2 and intersecting at the unique $\mathbb{S}^1$-fixed point $p_\pm \subset \pi_2^{-1}(\pm 1)$

the equation $u^2 + v^2 + z^2 = 1$, on which $\mathbb{S}^1$ acts by the restriction of the direct sum of the representation $\mathbb{S}^1 \to SO_2(\mathbb{R})$ with the trivial 1-dimensional representation. By construction, the inclusion $(S_2, \sigma_2) \hookrightarrow (Q_\mathbb{C}, \sigma_Q)$ is an $\mathbb{S}^1$-equivariant birational diffeomorphism.

The real quotient morphism of the $\mathbb{S}^1$-action on (S_2, σ_2) is the projection

$$\pi_2 = \mathrm{pr}_z : (S_2, \sigma_2) \to (C_2, \tau_2) = (\mathbb{A}^1_\mathbb{C} = \mathrm{Spec}(\mathbb{C}[z], \sigma_{\mathbb{A}^1_\mathbb{R}})$$

and a corresponding real DPD-pair on (C_2, τ_2) is $(D_2, h_2) = (0, 1 - z^2)$, where 0 denotes the trivial Weil divisor. The image by π_2 of the real locus of (S_2, σ_2) is the segment $[-1, 1]$ of the real locus $\mathbb{R}$ of $(\mathbb{A}^1_\mathbb{C}, \sigma_{\mathbb{A}^1_\mathbb{R}})$. The restriction of π_2 over the principal real affine open subset $(\mathbb{A}^1_\mathbb{C} \setminus \{\pm 1\}, \sigma_{\mathbb{A}^1_\mathbb{R}}|_{\mathbb{A}^1_\mathbb{C} \setminus \{\pm 1\}})$ is a nontrivial $\mathbb{S}^1$-torsor. The fibers of π_2 over the real points ± 1 of (C_2, τ_2) are of type c) in Theorem 20. Their respective real loci consist of a unique point $p_\pm = (0, 0, \pm 1)$, which is an $\mathbb{S}^1$-fixed point (Fig. 1).

4.1.3 Equivariant Rational Models of the Projective Plane

The group S^1 acts on $\mathbb{RP}^2$, viewed as the projective compactification of $\mathbb{R}^2$ by adding a "line at infinity" $\mathbb{RP}^1 \cong S^1$, by the extension of the linear action of $S^1 = SO(2)$ on $\mathbb{R}^2$ to an action on $\mathbb{RP}^2$ leaving the line at infinity invariant. All the orbits are principal, except for two of them: one is a fixed point corresponding to the origin of $\mathbb{R}^2$ and the other is the line at infinity, equivariantly isomorphic to S^1 on which S^1 act with stabilizer μ_2. An S^1-invariant tubular neighborhood of this second non principal orbit is isomorphic to the quotient $(S^1 \times \mathbb{R})/((z, u) \sim (-z, -u))$, that is, to an open Moebius band B, endowed with the induced S^1-action. The orbit space is a closed interval, each of its ends corresponding to a non-principal orbit.

A rational projective model is the complexification $(\mathbb{P}^2_\mathbb{C} = \mathrm{Proj}(\mathbb{C}[u, v, z]), \sigma_{\mathbb{P}^2_\mathbb{R}})$ of $\mathbb{P}^2_\mathbb{R}$ endowed with the $\mathbb{S}^1$-action defined by the projective representation induced by the direct sum of the representation $\mathbb{S}^1 \to SO_2(\mathbb{R})$ with the trivial 1-dimensional representation. The smooth conic Δ in $\mathbb{P}^2_\mathbb{C}$ with equation $u^2 + v^2 + z^2 = 0$ is $\sigma_{\mathbb{P}^2_\mathbb{R}}$-

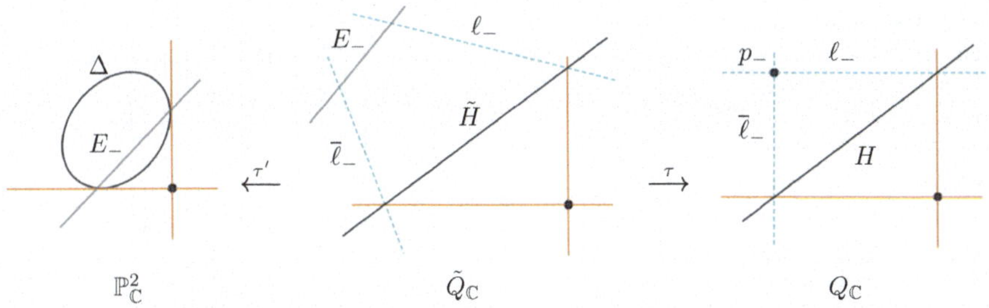

Fig. 2 A real birational maps between $(\mathbb{P}^2_{\mathbb{C}}, \sigma_{\mathbb{P}^2_{\mathbb{R}}})$ and $(Q_{\mathbb{C}}, \sigma_Q)$

invariant and $\mathbb{S}^1$-invariant and has empty real locus. Its complement $S_3 = \mathbb{P}^2_{\mathbb{C}} \setminus \Delta$ endowed with the restriction σ_3 of $\sigma_{\mathbb{P}^2_{\mathbb{R}}}$ is thus an affine model of $\mathbb{RP}^2$. By construction, the inclusion $(S_3, \sigma_3) \hookrightarrow (\mathbb{P}^2_{\mathbb{C}}, \sigma_{\mathbb{P}^2_{\mathbb{R}}})$ is an $\mathbb{S}^1$-equivariant birational diffeomorphism.

Another rational projective model of $\mathbb{RP}^2$ with $\mathbb{S}^1$-action is obtained by blowing-up the smooth quadric

$$Q_{\mathbb{C}} = \{u^2 + v^2 + z^2 - w^2 = 0\} \subset \mathbb{P}^3_{\mathbb{C}},$$

endowed with the real structure σ_Q and the $\mathbb{S}^1$-action defined in Sect. 4.1.2, at the real $\mathbb{S}^1$-fixed point $p_- = [0 : 0 : -1 : 1]$. Letting $\tau : (\tilde{Q}_{\mathbb{C}}, \sigma_{\tilde{Q}}) \to (Q_{\mathbb{C}}, \sigma_Q)$ be the blow-up morphism, with exceptional divisor $E_- \cong (\mathbb{P}^1_{\mathbb{C}}, \sigma_{\mathbb{P}^1_{\mathbb{R}}})$, the $\mathbb{S}^1$-action on $(Q_{\mathbb{C}}, \sigma_Q)$ lifts to an action on $(\tilde{Q}_{\mathbb{C}}, \sigma_{\tilde{Q}})$ for which the real locus of $(\tilde{Q}_{\mathbb{C}}, \sigma_{\tilde{Q}})$ endowed with the induced S^1-action is equivariantly diffeomorphic to the S^1-equivariant connected sum $\mathbb{RP}^2 \sharp_{S^1} S^2 \simeq \mathbb{RP}^2$ endowed with the S^1-action defined above. The proper transforms in $(\tilde{Q}_{\mathbb{C}}, \sigma_{\tilde{Q}})$ of the curves $\ell_- = \{u + iv = z + w = 0\}$ and $\overline{\ell}_- = \{u - iv = z + w = 0\}$ in $(Q_{\mathbb{C}}, \sigma_Q)$ are a pair of non-real disjoint smooth rational curves with self-intersection -1, exchanged by the real structure $\sigma_{\tilde{Q}}$. Their union is an $\mathbb{S}^1$-invariant real closed subset F_- of $(\tilde{Q}_{\mathbb{C}}, \sigma_{\tilde{Q}})$ and the contraction of F_- is an $\mathbb{S}^1$-equivariant birational diffeomorphism $\tau' : (\tilde{Q}_{\mathbb{C}}, \sigma_{\tilde{Q}}) \to (\mathbb{P}^2_{\mathbb{C}}, \sigma_{\mathbb{P}^2_{\mathbb{R}}})$ which maps the proper transform $\tilde{H} \subset \tilde{Q}_{\mathbb{C}}$ of the curve $H = \{w = 0\} \subset Q_{\mathbb{C}}$ onto the curve $\Delta \subset \mathbb{P}^2_{\mathbb{C}}$ (see Fig. 2).

We obtain a diagram

$$(S_3, \sigma_3) \cong (\mathbb{P}^2_{\mathbb{C}} \setminus \Delta, \sigma_{\mathbb{P}^2_{\mathbb{R}}}) \xleftarrow{\tau'} (\tilde{Q}_{\mathbb{C}} \setminus (\tilde{H} \cup F_-), \sigma_{\tilde{Q}}) \xrightarrow{\tau} (Q_{\mathbb{C}} \setminus H, \sigma_Q) \cong (S_2, \sigma_2)$$

in which the left hand side induced morphism τ' is an $\mathbb{S}^1$-equivariant real isomorphism. The right hand side morphism τ realizes $(\tilde{Q}_{\mathbb{C}} \setminus (\tilde{H} \cup F_-), \sigma_{\tilde{Q}})$ as the $\mathbb{S}^1$-equivariant real affine modification of (S_2, σ_2) obtained by blowing-up p_- and

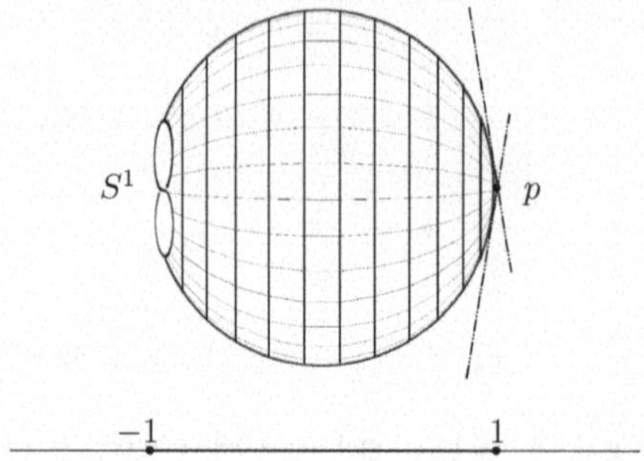

Fig. 3 Projection of the real locus $\mathbb{RP}^2$ of $(S_3, \sigma_3) = (\mathbb{S}^2_{\mathbb{C}}, \sigma_{\mathbb{S}^2})/\mathbb{Z}_2$ onto the interval $[-1, 1]$. The dashed lines represent the closures $\overline{O^{\pm}}$ of the two principal $\mathbb{G}_{m,\mathbb{C}}$-orbits in $\pi_3^{-1}(1)$ exchanged by σ_3 and intersecting at the unique $\mathbb{S}^1$-fixed point $p \subset \pi_3^{-1}(1)$

removing the proper transforms of the closures $\overline{O^{\pm}}$ of the two principal $\mathbb{G}_{m,\mathbb{C}}$-orbits in $\pi_2^{-1}(-1)$ exchanged by σ_2 and intersecting at p_- (see Fig. 1).

A real DPD-presentation of (S_3, σ_3) can be determined as follows. The smooth real quadric $(Q_{\mathbb{C}}, \sigma_Q)$ is isomorphic to the Galois double cover of $(\mathbb{P}^2_{\mathbb{C}}, \sigma_{\mathbb{P}^2_{\mathbb{R}}})$ branched along the real conic Δ. The commutative diagram

$$\begin{array}{ccc} S_2 = Q_{\mathbb{C}} \setminus \{w = 0\} & \longrightarrow & Q_{\mathbb{C}} \\ \downarrow & & \downarrow {\scriptstyle [u:v:z:w] \mapsto [u:v:z]} \\ S_3 = \mathbb{P}^2_{\mathbb{C}} \setminus \Delta & \longrightarrow & \mathbb{P}^2_{\mathbb{C}} \end{array}$$

then identifies S_3 with the quotient of $S_2 \cong \{u^2 + v^2 + z^2 = 1\} \subset \mathbb{A}^3_{\mathbb{C}}$ by the antipodal involution $(u, v, z) \mapsto (-u, -v, -z)$. This involution commutes with the real structure σ_2 on S_2 and the quotient morphism $(S_2, \sigma_2) \to (S_3, \sigma_3) \cong (S_2, \sigma_2)/\mathbb{Z}_2$ is a real morphism, which is equivariant for the $\mathbb{S}^1$-actions on (S_2, σ_2) and (S_3, σ_3). The real quotient morphism $\pi_2 : (S_2, \sigma_2) \to (C_2, \tau_2)$ thus descends to a real morphism

$$\pi_3 : (S_3, \sigma_3) \cong (S_2, \sigma_2)/\mathbb{Z}_2 \to (C_3, \tau_3) = (C_2, \tau_2)/\mathbb{Z}_2 \cong (\mathbb{A}^1_{\mathbb{C}} = \mathrm{Spec}(\mathbb{C}[Z], \sigma_{\mathbb{A}^1_{\mathbb{R}}}),$$

where $Z = 2z^2 - 1$, which is the real quotient morphism of the induced $\mathbb{S}^1$-action on (S_3, σ_3). With this choice of coordinate, a direct calculation shows that a real DPD-pair on (C_3, τ_3) corresponding to (S_3, σ_3) is $(D_3, h_3) = (\frac{1}{2}\{-1\}, 1 - Z^2)$.

The image by π_3 of the real locus $\mathbb{RP}^2$ of (S_3, σ_3) is the segment $[-1, 1]$ of the real locus $\mathbb{R}$ of (C_3, τ_3). The restriction of π_3 over the principal real affine open subset $(\mathbb{A}^1_{\mathbb{C}} \setminus \{\pm 1\}, \sigma_{\mathbb{A}^1_{\mathbb{R}}}|_{\mathbb{A}^1_{\mathbb{C}} \setminus \{\pm 1\}})$ of (C_3, τ_3) is a nontrivial $\mathbb{S}^1$-torsor. The fibers of π_3 over the real points -1 and 1 of (C_3, τ_3) are respectively of type (b) and (c) in Theorem 20. Their real loci consist respectively of a copy of S^1 on which S^1 acts with stabilizer μ_2 and a unique point p, which is a fixed point of the induced S^1-action on the real locus of (S_3, σ_3) (Fig. 3).

4.1.4 Equivariant Rational Models of the Klein Bottle

The Klein bottle K with its S^1-action is the S^1-equivariant connected sum $\mathbb{RP}^2\sharp_{S^1}\mathbb{RP}^2$ of two copies of $\mathbb{RP}^2$ endowed with the S^1-action defined in the previous subsection. Namely, $\mathbb{RP}^2\sharp_{S^1}\mathbb{RP}^2$ is obtained by removing on each copy of $\mathbb{RP}^2$ an S^1-invariant open disc containing the unique S^1-fixed point and gluing together the resulting boundary circles in an S^1-equivariant way. Equivalently, $\mathbb{RP}^2\sharp_{S^1}\mathbb{RP}^2$ is obtained by the S^1-equivariant gluing of two closed Moebius bands

$$\overline{B} = (S^1 \times [-1, 1])/((z, u) \sim (-z, -u))$$

with the S^1-action as in the previous subsection along their boundary circles. The resulting S^1-action on K has two non principal orbits isomorphic to S^1 with stabilizer μ_2, and with S^1-invariant tubular neighborhoods diffeomorphic to open Moebius bands. The orbit space is again a closed interval, each of its ends corresponding to a non-principal orbit.

A projective rational model of K is obtained as follows: let $(\mathbb{P}^2_{\mathbb{C}} = \mathrm{Proj}(\mathbb{C}[u, v, z]), \sigma_{\mathbb{P}^2_{\mathbb{R}}})$ be endowed with the $\mathbb{S}^1$-action defined by the projective representation induced by the direct sum of the representation $\mathbb{S}^1 \to SO_2(\mathbb{R})$ with the trivial 1-dimensional representation. The blow-up of $(\mathbb{P}^2_{\mathbb{C}}, \sigma_{\mathbb{P}^2_{\mathbb{R}}})$ at the real $\mathbb{S}^1$-fixed point $[0 : 0 : 1]$ is the real Hirzebruch surface $(\mathbb{F}_{1,\mathbb{C}} = \mathbb{P}(\mathcal{O}_{\mathbb{P}^1_{\mathbb{R}}} \oplus \mathcal{O}_{\mathbb{P}^1_{\mathbb{R}}}(-1)), \sigma_{\mathbb{F}_1})$ in which the exceptional divisor $E \cong (\mathbb{P}^1_{\mathbb{C}}, \sigma_{\mathbb{P}^1_{\mathbb{R}}})$ is the section with self-intersection -1 of the $\mathbb{P}^1$-bundle structure. The $\mathbb{S}^1$-action on $(\mathbb{P}^2_{\mathbb{C}}, \sigma_{\mathbb{P}^2_{\mathbb{R}}})$ lifts to an action on $(\mathbb{F}_{1,\mathbb{C}}, \sigma_{\mathbb{F}_1})$ and the real locus of $(\mathbb{F}_{1,\mathbb{C}}, \sigma_{\mathbb{F}_1})$ endowed with the induced S^1-action is diffeomorphic to $K \simeq \mathbb{RP}^2\sharp_{S^1}\mathbb{RP}^2$ endowed with the S^1-action defined above.

An affine model of K is in turn obtained from $(\mathbb{F}_{1,\mathbb{C}}, \sigma_{\mathbb{F}_1})$ by removing the union of the proper transform of the conic $\Delta = \left\{u^2 + v^2 + z^2 = 0\right\} \subset \mathbb{P}^2_{\mathbb{C}}$ and the proper transforms of the pair of non-real lines

$$\ell = \{u + iv = 0\} \quad \text{and} \quad \sigma_{\mathbb{P}^2_{\mathbb{R}}}(\ell) = \{u - iv = 0\}$$

of $(\mathbb{P}^2_{\mathbb{C}}, \sigma_{\mathbb{P}^2_{\mathbb{R}}})$ passing through $[0 : 0 : 1]$. Indeed, $\Delta \cup \ell \cup \sigma_{\mathbb{P}^2_{\mathbb{R}}}(\ell)$ is a real $\mathbb{S}^1$-invariant closed subset of $(\mathbb{P}^2_{\mathbb{C}}, \sigma_{\mathbb{P}^2_{\mathbb{R}}})$ with $[0 : 0 : 1]$ as a unique real point, whose proper transform B in $(\mathbb{F}_{1,\mathbb{C}}, \sigma_{\mathbb{F}_1})$ is an ample $\mathbb{S}^1$-invariant curve with empty real locus. So $(S_4, \sigma_4) = (\mathbb{F}_{1,\mathbb{C}} \setminus B, \sigma_{\mathbb{F}_1}|_{\mathbb{F}_{1,\mathbb{C}}\setminus B})$ endowed with the induced $\mathbb{S}^1$-action is a real affine surface whose real locus is diffeomorphic to K. By construction, the inclusion $(S_4, \sigma_4) \hookrightarrow (\mathbb{F}_{1,\mathbb{C}}, \sigma_{\mathbb{F}_1})$ is an $\mathbb{S}^1$-equivariant birational diffeomorphism.

An alternative projective model of K is obtained from the quadric $(Q_{\mathbb{C}}, \sigma_Q)$ of Sect. 4.1.2 by blowing-up the two real $\mathbb{S}^1$-fixed points $p_{\pm} = [0 : 0 : \pm 1 : 1]$ with respective exceptional divisors $E_{\pm} \cong (\mathbb{P}^1_{\mathbb{C}}, \sigma_{\mathbb{P}^1_{\mathbb{R}}})$. Letting $\beta : (\hat{Q}_{\mathbb{C}}, \sigma_{\hat{Q}}) \to (Q_{\mathbb{C}}, \sigma_Q)$ be the real blow-up morphism, the $\mathbb{S}^1$-action on $(Q_{\mathbb{C}}, \sigma_Q)$ lifts to an action on $(\hat{Q}_{\mathbb{C}}, \sigma_{\hat{Q}})$ for which the real locus of $(\hat{Q}_{\mathbb{C}}, \sigma_{\hat{Q}})$ endowed with the induced S^1-action is equivariantly diffeomorphic to the connected sum $\mathbb{RP}^2\sharp_{S^1} S^2\sharp_{S^1}\mathbb{RP}^2 \simeq$

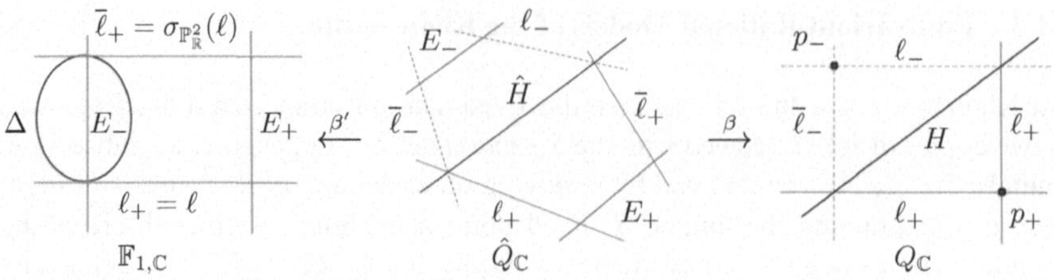

Fig. 4 A real birational maps between $(\mathbb{F}_{1,\mathbb{C}}, \sigma_{\mathbb{F}_1})$ and $(Q_{\mathbb{C}}, \sigma_Q)$

$\mathbb{RP}^2\sharp_{S^1}\mathbb{RP}^2$, hence to the Klein bottle K endowed with the S^1-action defined above. As in the previous subsection, the proper transforms in $(\hat{Q}_{\mathbb{C}}, \sigma_{\hat{Q}})$ of the curves $\ell_- = \{u + iv = z + w = 0\}$ and $\overline{\ell}_- = \{u - iv = z + w = 0\}$ in $(Q_{\mathbb{C}}, \sigma_Q)$ are a pair of non-real disjoint smooth rational curves with self-intersection -1, exchanged by the real structure $\sigma_{\hat{Q}}$. Their union is an $\mathbb{S}^1$-invariant real closed subset F_- of $(\hat{Q}_{\mathbb{C}}, \sigma_{\hat{Q}})$ and the contraction of F_- is an $\mathbb{S}^1$-equivariant birational diffeomorphism $\beta' : (\hat{Q}_{\mathbb{C}}, \sigma_{\hat{Q}}) \to (\mathbb{F}_{1,\mathbb{C}}, \sigma_{\mathbb{F}_1})$ which maps the proper transform $\hat{H} \subset \hat{Q}_{\mathbb{C}}$ of the curve $H = \{w = 0\} \subset Q_{\mathbb{C}}$ onto the proper transform in $(\mathbb{F}_{1,\mathbb{C}}, \sigma_{\mathbb{F}_1})$ of the conic $\Delta = \{u^2 + v^2 + z^2 = 0\} \subset \mathbb{P}^2_{\mathbb{C}}$. The proper transforms in $(\hat{Q}_{\mathbb{C}}, \sigma_{\hat{Q}})$ of the curves $\ell_+ = \{u + iv = z - w = 0\}$ and $\overline{\ell}_+ = \{u - iv = z - w = 0\}$ in $(Q_{\mathbb{C}}, \sigma_Q)$ are also a pair of non-real disjoint smooth rational curves with self-intersection -1, exchanged by the real structure $\sigma_{\hat{Q}}$. Their union is an $\mathbb{S}^1$-invariant real closed subset F_+ of $(\hat{Q}_{\mathbb{C}}, \sigma_{\hat{Q}})$ whose image by $\beta' : (\hat{Q}_{\mathbb{C}}, \sigma_{\hat{Q}}) \to (\mathbb{F}_{1,\mathbb{C}}, \sigma_{\mathbb{F}_1})$ is equal to the union of the proper transforms in $(\mathbb{F}_{1,\mathbb{C}}, \sigma_{\mathbb{F}_1})$ of the lines ℓ and $\sigma_{\mathbb{P}^2_{\mathbb{R}}}(\ell)$ of $(\mathbb{P}^2_{\mathbb{C}}, \sigma_{\mathbb{P}^2_{\mathbb{R}}})$ (Fig. 4).

We obtain a diagram

$$(S_4, \sigma_4) = (\mathbb{F}_{1,\mathbb{C}} \setminus B, \sigma_{\mathbb{F}_1}) \xleftarrow{\beta'} (\hat{Q}_{\mathbb{C}} \setminus (\hat{H} \cup F_- \cup F_+), \sigma_{\hat{Q}}) \xrightarrow{\beta} (Q_{\mathbb{C}} \setminus H, \sigma_Q) \cong (S_2, \sigma_2)$$

in which the left hand side induced morphism β' is an $\mathbb{S}^1$-equivariant real isomorphism. The right hand side morphism β realizes the real affine surface $(\hat{Q}_{\mathbb{C}} \setminus (\hat{H} \cup F_- \cup F_+), \sigma_{\hat{Q}})$ as the $\mathbb{S}^1$-equivariant real affine modification of (S_2, σ_2) obtained by blowing-up p_- and p_+ and removing the proper transforms of the closures $\overline{O^{\pm}_{\pm 1}}$ of the two principal $\mathbb{G}_{m,\mathbb{C}}$-orbits in $\pi_2^{-1}(\pm 1)$ exchanged by σ_2 and intersecting at $p_{\pm}$ (see Fig. 1).

The $\mathbb{S}^1$-equivariant affine modification $\beta : (S_4, \sigma_4) \to (S_2, \sigma_2)$ can be made explicit as follows. Let (Q_2, σ_2') be the smooth surface in $\mathrm{Spec}(\mathbb{C}[x, y, z])$ with equation $xy = 1 - z^2$, endowed with the real structure defined as the composition of the involution $(x, y, z) \mapsto (y, x, z)$ with the complex conjugation. The isomorphism

$$Q_2 \to S_2 \subset \mathrm{Spec}(\mathbb{C}[u, v, z]), \quad (x, y, z) \mapsto \left(\frac{x + y}{2}, \frac{x - y}{2i}, z\right)$$

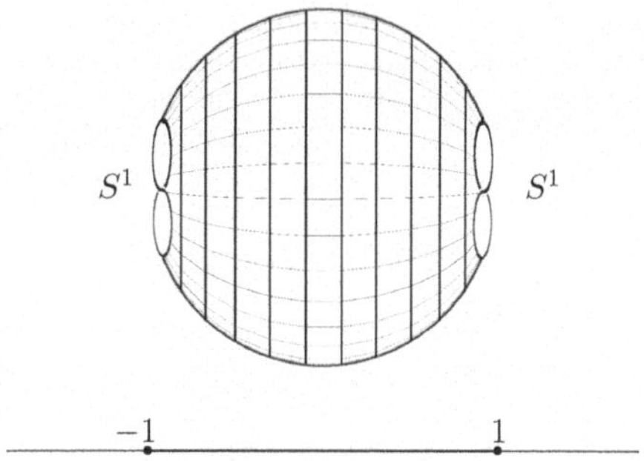

Fig. 5 Projection of the real locus K of (S_4, σ_4) onto the interval $[-1, 1]$

induces an real isomorphism $(Q_2, \sigma_2') \cong (S_2, \sigma_2)$ which is equivariant for the $\mathbb{S}^1$-action on (Q_2, σ_2') given by the hyperbolic $\mathbb{G}_{m,\mathbb{C}}$-action $\mu(t, (x, y, z)) = (tx, t^{-1}y, z)$. Via this isomorphism, the $\mathbb{S}^1$-equivariant real affine modification of (S_2, σ_2) described geometrically above coincides with the affine modification of (Q_2, σ_2') with center at the real $\mathbb{S}^1$-invariant closed subscheme with defining ideal $I = (x, y)^2$ and with real $\mathbb{S}^1$-invariant principal divisor $\mathrm{div}(xy)$. It follows that (S_4, σ_4) is $\mathbb{S}^1$-equivariantly isomorphic to the complex affine surface in $\mathrm{Spec}(\mathbb{C}[x^{\pm 1}, y, z])$ defined by the equation $xy^2 = 1 - z^2$, endowed with the real structure given by the composition of the involution $(x, y, z) \mapsto (x^{-1}, xy, z)$ with the complex conjugation, equipped with the $\mathbb{S}^1$-action given by the hyperbolic $\mathbb{G}_{m,\mathbb{C}}$-action $\mu(t, (x, y, z)) = (t^2x, t^{-1}y, z)$.

We deduce from this description that the real quotient morphism of (S_4, σ_4) is the projection

$$\pi_4 = \mathrm{pr}_z : (S_4, \sigma_4) \to (C_4, \tau_4) = (\mathbb{A}^1_{\mathbb{C}} = \mathrm{Spec}(\mathbb{C}[z], \sigma_{\mathbb{A}^1_{\mathbb{R}}})$$

and that a real DPD-pair on (C_4, τ_4) corresponding to (S_4, σ_4) is $(D_4, h_4) = (\frac{1}{2}\{-1\} + \frac{1}{2}\{1\}, 1 - z^2)$. The image by π_4 of the real locus of (S_4, σ_4) is the segment $[-1, 1]$ of the real locus $\mathbb{R}$ of (C_4, τ_4). The restriction of π_4 over the principal real affine open subset $(\mathbb{A}^1_{\mathbb{C}} \setminus \{\pm 1\}, \sigma_{\mathbb{A}^1_{\mathbb{R}}}|_{\mathbb{A}^1_{\mathbb{C}} \setminus \{\pm 1\}})$ of (C_4, τ_4) is a nontrivial $\mathbb{S}^1$-torsor. The fibers of π_4 over the real points ± 1 of (C_4, τ_4) are of type (b) in Theorem 20. Their real loci consist of a copy of S^1 on which S^1 acts with stabilizer μ_2 (Fig. 5).

4.2 *Uniqueness of Models up to Equivariant Birational Diffeomorphism*

This subsection is devoted to the proof of the following result which implies Theorem 1:

Table 1 Rational affine models of compact surfaces with S^1-actions. The notation $(Q_{1,\mathbb{C}}, \sigma_{Q_1})$ refers to the underlying real algebraic variety of $\mathbb{S}^1$, that is, the complexification of the smooth affine curve in $\mathbb{A}^2_{\mathbb{R}} = \mathrm{Spec}(\mathbb{R}[u, v])$ with equation $u^2 + v^2 = 1$

	$S^1 \times S^1$	S^2	$\mathbb{RP}^2$	K
Rational model	$S_1 = Q_{1,\mathbb{C}} \times \mathbb{S}^1_{\mathbb{C}}$	$S_2 = \mathbb{S}^2_{\mathbb{C}}$	$S_3 = \mathbb{S}^2_{\mathbb{C}}/\mathbb{Z}_2$	$S_4 = \{xy^2 = 1 - z^2\}$
Real categorical quotient	$(Q_{1,\mathbb{C}}, \sigma_{Q_1})$	$(\mathbb{A}^1_{\mathbb{C}}, \sigma_{\mathbb{A}^1_{\mathbb{R}}})$	$(\mathbb{A}^1_{\mathbb{C}}, \sigma_{\mathbb{A}^1_{\mathbb{R}}})$	$(\mathbb{A}^1_{\mathbb{C}}, \sigma_{\mathbb{A}^1_{\mathbb{R}}})$
Image of real locus	S^1	$[-1, 1]$	$[-1, 1]$	$[-1, 1]$
Real DPD-pair (D, h)	$(0, 1)$	$(0, 1 - z^2)$	$(\frac{1}{2}\{-1\}, 1 - z^2)$	$(\frac{1}{2}\{-1\} + \frac{1}{2}\{1\}, 1 - z^2)$

Proposition 24 *Every smooth rational quasi-projective real surface with an effective $\mathbb{S}^1$-action and whose real locus is a compact connected manifold of dimension 2 without boundary is $\mathbb{S}^1$-equivariantly birationally diffeomorphic to one of the affine models constructed in Sect. 4.1, summarized in Table 1.*

The scheme of the proof is the following. In Lemma 25 below, we first establish that every smooth rational quasi-projective model with $\mathbb{S}^1$-action of the torus T, or the sphere S^2, or the plane $\mathbb{RP}^2$ or the Klein bottle K is $\mathbb{S}^1$-equivariantly birationally diffeomorphic to an affine one. Then in Lemma 26, we split the study of the affine case into two subcases according to the nature of the image of the real locus by the real quotient morphism. These subcases are finally studied separately in Sects. 4.2.1 and 4.2.2.

Lemma 25 *Let (X, Σ) be a smooth rational real quasi-projective surface with an effective $\mathbb{S}^1$-action and whose real locus is a compact connected manifold of dimension 2 without boundary. Then (X, Σ) is $\mathbb{S}^1$-equivariantly birationally diffeomorphic to a smooth rational real affine surface (S, σ) with $\mathbb{S}^1$-action.*

Proof By Sumuhiro equivariant completion theorem [15] and equivariant desingularization results for normal surfaces with $\mathbb{G}_{m,\mathbb{C}}$-actions [12], there exists a smooth real projective surface $(\overline{X}, \overline{\Sigma})$ with $\mathbb{S}^1$-action and an $\mathbb{S}^1$-equivariant open embedding $(X, \Sigma) \hookrightarrow (\overline{X}, \overline{\Sigma})$. Since (X, Σ) is rational, so is $(\overline{X}, \overline{\Sigma})$. By a result of Comessatti [5, p. 257], the real locus of $(\overline{X}, \overline{\Sigma})$ is a connected compact smooth surface without boundary, either non-orientable, or diffeomorphic to T or S^2. Since the real locus of (X, Σ) is itself connected and compact, it follows that the real loci of (X, Σ) and $(\overline{X}, \overline{\Sigma})$ coincide, so that $(X, \Sigma) \hookrightarrow (\overline{X}, \overline{\Sigma})$ is a birational diffeomorphism. By [15, Theorem 1.6], there exists a very ample $\mathbb{S}^1$-linearized invertible sheaf $\mathcal{L}$ on $(\overline{X}, \overline{\Sigma})$. This yields in particular a representation of $\mathbb{S}^1$ into the group of linear automorphism of $H^0(\overline{X}, \mathcal{L})$. The underlying representation of $\mathbb{G}_{m,\mathbb{C}}$ splits as a direct sum of $n_1 \geq 1$ non trivial diagonal representations of the form

$$t \cdot (x_i, y_i) = (t^{m_i} x_i, t^{-m_i} y_i), \ m_i \in \mathbb{Z}_{>0}, \quad i = 0, \ldots, n_1 - 1$$

on which the real structure is given by the composition of the involution $(x_i, y_i) \mapsto (y_i, x_i)$ with the complex conjugation, and $n_2 \geq 0$ trivial 1-dimensional representations. The $\mathbb{G}_{m,\mathbb{C}}$-equivariant closed embedding

$$\overline{X} \hookrightarrow \mathbb{P}(H^0(\overline{X}, \mathcal{L})) \cong \mathrm{Proj}(\mathbb{C}[x_0, y_0, \ldots x_{n_1-1}, y_{n_1-1}, z_1, \ldots, z_{n_2}])$$

then becomes a real $\mathbb{S}^1$-equivariant closed embedding for the real structure $\overline{\Sigma}$ on $\overline{X}$ and the real structure on $\mathbb{P}(H^0(\overline{X}, \mathcal{L}))$ defined as the composition of the involution

$$(x_0, y_0, \ldots x_{n_1-1}, y_{n_1-1}, z_1, \ldots, z_{n_2}) \mapsto (y_0, x_0, \ldots y_{n_1-1}, x_{n_1-1}, z_1, \ldots, z_{n_2})$$

with the complex conjugation. The quadric $Q \subset \mathbb{P}(H^0(\overline{X}, \mathcal{L}))$ with equation $\sum_{i=0}^{n_1-1} x_i y_i + \sum_{j=1}^{n_2} z_j^2 = 0$ is a real ample $\mathbb{S}^1$-invariant divisor on $\mathbb{P}(H^0(\overline{X}, \mathcal{L}))$ with empty real locus. Since the real locus of $(\overline{X}, \overline{\Sigma})$ is not empty, $\overline{X}$ is not contained in Q. It follows that $(S, \sigma) = (\overline{X} \setminus Q, \overline{\Sigma}|_{\overline{X}\setminus Q})$ is a smooth rational real affine surface with $\mathbb{S}^1$-action. By construction, the open inclusion $(S, \sigma) \hookrightarrow (\overline{X}, \overline{\Sigma})$ is an $\mathbb{S}^1$-equivariant birational diffeomorphism.

The following lemma divides in turn the study of the affine case into two sub-cases:

Lemma 26 *Let (S, σ) be a smooth rational real affine surface with an effective $\mathbb{S}^1$-action and whose real locus is a connected compact surface without boundary. Let $\pi : (S, \sigma) \to (C, \tau)$ be the real quotient morphism for the $\mathbb{S}^1$-action. Then the following alternative holds:*

(a) (C, τ) is a real affine open subset of $(Q_{1,\mathbb{C}}, \sigma_{Q_1})$ and its real locus is equal to that of $(Q_{1,\mathbb{C}}, \sigma_{Q_1})$.

(b) (C, τ) is a real affine open subset of the real affine line $(\mathbb{A}^1_{\mathbb{C}}, \sigma_{\mathbb{A}^1_{\mathbb{R}}})$ and its real locus is a closed interval.

Proof The curve C is rational because S is rational. Since the real locus of (S, σ) is nonempty, connected and compact, its image by $\pi : (S, \sigma) \to (C, \tau)$ is a nonempty connected compact subset of the real locus of (C, τ). The smooth real projective model of (C, τ) is thus isomorphic to the real projective line $(\mathbb{P}^1_{\mathbb{C}}, \sigma_{\mathbb{P}^1_{\mathbb{R}}})$. If $\mathbb{P}^1_{\mathbb{C}} \setminus C$ contains a real point of $(\mathbb{P}^1_{\mathbb{C}}, \sigma_{\mathbb{P}^1_{\mathbb{R}}})$, then C is isomorphic to a real affine open subset of the real affine line. Being connected and compact, its real locus is then a closed interval. Otherwise, since the inclusion $(C, \tau) \hookrightarrow (\mathbb{P}^1_{\mathbb{C}}, \sigma_{\mathbb{P}^1_{\mathbb{R}}})$ is a real morphism and C is affine, $\mathbb{P}^1_{\mathbb{C}} \setminus C$ is not empty and consists of pairs of non-real points of $\mathbb{P}^1_{\mathbb{C}}$ which are exchanged by the real structure $\sigma_{\mathbb{P}^1_{\mathbb{R}}}$. Since the complement of a pair of such points q and $\sigma_{\mathbb{P}^1}(q)$ is isomorphic to the real affine quadric $(Q_{1,\mathbb{C}}, \sigma_{Q_1})$, it follows that (C, τ) isomorphic to a real affine open subset of $(Q_{1,\mathbb{C}}, \sigma_{Q_1})$, and since the real locus of $\mathbb{P}^1_{\mathbb{C}} \setminus C$ is empty, it follows that the real locus of (C, τ) is equal to that of $(Q_{1,\mathbb{C}}, \sigma_{Q_1})$. □

4.2.1 First Case: (C, τ) is a Real Affine Open Subset of $(Q_{1,\mathbb{C}}, \sigma_{Q_1})$ with Real Locus Equal to S^1

Proposition 27 *Let (S, σ) be a smooth rational real affine surface with an effective $\mathbb{S}^1$-action and whose real locus is a connected compact surface without boundary. Let $\pi : (S, \sigma) \to (C, \tau)$ be the real quotient morphism and assume that (C, τ) is a real affine open subset of $(Q_{1,\mathbb{C}}, \sigma_{Q_1})$ whose real locus is equal to that of $(Q_{1,\mathbb{C}}, \sigma_{Q_1})$. Then (S, σ) is $\mathbb{S}^1$-equivariantly birationally diffeomorphic to $(Q_{1,\mathbb{C}}, \sigma_{Q_1}) \times \mathbb{S}^1$ on which $\mathbb{S}^1$ acts by translations on the second factor.*

Proof Let (D, h) be a regular real DPD-pair on (C, τ) corresponding to (S, σ). Since h is a τ-invariant rational function on C, it is also a σ_{Q_1}-invariant rational function on $Q_{1,\mathbb{C}}$. We claim that by changing h for some rational function of the form $f\tau^* fh$, where f is a rational function on $Q_{1,\mathbb{C}}$, and changing D accordingly by $D + \mathrm{div}(f|_C)$, we can assume that h is the restriction to (C, τ) of a real regular function on $(Q_{1,\mathbb{C}}, \sigma_{Q_1})$ whose zero locus on $Q_{1,\mathbb{C}}$ consists of real points only. Indeed, since h is σ_{Q_1}-invariant, its poles on Q_1 are either real points of $(Q_{1,\mathbb{C}}, \sigma_{Q_1})$ or pairs of non-real points exchanged by σ_{Q_1}. So up to changing h for $f\tau^* fh$ and D for $D + \mathrm{div}(f|_C)$ for a suitable regular function f on $Q_{1,\mathbb{C}}$, we can assume from the very beginning that h is the restriction of a real regular function on $(Q_{1,\mathbb{C}}, \sigma_{Q_1})$. Let $q = (z_1, z_2)$ and $\sigma_{Q_1}(q) = (\overline{z}_1, \overline{z}_2) = \overline{q}$ be a pair of non-real points of $Q_{1,\mathbb{C}}$ at which h vanishes. The restrictions to $Q_{1,\mathbb{C}}$ of the regular functions

$$F_q = (v - z_2) - i(u - z_1) \quad \text{and} \quad F_{\overline{q}} = (v - \overline{z}_2) + i(u - \overline{z}_1)$$

on $\mathbb{A}^2_{\mathbb{C}} = \mathrm{Spec}(\mathbb{C}[u, v])$ are regular functions f_q and $f_{\overline{q}}$ on $Q_{1,\mathbb{C}}$ such that $\mathrm{div}(f_q) = q$ and $\mathrm{div}(f_{\overline{q}}) = \overline{q}$. Furthermore, since $\sigma^*_{Q_1} f_q = f_{\overline{q}}$ it follows that for $\delta = \mathrm{ord}_q(h) = \mathrm{ord}_{\overline{q}}(h)$, $f_q^{-\delta}\sigma^*_{Q_1} f_q^{-\delta} h$ is a real regular function on $(Q_{1,\mathbb{C}}, \sigma_{Q_1})$ which does not vanish at q and $\overline{q}$. The pair

$$(D', h') = (D - \delta \mathrm{div}(f_q|_C), f_q^{-\delta}\sigma^*_{Q_1} f_q^{-\delta} h)$$

is then a regular real DPD-pair on (C, τ) which defines a smooth real affine surface $\mathbb{S}^1$-equivariantly isomorphic to (S, σ) by Theorem 14(2). The desired regular real DPD-pair is then obtained by applying this construction to the finitely many pairs of non-real points of $(Q_{1,\mathbb{C}}, \sigma_{Q_1})$ exchanged by σ_{Q_1} at which h vanishes.

The set of non-real points q of (C, τ) such that either $D(q) \neq 0$ or $D(\tau(q)) \neq 0$ is a finite real subset Z of (C, τ). Its complement $(U, \tau|_U)$ is a real affine open subset of (C, τ) and the restriction $(D|_U, h|_U)$ of (D, h) is a regular real DPD-pair defining a smooth real affine surface $\mathbb{S}^1$-equivariantly isomorphic to $(\pi^{-1}(U), \sigma|_{\pi^{-1}(U)})$. Since Z consists of non-real points of (C, τ) only, the inclusion $(\pi^{-1}(U), \sigma|_{\pi^{-1}(U)}) \hookrightarrow (S, \sigma)$ is an $\mathbb{S}^1$-equivariant birational diffeomorphism. Replacing (C, τ) and (D, h) by $(U, \tau|_U)$ and $(D|_U, h|_U)$, we can therefore assume that $D(q) = \mathrm{ord}_q(h) = 0$ for every non-real point q of C. Now let $\tilde{D}$ be the Weil $\mathbb{Q}$-divisor on $Q_{1,\mathbb{C}}$ defined by $\tilde{D}(c) = D(c)$ if $c \in C$ and $\tilde{D}(c) = 0$ otherwise, and let

$\tilde{h} = h$. Since $(C, \tau) \subset (Q_{1,\mathbb{C}}, \sigma_{Q_1})$ is a real affine open subset with the same real locus as $(Q_{1,\mathbb{C}}, \sigma_{Q_1})$, $Q_{1,\mathbb{C}} \setminus C$ consists of finitely many pairs $\{q, \overline{q} = \sigma_{Q_1}(q)\}$ of non-real points of $Q_{1,\mathbb{C}}$ exchanged by the real structure σ_{Q_1}. For every such pair of points, we have by construction

$$\mathrm{ord}_q(\tilde{h}) = \mathrm{ord}_q(h) = \tilde{D}(q) + \tilde{D}(\sigma_{Q_1}(q)) = \tilde{D}(q) + \sigma^*_{Q_1}(\tilde{D})(q)$$

and similarly for $\overline{q}$. This implies that $(\tilde{D}, \tilde{h})$ is a real DPD-pair on $(Q_{1,\mathbb{C}}, \sigma_{Q_1})$ which is regular since (D, h) is regular. Let $(\tilde{S}, \tilde{\sigma})$ be the corresponding smooth real affine surface with $\mathbb{S}^1$-action and let $\tilde{\pi} : (\tilde{S}, \tilde{\sigma}) \to (Q_{1,\mathbb{C}}, \sigma_{Q_1})$ be its real quotient morphism. We then have a cartesian square of real algebraic varieties

$$\begin{array}{ccc} (S, \sigma) & \xrightarrow{\varphi} & (\tilde{S}, \tilde{\sigma}) \\ {\scriptstyle \pi}\downarrow & & \downarrow{\scriptstyle \tilde{\pi}} \\ (C, \tau) & \longrightarrow & (Q_{1,\mathbb{C}}, \sigma_{Q_1}) \end{array}$$

in which the top horizontal morphism φ is an $\mathbb{S}^1$-equivariant open embedding of (S, σ) as the complement of the fibers of $\tilde{\pi}$ over the points of $Q_{1,\mathbb{C}} \setminus C$. Since $Q_{1,\mathbb{C}} \setminus C$ consists of pairs of non-real points of $Q_{1,\mathbb{C}}$, the real loci of (S, σ) and $(\tilde{S}, \tilde{\sigma})$ coincide, which implies that $\varphi : (S, \sigma) \to (\tilde{S}, \tilde{\sigma})$ is a birational diffeomorphism.

By construction, $(\tilde{D}, \tilde{h})$ is a regular real DPD-pair on $(Q_{1,\mathbb{C}}, \sigma_{Q_1})$ such that the support of $\tilde{D}$ consists of real points of $(Q_{1,\mathbb{C}}, \sigma_{Q_1})$ and such that $\tilde{h}$ is a regular function whose zero locus consists of real points only. Furthermore, the image by $\tilde{\pi} : (\tilde{S}, \tilde{\sigma}) \to (Q_{1,\mathbb{C}}, \sigma_{Q_1})$ of the real locus of $(\tilde{S}, \tilde{\sigma})$ is equal to that of $(Q_{1,\mathbb{C}}, \sigma_{Q_1})$. By Lemma 28 below, $(\tilde{S}, \tilde{\sigma})$ is $\mathbb{S}^1$-equivariantly isomorphic to $(Q_{1,\mathbb{C}}, \sigma_{Q_1}) \times \mathbb{S}^1$ on which $\mathbb{S}^1$ acts by translations on the second factor. This completes the proof. □

In the proof of Proposition 27 above, we use the following auxiliary characterization of $(Q_{1,\mathbb{C}}, \sigma_{Q_1}) \times \mathbb{S}^1$ up to $\mathbb{S}^1$-equivariant real isomorphisms:

Lemma 28 *Let (D, h) be a regular real DPD pair on $(Q_{1,\mathbb{C}}, \sigma_{Q_1})$ such that the support of D consists of real points of $(Q_{1,\mathbb{C}}, \sigma_{Q_1})$ and such that h is a real regular function whose zero locus consists of real points of $(Q_{1,\mathbb{C}}, \sigma_{Q_1})$ only. Let (S, σ) be the corresponding smooth real affine surface with $\mathbb{S}^1$-action and let $\pi : (S, \sigma) \to (Q_{1,\mathbb{C}}, \sigma_{Q_1})$ be its real quotient morphism. Then the following are equivalent:*

(i) The image by π of the real locus of (S, σ) is equal to the real locus of $(Q_{1,\mathbb{C}}, \sigma_{Q_1})$.

(ii) The surface (S, σ) is $\mathbb{S}^1$-equivariantly isomorphic to $(Q_{1,\mathbb{C}}, \sigma_{Q_1}) \times \mathbb{S}^1$ on which $\mathbb{S}^1$ acts by translations on the second factor.

Proof The implication (ii)⇒(i) is clear. We now proceed to the proof of (i)⇒(ii). For every real point $c = (c_1, c_2)$ of $(Q_{1,\mathbb{C}}, \sigma_{Q_1})$, the restrictions to $Q_{1,\mathbb{C}}$ of the regular functions

$$F_c = (v - c_2) - i(u - c_1) \quad \text{and} \quad \overline{F}_c = (v - c_2) + i(u - c_2)$$

on $\mathbb{A}^2_{\mathbb{C}} = \mathrm{Spec}(\mathbb{C}[u, v])$ are regular functions f_c and $\overline{f}_c$ on $Q_{1,\mathbb{C}}$ such that $\mathrm{div}(f_c) = \mathrm{div}(\overline{f}_c) = c$. Furthermore, we have $\sigma^*_{Q_1} f_c = \overline{f}_c$ so that $\mathrm{div}(f_c \sigma^*_{Q_1} f_c) = 2c$. Arguing as in the proof of Lemma 18, we obtain that (S, σ) is $\mathbb{S}^1$-equivariantly isomorphic to a surface determined by a regular real DPD-pair (D', h') on $(Q_{1,\mathbb{C}}, \sigma_{Q_1})$ such that $\mathrm{Supp}(D')$ is contained in the real locus of $(Q_{1,\mathbb{C}}, \sigma_{Q_1})$, h' is regular and vanishes at real points only, and such that for every real point c of $(Q_{1,\mathbb{C}}, \sigma_{Q_1})$ exactly one of the following possibilities occurs

(a) $D'(c) = 0$ and $\mathrm{ord}_c(h') = 0$
(b) $D'(c) = \frac{1}{2}$ and $\mathrm{ord}_c(h') = 1$
(c) $D'(c) = 0$ and $\mathrm{ord}_c(h') = 1$.

Consider the restriction $h'|_{S^1} : S^1 \to \mathbb{R}$ of the real regular function h' to the real locus S^1 of $(Q_{1,\mathbb{C}}, \sigma_{Q_1})$. If c_0 is a real point of $(Q_{1,\mathbb{C}}, \sigma_{Q_1})$ of type (b) or (c) then $h'|_{S^1}$ is a continuous function on S^1 whose sign changes at c_0. It follows that there exists a real point c of $(Q_{1,\mathbb{C}}, \sigma_{Q_1})$ such that $D'(c) = 0$ and $h'(c) < 0$. But then, it follows from the proof of Theorem 20(1) that $(\pi^{-1}(c), \sigma|_{\pi^{-1}(c)})$ is $\mathbb{S}^1$-equivariantly isomorphic to the nontrivial $\mathbb{S}^1$-torsor $\hat{\mathbb{S}}^1$ which has empty real locus. This is impossible since by hypothesis the real locus of (S, σ) surjects onto that of $(Q_{1,\mathbb{C}}, \sigma_{Q_1})$. Thus $D'(c) = \mathrm{ord}_c(h') = 0$ for every real point c of $(Q_{1,\mathbb{C}}, \sigma_{Q_1})$. This implies in turn D' has empty support and that h' is a nowhere vanishing real regular function on $(Q_{1,\mathbb{C}}, \sigma_{Q_1})$. It follows that h' is constant, with positive value $\lambda \in \mathbb{R}^*_+$ at every point since the real locus of (S, σ) surjects onto that of $(Q_{1,\mathbb{C}}, \sigma_{Q_1})$. Writing $\lambda = \alpha\tau^*\alpha = \alpha^2$ for some real number α, we deduce from Theorem 14(2) that the surface (S, σ) is $\mathbb{S}^1$-equivariantly isomorphic to that defined by the real DPD-pair $(D', 1) = (0, 1)$ on $(Q_{1,\mathbb{C}}, \sigma_{Q,1})$. By Sect. 4.1.1, the latter is $\mathbb{S}^1$-equivariantly isomorphic to $(Q_{1,\mathbb{C}}, \sigma_{Q_1}) \times \mathbb{S}^1$ on which $\mathbb{S}^1$ acts by translations on the second factor. □

4.2.2 Second Case: (C, τ) is a Real Affine Open Subset of the Real Affine Line

Proposition 29 *Let (S, σ) be a smooth rational real affine surface with an effective $\mathbb{S}^1$-action and whose real locus is a connected compact surface without boundary. Let $\pi : (S, \sigma) \to (C, \tau)$ be its real quotient morphism and assume that (C, τ) is a real affine open subset of $(\mathbb{A}^1_{\mathbb{C}}, \sigma_{\mathbb{A}^1_{\mathbb{R}}})$. Then (S, σ) is $\mathbb{S}^1$-equivariantly birationally diffeomorphic to one of the affine surfaces (S_2, σ_2), (S_3, σ_3) and (S_4, σ_4) in Table 1.*

Proof Let $\mathbb{A}^1_{\mathbb{C}} = \mathrm{Spec}(\mathbb{C}[z])$ and let (D, h) be a regular real DPD-pair on $(C, \tau) \subseteq (\mathbb{A}^1_{\mathbb{C}}, \sigma_{\mathbb{A}^1_{\mathbb{R}}})$ corresponding to (S, σ). The image by π of the real locus of (S, σ) is a closed interval J of the real locus $\mathbb{R}$ of $(\mathbb{A}^1_{\mathbb{C}}, \sigma_{\mathbb{A}^1_{\mathbb{R}}})$. By Theorem 14(2), (S, σ) is $\mathbb{S}^1$-equivariantly isomorphic to the surface determined by a real DPD-pair of the form

$$(D + \mathrm{div}(f|_C), (f\tau^* f)|_C h)$$

on (C, τ), where f is any element of $\mathbb{C}(z)$. By choosing $f \in \mathbb{C}(z)$ suitably, we can assume from the very beginning that $h \in \mathbb{C}(z)$ is a real polynomial whose zero locus on $\mathbb{A}^1_{\mathbb{C}}$ is contained in J. Then arguing as in the proof of Proposition 27, we get that (S, σ) is $\mathbb{S}^1$-equivariantly birationally diffeomorphic to the smooth real affine surface $(\pi^{-1}(U), \sigma|_{\pi^{-1}(U)})$ determined by the real DPD pair $(D|_U, h|_U)$ on the open complement $(U, \tau|_U)$ in C of the set of non-real points q of (C, τ) such that either $D(q) \neq 0$ or $D(\tau(q)) \neq 0$. Then $(\pi^{-1}(U), \sigma|_{\pi^{-1}(U)})$ is in turn $\mathbb{S}^1$-equivariantly birationally diffeomorphic to the smooth real affine surface $(\tilde{S}, \tilde{\sigma})$ determined by the regular real DPD-pair $(\tilde{D}, \tilde{h}) = (\tilde{D}, h)$ on $(\mathbb{A}^1_{\mathbb{C}}, \sigma_{\mathbb{A}^1_{\mathbb{R}}})$, where $\tilde{D}$ is the Weil $\mathbb{Q}$-divisor defined by $\tilde{D}(c) = D(c)$ if $c \in J$ and $\tilde{D}(c) = 0$ otherwise.

By composing the real quotient morphism $\tilde{\pi} : (\tilde{S}, \tilde{\sigma}) \to (\mathbb{A}^1_{\mathbb{C}}, \sigma_{\mathbb{A}^1_{\mathbb{R}}})$ by a real automorphism of $(\mathbb{A}^1_{\mathbb{C}}, \sigma_{\mathbb{A}^1_{\mathbb{R}}})$, we can assume without loss of generality that J is equal to the interval $[-1, 1]$. For every real point c of J, $f_c = z - c$ is a real regular function f_c on $(\mathbb{A}^1_{\mathbb{C}}, \sigma_{\mathbb{A}^1_{\mathbb{R}}})$ such that $c = \mathrm{div}(f_c)$. Arguing as in the proof of Lemma 18, we obtain that $(\tilde{S}, \tilde{\sigma})$ is $\mathbb{S}^1$-equivariantly isomorphic to the surface $(\hat{S}, \hat{\sigma})$ determined by a real DPD-pair $(\hat{D}, \hat{h})$ on $(\mathbb{A}^1_{\mathbb{C}}, \sigma_{\mathbb{A}^1_{\mathbb{R}}})$ such that $\mathrm{Supp}(\hat{D})$ is contained in J, $\hat{h}$ is a regular function whose zero locus is contained in J and such that that for every $c \in J$ exactly one of the following possibilities occurs

(a) $\hat{D}(c) = 0$ and $\mathrm{ord}_c(\hat{h}) = 0$
(b) $\hat{D}(c) = \frac{1}{2}$ and $\mathrm{ord}_c(\hat{h}) = 1$
(c) $\hat{D}(c) = 0$ and $\mathrm{ord}_c(\hat{h}) = 1$.

By composing the real quotient morphism $\hat{\pi} : (\hat{S}, \hat{\sigma}) \to (\mathbb{A}^1_{\mathbb{C}}, \sigma_{\mathbb{A}^1_{\mathbb{R}}})$ by the real automorphism $z \mapsto -z$ of $(\mathbb{A}^1_{\mathbb{C}}, \sigma_{\mathbb{A}^1_{\mathbb{R}}})$, we can further assume without loss of generality that $\hat{D}(-1) \geq \hat{D}(1)$. By Theorem 20, for a real point c of $(\mathbb{A}^1_{\mathbb{C}}, \sigma_{\mathbb{A}^1_{\mathbb{R}}})$, the real locus of $(\hat{\pi}^{-1}(c), \hat{\sigma}|_{\hat{\pi}^{-1}(c)})$ is empty if and only if $\hat{D}(c) = \mathrm{ord}_c(\hat{h}) = 0$ and $\hat{h}(c) < 0$. It follows that $\hat{D}(c) = \mathrm{ord}_c(\hat{h}) = 0$ and $\hat{h}(c) < 0$ for every real point c of $(\mathbb{A}^1_{\mathbb{C}}, \sigma_{\mathbb{A}^1_{\mathbb{R}}})$ outside of J. On the other hand, the real locus of $(\hat{\pi}^{-1}(c), \hat{\sigma}|_{\hat{\pi}^{-1}(c)})$ being nonempty for every real point $c \in J$ by assumption, we have $\hat{h}(c) \geq 0$ for every $c \in J$. It follows that $\hat{h} \in \mathbb{R}[z] \subset \mathbb{C}[z]$ is a nonzero real polynomial with only simple real roots, whose restriction to the real locus $\mathbb{R}$ of $(\mathbb{A}^1_{\mathbb{C}}, \sigma_{\mathbb{A}^1_{\mathbb{R}}})$ is negative outside J and non-negative on J. This implies that $\hat{h} = \lambda(1 - z^2)$ for some $\lambda \in \mathbb{R}^*_+$ which can be further chosen equal to 1 by Theorem 14(2). It follows in turn that $\hat{D}(c) = 0$ for every real point c of $(\mathbb{A}^1_{\mathbb{C}}, \sigma_{\mathbb{A}^1_{\mathbb{R}}})$ other than -1 and 1, and since $\mathrm{ord}_{\pm 1}\hat{h} = 1$ and $\hat{D}(-1) \geq \hat{D}(1)$, the only remaining possibilities are the following:

(i) $\hat{D}(-1) = \hat{D}(1) = 0$
(ii) $\hat{D}(-1) = \frac{1}{2}$ and $\hat{D}(1) = 0$.
(iii) $\hat{D}(-1) = \hat{D}(1) = \frac{1}{2}$

These pairs $(\hat{D}, \hat{h})$ correspond respectively to the model (S_2, σ_2) of S^2, (S_3, σ_3) of $\mathbb{RP}^2$ and (S_4, σ_4) of K in Table 1. This completes the proof. □

References

1. Altmann, K., Hausen, J.: Polyhedral divisors and algebraic torus actions. Math. Ann. **334**, 557–607 (2006)
2. Audin, M.: Torus Actions on Symplectic Manifolds, Second revised edition, Progress in Mathematics, vol. 93. Birkhäuser Verlag, Basel (2004)
3. Biswas, I., Huisman, J.: Rational real algebraic models of topological surfaces. Doc. Math. **12**, 549–567 (2007)
4. Borel, A., Serre, J.-P.: Théorèmes de finitude en cohomologie galoisienne. Comment. Math. Helv. **39**, 111–164 (1964)
5. Comessatti, A.: Sulla connessione delle superfizie razionali reali. Annali di Math. **23**(3), 215-283 (1914)
6. Demazure, M.: Anneaux gradués normaux, Introduction à la théorie des singularités, II, 35–68, Travaux en Cours 37. Hermann, Paris (1988)
7. Dolgachev, I.V.: Automorphic forms and quasihomogeneous singularities. Func. Anal. Appl. **9**, 149–151 (1975)
8. Dubouloz, A., Liendo, A.: Normal Real Affine Varieties with Circle Actions. arXiv:1810.11712
9. Flenner, H., Zaidenberg, M.: Normal affine surfaces with $\mathbb{C}^*$-actions. Osaka J. Math. **40**(4), 981–1009 (2003)
10. Langlois, K.: Polyhedral divisors and torus actions of complexity one over arbitrary fields. J. Pure Appl. Algebra **219**, 2015–2045 (2015) and Erratum on "Polyhedral divisors and torus actions of complexity one over arbitrary fields" (in preparation)
11. Mumford, D., Fogarty, J., Kirwan, F.: Geometric invariant theory. Ergebnisse der Mathematik und ihrer Grenzgebiete (2), 34, 3rd edn. Springer, Berlin, New York (1994)
12. Orlik, P., Wagreich, P.: Isolated Singularities of Algebraic Surfaces with $\mathbb{C}^*$-*Action*, Annals of Mathematics Second Series, vol. 93, No. 2, pp. 205–228 (1971)
13. Pinkham, H.: Normal surface singularities with $\mathbb{C}^*$-action. Math. Ann. **227**, 183–193 (1977)
14. Serre, J.-P.: Géométrie algébrique et géométrie analytique. Ann. Inst. Fourier, Grenoble **6**, 1–42 (1955)
15. Sumihiro, H.: Equivariant completion II. J. Math. Kyoto Univ. **15**(3), 573–605 (1975)

The Super-Rank of a Locally Nilpotent Derivation of a Polynomial Ring

Gene Freudenburg

Abstract The super-rank of a k-derivation of a polynomial ring $k^{[n]}$ over a field k of characteristic zero is introduced. Like rank, super-rank is invariant under conjugation, and thus gives a way to classify derivations of maximal rank n. For each $m \geq 2$, we construct a locally nilpotent derivation of $k^{[m(m+1)]}$ with maximal super-rank $m(m+1)$.

Keywords Locally nilpotent derivation · Additive group action · Determinantal variety

2010 Mathematics Subject Classification 13A50 · 13N15 · 14M12 · 14R20

1 Introduction

For a field k of characteristic zero and positive integer n, let $B = k^{[n]}$ be the polynomial ring in n variables over k, and let $\Gamma(B)$ denote the set of coordinate systems of B, i.e., the set of $\gamma = (\gamma_1, \ldots, \gamma_n) \in B^n$ such that $B = k[\gamma_1, \ldots, \gamma_n]$. Let $\mathrm{Der}_k(B)$ denote the set of k-derivations $D : B \to B$, noting that $\mathrm{Der}_k(B)$ forms a Lie algebra over k. Given $D \in \mathrm{Der}_k(B)$ and $\gamma \in \Gamma(B)$, we say that D is **triangular** for γ if D respects the filtration of B by subalgebras given by

$$k[\gamma_1] \subset k[\gamma_1, \gamma_2] \subset \cdots \subset k[\gamma_1, \ldots, \gamma_n]$$

and we say that D is **linear** for γ if D restricts to the vector space $V = k\gamma_1 \oplus \cdots \oplus k\gamma_n$. For fixed $\gamma \in \Gamma(B)$, the triangular derivations and linear derivations each form a large and important Lie subalgebra of $\mathrm{Der}_k(B)$.

A linear k-derivation, when viewed as a linear operator on V, has a well-defined rank. This notion was generalized in [2]: Given $D \in \mathrm{Der}_k(B)$, the **nullity** of D is the

G. Freudenburg (✉)
Department of Mathematics, Western Michigan University, Kalamazoo, MI 49008, USA
e-mail: gene.freudenburg@wmich.edu

S. Kuroda et al. (eds.), *Polynomial Rings and Affine Algebraic Geometry*, Springer Proceedings in Mathematics & Statistics 319,
https://doi.org/10.1007/978-3-030-42136-6_5

largest integer $d, 0 \le d \le n$, such that, for some $\gamma \in \Gamma(B)$, the subring $k[\gamma_1, \dots, \gamma_d]$ is contained in the kernel of D. The **rank** of D equals $n - \text{nullity}(D)$. In this note, we introduce the following.

Definition 1.1 Let $B = k^{[n]}$. Given nonzero $D \in \text{Der}_k(B)$, the **super-rank** of D equals the least integer $d, 1 \le d \le n$, such that D restricts to $k[\gamma_1, \dots, \gamma_d]$ for some $\gamma \in \Gamma(B)$.

Note that both the rank and super-rank of a k-derivation of B are invariant under conjugation by automorphisms, and thus provide a means to classify derivations. Note also that super-rank does not distinguish between elements of submaximal rank: Given nonzero $D \in \text{Der}_k(k^{[n]})$, if $\text{rank}(D) < n$, then the super-rank of D equals one. Conversely, super-rank$(D) = n$ implies $\text{rank}(D) = n$. The super-rank is thus a tool intended to distinguish between derivations of maximal rank.

We are especially interested in k-derivations of B which are **locally nilpotent**, namely, those D in $\text{Der}_k(B)$ such that, for each $f \in B$, $D^n f = 0$ for $n \gg 0$. The set of locally nilpotent derivations of B is denoted by LND(B). Given $D \in \text{LND}(B)$ and $\gamma \in \Gamma(B)$, D is triangular for γ if and only if $D\gamma_{i+1} \in k[\gamma_1, \dots, \gamma_i]$ for $0 \le i \le n-1$, and D is linear for γ if and only if $D|_V$ is a nilpotent linear operator on $V = k\gamma_1 \oplus \cdots \oplus k\gamma_n$. Rentschler [6] showed that every $D \in \text{LND}(k^{[2]})$ is triangular for some $\gamma \in \Gamma(k^{[2]})$. In particular, every nonzero element of LND$(k^{[2]})$ has rank one. Then Bass [1] gave an example of $D \in \text{LND}(k^{[3]})$ such that, for all $\gamma \in \Gamma(k^{[3]})$, D is not triangular for γ. In Bass's example, the rank of D is two. Subsequently, elements of maximal rank n in LND$(k^{[n]})$ for each $n \ge 3$ were given in [3]. Rather than being exceptional, we now see that most elements of LND$(k^{[n]})$ are of maximal rank n when $n \ge 3$.

Recall that a subalgebra $A \subset B = k^{[n]}$ is a **coordinate subring** if there exists $\gamma \in \Gamma(B)$ such that $A = k[\gamma_1, \dots, \gamma_r]$ for some r with $1 \le r \le n$. Assume that $B = k^{[3]}$. Given $D \in \text{LND}(B)$, we have:

$$\text{rank}(D) = 3 \implies \text{super rank}(D) = 3$$

To see this, assume that the super-rank of D is 1. Then D restricts to a coordinate subring $k[f] \subset B$. Since $Df \neq 0$, we must have $Df \in k^*$ (units of k). In this case, there exist $g, h \in B$ such that $Dg = Dh = 0$ and $B = k[f, g, h]$. But then the rank of D is one, meaning this case cannot occur. If the super-rank of D is 2, then D restricts to a coordinate subring $k[f, g] \subset B$, where $k[f, g] \cong k^{[2]}$. By Rentschler's Theorem, $Dh = 0$ for some variable h of $k[f, g]$. But h is also a variable of B (since $k[f, g]$ is a coordinate subring), and again the rank of D is at most 2. So this case cannot occur.

But this implication does not generalize to higher dimensions. Example 8.16 of [4] gives $D \in \text{LND}(k^{[4]})$ of rank 4 which restricts to a coordinate subring $R \cong k^{[3]}$. By the foregoing discussion, it follows that the super-rank of D is 3.

Calculation of rank and super-rank can be difficult, owing to the fact that the automorphisms of $B = k^{[n]}$, and thus $\Gamma(B)$, are not well understood. In this note, we construct, for each $m \ge 2$, a locally nilpotent derivation of $B = k^{[m(m+1)]}$ with maximal super-rank $m(m+1)$. Each is homogeneous and its exponential is an automorphism of B of a type known as quasi-translation. In the case $m = 2$ we recover the derivation of DeBondt on $k^{[6]}$; see [4], 3.9.3.

Preliminaries. Throughout, k is a field of characteristic zero. For integer $n \geq 1$ and integral k-domain B, let $B^{[n]}$ be the polynomial ring in n variables over B. $\mathrm{Der}_k(B)$ is the set of k-derivations of B and $\mathrm{LND}(B)$ is the subset of locally nilpotent derivations. Note that any $D \in \mathrm{Der}_k(B)$ is determined by its images on a set of algebra generators of B.

For $D \in \mathrm{Der}_k(B)$, $\ker D$ denotes the kernel of D, which is a subalgebra of B, and DB denotes the image of D. Note that, if $D \in \mathrm{LND}(B)$ and $A = \ker D$, then $A \cap DB$ is an ideal of A, and if $D \neq 0$, then $A \cap DB \neq (0)$. If $R \subset B$ is a subalgebra, then $\mathrm{Der}_R(B)$ is the subset of $D \in \mathrm{Der}_k(B)$ with $DR = 0$. If $B = \bigoplus_i B_i$ is a $\mathbb{Z}$-grading, then a k-derivation D is said to be homogeneous if there exists $d \in \mathbb{Z}$ such that, for each $i \in \mathbb{Z}$, $DB_i \subset DB_{i+d}$.

In case k is algebraically closed, any $D \in \mathrm{LND}(k^{[n]})$ induces an algebraic action of the additive group $\mathbb{G}_a = (k, +)$ on affine space $\mathbb{A}^n_k$ via the exponential map $t \to \exp tD$. See [4] for details about locally nilpotent derivations and additive actions.

2 The Affine Cone over a Determinantal Variety

Assume that k is an algebraically closed field. Given positive integers m, n, let V be the vector space of $m \times n$ matrices over k, and let $\mathbb{P}_k(V) = \mathbb{P}^{mn-1}_k$ be the associated projective space with coordinates (x_{ij}). Given $0 \leq r \leq \min\{m, n\}$, let $M_r \subset \mathbb{P}^{mn-1}_k$ be the subset of matrices of rank at most r. We have the following.

(1) The $(r+1) \times (r+1)$ minors of the generic $m \times n$ matrix, which are homogeneous polynomials of degree $r+1$ in the x_{ij}, generate the ideal of M_r.
(2) M_r is irreducible of codimension $(m-r)(n-r)$ in $\mathbb{P}^{mn-1}_k$.
(3) $M_{r-1} \subset M_r$ is the set of singular points of M_r.

See Harris [5] for details.

Let $\mathcal{C}(M_r) \subset \mathbb{A}^{mn}_k$ be the affine cone over M_r.[1] Then $\mathcal{C}(M_{r-1})$ is the subvariety of singular points of $\mathcal{C}(M_r)$. Using the fact that any automorphism of a variety restricts to an automorphism of its singular subvariety, we obtain the following.

Lemma 2.1 *Every algebraic automorphism of $\mathcal{C}(M_r)$ respects the filtration:*

$$\{0\} = \mathcal{C}(M_0) \subset \mathcal{C}(M_1) \subset \cdots \subset \mathcal{C}(M_{r-1}) \subset \mathcal{C}(M_r)$$

In particular, every algebraic automorphism of $\mathcal{C}(M_r)$ fixes $\{0\}$.

Note that, if $k = \mathbb{C}$, then the lemma generalizes to all holomorphic automorphisms.

[1] In case $r = 0$, the cone over the empty set is a point.

3 Derivations of Maximal Super-Rank

3.1 *The Derivation* $\mathcal{D}_{(m,n)}$

Let $R = k[a_1, \ldots, a_m] = k^{[m]}$ for $m \geq 2$, and define $\mathbf{a} = (a_1, \ldots, a_m) \in R^m$. Set $\bar{\mathbf{a}} = \mathbf{a}M$, where M is an $m \times m$ skew-symmetric matrix over k of maximal rank, and write $\bar{\mathbf{a}} = (\bar{a}_1, \ldots, \bar{a}_m) \in R^m$. Note that $\langle \mathbf{a}, \bar{\mathbf{a}} \rangle = 0$. Form the $m \times m$ matrix $N = \mathbf{a}^T \bar{\mathbf{a}} = QM$ over R, where $Q = (a_i a_j)_{i,j}$. The entries of N are quadratic forms in $a_1, \ldots, a_m$. We have:

$$N^2 = (\mathbf{a}^T \bar{\mathbf{a}})(\mathbf{a}^T \bar{\mathbf{a}}) = \mathbf{a}^T (\bar{\mathbf{a}} \mathbf{a}^T) \bar{\mathbf{a}} = \mathbf{a}^T (0) \bar{\mathbf{a}} = 0$$

Let $\mathcal{B}_{(m,n)} = R[X] \cong R^{[mn]}$, where $n \geq m + 1$ and $X = (x_{ij})$ is an $m \times n$ matrix of indeterminates. Then $\mathcal{B}_{(m,n)}$ has the $\mathbb{Z}$-grading over k in which each a_i and x_{ij} is homogeneous, $\deg a_i = m$ and $\deg x_{ij} = 1$.

Define $\mathcal{D}_{(m,n)} \in \mathrm{Der}_R(\mathcal{B}_{(m,n)})$ by $\mathcal{D}_{(m,n)}(X) = NX = QMX$. To simplify notation, let us write:

$$\mathcal{B}_{(m,n)} = \mathcal{B} \quad \text{and} \quad \mathcal{D}_{(m,n)} = \mathcal{D}$$

Then $\mathcal{D}$ is homogeneous, and is locally nilpotent, since $\mathcal{D}^2(X) = N^2 X = 0$. Let X_j denote the jth column of X, $1 \leq j \leq n$. If $L_j \in \mathcal{B}$ is the R-linear form $L_j = \langle \bar{\mathbf{a}}, X_j \rangle$, then

$$\mathcal{D}(X_j) = \mathbf{a}^T (\bar{\mathbf{a}} X_j) = \mathbf{a}^T L_j$$

and:

$$\mathcal{D}(L_j) = \langle \bar{\mathbf{a}}, D(X_j) \rangle = \langle \bar{\mathbf{a}}, \mathbf{a}^T L_j \rangle = \langle \bar{\mathbf{a}}, \mathbf{a}^T \rangle L_j = 0$$

Let $p = \binom{n}{m}$ and let $f_1, \ldots, f_p$ be the $m \times m$ minors of X, noting that $p \geq n \geq 3$. Then for each i, $f_i \in k[X]$ and f_i is homogeneous of degree m. Specifically, assume that:

$$f_i = \det \left(X_1 \ \cdots \ \widehat{X_i} \ \cdots \ X_{m+1} \right) \ , \ 1 \leq i \leq m+1$$

Lemma 3.1 $f_i \in \ker \mathcal{D}$ *for each* $i = 1, \ldots, p$.

Proof By symmetry, we need only consider the minor $f_{m+1} = \det \tilde{X}$, where $\tilde{X} = (X_1 \ \cdots \ X_m)$. Let $Y = (Y_1, \ldots, Y_m)^T$ solve the equation:

$$\tilde{X} Y = \mathbf{a}^T$$

By Cramer's Rule we see that:

$$(\det \tilde{X}) Y_j = \det(X_1 \cdots \overset{(j)}{\mathbf{a}^T} \cdots X_m)$$

Note that $Y_j \in \mathrm{frac}(\mathcal{B})$ for each j. Applying $\mathcal{D}$ to $\det \tilde{X}$ yields:

$$
\begin{aligned}
\mathcal{D}(\det \tilde{X}) &= \sum_j \det(X_1 \cdots \overset{(j)}{\mathcal{D}X_j} \cdots X_m) \\
&= \sum_j \det(X_1 \cdots \overset{(j)}{\mathbf{a}^T L_j} \cdots X_m) \\
&= \sum_j L_j \det(X_1 \cdots \overset{(j)}{\mathbf{a}^T} \cdots X_m) \\
&= (\det \tilde{X}) \sum_j L_j Y_j \\
&= (\det \tilde{X}) \sum_j \langle \bar{\mathbf{a}}, X_j \rangle Y_j
\end{aligned}
$$

In matrix form this becomes: $\mathcal{D}(\det \tilde{X}) = (\det \tilde{X})\bar{\mathbf{a}}\tilde{X}Y = (\det \tilde{X})\bar{\mathbf{a}}\mathbf{a}^T = 0$.

3.2 The Derivation $D_{(m,n)}$

Let integers m, n be given, $m \geq 2$ and $n \geq m+1$. Let $H \subset \mathcal{B}_{(m,n)}$ be the homogeneous ideal $H = (f_1 - a_1, \ldots, f_m - a_m)$ and let

$$B_{(m,n)} = \mathcal{B}_{(m,n)}/H = k[X] \cong k^{[mn]}$$

with the induced $\mathbb{Z}$-grading from $\mathcal{B}_{(m,n)}$. Let $\pi : \mathcal{B}_{(m,n)} \to B_{(m,n)}$ be the standard projection. Since $a_i - f_i$ is a homogeneous element of $\ker \mathcal{D}_{(m,n)}$ for each i, it follows that

$$D_{(m,n)} := \mathcal{D}_{(m,n)}/H$$

is a well-defined homogeneous locally nilpotent derivation of $B_{(m,n)}$. Specifically, if $F = \pi(Q) = (f_i f_j)_{i,j}$, then $(FM)^2 = 0$, $D_{(m,n)}(FM) = 0$ and $D_{(m,n)}X = FMX$.

Let $A = \ker D_{(m,n)}$ and define $\mathbf{f}, \bar{\mathbf{f}} \in B^m_{(m,n)}$ by $\mathbf{f} = \pi(\mathbf{a})$, $\bar{\mathbf{f}} = \pi(\bar{\mathbf{a}})$, and $\lambda_j = \pi(L_j)$, $1 \leq j \leq n$. For $1 \leq j \leq n$ we have:

(1) $\lambda_j = \langle \bar{\mathbf{f}}, X_j \rangle \in A$
(2) $D_{(m,n)}X_j = \mathbf{f}^T \lambda_j$ and $D^2_{(m,n)}X_j = 0$

By Lemma 3.1, $f_i \in A$ for each i. Define A-ideals

$$I = f_1 A + \cdots + f_p A \quad \text{and} \quad J = \lambda_1 A + \cdots + \lambda_n A$$

and $B_{(m,n)}$-ideals $\tilde{I} = I B_{(m,n)}$ and $\tilde{J} = J B_{(m,n)}$. For each component $\bar{a}_i$ of $\bar{\mathbf{a}}$, we have $\bar{a}_i \in a_1 R + \cdots + a_m R$. Therefore, for each component $\bar{f}_i$ of $\bar{\mathbf{f}}$ we have $\bar{f}_i \in I$, which implies $\lambda_j \in I$ for each j. So the following properties hold for these ideals.

(1) $J \subset I$ and $\tilde{J} \subset \tilde{I}$
(2) $A \cap D_{(m,n)}B_{(m,n)} \subset IJ$ and $(D_{(m,n)}B_{(m,n)}) \subset \tilde{I}\tilde{J}$

Here, $(D_{(m,n)}B_{(m,n)})$ is the ideal of $B_{(m,n)}$ generated by the image $D_{(m,n)}B_{(m,n)}$.

In order to simplify the proof of the following result, let us assume that:

$$M = \begin{pmatrix} 0 & 1 & 0 & \cdots & 0 & 0 \\ -1 & 0 & 1 & \cdots & 0 & 0 \\ 0 & -1 & 0 & \cdots & 0 & 0 \\ \vdots & \vdots & \vdots & \ddots & \vdots & \vdots \\ 0 & 0 & 0 & \cdots & 0 & 1 \\ 0 & 0 & 0 & \cdots & -1 & 0 \end{pmatrix}_{m \times m}$$

Theorem 3.2 *The rank of $D_{(m,n)}$ equals mn.*

Proof Let $k[t] \cong k^{[1]}$ and define $\phi : B_{(m,n)} \to k[t]$ by:

$$\phi(x_{ij}) = (\phi(X))_{ij} = \begin{cases} t^{(j-1)m+i} & \text{if } i \neq j \\ t^{(j-1)m+i} - 1 & \text{if } i = j \end{cases}$$

Then $\phi(X_j) = t^{(j-1)m}T - I_j$, where I is the $m \times m$ identity matrix and:

$$T = \begin{pmatrix} t \\ t^2 \\ \vdots \\ t^m \end{pmatrix}_{m \times 1}$$

Therefore:

$$\begin{aligned} \phi(f_j) &= \det\left(T - I_1 \quad t^m T - I_2 \ \cdots\ \widehat{(t^{(j-1)m}T - I_j)} \ \cdots\ t^{(m-1)m}T - I_m \quad t^{m^2}T\right) \\ &= t^{m^2} \det\left(T - I_1 \quad t^m T - I_2 \ \cdots\ \widehat{(t^{(j-1)m}T - I_j)} \ \cdots\ t^{(m-1)m}T - I_m \quad T\right) \\ &= (-1)^{m-1} t^{m^2} \det\left(I_1 \quad I_2 \ \cdots\ \widehat{I_j} \ \cdots\ I_m \quad T\right) \\ &= (-1)^{2m+j} t^{m^2} t^j \\ &= (-1)^j t^{m^2+j} \end{aligned}$$

Since

$$(FM)_{ij} = \sum_{s=1}^{m} f_i f_s M_{sj} = \begin{cases} -f_i f_2 & (j = 1) \\ f_i(f_{j-1} - f_{j+1}) & (1 < j < m) \\ f_i f_{m-1} & (j = m) \end{cases}$$

it follows that

$$(\phi(F)M)_{ij} = \begin{cases} (-1)^{i+2}t^{2m^2+i+2} & (j=1) \\ (-1)^{i+j-1}t^{m^2+i}(t^{m^2+j-1} - t^{m^2+j+1}) & (1<j<m) \\ (-1)^{i+m-1}t^{2m^2+i+m-1} & (j=m) \end{cases}$$

and

$$(\phi(F)M)_{is}\phi(X)_{sj} = \begin{cases} (-1)^{i+2}t^{2m^2+i+2}(t^{(j-1)m+1} - \delta_{1j}) & (s=1) \\ (-1)^{i+s-1}t^{m^2+i}(t^{m^2+s-1} - t^{m^2+s+1})(t^{(j-1)m+s} - \delta_{sj}) & (1<s<m) \\ (-1)^{i+m-1}t^{2m^2+i+m-1}(t^{jm} - \delta_{mj}) & (s=m) \end{cases}$$

where δ_{ij} is the Kronecker function. Therefore, the highest degree term of

$$\phi(D_{(m,n)}x_{ij}) = (\phi(F)M\phi(X))_{ij} = \sum_{s=1}^{m}(\phi(F)M)_{is}\phi(X)_{sj}$$

equals $2(-1)^{i+m-1}t^{2m^2+i+(j+1)m-1}$. If $2m^2+i+(j+1)m-1 = 2m^2+i'+(j'+1)m-1$ for pairs (i,j) and (i',j'), then $i \equiv i' (\text{mod } m)$, and since $1 \le i, i' \le m$ we see that $i = i'$, which implies $j = j'$. Therefore, the degrees $\deg_t \phi(D_{(m,n)}x_{ij})$ are distinct for distinct x_{ij}. Consequently, the images $D_{(m,n)}x_{ij}$ are linearly independent, $1 \le i \le m$, $1 \le j \le n$.

Let $v \in \ker D_{(m,n)}$, and let v_1 be the linear (degree one) term of v. By homogeneity, $D_{(m,n)}v_1 = 0$. Since the images $D_{(m,n)}x_{ij}$ are linearly independent, it follows that $v_1 = 0$. Since any variable of $B_{(m,n)}$ has a nonzero linear term, v cannot be a variable.

Theorem 3.3 *Assume that k is algebraically closed. For $m \ge 2$, the super-rank of $D_{(m,m+1)}$ is m^2+m.*

Proof Given $m \ge 2$, let $B = B_{(m,m+1)}$ and $D = D_{(m,m+1)}$. Let d be the super-rank of D, $1 \le d \le m^2+m$, and let $\gamma \in \Gamma(B)$ be such that D restricts to $k[\gamma_1, \dots, \gamma_d]$. Let Δ denote the restriction of D to $k[\gamma_1, \dots, \gamma_d]$. Since the rank of D equals m^2+m by Theorem 3.2, it follows that the rank of Δ equals d. In particular, $\Delta \ne 0$. Consequently, there exists nonzero $h \in k[\gamma_1, \dots, \gamma_d] \cap A \cap DB$, which implies $h \in IJ \subset I^2$.

Let $\mathcal{V}(I), \mathcal{V}(h) \subset \mathbb{A}_k^{m^2+m}$ be the varieties defined by I and h, respectively, and let $\text{Sing}(\mathcal{V}(h))$ be the singular set of $\mathcal{V}(h)$. Since $h \in I^2 \subset \tilde{I}^2$, it follows[2] that $\partial h/\partial x_{ij} \in \tilde{I}$ for each i, j. Therefore:

$$\mathcal{V}(\tilde{I}) \subset \text{Sing}(\mathcal{V}(h)) \subset \mathcal{V}(h)$$

Now $\mathcal{V}(\tilde{I})$ is precisely the affine cone over the projective variety of $m \times (m+1)$ matrices of rank at most $m-1$. Therefore, $\mathcal{V}(\tilde{I})$ is irreducible and the codimension of $\mathcal{V}(\tilde{I})$ in $\mathbb{A}_k^{m^2+m}$ equals:

---

[2]For any commutative k-domain S, ideal $L \subset S$, and $\delta \in \text{Der}_k(S)$, we have $\delta(L^{n+1}) \subset L^n$, $n \ge 0$.

$$(m-(m-1))((m+1)-(m-1))=2$$

Since $\mathrm{Sing}(\mathcal{V}(h))$ has codimension at least 2, we conclude that $\mathcal{V}(\tilde{I})$ is an irreducible component of $\mathrm{Sing}(\mathcal{V}(h))$.

Assume that $d < m(m+1)$. Then $h \in k[\gamma_1, \ldots, \gamma_d]$ implies that $\mathcal{V}(h)$ is a cylindrical variety, which implies that $\mathrm{Sing}(\mathcal{V}(h))$ is a cylindrical variety, and that every irreducible component of $\mathrm{Sing}(\mathcal{V}(h))$ is a cylindrical variety. Write $\mathcal{V}(\tilde{I}) = Z \times \mathbb{A}^1_k$ for some variety Z. But then $\mathcal{V}(\tilde{I})$ admits a free $\mathbb{G}_a$-action, contradicting Lemma 2.1.

Therefore, $d = m(m+1)$. □

Remark 3.4 Geometrically, $D \in \mathrm{LND}(k^{[n]})$ of super-rank d corresponds to a $\mathbb{G}_a$-action on $\mathbb{A}^n_k$ for which $\mathbb{A}^n_k = \mathbb{A}^d_k \times \mathbb{A}^{n-d}_k$ and the action restricts to $\mathbb{A}^d_k$.

Question 3.5 *For $n \in \{4, 5\}$, does there exist $D \in \mathrm{LND}(k^{[n]})$ of super-rank n?*

In conclusion, we observe the following.

Criterion for Coordinate Subrings. Suppose that $B \cong k^{[n]}$ and $D \in \mathrm{LND}(B)$ has super-rank equal to n. If D restricts to the subalgebra $A \subsetneq B$, then A is not a coordinate subring of B.

For example, consider $B \cong k^{[6]}$ for k algebraically closed, and let $D \in \mathrm{LND}(B)$ be the derivation $D = D_{(2,3)}$ of super-rank 6 appearing in Theorem 3.3. Suppose that we can find $E \in \mathrm{LND}(B)$ such that $DE = ED$ and $Es = 1$ for some $s \in B$. Then D restricts to the subring $A = \ker E$, implying that A is not a coordinate subring. On the other hand, the Slice Theorem implies $B = A[s] \cong A^{[1]}$ (see [4], Corollary 1.26). This would yield a counterexample to the Cancellation Problem for Affine Spaces.

Acknowledgements The author wishes to thank Steve Mackey of Western Michigan University for his comments about an earlier version of this paper which led to a number of improvements in the final version.

References

1. Bass, H.: A non-triangular action of $\mathbb{G}_a$ on $\mathbb{A}^3$. J. Pure Appl. Algebra **33**, 1–5 (1984)
2. Freudenburg, G.: Triangulability criteria for additive group actions on affine space. J. Pure Appl. Algebra **105**, 267–275 (1995)
3. Freudenburg, G.: Actions of $\mathbb{G}_a$ on $\mathbb{A}^3$ defined by homogeneous derivations. J. Pure Appl. Algebra **126**, 169–181 (1998)
4. Freudenburg, G.: Algebraic Theory of Locally Nilpotent Derivations. Encyclopaedia of Mathematical Sciences, vol. 136, 2nd edn. Springer, Berlin, Heidelberg, New York (2017)
5. Harris, J.: Algebraic Geometry: A First Course, GTM, vol. 133. Springer (1992)
6. Rentschler, R.: Opérations du groupe additif sur le plan affine. C. R. Acad. Sc. Paris **267**, 384–387 (1968)

Affine Space Fibrations

Rajendra V. Gurjar, Kayo Masuda and Masayoshi Miyanishi

Abstract We discuss various aspects of affine space fibrations $f : X \to Y$ including the generic fiber, singular fibers and the case with a unipotent group action on X. The generic fiber X_η is a form of $\mathbb{A}^n$ defined over the function field $k(Y)$ of the base variety. Singular fibers in the case where X is a smooth (or normal) surface or a smooth threefold have been studied, but we do not know what they look like even in the case where X is a singular surface. The propagation of properties of a given smooth fiber to nearby fibers will be studied in the equivariant case of Abhyankar-Sathaye Conjecture in dimension three. We also treat the triviality of a form of $\mathbb{A}^n$ if it has a unipotent group action. Treated subjects are classified into the following four themes

1. Singular fibers of $\mathbb{A}^1$- and $\mathbb{P}^1$-fibrations,
2. Equivariant Abhyankar-Sathaye Conjecture in dimension three,
3. Forms of $\mathbb{A}^3$ with unipotent group actions,
4. Cancellation problem in dimension three.

Keywords Affine space fibration · Unipotent group action · Abhyankar-Sathaye conjecture

---

R. V. Gurjar (✉)
Department of Mathematics, Indian Institute of Technology Bombay, Powai,
Mumbai 400076, Maharashtra, India
e-mail: gurjar@math.iitb.ac.in

K. Masuda
Department of Mathematical Sciences, School of Science & Technology,
2-1 Gakuen, Sanda 669-1337, Japan
e-mail: kayo@kwansei.ac.jp

M. Miyanishi
Research Center for Mathematics and Data Science, Kwansei Gakuin University,
2-1 Gakuen, Sanda 669-1337, Japan
e-mail: miyanisi@kwansei.ac.jp

S. Kuroda et al. (eds.), *Polynomial Rings and Affine Algebraic Geometry*,
Springer Proceedings in Mathematics & Statistics 319,
https://doi.org/10.1007/978-3-030-42136-6_6

1 Introduction

In the present article, we will be interested in fibrations of algebraic varieties in order to study structures of higher-dimensional varieties. Fibration structure allows us to study varieties of smaller dimension and their interrelations. In particular, we will deal with fibrations whose fibers are affine spaces $\mathbb{A}^n$ or fibrations obtained as the quotient morphisms of varieties with unipotent group actions. There are long-standing problems which are still open in higher dimensions. We will discuss these problems.

Our base field is a field k of characteristic zero, which is algebraically closed throughout the article except section four. But, we assume that k is the complex number filed $\mathbb{C}$ whenever we use topological arguments. Indeed, there are cases where the Lefschetz principle is available. Given an algebro-geometric setting defined by finitely many equations, finitely many elements of the field k, say $\{a_\lambda\}_{\lambda\in\Lambda}$, are involved. Then one can embed the field $\mathbb{Q}(a_\lambda \mid \lambda \in \Lambda)$ into the complex field. Thus one can assume that the field k is replaced by $\mathbb{C}$. If the assertion is proved by a topological argument, one can use the descent of the field $\mathbb{C}$ to a subfield of k and then the ascent from the subfield to k.

Let $X \to Y$ be a dominant morphism of algebraic varieties X and Y. Let $k(X)$ and $k(Y)$ be the function fields of X and Y, respectively. Then f induces the inclusion of fields $k(Y) \hookrightarrow k(X)$. We thereby identify $k(Y)$ with a subfield of $k(X)$. We say that f is a *fibration* if $k(X)$ is a regular extension of $k(Y)$, or equivalently, if the generic fiber is geometrically integral. Further, the definition is equivalent to saying that $k(Y)$ is algebraically closed in $k(X)$. The following properties are well-known.

(1) If $f : X \to Y$ and $g : Y \to Z$ are fibrations then $g \circ f : X \to Z$ is a fibration.
(2) If $f : X \to Y$ is a fibration, then the base change $f_{Y'} : X \times_Y Y' \to Y'$ is a fibration for any algebraic variety Y' dominating Y.
(3) Let $f : X \to Y$ be a fibration. Then there exists a non-empty open set U of Y such that the fiber $X_y = X \times_Y \operatorname{Spec} k(y)$ is geometrically integral for every closed point y of U, where $k(y)$ denotes the residue field of Y at y.

We say that a fibration $f : X \to Y$ is an *F-fibration* for an algebraic variety F if X_y is isomorphic to F over k for general (closed) points y of Y. The F-fibration is *generically trivial* if there exists a non-empty open set U of Y such that $f^{-1}(U)$ is isomorphic to $U \times F$ over U. If $f^{-1}(U) \times_U U' \cong U' \times F$ over U' for an open set U of Y and a finite étale covering $U' \to U$, f is called *generically isotrivial*. If we can choose an open neighborhood U_y as U for each closed point y then X is a locally trivial *F-bundle* over Y in the Zariski topology of Y. If we can choose a finite étale covering U'_y of an open neighborhood U_y of y so that $f^{-1}(U_y) \times_{U_y} U'_y \cong F \times U'_y$, we say that X is a locally isotrivial *étale F-bundle* over Y. Let $X_\eta := X \times_Y \operatorname{Spec} k(Y)$ be the *generic fiber* of f. If f is a generically trivial F-fibration then X_η is isomorphic to F over the field $k(Y)$. Similarly, if f is generically isotrivial then X_η is isomorphic to F over a finite separable extension $k(U')$ of $k(U) = k(Y)$, while it is not necessarily isomorphic to F over $k(Y)$. It is a fundamental question about fibrations to ask if a fibration $f : X \to Y$ is generically trivial or generically isotrivial provided general closed fibers of f are isomorphic to an algebraic variety F (*the generic triviality or generic isotriviality problem*).

If k has infinite transcendence degree over $\mathbb{Q}$, F is an affine variety and f is an affine morphism, the generic isotriviality of f follows from the generic equivalence theorem of Kraft-Russell [33] which is stated as:

Generic Equivalence Theorem. *Let k be an algebraically closed field of infinite transcendence degree over the prime field. Let $p : S \to Y$ and $q : T \to Y$ be two affine morphisms where S, T and Y are k-varieties. Assume that for all closed points $y \in Y$ the two (schematic) fibers $S_y := p^{-1}(y)$ and $T_y := q^{-1}(y)$ are isomorphic. Then there is a dominant morphism of finite degree $\varphi : U \to Y$ and an isomorphism $S \times_Y U \cong T \times_Y U$ over U.*

We are interested in the case f is an affine morphism and general fibers are isomorphic to the affine space $\mathbb{A}^n$. An example of $\mathbb{A}^n$-fibration is obtained as the quotient morphism $q : X \to Y := X/G$ when a unipotent algebraic group G acts on an affine variety $X = \operatorname{Spec} A$ so that the ring of invariants $B = A^G$ is finitely generated over k. Then we set $Y = \operatorname{Spec} B$ and $q : X \to Y$ the morphism induced by the canonical inclusion $B \hookrightarrow A$.

The generic triviality problem for $F = \mathbb{A}^n$ is called the *Dolgachev-Weisfeiler problem* [10, Conjecture 3.8.5],[1] which was treated in many references including [2, 6, 25, 26, 28–30, 43, 49]. If $f : X \to Y$ is generically trivial, X contains an open set $f^{-1}(U) \cong U \times \mathbb{A}^n$. We call such an open set an $\mathbb{A}^n$-*cylinder* or a *cylinderlike open set* if $n = 1$. The answer to Dolgachev-Weisfeiler problem is positive if $n = 1, 2$[2] and not known for $n \geq 3$. The problem is related to the triviality of forms of $\mathbb{A}^n$ and the structure of the automorphism group $\operatorname{Aut}(\mathbb{A}^n)$.

A closed fiber X_y of an F-fibration $f : X \to Y$ is called a *singular fiber* if X_y is not isomorphic to F. There are two cases where the fiber X_y is not isomorphic to F.

(1) The fiber X_y is integral, i.e., irreducible and reduced, but not isomorphic to F.
(2) The fiber X_y is not integral. Hence either X_y has two or more irreducible components (*reducible fiber*), or X_y is irreducible but non-reduced (*multiple fiber*).

More precisely, the case (2) is divided into two subcases

(3) Each irreducible component Z_i of X_y has right dimension $\dim F$, but has multiplicity length $\mathcal{O}_{X_y, \xi_i}$ which is a multiple of some integer $d > 1$ (*multiple fiber*), where ξ is the generic point of Z_i.
(4) There exists some irreducible component Z_i of X_y of dimension bigger than $\dim F$.

---

[1]The original conjecture asserts that *if $f : X \to S$ is a flat affine morphism of smooth schemes with every fiber isomorphic (over the residue field) to an affine space $\mathbb{A}^n$, then f is locally trivial in the Zariski topology*. Taking credit of the conjecture, we call variants of the conjecture the *Dolgachev-Weisfeiler problem*.

[2]In fact, the answer is almost complete if $n = 1$ (see [29, 30]). However, the answer remains partial if $n = 2$. A key result is Sathaye's theorem [43], and a most general result is found in [27].

In section two, we consider first an $\mathbb{A}^1$-fibration $f : S \to C$ on an integral affine surface S over a curve C without assuming that S is smooth or normal, and consider singular fibers of f. This situation occurs when S is embedded into a bigger variety X with milder singularity as $S = \widetilde{f}^{-1}(C)$ with C embedded into Y, where $\widetilde{f} : X \to Y$ is an $\mathbb{A}^1$-fibration. If the $\mathbb{A}^1$-fibration f is obtained as the quotient morphism $q : S \to C$ of a G_a-action on S, we have a multiple fiber contained in the fixed-point locus S^{G_a}. But we have a chance to retrieve the G_a-action near the multiple fiber by looking at the infinitesimal neighborhoods of the multiple fiber (Lemmas 4 and 5). In [16], we observed singular fibers of the quotient morphism $q : X \to Y$ when G_a acts on a smooth affine threefold X. A singular fiber might contain a two-dimensional irreducible component S. We here, in Subsect. 2.3, show under a mild assumption that S contains a cylinderlike open set (Theorem 3). In Subsect. 2.4 we observe singular fibers of a $\mathbb{P}^1$-fibration on a normal surface S or a smooth algebraic threefold X. Our observation is limited to showing the simply-connectedness of the fiber F or its irreducible components.

Section three is devoted to an equivariant Abhyankar-Sathaye conjecture in dimension three. The conjecture in general asks if an embedded affine plane $\mathbb{A}^2$ in $\mathbb{A}^3$ is a fiber of an $\mathbb{A}^2$-bundle on $\mathbb{A}^3$. Contrary to the case of the affine line embedded into $\mathbb{A}^2$ (cf. Theorem of Abhyankar-Moh-Suzuki), this is far more difficult. We consider a G_a-action on $\mathbb{A}^3$ for which the embedded plane is stable. If we put two more conditions, the answer is positive (Theorem 6). These additional conditions will suggest what kind of observations must be made in the non-equivariant case.

Section four is for a study of forms of $\mathbb{A}^n$ with unipotent group actions. Consider a $\mathbb{A}^n$-fibration $f : X \to Y$. The question of generic isotriviality, that is to say, the generic fiber X_η isomorphic to $\mathbb{A}^n$ after a finite algebraic extension K of $k(Y)$, has been solved by the generic equivalence theorem of Kraft-Russell, a major questions in our mind is the generic triviality.

(1) Does the generic fiber X_η which is a $k(Y)$-form of $\mathbb{A}^n$, become isomorphic to $\mathbb{A}^n$ over $k(Y)$?

If this question is answered affirmatively, then f is generically trivial. In fact, if $n = 1, 2$ the answer of the question is positive. If $n \geq 3$, however, we do not know the answer to the question. Our purpose is to consider the question (1) with a suitable action of unipotent group. First we prove that a form X of $\mathbb{A}^3$ is trivial if there is a fixed-point free G_a-action on X (Theorem 7) or an effective action of a unipotent group of dimension two (hence necessarily commutative) (Theorem 8).[3] Second, we show that if X is a k-form of $\mathbb{A}^n$ with a *proper and q-tight* action of a commutative unipotent group G of dimension $n-2$ then the k-form is trivial, where k is a non-closed field of characteristic zero (Remark 4). But the proof is only given in the case $n = 4$. Assuming that a given action of a unipotent group G on a k-form be proper and q-tight is seemingly quite technical, but the properness and q-tightness imply nice conditions like the fixed-point freeness of the action and the irreducibility of each fiber of the algebraic quotient morphism if the fiber is not empty. Third, we

[3]The latter is a theorem of Daigle-Kaliman [5].

consider a k-form of $\mathbb{A}^n \times \mathbb{A}^1_*$, where $\mathbb{A}^1_*$ is the affine line with one point punctured. Then a k-form has a bundle structure over $\mathbb{A}^1_*$ or its nontrivial k-form with fiber isomorphic to a k-form of $\mathbb{A}^n$ (Lemma 31).

In section five, we apply the arguments in Subsect. 4.4 to prove a special case of the cancellation theorem in dimension three where the variety in question has a proper and q-tight G_a-action (Theorem 10).

2 Singular Fibers of $\mathbb{A}^1$- and $\mathbb{P}^1$-Fibrations

2.1 G_a-Actions on Affine Surfaces

Let $S = \text{Spec } A$ be an integral affine surface defined over k. Assume that S has a nontrivial G_a-action. Let $B = A^{G_a}$. Then $B = A \cap K^{G_a}$, where $K = Q(A)$. By Zariski's lemma [50], B is an affine domain of dimension one over k. For the readers convenience, we give a topological proof in this case.

Proposition 1 *Let A be a normal affine domain defined over $\mathbb{C}$ with the quotient field K, let L be a subfield of K with* $\text{tr.deg}_{\mathbb{C}} L = 1$ *and let $B = A \cap L$. Then B is finitely generated over $\mathbb{C}$.*

Proof If $B = \mathbb{C}$, we have nothing to show. So, we assume that $\mathbb{C} \subsetneqq B$. Since A is normal, B is integrally closed in K. Furthermore, since $\text{tr.deg}_{\mathbb{C}} L = 1$, the quotient field of B is equal to L. In fact, suppose that $Q(B) \neq L$. Let ξ be an element of $L \setminus Q(B)$. Since L is algebraic over $Q(B)$, there exist an element $b \in B \setminus \{0\}$ such that $b\xi$ is integral over B. Since it then follows that $b\xi \in A$, we have $b\xi \in B$. Hence $\xi \in Q(B)$. A contradiction. We can find elements $b_1, \ldots, b_r \in B$ such that the subring $B_1 := \mathbb{C}[b_1, \ldots, b_r]$ of B is normal and birational to B. Let $C_1 = \text{Spec } B_1$. Then C_1 is a smooth affine curve. Let $X = \text{Spec } A$ and let $q_1 : X \to C_1$ be the dominant morphism induced by the inclusion $B_1 \hookrightarrow A$. Then the homomorphism of homology groups $(q_1)_* : H_1(X; \mathbb{Q}) \to H_1(C_1; \mathbb{Q})$ is surjective. Suppose that $B_1 \subsetneqq B$. Then we find an affine subalgebra B_2 of B containing strictly B_1 such that B_2 is normal. Let $C_2 = \text{Spec } B_2$. Then the inclusion $B_1 \subsetneqq B_2$ induces a birational morphism $C_2 \to C_1$, which makes C_2 a strictly smaller open set of C_1, i.e., $C_2 \subsetneqq C_1$. The morphism $q_2 : X \to C_2$ induces a surjection $(q_2)_* : H_1(X; \mathbb{Q}) \to H_1(C_2; \mathbb{Q})$, where $\text{rank} H_1(C_2; \mathbb{Q}) > \text{rank} H_1(C_1; \mathbb{Q})$ because $C_2 \subsetneqq C_1$. Continuing this way, we obtain a sequence of smooth affine curves $C_1 \supsetneqq C_2 \supsetneqq C_3 \supsetneqq \cdots$, where $C_{i+1} \subsetneqq C_i$ is an open immersion. Hence $\text{rank} H_1(C_{i+1}; \mathbb{Q}) > \text{rank} H_1(C_i; \mathbb{Q})$. But, since $\text{rank} H_1(X; \mathbb{Q}) \geq H_1(C_i; \mathbb{Q})$ for every i, this decreasing series of open sets must stop at some i. This implies that B is finitely generated over $\mathbb{C}$. □

Let $C = \text{Spec } B$ and let $q : S \to C$ be the quotient morphism induced by the inclusion $B \hookrightarrow A$.

Problem 1 With the notations and assumptions as above, is every fiber $F := q^{-1}(P)$ for $P \in C$ a disjoint union of affine lines?

What we can say so far is the following result which will be proved below.

Theorem 1 *With the above notations and assumptions, the following assertions hold:*

(1) F is a disjoint union of the irreducible components, each of which is an affine rational curve with one place at infinity.
(2) If an irreducible component Z_i of F is reduced in F, then Z_i is isomorphic to the affine line.
(3) If S is normal, every irreducible component of F is isomorphic to the affine line.

We have the following related result [37, Theorem 2.1] (see also [11, Lemma 1.1]).

Lemma 1 *Let A be an affine k-algebra of dimension one with a nontrivial locally nilpotent derivation (lnd, in short) δ, and let $R = \text{Ker}\ \delta$. Assume that the associated G_a-action has no fixed points on* Spec A*, i.e., $\delta(A)A = A$. Then $A \cong R[t]$, where R is an Artin ring.*

Lemma 1 implies that if the G_a-action on an irreducible component Z_i of F is nontrivial, then the scheme structure of F on Z_i is $(\text{Spec}\ R_i) \times \mathbb{A}^1$, where R_i is an Artin local ring and represents the "thickening" of Z_i in F. Hence length(R_i) is the multiplicity of Z_i in F.

We begin with the following result [38, Chap. 3, Lemma 1.4.2] and [36].

Lemma 2 *Let $\widetilde{S}$ be a normal affine surface with a nontrivial G_a-action and let $\widetilde{q} : \widetilde{S} \to C$ be the quotient morphism, where C is a normal affine curve. Let $P \in C$ be a closed point and let $\widetilde{F} = \widetilde{q}^{-1}(P)$. Then $\widetilde{F}$ is a disjoint union of affine lines.*

In [38], it is not shown that each irreducible component of $\widetilde{F}$ is isomorphic to the affine line. But this follows from the fact that the irreducible component (plus one smooth point) is obtained from a part of a degenerate fiber of a $\mathbb{P}^1$-fibration on a smooth projective surface by contracting a linear chain such that it meets the (proper transform) of the irreducible component at an end component of the linear chain and that the linear chain is the exceptional locus of a cyclic quotient singularity. The assertion (3) of Theorem 1 follows from this result.

In order to prove the other two assertions of Theorem 1, we set our notations as follows. With A as given above, let $\widetilde{A}$ be the normalization of A and let $\widetilde{S} = \text{Spec}\ \widetilde{A}$.

Lemma 3 *The following reduction of the settings is possible.*

(1) We may assume that the curve $C =$ Spec B is normal.

(2) *The G_a-action on $S = \text{Spec } A$ lifts to $\widetilde{S} = \text{Spec } \widetilde{A}$. Namely, the lnd δ associated with the G_a-action on S extends to an lnd $\widetilde{\delta}$ on $\widetilde{A}$ in such a way that* $\text{Ker } \delta = \text{Ker } \widetilde{\delta}$. *Hence we have a commutative diagram*

$$\begin{array}{ccc} \widetilde{S} & \xrightarrow{\nu} & S \\ {\scriptstyle \widetilde{q}}\downarrow & & \downarrow{\scriptstyle q} \\ C & = & C \end{array}$$

where ν is the normalization morphism and $\widetilde{q}$ is the quotient morphism.

Proof (1) Note that the function field K is a regular extension of k, hence $L := Q(B)$ is an algebraic function field of dimension one over k. Let $\widetilde{C}$ be the normalization of C in L and let $\widetilde{P}$ be a point of $\widetilde{C}$ lying over P. Then the residue fields of P and $\widetilde{P}$ coincide with k. Let $\overline{S} = S \times_C \widetilde{C}$. Then $\overline{S} = \text{Spec } (\overline{A})$, where $\overline{A} = A \otimes_B \widetilde{B}$ and $\widetilde{B}$ is the integral closure of B in L. Since $\widetilde{C}$ is birational to C, taking the tensor product $\otimes_B \widetilde{B}$ is equivalent to adjoining finitely many fractions of B. Hence $\overline{A} = A[\widetilde{B}]$. Let $\mathfrak{m}$ (resp. $\widetilde{\mathfrak{m}}$) be the maximal ideal of B (resp. $\widetilde{B}$) corresponding to the point P (resp. $\widetilde{P}$). Then we have

$$\begin{aligned}\overline{A}/\widetilde{\mathfrak{m}}\overline{A} &= (A \otimes_B \widetilde{B}) \otimes_{\widetilde{B}} (\widetilde{B}/\widetilde{\mathfrak{m}}) \\ &= (A \otimes_B (B/\mathfrak{m})) \otimes_{B/\mathfrak{m}} (\widetilde{B}/\widetilde{\mathfrak{m}}) \cong A \otimes_B (B/\mathfrak{m}) = A/\mathfrak{m}.\end{aligned}$$

Namely, we have $f^{-1}(P) \cong \overline{f}^{-1}(\widetilde{P})$, where $\overline{f} : \overline{S} \to \widetilde{C}$ is the base change of S. Hence we may replace C by $\widetilde{C}$ and assume that C is normal.

(2) It is known by Seidenberg [46] that δ extends to a derivation $\widetilde{\delta}$ of $\widetilde{A}$. Namely, $\widetilde{\delta}(\widetilde{A}) \subseteq \widetilde{A}$. Assume that C is normal by (1). Since S contains a cylinderlike open set $U \times \mathbb{A}^1$ with an open set $U = D(b) = \{b \neq 0\}$ of C for $b \in B$, we have $\widetilde{\delta}|_{U \times \mathbb{A}^1} = \delta|_{U \times \mathbb{A}^1}$. Since $\widetilde{\delta}$ defines a G_a-action on this cylinderlike open set, $\widetilde{\delta}$ is an lnd on $\widetilde{A}$. □

Proof of Theorem 1 (1) Let $\widetilde{F} = \widetilde{q}^{-1}(P)$. Then $\widetilde{F} = \coprod_j \widetilde{Z}_j$, where $\widetilde{Z}_j \cong \mathbb{A}^1$ by Lemma 2. Note that $\nu|_{\widetilde{F}} : \widetilde{F} \to F$ is a finite morphism. Suppose that $Q \in Z_i \cap Z_{i'}$. Let $\nu^{-1}(Z_i) = \widetilde{Z}_{ij_1} \coprod \cdots \coprod \widetilde{Z}_{ij_r}$ and $\nu^{-1}(Z_{i'}) = \widetilde{Z}_{i'j_1} \coprod \cdots \coprod \widetilde{Z}_{i'j_s}$. Then a point $\widetilde{Q} \in \nu^{-1}(Q)$ lies in the intersection of $\widetilde{Z}_{ij}$ and $\widetilde{Z}_{i'j'}$ for some j and j'. Hence the components $\widetilde{Z}_{ij}$ and $\widetilde{Z}_{i'j'}$ meet each other. But this contradicts Lemma 2. So, F is a disjoint union $\coprod_{i=1}^r Z_i$. Note that, for each $1 \leq i \leq r$, a multiple of Z_i is locally defined by $t = 0$. By Nagata [40, p. 65, footnote], $S - (F \setminus Z_i)$ is affine, and $S - (F \setminus Z_i)$ has a nontrivial G_a-action since $F \setminus Z_i$ is G_a-stable. Thus we may assume that F consists of a single irreducible component Z_i. Then Z_i is a surjective image of the affine line $\widetilde{Z}_j$ for some j. Hence Z_i is an affine rational curve with one place at infinity.

(2) Let t be a generator of the maximal ideal of $\mathcal{O}_{C,P}$. By replacing C by an open neighborhood U of P such that t is regular on U and P is the only zero of t on U, we may assume that $\mathfrak{m}$ is a principal ideal tB of B. Furthermore, we may assume that $S \setminus F \cong \widetilde{S} \setminus \widetilde{F}$, i.e., $A[t^{-1}] = \widetilde{A}[t^{-1}]$. Then we know that $\widetilde{A}$ is obtained from A

by adjoining element of $A[t^{-1}]$ of the form a/t^m with $a \in A$ and $m > 0$. Take such an element a/t^m of $\widetilde{A}$. Assume that $a \notin tA$. Then it satisfies a monic relation

$$(a/t^m)^n + a_1(a/t^m)^{n-1} + \cdots + a_{n-1}(a/t^m) + a_n = 0, \quad a_i \in A .$$

Then we have

$$a^n + t^m a_1 a^{n-1} + \cdots + t^{m(n-1)} a_{n-1} a + t^{mn} a_n = 0.$$

Hence $a^n \in t^m A$. By the assertion (1), we now assume that F is irreducible. Let $R = A/tA$ and let $\mathfrak{p} = \sqrt{0}$ which represents the "thickening" of the component Z_i in F. If Z_i is a reduced component, $\mathfrak{p} = (0)$. Meanwhile, $\mathfrak{p} = \sqrt{tA}/tA$ and $a \in \sqrt{tA}$. Since $\mathfrak{p} = (0)$, $a \in tA$. This is a contradiction to the above assumption. Hence $A = \widetilde{A}$, and $Z_i \cong \mathbb{A}^1$. □

We need the following result in Subsect. 2.3.

Theorem 2 *Let S be an affine algebraic surface and let $f : S \to C$ be a dominant morphism to an affine curve C. Assume that, for every closed point $P \in C$, the fiber $f^{-1}(P)$ is a disjoint union of affine lines. Then the following assertions hold.*

(1) There exists a cylinderlike affine open set $Z = U \times \mathbb{A}^1$ of S such that every fiber of the projection $p_1 : Z \to U$ is a fiber component of f.
(2) There exists a nontrivial G_a-action on S such that the morphism f is factored by the quotient morphism $q : S \to S/G_a$ as

$$f : S \xrightarrow{q} S/G_a \xrightarrow{g} C,$$

where g is a quasi-finite morphism.

Proof (1) Restricting C to the smooth part of C, we may assume that C is smooth. Let $\nu : \widetilde{S} \to S$ be the normalization morphism and let $\widetilde{f} = f \circ \nu$. Removing the fibers of $\widetilde{f}$ which pass through the singular points of $\widetilde{S}$, we may assume that $\widetilde{S}$ is smooth. On the other hand, if the singular locus $\mathrm{Sing}(S)$ contains a fiber component of f, then we throw away the complete fiber containing the component. So, we may assume that if $\mathrm{Sing}(S)$ has dimension one, all the components of dimension one are horizontal to the fibration f. Let $\widetilde{F}$ be a fiber of $\widetilde{f}$. Since ν is birational, every irreducible component, say $\widetilde{F}_i$ of $\widetilde{F}$ is a rational curve, which surjects birationally onto the irreducible component F_i of $F := \nu(\widetilde{F})$. Since $F_i \cong \mathbb{A}^1$ by the assumption, it follows that $\widetilde{F}_i \cong F_i \cong \mathbb{A}^1$. Thus the fiber $\widetilde{F}$ is isomorphic to F. This implies that $\mathrm{Sing}(S) = \emptyset$, and S is smooth. Now take the normalization $\widetilde{C}$ of C in the function field of S and consider the factorization

$$f : S \xrightarrow{\widehat{f}} \widetilde{C} \xrightarrow{g} C \quad \text{(Stein factorization)}.$$

Then the morphism $\widehat{f}$ is an $\mathbb{A}^1$-fibration. We may assume that this fibration is trivial, i.e., $S \cong \widetilde{C} \times \mathbb{A}^1$.

(2) Let $S = \operatorname{Spec} A$ and $C = \operatorname{Spec} B$. We show that there exists an element $b \in B$ such that $A[b^{-1}] = \widetilde{B}[b^{-1}][t]$, where $\widetilde{B}$ is the integral closure of B in $Q(A)$. In fact, by the construction in (1), there exists an element $\widetilde{b} \in \widetilde{B}$ such that $Z = \widehat{f}^{-1}(U)$, where $U = D(\widetilde{b})$ in $\widetilde{C}$. Let

$$\widetilde{b}^n + b_{n-1}\widetilde{b}^{n-1} + \cdots + b_1\widetilde{b} + b_0 = 0, \quad b_i \in B, \ b_0 \neq 0$$

be the monic equation for $\widetilde{b}$ over B. Since $g^{-1}(D(b_0)) \subseteq D(\widetilde{b})$, we may take $U = g^{-1}(D(b_0))$ in the assertion (1). Let $b = b_0$. Then $A[b^{-1}] = \widetilde{B}[b^{-1}][t]$. Let $\delta = b^N(\partial/\partial t)$, where we choose N a positive integer such that if $A = k[a_1, \ldots, a_r]$ then $\delta(a_i) \in A$ for all i. Then δ is an lnd of A such that $\widetilde{B} \subseteq \operatorname{Ker} \delta$. Consider the associated G_a-action on S and the quotient morphism $q : S \to \operatorname{Spec} \operatorname{Ker} \delta$. Then $g : S/G_a \to C$ is the morphism associated with the inclusion $B \hookrightarrow \widetilde{B} \hookrightarrow \operatorname{Ker} \delta$. □

2.2 *Hidden G_a-Actions on Multiple Fiber Components*

With the notations in subsection one, we consider the case where all points of the fiber F are fixed by the G_a-action. Since the G_a-action on $S = \operatorname{Spec} A$ is non-trivial, the following lemma shows that the G_a-action on the fiber F is hidden in the "thickening" part of F. To simplify the situation, we assume below that *the fiber F is irreducible* and that *the G_a-action on F_{red} is trivial*. We denote F_{red} by Z.

Lemma 4 *If F is reduced, then the G_a-action induces a nontrivial action on the mth infinitesimal neighborhood J_m of F for some $m > 0$, where $J_m := t^m A/t^{m+1} A$ which is isomorphic to A/tA as an A-module.*

Proof Suppose that F is reduced. By Theorem 1, Z is isomorphic to $\mathbb{A}^1 = \operatorname{Spec} k[\overline{v}]$. Let Q be a closed point of Z. Choose an element v of A such that $\overline{v} = v \pmod{tA}$ and $v(Q) = 0$. Then S is smooth at Q and $\{t, v\}$ is a system of local parameters at Q. We show that $\delta(v) = bt^m$ for $m > 0$ and $b \in A$ such that $b \notin tA$ and $b \pmod{tA} \in k^*$. Since any element a is written as $a - f(v) \in tA$ for some $f(v) \in k[v]$, A is a subalgebra of the tA-adic completion $\widehat{A} = \varprojlim_n A/t^n A \cong k[v][[t]]$. Hence we can write

$$\delta(v) = t^m \left(\sum_{i \geq 0} f_i(v) t^i \right), \quad f_i(v) \in k[v], \ f_0(v) \neq 0.$$

Write

$$f_0(v) = c_0 v^n + c_1 v^{n-1} + \cdots + c_{n-1} v + c_n, \quad c_i \in k, \ c_0 \neq 0.$$

We may assume that $c_0 = 1$. Suppose $n > 0$. Then, by a straightforward computation, we have

$$\delta^\ell(v) = g_0(v) t^{\ell m} + g_1(v) t^{\ell m+1} + \cdots,$$

where the highest degree v-term of $g_0(v)$ is

$$\left(\prod_{i=1}^{\ell-1}(in-(i-1))\right)\cdot v^{\ell n-(\ell-1)},$$

which is nonzero for every $\ell > 0$. Since $\delta^\ell(v) = 0$ for some ℓ, we have $n = 0$. Thus the representation of $\delta(v)$ on $J_m = t^m A/t^{m+1}A \cong k[v]$ is given by $\delta(v) = \Delta(v)t^m \pmod{t^{m+1}A}$, where $\Delta(v) = 1$. By the construction, one can verify that Δ is an lnd on A/tA. So, Δ gives a nontrivial G_a-action on J_m. If $m = 0$, this Δ coincides with the lnd induced on A/tA by δ. Since $\Delta(v) = 1$, the action of G_a on A/tA is nontrivial. It contradicts the assumption that $F = Z$ is contained in the fixed point locus. So, $m > 0$. □

Next, we consider the case F is non-reduced. Let $R = A/tA$ and let $\mathfrak{p} = \sqrt{0}$ in R. By the assumption, $\mathfrak{p} \neq (0)$, $\mathfrak{p}^{N-1} \neq (0)$ and $\mathfrak{p}^N = (0)$. Since $\mathfrak{p}$ is a δ-ideal, i.e., $\delta(\mathfrak{p}) \subseteq \mathfrak{p}$, where we denote the induced lnd on R by the same letter δ. We assume that $\delta(\mathfrak{p}) \neq (0)$. Then there exists an integer d such that $\delta(\mathfrak{p}) \subseteq \mathfrak{p}^d$ and $\delta(\mathfrak{p}) \not\subseteq \mathfrak{p}^{d+1}$, where $d < N$. We call the integer d the *depth* of δ at F and denote $d = \mathrm{depth}_F(\delta)$.

We have the following result.

Lemma 5 *Assume that A is normal and F is irreducible and non-reduced. Let $R_0 = R/\mathfrak{p}$. Let $M = \mathfrak{p}^{d-1}/\mathfrak{p}^d$ and let M^{**} be the double dual of the R_0-module M. Then $M^{**} \cong R_0 e$ with a free generator e, M^{**} is a δ-module, i.e., $\delta(M^{**}) \subseteq M^{**}$ and the representation of δ on M^{**} is given by $\delta(a) = \Delta(a)e$ and Δ is a nonzero lnd on R_0.*

Proof Let Q be a general smooth point of $Z \cong \mathbb{A}^1$. Then Q is a smooth point of S. Hence we can choose local parameters $\{u, v\}$ of S such that Z is defined by $u = 0$ and v is a fiber coordinate of Z. Then $t = u^N\xi$, where $\xi \in \mathcal{O}_{S,Q}^*$. Let $\mathfrak{p}_Q = \mathfrak{p} \otimes \mathcal{O}_{S,Q}$ and $R_Q = R \otimes \mathcal{O}_{S,Q}$. Then $\mathfrak{p}_Q = uR_Q$ and $\mathfrak{p}_Q^d = u^d R_Q$. We take the element u from A.

Here we remark that $\delta(R) \subseteq \mathfrak{p}^{d-1}$. In fact, for any element $a \in A$, we have $au \pmod{tA} \in \mathfrak{p}$. Since $\mathfrak{p}_Q = uR_Q$, we have $\delta(au) \pmod{tA} \in \mathfrak{p}^d$. Since $\mathfrak{p}_Q^d = u^d R_Q$, $s\delta(au) = u^d z$ in A for $s, z \in A$ and $s(Q) \neq 0$. We can write the last equation as $su\delta(a) = u^d z - sa\delta(u)$. Since $\delta(\mathfrak{p}_Q) \subseteq \mathfrak{p}_Q^d = u^d R_Q$, we have

$$ss'u\delta(a) = u^d(s'z - saw) \text{ in } A$$

for $s', w \in A$ with $s'(Q) \neq 0$. Since A is an integral domain, we have $ss'\delta(a) = u^{d-1}(s'z - saw)$ in A. This implies that $\delta(a) \in \mathfrak{p}_Q^{d-1} \cap R = \mathfrak{p}^{d-1}$. By the definition of d, we have an induced k-module homomorphism

$$\delta : R_0 := R/\mathfrak{p} \longrightarrow M := \mathfrak{p}^{d-1}/\mathfrak{p}^d, \quad a \mapsto \delta(a) \pmod{\mathfrak{p}^d}. \qquad (*)$$

Define

$$\widetilde{\delta} := \rho \circ \delta : R_0 \xrightarrow{\delta} M \xrightarrow{\rho} M^{**},$$

where $\rho : M \to M^{**}$ is the canonical R_0-module homomorphism. Since M^{**} is a torsion-free module over $R_0 \cong k[z]$, M^{**} is a free R_0-module. Write $M^{**} = R_0 e$. For $a_1, a_2 \in R$, we have

$$\begin{aligned}\widetilde{\delta}(a_1 a_2) &= \rho \circ \delta(a_1 a_2) = \rho(a_1 \delta(a_2) + a_2 \delta(a_1)) \\ &= a_1(\rho \circ \delta)(a_2) + a_2(\rho \circ \delta)(a_1) = a_1 \widetilde{\delta}(a_2) + a_2 \widetilde{\delta}(a_1).\end{aligned}$$

If we write $\overline{a}_i = a_i \pmod{\mathfrak{p}}$ $(i = 1, 2)$ and $\widetilde{\delta}(\overline{a}_i) = \Delta(\overline{a}_i)e$, we have

$$\Delta(\overline{a}_1 \overline{a}_2) = \overline{a}_1 \Delta(\overline{a}_2) + \overline{a}_2 \Delta(\overline{a}_1).$$

It is easy to show that $\Delta : R_0 \to R_0$ is an lnd. Hence Δ gives a required representation on M^{**}. □

If $\delta(R) = 0$ then $\delta(A) \subseteq tA$, and we replace δ by $t^{-1}\delta$. Hence we may assume that $\delta(R) \neq 0$. If $\delta(\mathfrak{p}) = 0$ then Lemma 5 implies that $\delta(R) \subseteq \mathfrak{p}^{N-1}$, and we can take M^{**} as the double dual of the R_0-module $\mathfrak{p}^{N-1}$ to obtain the hidden lnd on $R_0 = R/\mathfrak{p}$. If A is not normal, the argument of Lemma 5 does not work. The following example reflects well the above situation.

Example 1 Let S be an affine hypersurface $t^2 z - y^n = 0$ in $\mathbb{A}^3$, where $n \geq 2$. Define an lnd δ by $\delta(t) = 0$, $\delta(y) = t^2$ and $\delta(z) = ny^{n-1}$. Then S is non-normal with $\mathrm{Sing}(S) = \{t = 0\}$, $A = k[t, y, z]$ and $R = k[z][\varepsilon]$, where $\varepsilon = y \pmod{tA}$ and $\varepsilon^n = 0$. Then $\mathrm{depth}_F(\delta) = n$ and $M = k[z]\varepsilon^{n-1}$. Here we can argue as in the proof of Lemma 5 though S is not normal. Then the lnd Δ on $R_0 = k[z]$ is given by $\Delta(z) = n$.

2.3 G_a-Actions on Affine Threefolds

Let X be a smooth affine threefold with a nontrivial G_a-action and let $q : X \to Y$ be the quotient morphism. Assume that the quotient surface $Y := X/G_a$ is smooth. Let $P \in Y$ and let $F = q^{-1}(P)$. We proved in [16] that every irreducible component of F of dimension one is isomorphic to $\mathbb{A}^1$ and disjoint from other irreducible components. Here we consider a two-dimensional component S and show that S contains a cylinderlike open set. We consider a linear pencil Λ of hyperplane sections of Y passing through the point P. Namely, if Y is embedded into an affine space $\mathbb{A}^N$ with a system of coordinates $\{x_1, \ldots, x_N\}$ so that P corresponds to $(0, \ldots, 0)$, then let $L_\lambda = Y \cap H_\lambda$, where H_λ is a hyperplane $a_1 x_1 + \cdots + a_N x_N = 0$ and the set $\{\lambda = (a_1, a_2, \ldots, a_N)\}$ corresponds to a line Λ in the dual projective space $\mathbb{P}^{N-1}$, which we take to be sufficiently general. Let T_λ be the closure in X of $q^{-1}(L_\lambda \setminus \{P\})$. *We assume the hypothesis* (H) *that* $T_\lambda \cap S \neq \emptyset$ *for general* $\lambda \in \Lambda$.

Theorem 3 *With the same notations and assumptions as above, the component S contains a cylinderlike open set $U \times \mathbb{A}^1$.*

Proof (1) Let $q_\lambda : T_\lambda \to L_\lambda$ be the restriction of q onto the affine surface T_λ. If λ is general in Λ, then T_λ is smooth by Bertini's second theorem and q_λ is an $\mathbb{A}^1$-fibration. Let $F_\lambda = S \cap T_\lambda$. Then F_λ is a disjoint union of affine lines by Lemma 2.

(2) Let $\sigma : \widehat{Y} \to Y$ be the blowing-up of the point P, hence $E := \sigma^{-1}(P) \cong \mathbb{P}^1$. Consider the fiber product $X \times_Y \widehat{Y}$ and the base change $q_{\widehat{Y}} = q \times_Y \widehat{Y}$. Then $X \times_Y \widehat{Y}$ contains three-dimensional components whose underlying sets are isomorphic to $S_i \times E$, where the S_i are two-dimensional components of $F = q^{-1}(P)$. The open set $X \setminus q^{-1}(P)$ is contained in $X \times_Y \widehat{Y}$ as an open set. Let $\widehat{X}$ be the closure of this open set in $X \times_Y \widehat{Y}$ [4] and let $\widehat{q} : \widehat{X} \to \widehat{Y}$ be the restriction of $q_{\widehat{Y}}$ to $\widehat{X}$. By the definition, $\widehat{q}$ coincides with q on the open set $\widehat{X} \setminus \widehat{q}^{-1}(E)$. The closed set $\widehat{q}^{-1}(E)$ contains the *proper* transform $\widehat{S}$ which is biregular to S.

(3) Let $\widehat{L}_\lambda$ be the proper transform of L_λ on $\widehat{Y}$. Then $\widehat{L}_\lambda \cap \widehat{L}_\mu = \emptyset$ if $\lambda \neq \mu$. Let $\widehat{T}_\lambda$ be the proper transform of T_λ in $\widehat{X}$. Then $\widehat{T}_\lambda$ meets $\widehat{S}$ along a curve $\widehat{F}_\lambda$ because $T_\lambda \cap S \neq \emptyset$ by the hypothesis (H). Since $\widehat{S}$ is isomorphic to S, $\widehat{F}_\lambda$ is isomorphic F_λ, which is a disjoint union of affine lines.

(4) By Theorem 2 and its proof in the step (1) applied to the morphism $\widehat{q}|_{\widehat{S}} : \widehat{S} \to E$, we conclude that $\widehat{S}$, hence S, contains a cylinderlike open set. Note that in Theorem 2 the base curve C is assumed to be affine, whereas E is a complete curve. □

The following is a simple example which illustrates our present situation, which is Example 3.14 in [16].

Example 2 Let X be a smooth hypersurface in $\mathbb{A}^4 = \operatorname{Spec} k[x, y, u, z]$ defined by $xu - y^2 z = y$. Then X has a G_a-action defined by an lnd

$$\delta = x\frac{\partial}{\partial z} + y^2\frac{\partial}{\partial u}.$$

Then $\operatorname{Ker}\delta = k[x, y]$ and the quotient morphism $q : X \to Y \cong \mathbb{A}^2$ is the projection $(x, y, z, u) \mapsto (x, y)$. Hence $F := q^{-1}(0, 0) = \mathbb{A}^2 = \operatorname{Spec} k[z, u]$ and $q^{-1}(\alpha, \beta) \cong \mathbb{A}^1$ if $(\alpha, \beta) \neq (0, 0)$. With the above notations, let $L_\lambda = \{y = \lambda x\}$. Then $T_\lambda = q^{-1}(L_\lambda)$ meets F along the line $u = \lambda$. □

Remark 1 The hypothesis (H) holds if the fiber $F := q^{-1}(P)$ consists of only one component S of dimension two. In fact, $T_\lambda \cap S \neq \emptyset$ clearly, and then $\dim(T_\lambda \cap S)=1$.

---

[4] Let M be the maximal ideal of $B := \Gamma(Y, \mathcal{O}_Y)$ corresponding to the point P. Then $\widehat{X}$ is the blow-up of X with respect to MA, i.e., $\widehat{X} \cong \operatorname{Proj}_X(\oplus_{n\geq 0}(MA)^n)$. Let $\tau : \widehat{X} \to X$ be the canonical morphism. Then $\tau^{-1}(C_j) \cong C_j \times E$ for every one-dimensional component C_j of F. But $\tau^{-1}(S_i) \cong S_i$ for every two-dimensional component S_i of F. We note here that $S_i \cap C_j = \emptyset$ for all i and j.

2.4 Singular Fibers of $\mathbb{P}^1$-Fibrations on Smooth Projective Threefolds

Let $f : X \to Y$ be a $\mathbb{P}^1$-fibration with smooth algebraic varieties X and Y of dimension n and $n-1$, respectively. Namely, f is a projective morphism whose general fibers as well as the generic fiber $X_\eta := X \times_Y \operatorname{Spec} k(Y)$ are isomorphic to $\mathbb{P}^1$. In general, the general fibers being isomorphic to $\mathbb{P}^1$ implies that the generic fiber X_η is a form of $\mathbb{P}^1$, i.e., $X_\eta \otimes_{k(Y)} K$ is K-isomorphic to $\mathbb{P}^1$ for an algebraic extension K of $k(Y)$. The form X_η is trivial if $n = 2$ by Tsen's theorem. Meanwhile, if $n > 2$, this is not the case because the Brauer group of $k(Y)$ is not necessarily trivial.

Lemma 6 *Let $f : X \to Y$ be a $\mathbb{P}^1$-fibration with smooth algebraic varieties X and Y. Then the following assertions hold.*

(1) Let S be the closure of the set of points $Q \in Y$ such that either the scheme-theoretic fiber $F_Q := X \times_Y \operatorname{Spec} k(Q)$ has an irreducible component of dimension > 1 or every irreducible component F_i of F_Q has multiplicity > 1, i.e., $\operatorname{length} \mathcal{O}_{F_Q, F_i} > 1$. Then $\operatorname{codim}_Y S > 1$.
(2) Let $n = 3$. Then every fiber F_Q is simply-connected.

Proof (1) Suppose that $\operatorname{codim}_Y S = 1$. Then there exists an irreducible subvariety Z of codimension 1 of Y such that, for a general point Q of Z, either the fiber F_Q has an irreducible component of dimension > 1 or every irreducible component has multiplicity > 1. The first case is impossible. In fact, it then occurs that $\dim f^{-1}(Z) \geq \dim Z + 2 = n - 2 + 2 = n$ and $f^{-1}(Z) = X$. This is a contradiction. Consider the second case. Suppose that the fiber F_Q has only irreducible components of dimension one whose multiplicity is greater than one. We may assume that X and Y are projective. Let H be a hyperplane section of Y through the point Q. If H is general, the inverse image $X_H := f^{-1}(H)$ is smooth with the induced $\mathbb{P}^1$-fibration. In fact, by Bertini's second theorem, X_H has singularity along the base locus of the linear system $L := \{H \mid Q \in H\}$. Consider the blowing-up of X along the center $f^{-1}(Q)$. Since f is equi-dimensional locally over the point Q and since X and Y are smooth, f is a flat morphism locally over Q. Then the blowing-up of X along $f^{-1}(Q)$ is isomorphic to the fiber product $X \times_Y \widehat{Y}$ by the reasoning as in the footnote to the proof of Theorem 3, where $\widehat{Y} \to Y$ is the blowing-up of the point Q. Then the proper transforms of $f^{-1}(H)$ get separated from each other and have the same fiber $f^{-1}(Q)$ as the intersection with the exceptional divisor. Furthermore, we note that the blowing-up $B\ell_{f^{-1}(Q)_{\mathrm{red}}} X$ of X with respect to the center $f^{-1}(Q)_{\mathrm{red}}$ which is defined by the Ideal $\sqrt{\mathcal{I}_{f^{-1}(Q)}}$ is smooth since the fiber $f_{\widehat{Y}}^{-1}(\widehat{Q})$ of $f_{\widehat{Y}} : X \times_Y \widehat{Y} \to \widehat{Y}$ is locally isomorphic to the fiber $f^{-1}(Q)$, where $\widehat{Q}$ is a point of the exceptional locus of $\widehat{Y} \to Y$. Since $\mathcal{I}_{f^{-1}(Q)_{\mathrm{red}}} = \sqrt{\mathcal{I}_{f^{-1}(Q)}}$, we have $B\ell_{f^{-1}(Q)_{\mathrm{red}}} X = B\ell_{f^{-1}(Q)} X$. Hence $B\ell_{f^{-1}(Q)} X$ is smooth. By reducing dimension of Y by iterated hyperplane sections, we may assume that $\dim Y = 1$. Then $f^{-1}(H)$ is a smooth projective surface with a $\mathbb{P}^1$-fibration over a smooth projective curve H. Then it is well-known that the fiber $f^{-1}(Q)$ has a reduced component, which is a contradiction to the beginning hypothesis.

(2) By the assertion (1), the closed set S is a finite set. Let Q be a point of Y. We choose a small open disk D of Q in Y so that $D \cap S = \emptyset$ if $Q \notin S$ or $D \cap (S \setminus \{Q\}) = \emptyset$ if $Q \in S$. Then we may assume that the fiber F_Q is a strong deformation retract of $U := f^{-1}(D)$. Hence $\pi_1(U) = \pi_1(F_Q)$. On the other hand, we can apply Nori's result [16, Lemma 1.2.1] to $U \setminus F_Q \to D \setminus \{Q\}$ to obtain an exact sequence

$$\pi_1(\mathbb{P}^1) \to \pi_1(U \setminus F_Q) \to \pi_1(D \setminus \{Q\}) \to (1),$$

where $\pi_1(\mathbb{P}^1) = \pi_1(D \setminus \{Q\}) = (1)$. Hence $\pi_1(U \setminus F_Q) = (1)$. Since $U \setminus F_Q$ is an open set of U, we have a surjection $\pi_1(U \setminus F_Q) \to \pi_1(U)$. Hence we have $\pi_1(F_Q) = \pi_1(U) = (1)$. Hence F_Q is simply-connected. □

If $n = 3$, write the fiber F_Q as F_0 for the sake of simplicity, which may contain irreducible components of dimension 2. Write $F_0 = S \cup C$, where S (resp. C) is the sum of irreducible components of dimension 2 (resp. 1).

Lemma 7 *With the above notations and assumptions, if $n = 3$, we have the assertions.*

(1) $H_1(S; \mathbb{Z}) = H_1(C; \mathbb{Z}) = 0$.
(2) Each component of C is a rational curve, and each component of S is a rational surface or a rationally ruled surface.

Proof (1) As the Mayer-Vietoris exact sequence applied to $F_0 = S \cup C$, we have an exact sequence of integral homology groups

$$\cdots \longrightarrow H_2(F_0) \longrightarrow H_1(S \cap C) \longrightarrow H_1(S) \oplus H_1(C) \longrightarrow H_1(F_0) \longrightarrow \cdots,$$

where $H_1(F_0) = 0$ by the assertion (2) of Lemma 6 and $H_1(S \cap C) = 0$ because $S \cap C$ is a non-empty finite set. Hence $H_1(S) \oplus H_1(C) = 0$, whence $H_1(S) = H_1(C) = 0$.

(2) For the irreducible components of C, they have trivial first homology groups by the Mayer-Vietoris sequence. This implies that each irreducible component of C is at worst a cuspidal rational curve, i.e., a rational curve with only unibranch singularities. In order to show that each irreducible component of S is a rational surface or a rationally ruled surface, we apply Hironaka's flattening theorem [19]. We can thereby find a proper birational morphism $g : Y' \to Y$ with a smooth Y' such that the fiber product $X \times_Y Y'$ has an irreducible component X' with a flat $\mathbb{P}^1$-fibration $f' : X' \to Y'$. Let Q' be a point of Y', possibly in the exceptional locus of g as g is a composite of blowing-ups with centers at points. Let H' be a general hyperplane section of Y' passing through Q' and let V be the inverse image $f'^{-1}(H')$. Then $f'|_V : V \to H'$ is a $\mathbb{P}^1$-fibration. Let $\widetilde{V}$ be the normalization of V. Then $\widetilde{f} = (f'|_V) \circ \nu : \widetilde{V} \to H'$ is also a $\mathbb{P}^1$-fibration, where $\nu : \widetilde{V} \to V$ is the normalization morphism. Then every fiber is a sum of smooth rational curves. In fact, the smoothness of each rational curve follows from [31, (2.8.6.3), p. 107]. This implies that every fiber of $f'|_V$ is a sum of rational curves. Let S_i be an irreducible component of S. Then there exists a closed subvariety S_i' of X' such that S_i is dominated by S_i'. Note

that S_i' is a rationally ruled surface. If the image of the ruling on S_i has a base point, S_i is a rational surface. Otherwise, S_i is a rationally ruled surface. □

We occasionally have to consider an algebraic surface S with a $\mathbb{P}^1$-fibration like a subvariety of a smooth algebraic threefold X with a $\mathbb{P}^1$-fibration which has the induced $\mathbb{P}^1$-fibration. In general, S has singularity worse than normal singularity. We consider first the case of a normal surface.

Lemma 8 *Let S be a normal projective surface with a $\mathbb{P}^1$-fibration $f : S \to C$, where C is a smooth projective curve. Then we have:*

(1) S has only rational singularities, whose resolution graph is a tree of smooth rational curves and is a part of a degenerate fiber of $\mathbb{P}^1$.
(2) Every fiber F of f is a union of smooth rational curves, and its intersection dual graph is a tree in the sense that the dual graph of the inverse image of F in a minimal resolution of singularity of S is a tree.
(3) $H_1(F; \mathbb{Z}) = 0$.

Proof (1) Let $\sigma : \widehat{S} \to S$ be a minimal resolution of singularities. Then the composite $f \circ \sigma : \widehat{S} \to C$ is a $\mathbb{P}^1$-fibration. Since $\sigma^{-1}(f^{-1}(U)) = \widehat{f}^{-1}(U)$ for an open set U of C, each connected component of the exceptional locus of σ is a part of a degenerate fiber of $\widehat{f}$. Then it is well-known that this part contracts to a rational singular point. In fact, by using the five-term exact sequence associated to a spectral sequence $E_2^{pq} = R^q f_* R^p \sigma_* \mathcal{O}_{\widehat{S}} \Rightarrow R^n \widehat{f}_* \mathcal{O}_{\widehat{S}}$, we obtain an exact sequence

$$0 \to f_* R^1 \sigma_* \mathcal{O}_{\widehat{S}} \to R^1 \widehat{f}_* \mathcal{O}_{\widehat{S}} \to R^1 f_* \sigma_* \mathcal{O}_{\widehat{S}},$$

where $R^1 \widehat{f}_* \mathcal{O}_{\widehat{S}} = 0$ by [31, (2.8.6.2), p. 107]. Hence $f_* R^1 \sigma_* \mathcal{O}_{\widehat{S}} = 0$, whence $R^1 \sigma_* \mathcal{O}_{\widehat{S}} = 0$ because it is supported by a finite set of S.

(2) Let $\widehat{F} = \sigma^{-1}(F)$. Then F is the surjective image of $\widehat{F}$. Since $\widehat{F}$ is a union of rational curves, so is the image F. More precisely, $H^1(F_{\text{red}}, \mathcal{O}_{F_{\text{red}}}) = 0$ by [31, (2.8.6.3), p. 107]. If $F_{\text{red}} = \cup_{i=1}^{r} C_i$ is the irreducible decomposition, it follows that $H^1(C_i, \mathcal{O}_{C_i}) = 0$ for every i. This in turn implies that $C_i \cong \mathbb{P}^1$ for every i.

(3) The fiber F is obtained topologically from $\widehat{F}$ by removing the sum of the exceptional loci of singular points $Q_1, \ldots, Q_n$, which are rational singularities and adding back the points $T = \{Q_1, \ldots, Q_n\}$. As long exact sequences of integral homology (cohomology) groups for the pairs $(\widehat{F}, \Gamma)$ and (F, T), we have

$$\begin{array}{l} H_1(\Gamma) \to H_1(\widehat{F}) \to H^1(\widehat{F} \setminus \Gamma) \to H_0(\Gamma) \to H_0(\widehat{F}) \\ H_1(T) \to H_1(F) \to H^1(F \setminus T) \to H_0(T) \to H_0(F), \end{array}$$

where $H_1(\Gamma) = H_1(T) = 0$, $\widehat{F} \setminus \Gamma$ is homeomorphic to $F \setminus T$ and $H_0(\Gamma) \cong H_0(T) \cong \mathbb{Z}^{\oplus \#(T)}$. Hence $H_1(F; \mathbb{Z}) \cong H_1(\widehat{F}; \mathbb{Z}) = 0$. □

Remark 2 Let $f : S \to C$ be a $\mathbb{P}^1$-fibration as in Lemma 8. If $\text{Sing}(S) \neq \emptyset$, we may have either a multiple fiber or a fiber with more than two irreducible fiber components meeting in one point. We exhibit these phenomena by the following examples.

(1) Let L be a smooth fiber of a $\mathbb{P}^1$-fibration on a smooth surface which is isomorphic to $\mathbb{P}^1$. Blow up a point P_0 on L to obtain the exceptional curve E_1. Then blow up the intersection point P_1 of E_1 and the proper transform of L. Let E_2 be the exceptional curve. Then the inverse image of L is $L' + 2E_2 + E_1'$ with $(L')^2 = (E_1')^2 = -2$, where L' and E_1' are the proper transforms of L and E_1. Now contract L' and E_1' to rational double points. Let $\overline{E}_2$ be the image of E_2. Then $2\overline{E}_2$ is a fiber of the induced $\mathbb{P}^1$-fibration.

(2) In the above example, let P_2 be a point on E_2 which is different from the intersection points $L' \cap E_2$ and $E_1' \cap E_2$. Blow up P_2 and let E_3 be the exceptional curve. Then the obtained fiber is $L' + E_1' + 2E_2 + 2E_3$, where we denote the proper transforms of L', E_1', E_2 by the same letters. We have $(L')^2 = (E_1')^2 = (E_2)^2 = -2$. We contract the (-2)-curve E_2 to a rational double point. Then the images $\overline{L}'$, $\overline{E}_1'$, $\overline{E}_3$ of L', E_1', E_3 meet in the contracted singular point. According to [42], the intersection numbers are given as follows: $(\overline{E}_1')^2 = (\overline{L}')^2 = -\frac{3}{2}$, $(\overline{E}_3)^2 = -\frac{1}{2}$, $(\overline{E}_1' \cdot \overline{L}') = (\overline{E}_3 \cdot \overline{L}') = (\overline{E}_1' \cdot \overline{E}_3) = \frac{1}{2}$.

Lemma 9 *Let S be an algebraic surface, i.e., an integral k-scheme of finite type of dimension 2, and let $f : S \to C$ be a $\mathbb{P}^1$-fibration, i.e., f is a proper morphism whose general fibers are isomorphic to $\mathbb{P}^1$. Let F be a scheme-theoretic fiber of f over a closed point P. Then the following assertions hold.*

(1) Let $\widetilde{C}$ be the normalization of C and $\widetilde{f} : S \times_C \widetilde{C} \to \widetilde{C}$ be the base change of f. Let $\widetilde{P}$ be a point of $\widetilde{C}$ lying over P and let $\widetilde{F}$ be the scheme-theoretic fiber of $\widetilde{f}$ over $\widetilde{P}$. Then $\widetilde{F}$ is isomorphic to F as k-schemes. Hereafter we assume that C is normal.
(2) The singular locus of S is contained in the union of finitely many fibers of f.
(3) F is a connected union of rational irreducible components.
(4) $\pi_1(F)$ is a cyclic group. If S is normal, F is simply-connected.

Proof (1) The proof for the assertion (1) of Lemma 3 applies.

(2) There exists an open set U of C such that $f^{-1}(U) \cong U \times \mathbb{P}^1$. Let $\mathrm{Sing}(S)$ be the singular locus of S which is a closed set. Suppose that $\mathrm{Sing}(S)$ contains an irreducible component T of dimension one. Then T cannot lie horizontally to the fibration f because $\mathrm{Sing}(S) \cap f^{-1}(U) = \emptyset$. Hence T lies vertically to f. Namely it is contained in a fiber.

(3) Let $\nu : \widehat{S} \to S$ be the normalization morphism. Then $\widehat{f} = f \circ \nu : \widehat{S} \to C$ is a $\mathbb{P}^1$-fibration. Let $\widehat{F} := \widehat{f}^*(P)$ be the fiber of $\widehat{f}$ over P. Then $\widehat{F}$ is a union of irreducible components isomorphic to $\mathbb{P}^1$ by Lemma 8. Hence the fiber F is a connected union of rational curves.

(4) Let Δ be a small open disk (analytic neighborhood) of P and let $\Delta^* = \Delta \setminus \{P\}$. By the assertion (3), we can take Δ so small that $f^{-1}(\Delta^*) \cong \Delta^* \times \mathbb{P}^1$. Since F is a strong deformation retract of $f^{-1}(\Delta)$, we have a surjection $\pi_1(f^{-1}(\Delta^*)) \to \pi_1(f^{-1}(\Delta)) \to (1)$. Since $\pi_1(f^{-1}(\Delta^*)) \cong \pi_1(\Delta^*) \cong \mathbb{Z}$ and $\pi_1(f^{-1}(\Delta)) \cong \pi_1(F)$, it follows that $\pi_1(F)$ is a cyclic group. If S is normal, then $H_1(F) = 0$ by Lemma 8, (2). Hence F is simply connected. □

Lemma 10 *Let $f : X \to C$ be a projective morphism which is a $\mathbb{P}^1$-fibration over a smooth algebraic curve C and let F_0 be a singular fiber of f. Then F_0 is simply connected. In particular, every irreducible component is homeomorphic to $\mathbb{P}^1$.*

Proof We can replace C by its smooth complete model and assume that C is a smooth projective curve and X is a projective surface. Further, we may assume that the fiber F_0 is a unique singular fiber of f. By Lemma 9, F_0 is a connected union of rational irreducible components. Further, f has a cross-section T. For the existence of a cross-section, note that the generic fiber X_η of f is isomorphic to a smooth conic in $\mathbb{P}^2$ defined over the function field $k(C)$, where $k(C)$ is a C_1-field by Tsen's theorem. Hence X_η has a $k(C)$-rational point whose closure in X gives a cross-section of f. Since $k(C)$ is an infinite field, f has infinitely many cross-sections.

Suppose that F_0 is not simply-connected. We can assume, by blowing up points on F_0, that there is a loop in the dual graph of F_0. Let E be an irreducible component contained in a loop. We can assume that the section T meets some irreducible component of F_0 other than E. Consider the union D of T and all the components of F_0 except E. We may assume that D is connected. In fact, if F_0 contains an irreducible component E_1 which is not connected to the section T by a chain of irreducible components of F_0 unless it contains E, then we replace E by the exceptional curve arising by the blowing-up of the intersection point of E and an adjacent component of the loop. Now $E - D$ has at least two places at infinity.

Let $\widetilde{X}$ be the normalization of X and let $\widetilde{D}$, $\widetilde{E}$ be the inverse images of D, E in $\widetilde{X}$ respectively. Then $\widetilde{D}$ is connected. In fact, the section T is smooth because it dominates birationally the smooth curve C. If F_1 is an irreducible component of D meeting T, every irreducible component of its inverse image $\widetilde{F_1}$ in $\widetilde{X}$ passes through the unique point of the inverse image $\widetilde{T}$ of T lying over $T \cap F_0$, whence $\widetilde{F_1}$ is connected. Since every irreducible component of the inverse image of $F_0 \cap D$ is connected to an irreducible component of $\widetilde{F_1}$, we know that $\widetilde{D}$ is connected. Note that $\widetilde{E}$ might be reducible.

Now consider the relative cohomology exact sequence of a pair $(\widetilde{X}, \widetilde{D})$ with integral coefficients,

$$\longrightarrow H^0(\widetilde{X}) \longrightarrow H^0(\widetilde{D}) \longrightarrow H^1(\widetilde{X}, \widetilde{D}) \longrightarrow H^1(\widetilde{X}) \longrightarrow H^1(\widetilde{D}) \longrightarrow .$$

Since $\widetilde{X}$ is normal, $\widetilde{X} \setminus \widetilde{F_0}$ is a $\mathbb{P}^1$-bundle and $\widetilde{F_0}$ is simply-connected by Lemma 9, we know that $H^1(\widetilde{X}) \cong H^1(C)$. Since $\widetilde{D}$ is contracted to C, we know that $H^1(\widetilde{D}) \cong H^1(C)$. Hence $H^1(\widetilde{X}) \to H^1(\widetilde{D})$ is an isomorphism. Also, the map $H^0(\widetilde{X}) \to H^0(\widetilde{D})$ is an isomorphism since both $\widetilde{X}$ and $\widetilde{D}$ are connected. The above exact sequence shows that $H^1(\widetilde{X}, \widetilde{D}) = 0$. Hence we have $H_3(\widetilde{X} - \widetilde{D}; \mathbb{Q}) = 0$ by Lefschetz duality.

We note that $\widetilde{X}$ has at worst rational singular points. Hence Poincare (and Lefschetz) duality with rational coefficients is valid as here and used below.

Consider the relative $\mathbb{Q}$-coefficient cohomology exact sequence with compact support for $(\widetilde{X} - \widetilde{D}, \widetilde{E} - \widetilde{D})$.

$$\longrightarrow H_c^0(\widetilde{X}-\widetilde{D}) \longrightarrow H_c^0(\widetilde{E}-\widetilde{D}) \longrightarrow H_c^1(\widetilde{X}-\widetilde{D},\widetilde{E}-\widetilde{D})$$
$$\longrightarrow H_c^1(\widetilde{X}-\widetilde{D}) \longrightarrow H_c^1(\widetilde{E}-\widetilde{D}) \longrightarrow H_c^2(\widetilde{X}-\widetilde{D},\widetilde{E}-\widetilde{D}) \longrightarrow .$$

By duality, $H_c^1(\widetilde{X}-\widetilde{D},\widetilde{E}-\widetilde{D})$ is isomorphic to $H_3(\widetilde{X}-(\widetilde{D}\cup\widetilde{E}))=0$, since $\widetilde{X}-(\widetilde{D}\cup\widetilde{E})$ is an $\mathbb{A}^1$-bundle over C_0 and hence $H_3(\widetilde{X}-(\widetilde{D}\cup\widetilde{E}))\cong H_3(C_0)=0$. Similarly, $H_c^2(\widetilde{X}-\widetilde{D},\widetilde{E}-\widetilde{D})\cong H_2(C_0)=0$ because C_0 is an affine curve. Further, by the duality, $H_c^1(\widetilde{X}-\widetilde{D};\mathbb{Q})\cong H_3(\widetilde{X}-\widetilde{D};\mathbb{Q})=0$ as seen above. Hence $H_c^1(\widetilde{E}-\widetilde{D};\mathbb{Q})=0$. This is a contradiction since $E-D$ has at least two places at infinity and $H_c^1(\widetilde{E}-\widetilde{D};\mathbb{Q})\cong H^1(\widetilde{E}-\widetilde{D};\mathbb{Q})$. In fact, write $\widetilde{E}-\widetilde{D}=Z_1\coprod\cdots\coprod Z_r$, where each Z_i is completed to a tree of smooth rational curves $\widetilde{Z}_i$. If none of Z_i has an irreducible component with two or more places missing, an invertible function on Z_i is a constant. Since $\Gamma(E-D,\mathcal{O}^*)\subseteq\Gamma(\widetilde{E}-\widetilde{D},\mathcal{O}^*)$ and $\Gamma(E-D,\mathcal{O}^*)$ is not a finite sum of k, we have a contradiction. Hence some connected component Z_i has an irreducible component with at least two places missing. This irreducible component gives a non-zero element in $H^1(\widetilde{E}-\widetilde{D};\mathbb{Q})\cong H_1(\widetilde{E}-\widetilde{D};\mathbb{Q})$. □

Theorem 4 *Let $f: X\to Y$ be an equi-dimensional $\mathbb{P}^1$-fibration over a smooth projective variety Y and let $F_0:=f^{-1}(P)$ be a closed fiber of f. Then F_0 is simply-connected.*

Proof Suppose that $\dim Y\geq 2$. Let H be a hyperplane section of Y passing through the point P and let $X_H:=f^{-1}(H)$. Then H is a smooth projective variety and $f_H:=f|_{X_H}:X_H\to H$ is an equi-dimensional $\mathbb{P}^1$-fibration with the fiber F_0 over the point P. By taking hyperplane sections through P repeatedly, we are reduced to the case where $\dim Y=1$. Then Lemma 10 gives the result. □

2.5 *Freeness Conjecture of G. Freudenburg*

In his talk at the international conference "Affine Geometry, hyperbolicity, complex spaces", October 4–28, 2016 in Grenoble, G. Freudenburg made the following:

Conjecture 1 Let $A=k[x,y,z]$ be a polynomial ring in three variables with a nontrivial locally nilpotent derivation (lnd, for short) D and let $B=\mathrm{Ker}\,D$. Then A is a free B-module.

We denote by $k^{[n]}$ a polynomial ring in n variables with n generating variables not specified. Let $A=k^{[n]}$ with a nontrivial lnd D and let $B=\mathrm{Ker}\,D$. We can ask a similar question for $k^{[n]}$ as in Conjecture 1. If $n\geq 4$, then A is not a free B-module. Even less, A is not B-flat. A counterexample is $A=k[x_1,x_2,x_3,x_4]$ and

$$D=x_1\frac{\partial}{\partial x_2}+x_2\frac{\partial}{\partial x_3}+(x_2^2-2x_1x_3-1)\frac{\partial}{\partial x_4}.$$

Then B is a hypersurface $\xi_1\xi_4=\xi_3^2-\xi_2(\xi_2-1)^2$. If one checks the singular fibers of the quotient morphism $q:X\to Y$ with $X=\mathrm{Spec}\,A$ and $Y=\mathrm{Spec}\,B$, there is a unique singular fiber $\mathbb{A}^2 + \mathbb{A}^2$ (see [16, Example 4.17]). Hence q is not equi-dimensional. This implies that A is not B-flat.

If $n = 2$, we can choose variables x, y of A so that $A = k[x, y]$, $B = k[x]$ and $D(y) = f(x)$ with $f(x) \in A$. Hence A is a free B-module with a free basis $\{y^i \mid i \geq 0\}$. Hence the case $n = 3$ is the only case remained open. Even in this case, it is known by [3, 39] that $Y \cong \mathbb{A}^2$ and $q : X \to Y$ is equi-dimensional. Furthermore, every singular fiber is a disjoint union of the curves isomorphic to $\mathbb{A}^1$ by [16]. Hence we know that A is B-flat and that A is the inductive limit of a directed set of finitely generated free B-submodules by [34]. We can replace $A = k[x, y, z]$ in Conjecture1 by an affine domain of dimension $n = 2$ or 3. We consider the case $n = 2$.

Theorem 5 *Let X be an irreducible affine surface defined over k with an $\mathbb{A}^1$-fibration $f : X \to C$ onto a smooth affine curve C. Let A and B be the coordinate rings of X and C, respectively. Then A is a free B-module.*

Proof We reduce to the case when X is normal as follows. Let $\widetilde{A}$ be the normalization of A in its quotient field. Then $\widetilde{A}$ is the coordinate ring of the normalization $\widetilde{X}$ of X in its function field. Suppose that $\widetilde{A}$ is a free B-module. Now A is an B-submodule of $\widetilde{A}$. For any localization $B_{\mathfrak{m}}$ of B with respect to a maximal ideal $\mathfrak{m}$ of B, the localization $A_{\mathfrak{m}} := A \otimes_B B_{\mathfrak{m}}$ is an $B_{\mathfrak{m}}$-submodule of $\widetilde{A}_{\mathfrak{m}}$. Since $B_{\mathfrak{m}}$ is a PID, $A_{\mathfrak{m}}$ is a free $B_{\mathfrak{m}}$-module because a submodule of a free module over a PID is free. This proves that A is a projective B-module of infinite rank. By a result of Bass [1, Theorem 4.3], A is B-free. Now we can assume that X, and hence A, is normal.

If f is an $\mathbb{A}^1$-bundle then A can be seen to be a projective B-module because A is locally a polynomial ring in one variable over B, hence also a free B-module by Bass' result above. We will reduce to the $\mathbb{A}^1$-bundle case by the usual ramified covering trick. There exists a finite covering $C' \to C$ of C with smooth C' such that

(1) C' is ramified with a suitable ramification index over every point P of C with $f^{-1}(P)$ containing no reduced irreducible components,
(2) the normalization of the fiber product $X \times_C C'$, say X', has a surjective $\mathbb{A}^1$-fibration $f' : X' \to C'$ such that every fiber of f' has a reduced irreducible component.

By removing all the irreducible components from all the singular fibers except one reduced irreducible component, we obtain an open subset X'' of X' such that the induced $\mathbb{A}^1$-fibration $f'' : X'' \to C'$ has no singular fibers, i.e., no fibers which are not isomorphic to $\mathbb{A}^1$. Then X'' is smooth and is an $\mathbb{A}^1$-bundle. The coordinate ring of X'' is a free B'-module, where B' is the coordinate ring of C'. Note that B' is a projective B-module because B' is a finitely generated, torsion-free module over a Dedekind domain B. Since the coordinate ring A' of X' is a B'-submodule of the coordinate ring of X'', A' is a projective B'-module, hence a projective B-module. Then the result follows from the Bass' theorem above. □

3 Equivariant Abhyankar-Sathaye Conjecture in Dimension Three

In this section, the Abhyankar-Sathaye Conjecture in dimension three is affirmatively proved if the given polynomial f is invariant under a nontrivial action of the additive group G_a on Spec $k[x, y, z]$ and some technical conditions on f are satisfied.

3.1 Arguments on Singular and Plinth Loci

Let $A = k[x, y, z]$ be a polynomial ring in three variables. Let f be a non-constant element of A. Then the Abhyankar-Sathaye Conjecture in dimension three asserts that *if the affine hypersurface* $X_0 = \{f = 0\}$ *in* $X = \operatorname{Spec} A \cong \mathbb{A}^3$ *is isomorphic to the affine plane* $\mathbb{A}^2$ *then so is the hypersurface* $X_c = \{f = c\}$ *for every* $c \in k$. The conjecture implies that f is a coordinate, that is to say, $A = k[f, g, h]$ with certain elements $g, h \in A$, and *vice versa*.

It is noteworthy that by a result of Kaliman [21], if $X_c = \{f = c\}$ is isomorphic to $\mathbb{A}^2$ for general $c \in k$, then f is a coordinate of A. See also Daigle-Kaliman [5].

There are several partial results proving the conjecture in the general case with n variables, e.g., by Sathaye [44] and Popov [41]. We also add one partial affirmative result by assuming that the additive group G_a acts non-trivially on X and X_0 is G_a-stable. Let δ be a locally nilpotent derivation (*lnd*, in short) on A which corresponds to the G_a-action. Let $B = \operatorname{Ker} \delta$. Then B is a polynomial ring in two variables by [39], and B is factorially closed in A. For the terminology, see [12]. Let $Y = \operatorname{Spec} B$. Then the inclusion $B \hookrightarrow A$ defines the quotient morphism $q : X \to Y$, which is equi-dimensional and surjective by [3]. Assume that $f \in B$.

Since $q : X \to Y$ is an $\mathbb{A}^1$-fibration, we define the *singular locus* of q by $\operatorname{Sing}(q) = \{Q \in Y \mid q^{-1}(Q) \not\cong \mathbb{A}^1\}$, where $q^{-1}(Q)$ is the scheme-theoretic fiber over a point $Q \in Y$. Then, by [13, Lemma 3.1], $\operatorname{Sing}(q)$ is a closed set of Y. By Dutta [6], $\operatorname{Sing}(q)$ has pure dimension one.

We define the *plinth ideal* of δ by $\operatorname{pl}(\delta) = B \cap \delta(A)$ which is an ideal of B, and the *plinth locus* of q by $\operatorname{pl}(q) = V(\operatorname{pl}(\delta))$. Then we have the following result which we state in a slightly more general situation.

Lemma 11 *Let A be a regular factorial affine domain of dimension* 3 *and let δ be a nonzero lnd on A. With the same notations as above,* $\operatorname{Sing}(q)$ *is the codimension one part of* $\operatorname{pl}(q)$.

Proof (1) Let S be an irreducible component of $\operatorname{Sing}(q)$. Then it is defined by a prime element p of B. Suppose that $S \not\subset \operatorname{pl}(q)$. Then there exists a maximal ideal $\mathfrak{m}$ of B such that $p \in \mathfrak{m}$ but $\operatorname{pl}(\delta) \not\subset \mathfrak{m}$. Then there exist elements $b \in B \setminus \{0\}$ and $u \in A$ such that $b = \delta(u) \notin \mathfrak{m}$ and hence $A[b^{-1}] = B[b^{-1}][u]$. Since $b^{-1} \in B_{\mathfrak{m}}$, it follows that $A_{\mathfrak{m}} := A \otimes_B B_{\mathfrak{m}} = B_{\mathfrak{m}}[u]$. Namely, the point Q defined by $\mathfrak{m}$ is not a point of $\mathrm{Sing}(q)$. This is a contradiction because $Q \in V(pB) \subset \mathrm{Sing}(q)$. Hence $\mathrm{Sing}(q) \subseteq \mathrm{pl}(q)$.

(2) Take an irreducible component $V(pB)$ of $\mathrm{pl}(q)$, where p is a prime element of B. Suppose that $V(pB) \not\subset \mathrm{Sing}(q)$. Take a closed point Q of $V(pB)$ such that $Q \notin \mathrm{Sing}(q)$. Let $\mathfrak{m}$ be the maximal ideal of B corresponding to Q. Then $A_{\mathfrak{m}} = B_{\mathfrak{m}}[u]$ since the quotient morphism $q : X \to Y$ is an $\mathbb{A}^1$-bundle in a small open neighborhood of Q. We can take u to be an element of A such that $\delta(u) = b \in B \setminus \mathfrak{m}$. Hence $b \in \mathrm{pl}(\delta)$, and $\mathrm{pl}(\delta) \not\subset \mathfrak{m}$. This is a contradiction to the choice of $\mathfrak{m}$. Hence every irreducible component of $\mathrm{pl}(q)$ is contained in $\mathrm{Sing}(q)$. □

Let $f \in B$ be an element such that the hypersurface X_0 in $X = \mathbb{A}^3$ defined by the ideal fA is isomorphic to $\mathbb{A}^2$. Let $\varphi : X \to \mathbb{A}^1 = \mathrm{Spec}\, k[f]$ be the morphism defined by $P \mapsto f(P)$. The hypersurface X_c is the scheme-theoretic fiber $\mathrm{Spec}\, A/(f - c)A$ of the morphism φ over the point of $\mathbb{A}^1$ defined by $f = c$. The morphism φ is decomposed as

$$\varphi : X \xrightarrow{q} Y \xrightarrow{p} \mathbb{A}^1,$$

where p is induced by the inclusion $k[f] \hookrightarrow B$. The G_a-equivariant Abhyankar-Sathaye conjecture will be proved if the following two assertions hold true.

(1) The curve $Y_c = \mathrm{Spec}\, B/(f - c)B$ in Y is the affine line in the affine plane Y.
(2) The restriction of the quotient morphism $q|_{X_c} : X_c \to Y_c$ is an $\mathbb{A}^1$-bundle.

In fact, if these two assertions are proved, then X_c is an $\mathbb{A}^1$-bundle over $\mathbb{A}^1$, which is trivial. Thus, $X_c \cong \mathbb{A}^2$ for every $c \in k$. Then X is an $\mathbb{A}^2$-bundle over $\mathbb{A}^1$ by [43], where the local triviality in the sense of Zariski topology follows from [15, Theorem 3.10] (see also [27]). Then this $\mathbb{A}^2$-bundle is trivial by [2], i.e., $X \cong \mathbb{A}^2 \times \mathbb{A}^1$. This implies that f is a coordinate of A. We call (1) and (2) the requirements.

We show that the requirement (1) always holds true in the following lemma. We may assume that δ is irreducible (cf. [12]). Then the induced G_a-action on X_c is non-trivial. Hence $q|_{X_c}$ is decomposed as

$$q|_{X_c} : X_c \xrightarrow{q_c} X_c//G_a \xrightarrow{r_c} Y_c,$$

where q_c is the quotient morphism and $X_c//G_a$ is the algebraic quotient.

Lemma 12 *The following assertions hold.*

(1) For every $c \in k$, the element $f - c$ is irreducible in B.
(2) The curve Y_c is isomorphic to $\mathbb{A}^1$ for every $c \in k$.

Proof (1) Consider $q|_{X_0} : X_0 \to Y_0$. Since $X_0 \cong \mathbb{A}^2$ by the assumption, it follows that $X_0//G_a \cong \mathbb{A}^1$. Then the morphism $r_0 : X_0//G_a \to Y_0$ is a finite morphism (Stein factorization of $q|_{X_0}$). Hence Y_0 is an irreducible curve in $Y \cong \mathbb{A}^2$ with only one place at infinity. Then by the irreducibility theorem [35, p. 89], the curve Y_c is an irreducible curve with only one place at infinity. Hence $f - c$ is an irreducible element of B.

(2) We show that Y_0 is a smooth curve. By the initial assumption that $X_0 \cong \mathbb{A}^2$, Y_0 is a rational affine curve with one place at infinity. Hence if it is smooth, then $Y_0 \cong \mathbb{A}^1$. Now note that $B = k[v, w]$ is a polynomial ring over k. Then we have

$$\frac{\partial f}{\partial x} = \frac{\partial f}{\partial v}\frac{\partial v}{\partial x} + \frac{\partial f}{\partial w}\frac{\partial w}{\partial x}$$
$$\frac{\partial f}{\partial y} = \frac{\partial f}{\partial v}\frac{\partial v}{\partial y} + \frac{\partial f}{\partial w}\frac{\partial w}{\partial y}$$
$$\frac{\partial f}{\partial z} = \frac{\partial f}{\partial v}\frac{\partial v}{\partial z} + \frac{\partial f}{\partial w}\frac{\partial w}{\partial z}$$

If Q is a singular point of Y_0 then $\frac{\partial f}{\partial v}(Q) = \frac{\partial f}{\partial w}(Q) = f(v(Q), w(Q)) = f(P) = 0$, where P is a point of X such that $Q = q(P)$. Hence P is a singular point of X_0 which contradicts the smoothness of X_0. Hence Y_0 is smooth, and isomorphic to $\mathbb{A}^1$. By Abhyankar-Moh-Suzuki theorem, it follows that $Y_c \cong \mathbb{A}^1$ for all $c \in k$. □

Lemma 13 *The morphism $r_c : X_c//G_a \to Y_c$ for $c \in k$ is an isomorphism if the ideal $(f - c)B$ does not contain* $\mathrm{pl}(\delta)$.

By Lemma 11, the assumption that $(f - c)B \not\ni \mathrm{pl}(\delta)$ is equivalent to the condition that Y_c is not an irreducible component of $\mathrm{pl}(q)$, hence of $\mathrm{Sing}(q)$. This implies that $\mathrm{Sing}(q)$ lies horizontally along the fibration $p : Y \to \mathbb{A}^1$ if $\mathrm{Sing}(q) \neq \emptyset$.

Proof To simplify the notations, we consider the case $c = 0$. The proof is the same in the general case.

The morphism $q|_{X_0} : X_0 \to Y_0$ is induced by the natural inclusion $B/fB \hookrightarrow A/fA$. Let δ_0 be the lnd on A/fA induced by δ. Then we have an inclusion $B/fB \hookrightarrow \mathrm{Ker}\,\delta_0$, which gives the morphism r_0 in the above decomposition of $q|_{X_0}$.

In order to show that r_0 is an isomorphism, it suffices to show that if $\delta(a) \in fA$ then there exists an element $b \in B$ such that $a - b \in fA$.

Suppose first that δ has a slice u, i.e., $\delta(u) = 1$. Then $A = B[u]$, and we can write

$$a = b_0 + b_1 u + \cdots + b_n u^n, \quad b_0, \ldots, b_n \in B.$$

Since $\delta(a) = b_1 + 2b_2 u + \cdots + nb_n u^{n-1}$ is divisible by f in A, the coefficients $b_1, \ldots, b_n$ are divisible by f in B. Write

$$\delta(a) = f\left(b_1' + 2b_2' u + \cdots + nb_n' u^{n-1}\right).$$

Set

$$a' = b_1' u + b_2' u^2 + b_3' u^3 + \cdots + b_n' u^n.$$

Then $\delta(a) = f\delta(a')$, and hence $a - fa' \in B$. Set $b = a - fa'$. Then $a - b = fa' \in fA$.

Suppose next that u is a local slice. Namely $\alpha = \delta(u)$ is a nonzero element of B. Then δ extends to an lnd δ' of $A[\alpha^{-1}]$ and $B[\alpha^{-1}] = \mathrm{Ker}\,\delta'$. Since $\delta'(u/\alpha) = 1$, by the previous result, we have $a - (b/\alpha^r) \in fA[\alpha^{-1}]$. Hence, replacing b by $\alpha^s b$ with a suitable s, we have $\alpha^r a - b \in fA$.

Let $\mathfrak{p}$ be a prime ideal of B such that $f \in \mathfrak{p}$ and $\mathfrak{p} \notin V(\mathrm{pl}(\delta))$, i.e., $\mathfrak{p} \in Y_0 \setminus V(\mathrm{pl}(\delta))$. Such a prime ideal $\mathfrak{p}$ exists by the assumption. Let $\delta_\mathfrak{p}$ be the extension of δ to $A_\mathfrak{p} = A \otimes_B B_\mathfrak{p}$. Then we can find an element $\alpha \in \mathrm{pl}(\delta)$ such that $\alpha \notin \mathfrak{p}$. Let $(\delta_\mathfrak{p})_0$ be the restriction of $\delta_\mathfrak{p}$ onto $A_\mathfrak{p}/fA_\mathfrak{p}$. Then we have Ker $(\delta_\mathfrak{p})_0 = B_\mathfrak{p}/fB_\mathfrak{p}$ because u/α is a slice of $\delta_\mathfrak{p}$ if $\alpha = \delta(u)$. This implies that r_c is birational.

Since $Y_c \cong \mathbb{A}^1$ by Lemma 12, the birational morphism r_c is an isomorphism. □

We call (H1) the condition that $(f - c)B \not\supset \mathrm{pl}(\delta)$ for every $c \in k$.

Lemma 14 *Assume the condition* (H1) *and that* $Y_0 \cap \mathrm{Sing}(q) = \emptyset$. *Then* f *is a variable of* A, *i.e.,* $A = k[f, g, h]$ *for some* $g, h \in A$.

Proof Since $Y_0 \cap \mathrm{Sing}(q) = \emptyset$ by the assumption, Sing(q) is either the empty set or a finite disjoint union $\coprod_i Y_{c_i}$ with $c_i \in k$. Meanwhile, the condition (H1) implies that the last case does not occur. Hence Sing$(q) = \emptyset$. Therefore the requirements (1) and (2) are fulfilled. □

Lemma 15 *Assume that* $Y_c \cap \mathrm{Sing}(q) = \emptyset$ *for some* $c \in k$. *Then* f *is a variable of* A.

Proof Since $\mathrm{Sing}(q) \cap Y_c = \emptyset$, $q|_{X_c} : X_c \to Y_c$ is an $\mathbb{A}^1$-fibration which has no singular fibers. It implies, in particular, that $r_c : X_c//G_a \to Y_c$ is an isomorphism and $q_c : X_c \to Y_c$ is an $\mathbb{A}^1$-bundle. Furthermore, since $Y_c \cong \mathbb{A}^1$ as a parallel line by Lemma 12, it follows that $X_c \cong \mathbb{A}^2$, and Sing(q) is a disjoint sum of Y_{c_i} for $c_1, \ldots, c_r$ unless Sing$(q) = \emptyset$. This implies that almost all fibers of $\varphi : X \to \mathbb{A}^1 = \mathrm{Spec}\, k[f]$ are isomorphic to $\mathbb{A}^2$. Then f is a variable by Kaliman [21]. □

3.2 Statement of Theorem

We inherit the notations of Sect. 3.1, namely, δ is an lnd on $A = k[x, y, z]$ and $B = \mathrm{Ker}\, \delta$. Let $q : X = \mathrm{Spec}\, A \to \mathrm{Spec}\, B$ be the quotient morphism and let $f \in B$. We assume that the hypersurface $f = 0$ in $X \cong \mathbb{A}^3$ is isomorphic to $\mathbb{A}^2$.

Theorem 6 *With the above notation and assumption, the following assertions hold.*

(1) If $(f - c)B \not\supset \mathrm{pl}(\delta) = B \cap \delta(A)$ *for every* $c \in k$, *then the singular fibers of the quotient morphism* $q : X \to Y$ *are disjoint union of affine lines, and* $\mathrm{Sing}(q) = \{Q \in Y \mid q^{-1}(Q) \not\cong \mathbb{A}^1\}$ *lies horizontally along the* $\mathbb{A}^1$*-bundle* $p : Y \to \mathbb{A}^1 = \mathrm{Spec}\, k[f]$.

(2) f *is a variable of* A *if* $\mathrm{Sing}(q) \cap Y_c = \emptyset$ *for some* $c \in k$.

Proof The first assertion follows from Lemma 13, and the second assertion from Lemma 15. □

4 Forms of $\mathbb{A}^3$ with Unipotent Group Actions

We show that a form of $\mathbb{A}^3$ is trivial if it has either a fixed-point free G_a-action or an effective action of a unipotent algebraic group of dimension two.

4.1 Preliminary Results

In the present section, we let k be a field of characteristic zero and let $\overline{k}$ be an algebraic closure of k. A geometrically integral algebraic scheme over k is simply called a *k-variety*. A k-variety X is smooth if $Y \otimes_k \overline{k}$ is smooth. An algebraic k-variety X is called a *k-form* of $\mathbb{A}^n$ (or simply, a form of $\mathbb{A}^n$) if $X \otimes_k \overline{k}$ is isomorphic to $\mathbb{A}^n$ over $\overline{k}$. A k-form X of $\mathbb{A}^n$ is *trivial* if Y itself is isomorphic to $\mathbb{A}^n$ over k. It is well-known that, for $n = 1, 2$, the Galois cohomology $H^1_{et}(\overline{k}/k, \mathrm{Aut}_{\mathbb{A}^n/k})$ vanishes due to the structure theorem of the automorphism group $\mathrm{Aut}_{\mathbb{A}^n/k}$. Therefore, any k-form of $\mathbb{A}^n$ is trivial if $n = 1, 2$ (see [28] for the case $n = 2$).

Koras and Russell [32] proved that a k-form of $\mathbb{A}^3$ with a nontrivial G_m-action is trivial. We are motivated by this result to ask if a k-form of $\mathbb{A}^3$ (or $\mathbb{A}^n$ with $n \geq 3$) is trivial provided it has a nontrivial unipotent group action. We shall show that this is the case if a k-form of $\mathbb{A}^3$ has either a fixed-point free G_a-action or an effective action of a unipotent group of dimension 2. These are essentially due to Kaliman [22] and Daigle-Kaliman [5]. The third author was informed by Neena Gupta of her recent work [7] which asserts that a (separable) k-form $X = \mathrm{Spec}\, A$ of $\mathbb{A}^3$ is trivial provided A is endowed with a locally nilpotent k-derivation D such that $\mathrm{rank}(D \otimes 1_{\overline{k}}) \leq 2$.

Lemma 16 *Let X be a smooth k-variety such that $\overline{X} := X \otimes_k \overline{k}$ is affine and factorial and that $\Gamma(\overline{X}, \mathcal{O}^*_{\overline{X}}) = \overline{k}^*$. Then X is an affine factorial k-variety with $\Gamma(X, \mathcal{O}^*_X) = k^*$.*

Proof Let $\mathcal{F}$ be a quasi-coherent $\mathcal{O}_X$-sheaf. Then $\mathcal{F} \otimes_k \overline{k}$ is a quasi-coherent $\mathcal{O}_{\overline{X}}$-sheaf. Since $\overline{X}$ is affine, $H^i(\overline{X}, \mathcal{F} \otimes \overline{k}) \cong H^i(X, \mathcal{F}) \otimes_k \overline{k} = 0$ for all $i > 0$. Hence $H^i(X, \mathcal{F}) = 0$ for all $i > 0$, and X is affine by Serre's criterion of affineness. Write $A = \Gamma(X, \mathcal{O}_X)$. Then $\overline{A} := A \otimes_k \overline{k} = \Gamma(\overline{X}, \mathcal{O}_{\overline{X}})$. It is clear that $k^* \subseteq \Gamma(X, \mathcal{O}^*_X) \subseteq \Gamma(\overline{X}, \mathcal{O}^*_{\overline{X}}) = \overline{k}^*$. Since $\Gamma(X, \mathcal{O}_X)^*$ is invariant under the Galois group of $\overline{k}/k$, we obtain $\Gamma(X, \mathcal{O}^*_X) = k^*$.

Let D be a k-irreducible subvariety of codimension one in X and let $\overline{D} = D \otimes_k \overline{k}$. Let $\overline{D} = D_1 + \cdots + D_r$ be the irreducible decomposition of $\overline{D}$. If one fixes one irreducible component, say D_1, the other irreducible components are translates of D_1 under the Galois group of $\overline{k}/k$. Since $\overline{X}$ is factorial, D_1 is defined by an element f_1 of $\overline{A}$. Then the translates D_i of D_1 are defined by the translates f_i of f_1. Let $F = \prod_{i=1}^r f_i$. Then $g(F) = a(g)F$ for an element g of the Galois group $\mathfrak{G}$.[5] Then

[5]More precisely, we take a finite Galois extension k'/k such that $D \otimes_k k'$ splits into a sum of geometrically irreducible components and consider the Galois group $\mathrm{Gal}(k'/k)$.

$\{a(g)\}$ determines a cocycle of $\mathfrak{G}$ with values in $\overline{k}^*$. By Hilbert Theorem 90 (the multiplicative case), we find an element $b \in \overline{k}^*$ such that $a(g) = g(b)b^{-1}$. Then Fb^{-1} is invariant under $\mathfrak{G}$ and D is defined by $Fb^{-1} \in \Gamma(X, \mathcal{O}_X)$. □

An algebraic group G defined over k is called a *unipotent* group if so is $\overline{G} := G \otimes_k \overline{k}$.

Lemma 17 *Let G be a unipotent algebraic group defined over k. Then the following assertions hold.*

(1) If $\dim G = 1$ *then G is k-isomorphic to the additive group scheme* G_a*. Hence a G-action on an affine k-scheme* $X = \operatorname{Spec} A$ *is given by a locally nilpotent k-derivation (k-lnd, for short)* δ *on A.*
(2) If $\dim G = 2$ *then G is k-isomorphic to a direct product* $G_a \times G_a$*. Hence a G-action on an affine k-scheme X as above is given by two k-lnds* δ_1, δ_2 *on A such that* $\delta_1\delta_2 = \delta_2\delta_1$.

Proof (1) Since $\dim G = 1$, it follows that $\overline{G} \cong G_a$ over $\overline{k}$. Hence the underlying scheme of G is a k-form of $\mathbb{A}^1$, which is trivial. Thus we can write $G = \operatorname{Spec} k[x]$. Let Δ be the comultiplication of G. Then $\Delta(x) = x \otimes 1 + 1 \otimes x$ over $\overline{k}$ since this gives a unique group structure on $\overline{G} \cong G_{a,\overline{k}}$. Hence we have $\Delta(x) = x \otimes 1 + 1 \otimes x$ over k and $G \cong G_{a,k}$.

(2) Since $\dim G = 2$, it is known that $\overline{G}$ is $\overline{k}$-isomorphic to a direct product $G_a \times G_a$.[6] Hence G is commutative and the underlying scheme of G is k-isomorphic to $\mathbb{A}^2$. Let $L(G)$ be the Lie algebra of G. Then $L(G)$ is a k-vector space of rank 2 with bracket product. Hence there exist k-derivations Δ_1, Δ_2 of $R := \Gamma(G, \mathcal{O}_G)$ such that $L(G) = k\Delta_1 + k\Delta_2$. Since $L(\overline{G}) = L(G) \otimes_k \overline{k} = \overline{k}\partial_1 + \overline{k}\partial_2$ with $\overline{k}$-lnds ∂_1, ∂_2 of $\overline{R} = R \otimes_k \overline{k}$, then Δ_1, Δ_2 as elements of $L(\overline{G})$ are expressed as $\overline{k}$-linear combinations of ∂_1, ∂_2. Since $\partial_1\partial_2 = \partial_2\partial_1$, it is straightforward to verify that Δ_1, Δ_2 are k-lnds such that $\Delta_1\Delta_2 = \Delta_2\Delta_1$. This implies that G is k-isomorphic to $G_a \times G_a$. A G-action on X corresponds to an algebraic homomorphism $\rho : G \to \operatorname{Aut}_{X/k}$ whose associated homomorphism of Lie algebras is $L(\rho) : L(G) \to \operatorname{Der}_k(A)$. Then the images of Δ_1, Δ_2 by $L(\rho)$ are the lnds δ_1, δ_2 of A such that $\delta_1\delta_2 = \delta_2\delta_1$. □

Let G be a unipotent group of dimension 2 defined over k. A direct product decomposition $G = G_a \times G_a$ corresponds to the writing $L(G) = k\Delta_1 + k\Delta_2$ with the notations in the proof of Lemma 17, where Δ_1, Δ_2 are the k-lnds of $R = \Gamma(G, \mathcal{O}_G)$ with $\Delta_1\Delta_2 = \Delta_2\Delta_1$. In other words, $\Delta_1 = \partial/\partial x_1$ and $\Delta_2 = \partial/\partial x_2$, where x_1, x_2 are the coordinates of the underlying schemes of the direct factors G_a with $\Delta(x_i) = x_i \otimes 1 + 1 \otimes x_i$ $(i = 1, 2)$, where Δ is the comultiplication. Another splitting of G as a direct product $G = G_a \times G_a$ corresponds to the lnds Δ_1', Δ_2'

[6]The commutativity of a two-dimensional unipotent algebraic group G in the case of characteristic zero (and $k = \overline{k}$) follows from the triviality of the adjoint representation of G on the Lie algebra $L(G)$ (see [20] for the general facts). By V. Popov, this is not the case in positive characteristic p since $G = \left\{ \begin{pmatrix} 1 & a & b \\ 0 & 1 & a^p \\ 0 & 0 & 1 \end{pmatrix} \middle| \, a, b \in k \right\}$ is a two-dimensional non-commutative unipotent group.

with similar properties. Since $L(G) = k\Delta_1' + k\Delta_2'$, we have $\Delta_i' = \alpha_{i1}\Delta_1 + \alpha_{i2}\Delta_2$ for $i = 1, 2$, we have

$$\begin{pmatrix} \alpha_{11} & \alpha_{12} \\ \alpha_{21} & \alpha_{22} \end{pmatrix} \in \mathrm{GL}(2, k).$$

Fix a direct product decomposition $G = G_1 \times G_2$ corresponding to Δ_1, Δ_2. We denote the factor G_a corresponding to Δ_i by G_i for $i = 1, 2$.

Suppose, in general, that the additive group G_a acts on a factorial k-variety $X = \mathrm{Spec}\, A$ via a k-lnd δ. The fixed-point locus X^{G_a} is defined by the ideal $I := \sum_{a \in A} \delta(a)A$. Suppose that X^{G_a} contains an irreducible component F of codimension one. Then F is defined by $f \in A$ such that $\delta(a)$ is divisible by f for all $a \in A$. If we write $\delta = f\delta'$, then δ' is a k-lnd of A. By repeating this operation which changes the G_a-action only on the components of X^{G_a}, we may assume that X^{G_a} has no irreducible components of codimension one. Then we have the following result.

Lemma 18 *With the same notations and assumptions, we assume that X is a smooth, $\overline{k}$-factorial,*[7] *k-variety of dimension 3. Let $q : X \to Y$ be the quotient morphism, where $Y = \mathrm{Spec}\, B$ with $B = \mathrm{Ker}\, \delta$. Let C be an irreducible curve contained in X^{G_a}. Then C is a fiber component of q which is isomorphic to $\mathbb{A}^1$. Furthermore, X^{G_a} is a disjoint union of affine lines.*

Proof We may assume that $k = \overline{k}$. Since X is factorial, there are no fiber components of dimension 2 in the morphism q. Suppose that C is transversal to q, i.e., $q(C)$ is a curve. Let Z be an irreducible component of $q^{-1}(q(C))$ which contains the curve C. Since X^{G_a} contains no components of codimension one, Z is an algebraic surface which is G_a-stable but not contained in the fixed-point locus. The restriction of q to Z, $q_Z : Z \to q(C)$, is factored by an $\mathbb{A}^1$-fibration $\rho : Z \to C'$, which is the quotient morphism of the induced G_a-action on Z. Let F be a general fiber of ρ. Then G_a acts non-trivially on F, and F meeets the curve C. Then F is pointwise fixed by G_a. This is a contradiction. Hence C is contained in a fiber of q. By [16], C is isomorphic to $\mathbb{A}^1$ and X^{G_a} is a disjoint union of affine lines. □

Back to the original situation, we assume that a unipotent k-group G of dimension 2 acts on a smooth, $\overline{k}$-factorial k-variety $X = \mathrm{Spec}\, A$ of dimension 3. Let $G = G_1 \times G_2$ be a direct product decomposition with $G_1 \cong G_2 \cong G_a$. Let σ be the G-action on X and let σ_i be the restriction of G onto the factors G_i, where G_1 (resp. G_2) is identified with $G_1 \times \{1\}$ (resp. $\{1\} \times G_2$). Let $q_i : X \to Y_i$ be the quotient morphism of the G_i-action σ_i for $i = 1, 2$. We say that the G-action is *non-confluent* if there exists a decomposition $G = G_1 \times G_2$ as above such that the quotient morphisms q_1 and q_2 have no common irreducible components. Otherwise, the action is called *confluent*. In the rest of this section, we consider the tangential confluence in a similar setting.

Let $X = \mathrm{Spec}\, A$ be a k-variety of dimension n such that $\overline{X}$ is smooth and factorial. Suppose that G_a acts on X via a k-lnd δ and X^{G_a} has no components of codimension

[7] This means by definition that $\overline{X}$ is factorial.

one. Concerning the lift of the G_a-action on $\overline{X}$, we define the tangential direction of the orbit at a closed point P of $\overline{X}$. Let $(\mathcal{O}_P, \mathfrak{m}_P)$ be the local ring at P and let $\{\xi_1, \ldots, \xi_n\}$ be a system of local coordinates. Then the tangent space $T_{\overline{X},P}$ of $\overline{X}$ at P is the $\overline{k}$-vector space $\sum_{i=1}^n \overline{k}(\partial/\partial\xi_i)$. The tangential direction ℓ_P of the G_a-orbit is a line generated by $\sum_{i=1}^n \overline{\delta}(\xi_i)(P)(\partial/\partial\xi_i)$, where $\overline{\delta}$ is the lift of δ to $\overline{A}$ and $\overline{\delta}(\xi_i)(P)$ is the class of $\overline{\delta}(\xi_i)$ in the residue field $\mathcal{O}_P/\mathfrak{m}_P = \overline{k}$. The line ℓ_P is defined as a point of the projective space $\mathbb{P}(T_{\overline{X},P})$ unless $\overline{\delta}(\xi_i)(P) = 0$ for all i. The condition that $\overline{\delta}(\xi_i)(P) = 0$ for all i is equivalent to the condition that the point P is a fixed point. In fact, suppose that $\overline{\delta}(\xi_i)(P) = 0$ for all i. For an element $a \in \overline{A}$, there exists $\alpha \in \overline{k}$ such that $a - \alpha = \sum_{i=1}^n f_i\xi_i$ with $f_i \in \mathcal{O}_P$. Then we have

$$\overline{\delta}(a) = \sum_{i=1}^{n} \Big((\overline{\delta}(f_i)\xi_i + f_i\overline{\delta}(\xi_i) \Big) \in \mathfrak{m}_P.$$

This implies that $\sum_{a\in\overline{A}} \overline{\delta}(a)\overline{A} \subseteq \mathfrak{m}_P$. Hence P is the G_a-fixed point. Conversely, if P is a fixed point, then $\overline{\delta}(\overline{A}) \subseteq \mathfrak{m}_P$. Hence $\overline{\delta}(\xi_i) \in \mathfrak{m}_P$. Thus the tangential direction is not defined.

Given a unipotent group G of dimension 2 and its action on a smooth affine variety X, we consider a decomposition $G = G_1 \times G_2$ with G_1, G_2 isomorphic to G_a. With the previous notations, we consider the tangential directions of the G_i-action on $\overline{X}$.

Before going further, we give a remark on a nontrivial, non-effective $\overline{G}$-action on $\overline{X}$.

Lemma 19 *Assume that $k = \overline{k}$. Let $\overline{G} = G_1 \times G_2$ be a direct product decomposition with $G_1 \cong G_2 \cong G_a$ and let δ_1, δ_2 be the corresponding k-lnds. Then the following conditions on a nontrivial G-action on $X = \operatorname{Spec} A$ are equivalent.*

(i) The G-action is not effective, i.e., there exists an element $g \neq e$ of G which acts on X trivially.
(ii) There exists another direct product decomposition $G = G_1' \times G_2'$ such that G_2' acts trivially on X.
(iii) There exists an element $c \in k^*$ such that $\delta_1 = c\delta_2$.

Proof (i) $\Rightarrow$ (ii). Since the G-action is nontrivial but non-effective, the closure of the subgroup G_2' generated by an element g in the statement is a connected algebraic subgroup isomorphic to G_a, and the quotient G/G_2' is also isomorphic to G_a. In fact, by the argument using the Lie algebra of G in the proof of Lemma 17, we have a direct product decomposition $G = G_1' \times G_2'$ such that $G_1' \cong G/G_2'$.

(ii) $\Rightarrow$ (i). This is clear.

(ii) $\Rightarrow$ (iii). With the notations in the proof of Lemma 17, δ_i is the image of Δ_i by $L(\rho) : L(G) \to \operatorname{Der}_k(A)$ for $i = 1, 2$. Let δ_i' be the image of Δ_i' for $i = 1, 2$, where Δ_i' corresponds to the subgroup G_i'. We may assume that $\Delta_2' = \Delta_1 - c\Delta_2$ for $c \in k$. Since $\Delta_2' = 0$ as G_2' acts trivially on X, we have $\delta_1 = c\delta_2$.

(iii) $\Rightarrow$ (ii). Let $\Delta_2' := \Delta_1 - c\Delta_2 \in L(G)$. Then Δ_2' defines a subgroup G_2' of G which acts trivially on X. Further, $G = G_1' \times G_2'$ for some subgroup G_1' of G. □

For a closed point P of $\overline{X} \setminus (\overline{X}^{G_1} \cup \overline{X}^{G_2})$, we can consider the tangential directions $\ell_P^{(1)}$ and $\ell_P^{(2)}$ for the G_1 and G_2 actions.

Lemma 20 *With the above notations, suppose that $\ell_P^{(1)} = \ell_P^{(2)}$.*[8] *Then there exists another decomposition $\overline{G} = G_1 \times G_2'$ such that the G_a-action σ_2' on $\overline{X}$ induced by the G_2'-factor has the point P as a fixed point.*

Proof Suppose that the G_a-actions σ_1, σ_2 correspond to k-lnd δ_1, δ_2. Since $\ell_P^{(1)}$ and $\ell_P^{(2)}$ are defined and $\ell_P^{(1)} = \ell_P^{(2)}$, there exists $c \in \overline{k}^*$ such that

$$\overline{\delta}_2(\xi_i)(P) = c\overline{\delta}_1(\xi_i)(P), \quad (1 \le i \le n).$$

Hence $(\overline{\delta}_2 - c\overline{\delta}_1)(\mathcal{O}_P) \subseteq \mathfrak{m}_P$. Set $\delta_2' = \delta_2 - c\delta_1$. If $\delta_2' = 0$, Lemma 19 implies that the subgroup G_2' acts on $\overline{X}$ trivially. Suppose that $\delta_2' \neq 0$. Since δ_1, δ_2 commute, δ_2' ia a $\overline{k}$-lnd and $\delta_1\delta_2' = \delta_2'\delta_1$. Hence there is a direct product decomposition $\overline{G} = G_1 \times G_2'$ corresponding to $L(\overline{G}) = \overline{k}\Delta_1 + \overline{k}\Delta_2'$, where $\Delta_2' := \Delta_2 - c\Delta_1$. Now $\delta_2' = L(\overline{\rho})(\Delta_2')$ for the k-group homomorphism $\overline{\rho} : \overline{G} \to \mathrm{Aut}_{\overline{X}/\overline{k}}$ associated to the given $\overline{G}$-action. It is clear that the point P is a fixed point with respect to the σ_2'-action. Note that if $c \in k$ the above decomposition is over k as $G = G_1 \times G_2'$. □

Together with the assumptions in Lemma 20, we further assume that X is $\overline{k}$-factorial and has dimension 3. Concerning a subgroup of G which is isomorphic to G_a, every fiber component C of the quotient morphism $\overline{q} : \overline{X} \to \overline{Y} := \overline{X}//G_a$ is isomorphic to $\mathbb{A}^1$ if considered with a reduced structure (see [16]). If this curve C passes through a point P of $\overline{X}$, the tangential direction ℓ_P of C at P is well-defined since $\overline{X}$ is smooth. We say that ℓ_P is the tangential direction at P even if P is a fixed point. We say that the G-action is *tangentially non-confluent* if there exists a decomposition $G = G_1 \times G_2$ as above such that the tangential directions $\ell_P^{(1)}$ is different from $\ell_P^{(2)}$ for every closed point P of $\overline{X}$. It is obvious that the G-action is non-confluent if it is tangentially non-confluent.

At the end of the section we give an example of a unipotent algebraic group of dimension 2 being decomposed into a product $G_a \times G_a$.

Example 3 We show that a unipotent group G of dimension two defined by

$$G = \left\{ \begin{pmatrix} 1 & a & b \\ 0 & 1 & a \\ 0 & 0 & 1 \end{pmatrix} \;\middle|\; a, b \in k \right\} \subset \mathrm{GL}(3, k)$$

is decomposed as a direct product $G = G_a \times G_a$. In fact, we change an expression of element of G by an automorphism

$$\begin{pmatrix} 1 & a & b \\ 0 & 1 & a \\ 0 & 0 & 1 \end{pmatrix} \mapsto \begin{pmatrix} 1 & a & b + \frac{1}{2}a^2 \\ 0 & 1 & a \\ 0 & 0 & 1 \end{pmatrix}.$$

---

[8]This implies that the G_1-orbit G_1P and G_2-orbit G_2P are tangent at P.

Then we have

$$\begin{pmatrix} 1 & a & b+\frac{1}{2}a^2 \\ 0 & 1 & a \\ 0 & 0 & 1 \end{pmatrix} = \begin{pmatrix} 1 & a & \frac{1}{2}a^2 \\ 0 & 1 & a \\ 0 & 0 & 1 \end{pmatrix} \cdot \begin{pmatrix} 1 & 0 & b \\ 0 & 1 & 0 \\ 0 & 0 & 1 \end{pmatrix},$$

where

$$\left\{ \begin{pmatrix} 1 & a & \frac{1}{2}a^2 \\ 0 & 1 & a \\ 0 & 0 & 1 \end{pmatrix} \;\middle|\; a \in k \right\} \quad \text{and} \quad \left\{ \begin{pmatrix} 1 & 0 & b \\ 0 & 1 & 0 \\ 0 & 0 & 1 \end{pmatrix} \;\middle|\; b \in k \right\}$$

are isomorphic to G_a and commute each other.

4.2 Case of a Fixed-Point Free G_a-Action

Let $X = \operatorname{Spec} A$ be an affine k-variety and let $\sigma : G_a \times X \to X$ be an action defined over k, which is induced by a k-lnd δ on the k-algebra A. Let $B = \operatorname{Ker} \delta$ and $Y = \operatorname{Spec} B$. Hereafter we assume that X is a k-form of $\mathbb{A}^3$. Then Y is the (algebraic) quotient variety of X with respect to σ. We denote by $q : X \to Y$ the quotient morphism. We say that the G_a-action σ is *free* if $X^{G_a} = \emptyset$, i.e., $\delta(A)A = A$.[9]

Lemma 21 *With the above notations, Y is k-isomorphic to $\mathbb{A}^2$. Namely, B is a polynomial ring $k[x_1, x_2]$.*

Proof Let $\overline{A} = A \otimes_k \overline{k}$, which is isomorphic to a polynomial ring in dimension three over $\overline{k}$. Then the k-lnd δ lifts to a $\overline{k}$-lnd $\overline{\delta}$ on $\overline{A}$. On the other hand, the following sequence of k-vector spaces is exact

$$0 \longrightarrow B \longrightarrow A \overset{\varphi - \mathrm{id}}{\longrightarrow} A[t],$$

where $\varphi : A \to A[t]$ is defined by $\varphi(a) = \sum_{i \geq 0} \frac{1}{i!} \delta^i(a) t^i$. Then the tensor product $\otimes_k \overline{k}$ gives rise to an exact sequence of $\overline{k}$-vector spaces

$$0 \longrightarrow B \otimes_k \overline{k} \longrightarrow \overline{A} \overset{\overline{\varphi} - \mathrm{id}}{\longrightarrow} \overline{A}[t],$$

where $\overline{\varphi}$ is the $\overline{k}$-algebra homomorphism $\overline{A} \to \overline{A}[t]$ defined by $\overline{\delta}$ as φ for δ. Hence $B \otimes_k \overline{k} = \operatorname{Ker} \overline{\delta}$. By [38], $B \otimes_k \overline{k}$ is a polynomial ring of dimension two over $\overline{k}$.

[9] For a G_a-action $\sigma : G_a \times X \to X$, let $\Psi_X := (\sigma, p_2) : G_a \times X \to X \times X$ be the *graph morphism* of σ. We say that the action σ is *free* (resp. *proper*) if the graph morphism Ψ_X is a closed immersion (resp. a proper morphism). Then σ is free (resp. proper) if and only if $\overline{\sigma} := \sigma \otimes_k \overline{k}$ is free (resp. proper). By an argument as in subsection 4.4, we have the implications: σ is free $\Rightarrow$ σ is proper $\Rightarrow$ σ is fixed-point free. If X is a k-form of $\mathbb{A}^3$, a theorem of Kaliman [22] implies that these three conditions are equivalent. Hence we can use this terminology without confusion.

Since there is no nontrivial k-form of $\mathbb{A}^2$ (cf. [28]), it follows that B is a polynomial ring of dimension two over k. □

Let $\overline{Y} = Y \otimes_k \overline{k}$ and let $\overline{q} : \overline{X} \to \overline{Y}$ be the quotient morphism of the G_a-action $\overline{\sigma} = \sigma \otimes_k \overline{k}$. We note that $\overline{\delta}(\overline{A}) = \delta(A) \otimes_k \overline{k}$ and hence $\overline{\delta}(\overline{A})\overline{A} = \delta(A)\overline{A} = \delta(A)A \otimes_k \overline{k}$. This implies that $\overline{X}^{G_a} = X^{G_a} \otimes_k \overline{k}$ and hence that $\overline{\sigma}$ is free if and only if σ is free. This remark yields the following result in view of Kaliman's theorem [22] which asserts the following:

Lemma 22 *Let $\overline{\sigma} : G_a \times \mathbb{A}^3 \to \mathbb{A}^3$ be a fixed-point free G_a-action. Then the quotient morphism $\overline{q} : \mathbb{A}^3 \to \overline{Y}$ defines an $\mathbb{A}^1$-bundle over $\overline{Y} \cong \mathbb{A}^2$ and the G_a-action is a translation along the fibers. In particular, the quotient morphism $\overline{q}$ is surjective and the fiber $\overline{q}^{-1}(\overline{Q})$ is isomorphic to $\mathbb{A}^1$ with multiplicity 1 for every (closed and non-closed) point $\overline{Q}$ of $\overline{Y}$.*

Lemma 23 *Assume that the action σ is free. Then the following assertions hold.*

(1) Any closed fiber of the quotient morphism $q : X \to Y$ is isomorphic to $\mathbb{A}^1$.
(2) Let $\mathfrak{p}$ be a height one prime ideal of B and let $\kappa(\mathfrak{p})$ be the quotient field of $B/\mathfrak{p}$. Then the fiber of q over $\mathfrak{p}$ is isomorphic to $\mathbb{A}^1$ over $\kappa(\mathfrak{p})$. Namely, $(A/\mathfrak{p}A) \otimes_{B/\mathfrak{p}} \kappa(\mathfrak{p})$ is a polynomial ring of dimension one over $\kappa(\mathfrak{p})$.
(3) Let K be the quotient field of B. Then the generic fiber of q, $X_K = \mathrm{Spec}\,(A \otimes_A K)$, is isomorphic to $\mathbb{A}^1$ over K.

Proof (1) Let $\mathfrak{m}$ be the maximal ideal of B corresponding to the closed point of Y. Then $(A/\mathfrak{m}A) \otimes_k \overline{k} = \bigoplus_{i=1}^r \overline{A}/\mathfrak{m}_i\overline{A}$, where $\mathfrak{m}\overline{B} = \bigcap_{i=1}^r \mathfrak{m}_i$ and r is the extension degree of $B/\mathfrak{m}$ over k. Let $\lambda_i : B/\mathfrak{m} \to \overline{k}$ be an embedding of $B/\mathfrak{m}$ to a subfield of $\overline{k}$. Then we have $(A/\mathfrak{m}A) \otimes_{B/\mathfrak{m}} (\overline{k}, \lambda_i)$ is a polynomial ring of dimension one over $\overline{k}$ by Kaliman's theorem. Since there is no nontrivial form of $\mathbb{A}^1$, it follows that $A/\mathfrak{m}A$ is a polynomial ring of dimension one over $B/\mathfrak{m}$.

(2) Let $\mathfrak{p}$ be a height one prime ideal of B. Since B is a polynomial ring, $\mathfrak{p}$ is principal. Write $\mathfrak{p} = (f)$ with a k-irreducible element f of B. With f considered in $\overline{B}$, it is a product $f = \prod_{i=1}^s f_i$, where the f_i are conjugate of each other under the Galois group of the splitting field of f over k. We have $\overline{A}/f_i\overline{A} = (A/fA) \otimes_{B/fB} (\overline{B}/f_i\overline{B})$, which is geometrically integral. Now the restriction of $\overline{q}$ onto the hypersurface $V(f_i\overline{A}) = \mathrm{Spec}\,\overline{A}/f_i\overline{A}$ is a fibration over the curve $V(f_i\overline{B}) = \mathrm{Spec}\,\overline{B}/f_i\overline{B}$ whose closed fibers are isomorphic to $\mathbb{A}^1$ and whose generic fiber is geometrically integral. By [29, Theorem 2], it is an $\mathbb{A}^1$-bundle. Hence the generic fiber of $\mathrm{Spec}\,\overline{A}/f_i\overline{A}$ is isomorphic to $\mathbb{A}^1$ over $Q(\overline{B}/f_i\overline{B})$.[10] Since $Q(\overline{B}/f_i\overline{B})$ is an algebraic extension of $Q(B/fB)$, it follows that $(A/fA) \otimes_{B/fB} Q(B/fB)$ is a polynomial ring of dimension one over $Q(B/fB)$.

(3) By the slice theorem, the generic fiber $\overline{X}_{\overline{\eta}} = \mathrm{Spec}\,(\overline{A} \otimes_{\overline{B}} Q(\overline{B}))$ is isomorphic to $\mathbb{A}^1$ over $Q(\overline{B})$. Since $\overline{A} \otimes_{\overline{B}} Q(\overline{B}) = (A \otimes_B Q(B)) \otimes_{Q(B)} Q(\overline{B})$ and since

[10] This follows easily from the fact that $\overline{q} : \overline{X} \to \overline{Y}$ is an $\mathbb{A}^1$-bundle. But we would like to make a detour to give an argument which can be applied in case we drop the hypothesis that the action σ is free.

there is no nontrivial form of $\mathbb{A}^1$, it follows that $A \otimes_B Q(B)$ is a polynomial ring of dimension one over $Q(B)$. □

Now the following theorem is an easy consequence of a result of Kambayashi-Wright [30].

Theorem 7 *Let X be a k-form of $\mathbb{A}^3$ having a free G_a-action σ. Then X is k-isomorphic to $\mathbb{A}^3$.*

Proof By Lemma 23, every closed (and non-closed) fiber of the quotient morphism $q : X \to Y$ is isomorphic to $\mathbb{A}^1$ over the respective residue field. Hence q is an $\mathbb{A}^1$-bundle in the Zariski topology by [30]. Since A is factorial, it follows that this $\mathbb{A}^1$-bundle is trivial. Hence X is isomorphic to $\mathbb{A}^3$. □

4.3 Case of an Effective $G_a \times G_a$-Action

We consider the following result.

Lemma 24 *Let A be an affine k-domain of dimension 2. Suppose that A is geometrically integral over k, i.e., $Q(A)$ is a regular extension of k, and that A has two nonzero k-lnds δ_1, δ_2. Suppose further that $\delta_1\delta_2 = \delta_2\delta_1$ and $\operatorname{Ker} \delta_1 \neq \operatorname{Ker} \delta_2$. Then A is k-isomorphic to $k[t_1, t_2]$.*

Proof Let $A_1 = \operatorname{Ker} \delta_2$, $A_2 = \operatorname{Ker} \delta_1$ and $B = A_1 \cap A_2$. Then $\dim A_i = 1$ for $i = 1, 2$ and $\dim B = 0$. In particular, B is a field extension of k. Since $B \subset Q(A)$ and k is algebraically closed in $Q(A)$ by the hypothesis, it follows that $B = k$. Since $\delta_1\delta_2 = \delta_2\delta_1$, the lnd δ_1 induces a k-lnd on A_1 which we denote by the symbol Δ_1. If the induced Δ_1 is zero on A_1, then $A_2 \supseteq A_1$. Since $\dim A_2 = \dim A_1 = 1$, it follows that $Q(A_2)$ is an algebraic extension of $Q(A_1)$. Then the induced derivation δ_2 is trivial on $Q(A_2)$ since δ_2 is trivial on $Q(A_1)$. Then $A_1 = \operatorname{Ker} \delta_2 \supseteq A_2$. Hence $A_1 = A_2$ which contradicts the hypothesis. Thus the induced Δ_1 on A_1 is nontrivial. Similarly, δ_2 induces a nontrivial k-lnd Δ_2 on A_2. Then $\operatorname{Ker} \Delta_1 = \operatorname{Ker} \Delta_2 = k$. This implies that $A_1 = k[t_1]$ and $A_2 = k[t_2]$. We can choose t_1, t_2 so that $\Delta_1(t_1) = 1$ and $\Delta_2(t_2) = 1$. Then t_1 and t_2 are slices of the lnds δ_1 and δ_2 in A. Then $A = A_2[t_1] = A_1[t_2] = k[t_1, t_2]$. □

Let $X = \operatorname{Spec} A$ be an affine k-variety of dimension 3. By Lemma 17, a $G_a \times G_a$-action σ on X is given by two k-lnds δ_1, δ_2 on A such that $\delta_1\delta_2 = \delta_2\delta_1$. We prove the following result of Daigle-Kaliman [5].

Theorem 8 *Let $X = \operatorname{Spec} A$ be a k-form of $\mathbb{A}^3$. Suppose that X has an effective action of a unipotent group G of dimension 2. Then Y is k-isomorphic to $\mathbb{A}^3$.*

Proof Our proof consists of three steps.

(I) There exists a direct product decomposition $G = G_1 \times G_2$ such that the two G_a-actions induced by G_1 and G_2 are non-confluent. By Lemma 16, A is a factorial

k-domain of dimension 3 with $A^* = k^*$. Let $A_1 = \text{Ker}\,\delta_2$, $A_2 = \text{Ker}\,\delta_1$, and $B = A_1 \cap A_2$. Then B, A_1, A_2 are factorial affine k-domains of respective dimensions $1, 2, 2$. The units of these domains are all constants. By what we have discussed, $\overline{A_i} = A_i \otimes_k \overline{k}$ is a polynomial ring of dimension 2 over $\overline{k}$, and $\overline{B} = B \otimes_k \overline{k}$ is a polynomial ring of dimension 1 over $\overline{k}$. Since all k-forms of $\mathbb{A}^1$ and $\mathbb{A}^2$ are trivial, it follows by the construction of A_i, B that $B = k[u]$, $A_1 = k[u, t_1]$, and $A_2 = k[u, t_2]$ with $\delta_1(t_1) = f_1(u)$ and $\delta_2(t_2) = f_2(u)$ for nonzero elements $f_1(u), f_2(u)$ of $k[u]$.

(II) Let $A_0 = A_1 \otimes_B A_2 = k[u, t_1, t_2]$, which is a $k[u]$-subalgebra of A. Let $X_0 = \text{Spec}\,A_0$ and $Z = \text{Spec}\,B$. The inclusion $A_0 \hookrightarrow A$ defines a Z-morphism $\varphi : X \to X_0$. Let $K = Q(B)$. Then $A_K := A \otimes_B K = K[t_1, t_2]$ by Lemma 24. Hence φ is birational because

$$\begin{aligned} A[f_1(u)^{-1}, f_2(u)^{-1}] &= (A[f_1(u)^{-1}])[f_2(u)^{-1}] \\ &= (A_2[f_1(u)^{-1}][t_1])[f_2(u)^{-1}] \\ &= (A_2[f_2(u)^{-1}])[f_1(u)^{-1}, t_1] \\ &= (B[f_2(u)^{-1}, t_2])[f_1(u)^{-1}, t_1] \\ &= (B[f_1(u)^{-1}, f_2(u)^{-1}])[t_1, t_2]. \end{aligned}$$

(III) The morphism $\varphi : X \to X_0$ is G-equivariant by the construction. Since $\overline{X} := X \otimes_k \overline{k} \cong \mathbb{A}^3$ and $X_0 \cong \mathbb{A}^3 = \text{Spec}\,k[u, t_1, t_2]$, the birational morphism $\overline{\varphi} : \overline{X} \to \overline{X}_0$ is a polynomial morphism

$$(x, y, z) \mapsto (f_1(x, y, z), f_2(x, y, z), f_3(x, y, z)).$$

Let $J = \det(\partial(f_1, f_2, f_3)/\partial(x, y, z))$ be the Jacobian determinant. Then $J = 0$ defines the locus E of exceptional varieties of $\overline{\varphi}$, which is G-stable. Hence J is G-invariant, i.e., $J \in B$. This implies that the subvarieties of $\overline{X}$ which are contracted to curves or points by $\overline{\varphi}$ are the fibers of $\overline{p} : \overline{X} \to \overline{C}$, where $\overline{p}$ is the base change of $p : X \to C = \text{Spec}\,k[u]$ defined by the inclusion $B = k[u] \hookrightarrow A$. Furthermore, $\overline{\varphi} : \overline{X} \to \overline{X}_0$ is an isomorphism on $\overline{X} \setminus E$, where $E = \overline{p}^{-1}(V(J))$. In particular, the generic fiber of $p : X \to C$ is isomorphic to $\mathbb{A}^2$ because a form of $\mathbb{A}^2$ is trivial, and all closed fibers $p^{-1}(P)$ are isomorphic to $\mathbb{A}^2$ for all $P \in C \setminus V(J)$. It follows then from Kaliman [21] that u is a coordinate of $\overline{X} \cong \mathbb{A}^3$ and hence the fibers $\overline{p}^{-1}(\overline{P}) \cong \mathbb{A}^2$ for $\overline{P} \in V(J)$. Then $X \cong \mathbb{A}^3$ by Sathaye [43]. □

The morphism $\varphi : X \to X_0$ in the above proof is not an isomorphism in general as shown by the following example.

Example 4 Consider the following lnds on a polynomial ring $A = k[x, y, z]$,

$$\delta_1 = x\frac{\partial}{\partial y} + \frac{\partial}{\partial z}, \quad \delta_2 = x^2\frac{\partial}{\partial y} + \frac{\partial}{\partial z}.$$

Then $\delta_1\delta_2 = \delta_2\delta_1$. We can compute $A_1 = \text{Ker}\,\delta_1$, $A_2 = \text{Ker}\,\delta_2$ and $B = A_1 \cap A_2$ as follows.

$$A_1 = k[x, y - xz], \quad A_2 = k[x, y - x^2z], \quad B = k[x] .$$

Hence $A_0 = k[x, y - xz, y - x^2z]$ and $\varphi : X \to X_0$ is given by

$$(x, y, z) \mapsto (x, y - xz, y - x^2z).$$

The Jacobian determinant J is $x(1 - x)$. On the fibers $p^{-1}(x = 0)$ and $p^{-1}(x = 1)$, the morphism φ is given by

$$(0, y, z) \mapsto (0, y, y), \quad (1, y, z) \mapsto (1, y - z, y - z) .$$

In this example the planes $\{x = 0\}$ and $\{x = 1\}$ are mapped to lines. □

Remark 3 (1) With the notations of the proof of Lemma 24, the case where a fiber $\overline{p}^{-1}(P)$ is contracted to a point by $\overline{\varphi}$ is impossible. We may assume that $x = u$ and the fiber $\overline{p}^{-1}(P_0)$ is sent to the point $(x, t_1, t_2) = (0, 0, 0)$. The morphism $\overline{\varphi}$ is given by a polynomial morphism

$$(x, y, z) \mapsto (x, t_1, t_2) = (x, f_2(x, y, z), f_3(x, y, z)) ,$$

where $f_2 \in \mathrm{Ker}\,\delta_2$ and $f_3 \in \mathrm{Ker}\,\delta_1$. By the assumption, $f_2(0, y, z) = 0$ and $f_3(0, y, z)=0$. Then $f_2(x, y, z)$ and $f_3(x, y, z)$ are divisible by x, and hence $f_2(x, y, z)/x \in k[x, f_2(x, y, z)]$. This is a contradiction.

(2) If the action of $G = G_a \times G_a$ is non-confluent, then the birational morphism φ is an isomorphism. In fact, if D is an irreducible curve in $\overline{X}$ such that $\overline{\varphi}(D)$ is a point of $\overline{X}_0$. Then D is a fiber component of both quotient morphism $\overline{q}_i : \overline{X} \to \overline{X}_i$ for $i = 1, 2$. Such a curve does not exist if the G-action is non-confluent. Hence $\overline{\varphi}$ is a quasi-finite morphism. Then $\overline{\varphi}$ is an isomorphism by Zariski main theorem. Hence φ is an isomorphism. □

4.4 A k-Form of $\mathbb{A}^4$ with a Proper Action of a Unipotent Group of Dimension 2

Let $X = \mathrm{Spec}\,A$ be now an affine algebraic k-variety of dimension 4 such that $X \otimes_k \overline{k}$ is $\overline{k}$-isomorphic to $\mathbb{A}^4$. Let G be a two-dimensional unipotent group. By Lemma 17, G is k-isomorphic to $G_a \times G_a$. Assume that X has a proper G-action. Let B be the ring of G-invariant elements of A. By a lemma of Zariski, $B = A \cap K$ is an affine k-domain, where K is the field of G-invariant elements of the function field $k(X) = Q(A)$. Let $Y = \mathrm{Spec}\,B$ and let $q : X \to Y$ be the quotient morphism. We denote by $\overline{q} : \overline{X} \to \overline{Y}$ the base change of q by the field extension $\overline{k}/k$. A G-action $\sigma : G \times X \to X$ is called *q-tight* if $A \otimes_B A$ is an integral domain for $B = A^G$. If $\Psi_X = (\sigma, p_2) : G \times X \to X \times X$ is the graph morphism of the action, $\Psi_X^* : A \otimes_k A \to A[G] := A \otimes_k k[G]$ splits as

$$\Psi_X^* : A \otimes_k A \xrightarrow{\iota^*} A \otimes_B A \xrightarrow{\varphi_B} A[G],$$

where $k[G]$ is the coordinate ring of G isomorphic to $k[t_1, t_2]$. Since ι^* is surjective, σ is q-tight if and only if φ_B is injective. Use the fact that $\dim_K (A \otimes_B A) \otimes_B K = 4$ for $K = Q(B)$. Further, we can show that σ is q-tight if and only if $\overline{\sigma} = \sigma \otimes_k \overline{k}$ is q-tight (see the argument in [17, Lemma 2.5]). Hereafter in this subsection, *we assume that the G-action σ on X is q-tight.*

Lemma 25 *Y is k-isomorphic to $\mathbb{A}^2$.*

Proof It suffices to show that $\overline{Y}$ is isomorphic to $\mathbb{A}^2$. Since $\overline{Y} \cong \mathbb{A}^4/\overline{G}$ with $\overline{G} = G \otimes_k \overline{k}$, $\overline{B} = B \otimes_k \overline{k}$ is factorial and $\overline{B}^* = \overline{k}^*$. By the argument as in [39, Sect. 5], $\overline{Y}$ is isomorphic to $\mathbb{A}^2$. In fact, $\mathrm{Sing}(\overline{Y})$ is a finite set. Let $\overline{Y}' = \overline{Y} \setminus \mathrm{Sing}(\overline{Y})$. Since $\overline{X} \cong \mathbb{A}^4$, there is a linear plane $\overline{L}$ in $\overline{X}$ such that $\overline{q}|_{\overline{L}} : \overline{L} \to \overline{Y}$ is a dominant morphism and $\dim \overline{q}^{-1}(\overline{P}) \cap \overline{L} \leq 0$ for every point $\overline{P}$ of $\mathrm{Sing}(\overline{Y})$. Then $\overline{\kappa}(\overline{Y}') = -\infty$. Hence $\overline{Y}'$ is either affine 1-ruled, i.e., $\overline{Y}'$ has an $\mathbb{A}^1$-fibration, or contains a Platonic $\mathbb{C}^*$-fiber space (we may assume that $\overline{k} = \mathbb{C}$). In the first case, $\overline{Y} \cong \mathbb{A}^2$. In the second case, argue exactly in the same fashion as in [39]. □

Lemma 26 *The following assertions hold.*

(1) The graph morphism $\Psi_{\overline{X}/\overline{Y}} := (\overline{\sigma}, p_2) : G \times \overline{X} \to \overline{X} \times_{\overline{Y}} \overline{X}$ is a surjective morphism. Further, the generic fiber $X_\eta := X \times_Y \mathrm{Spec}\, k(Y)$ is $k(Y)$-isomorphic to $\mathbb{A}^2$.

(2) The $\overline{G}$-action on $\overline{X}$ is fixpoint-free, and each fiber of $\overline{q}$ is a $\overline{G}$-orbit, hence isomorphic to $\mathbb{A}^2$ if considered with reduced structure.

(3) For each closed point P of Y, the fiber $q^{-1}(P)_{\mathrm{red}}$ is isomorphic to $\mathbb{A}^2$.

Proof (1) The graph morphism $\Psi_{\overline{X}} : \overline{G} \times \overline{X} \to \overline{X} \times \overline{X}$ splits as

$$\Psi_{\overline{X}} : \overline{G} \times \overline{X} \xrightarrow{\Psi_{\overline{X}/\overline{Y}}} \overline{X} \times_{\overline{Y}} \overline{X} \xrightarrow{\overline{\iota}} \overline{X} \times \overline{X},$$

where $\overline{\iota}$ is a closed immersion. Hence $\overline{\iota}$ is a separated morphism and the morphism $\Psi_{\overline{X}/\overline{Y}}$ is a proper morphism since so is the composite $\overline{\iota} \circ \Psi_{\overline{X}/\overline{Y}}$. On the other hand, let ζ (resp. η) be the generic point of X (resp. Y). Let $X_\eta := X \times_Y \mathrm{Spec}\, k(\eta)$ be the generic fiber of q. By taking the fiber product with $\mathrm{Spec}\, k(\zeta)$ over X, the graph morphism $\Psi_{X/Y} : G \times X \to X \times_Y X$ induces a $k(\zeta)$-morphism $\Psi_\zeta : G \otimes_k k(\zeta) \to X_\eta \otimes_{k(\eta)} k(\zeta)$ which is proper. Hence it is a finite morphism because Ψ_ζ is an affine morphism as well. Since G contains no finite subgroups, Ψ_ζ is birational. Since X_η is normal, Ψ_ζ is an isomorphism by Zariski's Main Theorem. This implies that X_η is $k(\eta)$-isomorphic to $\mathbb{A}^2$ because G is k-isomorphic to $\mathbb{A}^2$ and a form of $\mathbb{A}^2$ is trivial, and that $\Psi_{X/Y} : G \times X \to X \times_Y X$ is a dominant morphism because $X \times_Y X$ is integral and hence $\varphi_B : A \otimes_B A \to A[G]$ is injective. Hence $\Psi_{\overline{X}/\overline{Y}}$ is also dominant and proper. This implies that $\Psi_{\overline{X}}$ is surjective.

(2) Hence each fiber of $\overline{q}$ is a $\overline{G}$-orbit and two-dimensional by the upper-semicontinuity of dimension of fibers, i.e., each closed fiber of $\overline{q}$ has dimension

not less than 2, and the $\overline{G}$-action is fixed-point free because proper subgroups of G have dimension one.

(3) As in the proof of Lemma 23, let $\lambda_i : k(P) \to \overline{k}$ be an embedding of fields. Let $F := q^{-1}(P)_{\text{red}}$. Then $F \otimes_{k(P)} (\overline{k}, \lambda_i) \cong \mathbb{A}^2_{\overline{k}}$. Namely, F is a $k(P)$-form of $\mathbb{A}^2$. Hence F is isomorphic to $\mathbb{A}^2$ over $k(P)$. □

Lemma 27 *$\overline{q}$ has no singular fibers over codimension one points of $\overline{Y}$. Hence all singular fibers are either empty fibers or irreducible multiple fibers whose reduced form is isomorphic to $\mathbb{A}^2$. Let $\overline{S}$ be the set of points $\overline{P}$ of $\overline{Y}$ such that $\overline{q}^{-1}(\overline{P})$ is either the empty set or a singular fiber of $\overline{q}$. Then $\overline{S}$ is a finite set.*

Proof Note that $\overline{B}$ is factorial. If $\overline{p}$ is a prime element of $\overline{B}$, $\overline{p}$ is a prime element in $\overline{A}$. In fact, $\overline{p}$ is a prime element of $\overline{A}^{\overline{G}_1}$ which is factorial because $\overline{B} = \left(\overline{A}^{\overline{G}_1}\right)^{\overline{G}_2}$, hence a prime element of $\overline{A}$, where $\overline{G} = \overline{G}_1 \times \overline{G}_2$ with $\overline{G}_1 \cong \overline{G}_2 \cong G_a$. Let $\overline{C}$ be an irreducible curve on $\overline{Y}$ defined by $\overline{p} = 0$. Then $\overline{S} := \overline{q}^{-1}(\overline{C})$ is an irreducible threefold defined by $\overline{p} = 0$ in $\overline{X}$ which is G-stable, and the restriction of $\overline{q}$ onto $\overline{S}$ is factored by the quotient morphism, $\overline{q}|_{\overline{S}} : \overline{S} \to \overline{S}/G \to \overline{C}$. If $\overline{q}|_{\overline{S}} : \overline{S} \to \overline{C}$ is not generically integral, then the general fiber is a disjoint union of irreducible components which are isomorphic to $\mathbb{A}^2$. This contradicts the properness hypothesis of the $\overline{G}$-action by Lemma 26. Hence if $\overline{q}$ has singular fibers, they are irreducible multiple fibers over isolated points of $\overline{Y}$. □

Lemma 28 *The quotient morphism $\overline{q} : \overline{X} \to \overline{Y}$ has no multiple fibers. Hence the set $\overline{S}$ consists of the points of $\overline{Y}$ over which the fiber is the empty set.*

Proof Let $\overline{Q}$ be a point of $\overline{Y}$ such that $\overline{q}^{-1}$ is a multiple fiber, i.e., a non-reduced fiber. Take a sufficiently general point $\overline{P} \in \overline{q}^{-1}(\overline{Q})$ and take a general hyperplane section H of $\overline{X}$ which passes through the point $\overline{P}$. By the Lefschetz principle, we can work over the complex number field $\mathbb{C}$. We take a small Euclidean neighborhood T of $\overline{P}$ in H and call T a slice transverse to $\overline{q}^{-1}(\overline{Q})$. By Seidenberg [45], $\overline{q}|_T$ is smooth at a point $\overline{P}'$ near $\overline{P}$ if and only if $\overline{q}$ is smooth at $\overline{P}'$. By Lemma 27, $\overline{q}|_T : T \to \overline{Y}$ is unramified outside a closed subset of T of codimension ≥ 1. By purity of branch locus, $\overline{q}|_T : T \to \overline{Y}$ is also smooth at $\overline{P}$. This is a contradiction because $\overline{q}^{-1}(\overline{Q})$ is a multiple fiber. Hence there are no multiple fibers.

By Seshadri [47, Theorem 6.1] and [17, Lemma 2.6],[11] there exists a geometric quotient U of $\overline{X}$ by G. Let $\rho : \overline{X} \to U$ be the quotient morphism. Then ρ is a locally trivial G-torsor and $\overline{q}$ is decomposed by π as

$$\overline{q} : \overline{X} \xrightarrow{\rho} U \xrightarrow{\pi} \overline{Y}.$$

Then π is one-to-one by the correspondence of G-orbits and hence birational. Since U and $\overline{Y}$ are normal, π is an open immersion by Zariski's main theorem. Then ρ is a principal G-bundle, and hence any fiber of $\overline{q}$ is smooth unless it is empty. □

[11] The same argument in the proof of Lemma 2.6 works for $G \cong G_a \times G_a$.

Lemma 29 *Let* $\overline{Y}^0 := \overline{Y} \setminus \overline{S}$ *and let* $\overline{X}^0 := \overline{q}^{-1}(\overline{Y}^0)$. *Then* $\overline{X}^0 = \overline{X}$ *and* $\overline{X}^0$ *is a* G*-torsor over* $\overline{Y}^0$.

Proof The first assertion follows from Lemma 28. It suffices to show that $\overline{X}^0 \times_{\overline{Y}^0} \overline{X}^0$ is smooth. Then $\overline{\Psi}_{\overline{X}/\overline{Y}}$ restricted onto $G \times \overline{X}^0$ induces an isomorphism $G \times \overline{X}^0 \cong \overline{X}^0 \times_{\overline{Y}^0} \overline{X}^0$. Let $\overline{P}, \overline{P}'$ be closed points of $\overline{X}^0$ such that $\overline{q}(\overline{P}) = \overline{q}(\overline{P}') = \overline{Q}$. Then there exists $\overline{g} \in \overline{G}(\overline{k})$ such that $\overline{P}' = {}^{\overline{g}}\overline{P}$. Then $\overline{g} \times \mathrm{id}_{\overline{X}^0}$ induces an isomorphism between local rings $\mathcal{O}_{\overline{Z},(\overline{P},\overline{P})}$ and $\mathcal{O}_{\overline{Z},(\overline{P}',\overline{P})}$, where $\overline{Z} = \overline{X}^0 \times_{\overline{Y}^0} \overline{X}^0$. Hence it suffices to show that $\mathcal{O}_{\overline{Z},(\overline{P},\overline{P})}$ is a regular local ring. Let $\{x_1, x_2\}$ be a regular system of parameters of the maximal ideal of $\mathcal{O}_{\overline{Y}^0,\overline{Q}}$ and let $\{t_1, t_2\}$ be elements of $\mathcal{O}_{\overline{X}^0,\overline{P}}$ whose images in $\mathcal{O}_{\overline{q}^{-1}(\overline{Q}),\overline{P}}$ form a regular system of parameters. Then it is clear that $\{x_1, x_2, t_1, t_2, t_1', t_2'\}$ is a regular system of parameters of the local ring $\mathcal{O}_{\overline{Z},(\overline{P},\overline{P})}$, where $\{t_1', t_2'\}$ is a copy of $\{t_1, t_2\}$. Hence $\overline{Z}$ is smooth at the point $(\overline{P}, \overline{P})$. □

We can now answer the triviality question of a k-form of $\mathbb{A}^4$ as follows.

Theorem 9 *Let* $X = \operatorname{Spec} A$ *be a* k*-form of* $\mathbb{A}^4$ *equipped with a proper and* q*-tight* G*-action, where* G *is a unipotent group of dimension* 2. *Then* X *is* k*-isomorphic to* $\mathbb{A}^4$.

Proof If $\overline{S} = \emptyset$ then $\overline{X}$ is a $\overline{G}$-torsor over $\overline{Y}$. Since the G-action is defined over the field k, X is a G-torsor over Y. Write $G = G_1 \times G_2$ with $G_1 \cong G_2 \cong G_a$. Let $X_1 = X/G_1$.[12] Then X_1 is a G-torsor over Y. Since the set of isomorphism classes of G_a-torsors over Y is bijective to $H^1_{\text{fl}}(Y, G_a) \cong H^1(Y, \mathcal{O}_Y) = 0$, we know that $X_1 \cong Y \times G_a$. Similarly, we have $X \cong X_1 \times G_a$. Hence $X \cong \mathbb{A}^2 \times G \cong \mathbb{A}^4$.

We shall show that $\overline{S} = \emptyset$. Let $\overline{q} : \overline{X} \to \overline{Y}$ be the same as in Lemma 28. Then $\overline{X}$ is a $\overline{G}$-torsor over $\overline{Y}^0$. By Lefschetz principle, we may assume that $\overline{k} = \mathbb{C}$. By a long exact sequence of homotopy groups for a fiber bundle, we have $\pi_i(\overline{Y}^0) \cong \pi_i(\overline{X}) \cong \pi_i(\mathbb{A}^4) = 0$ for every $i > 0$. Since $\pi_1(\overline{Y}^0) = (1)$, we have $H_i(\overline{Y}^0; \mathbb{Z}) \cong H_i(\overline{X}; \mathbb{Z}) = 0$ for every $i > 0$ by Hurwicz's isomorphism theorem. But, if $\overline{S} \neq \emptyset$, then $H_3(\overline{Y}^0; \mathbb{Z}) \cong \mathbb{Z}^{\oplus \#(\overline{S})} \neq 0$ by a long exact sequence of integral cohomologies for a pair $(\overline{Y}, \overline{S})$. This is a contradiction. □

---

[12] The algebraic quotient morphism $q : X \to Y = X/\!/G$ splits as a composite $X \xrightarrow{q_1} Y_1 := X/\!/G_1 \xrightarrow{q_2} Y$. If the G-action is q-tight, $\overline{X}$ is a G-torsor over $U := \overline{q}(\overline{X})$ by Lemma 29. This implies that $\overline{q}_1 : \overline{X} \to \overline{Y}_1$ is surjective and a G_1-torsor. Hence $\overline{Y}_1$ is the geometric quotient of $\overline{X}$ by G_1. In particular, the G_1-action on X is q-tight. Since $\dim A = 4$ and $A_1 = \operatorname{Ker} \Delta$ with an lnd Δ on A associated to the G_1-action, we need to show that A_1 is finitely generated over k. By Seshadri [47], X is a locally trivial G-principal fiber bundle. Hence there exists an open covering $\mathcal{U} = \{U_i\}_{i \in I}$ with a finite index set I such that $q^{-1}(U_i) \cong U_i \times G$ which is a G-equivalent isomorphism over U_i. We can take the open sets U_i of the form $D(b_i)$ with $b_i \in B$. Then $A_1[b_i^{-1}] \cong A[b_i^{-1}]^{G_1} \cong B[b_i^{-1}][t_i]$ since G_1 acts on G and $G/G_1 \cong G_a$. Hence $A_1[b_i^{-1}]$ is generated by a single element f_i over $B[b_i^{-1}]$. Since the positive powers of b_i generate the unitary ideal in B, it is then easy to show that A_1 is generated by $\{f_i\}_{i \in I}$ over B.

Remark 4 Let n be an integer greater than 2. Let G be a commutative unipotent group of dimension $n-2$. Since G has a subnormal series whose subquotient is isomorphic to G_a, it follows that the underlying space of G is isomorphic to $\mathbb{A}^{n-2}$. Let $X = \mathrm{Spec}\, A$ be a k-form of $\mathbb{A}^n$ with a proper and q-tight G-action. Let $B = A^G$ and let $Y = \mathrm{Spec}\, B$. Then B is an affine domain of dimension 2, and the inclusion $B \hookrightarrow A$ defines the quotient morphism. To avoid the difficulty arising from the non-triviality of forms of $\mathbb{A}^{n-2}$, we assume that the quotient morphism q has a cross-section if $n \geq 5$. Putting this assumption, every fiber of q will be a non-multiple fiber. For a closed point Q of Y, let $F = q^{-1}(Q)_{\mathrm{red}}$. Then F is defined over the residue field $k(Q)$ and given a proper G-action. Since $\dim F = n-2$ and the G-action on F has no fixed points, it follows that $F \cong \mathbb{A}^{n-2}$. Under this situation, Lemmas 25–29 hold with only $\mathbb{A}^2$ in the assertions replaced by $\mathbb{A}^{n-2}$. So, by pursuing each step of the proof of Theorem 9, we can show that X is k-isomorphic to $\mathbb{A}^n$.

Furthermore, we note that proper G_a-actions on smooth affine varieties of dimension 3 or 4 are studied by Kaliman [23, Theorem 0.1, Corollary 0.3]. □

4.5 Forms of $\mathbb{A}^n \times \mathbb{A}^1_*$

Let $X = \mathrm{Spec}\, A$ be a k-form of $\mathbb{A}^n \times \mathbb{A}^1_*$, where $\mathbb{A}^1_*$ is the affine line with one point punctured. Then there exists a finite Galois extension k' of k with Galois group $\mathfrak{G}$ such that $X \otimes_k k'$ is k'-isomorphic to $\mathbb{A}^n \times \mathbb{A}^1_*$. Namely $A_{k'} := A \otimes_k k' \cong k'[x_1, \ldots, x_n, y, y^{-1}]$. Since $A^*_{k'} = k'^* \times \{y^m \mid m \in \mathbb{Z}\}$, we have

$$^g y = \gamma(g) y^{m(g)}, \quad \gamma(g) \in k'^*, m(g) \in \mathbb{Z}, \quad g \in \mathfrak{G}$$

where

$$\gamma(g'g) = {}^{g'}\gamma(g)\gamma(g')^{m(g)}, \quad m(g'g) = m(g')m(g), \quad g', g \in \mathfrak{G}.$$

Since $\gamma(e) = 1$ and $m(e) = 1$ for the identity element e of $\mathfrak{G}$, we know that $m : \mathfrak{G} \to \mathbb{Z}^*$ is a multiplicative character. Hence $m(g) = \pm 1$ for $g \in \mathfrak{G}$. Let $N = \{g \in \mathfrak{G} \mid m(g) = 1\}$. Then N is a normal subgroup of $\mathfrak{G}$ such that $[\mathfrak{G} : N] \leq 2$.

Lemma 30 *Assume that $N = \mathfrak{G}$. Then there exists a k-form Y of $\mathbb{A}^n$ such that X is k-isomorphic to $Y \times \mathbb{A}^1_*$.*

Proof Since $m(g) = 1$ for every $g \in \mathfrak{G}$, the equality $\gamma(g'g) = {}^{g'}\gamma(g)\gamma(g')$ holds for $g', g \in \mathfrak{G}$. By Hilbert's theorem 90, there exists an element $c \in k'^*$ such that $\gamma(g) = {}^g c \cdot c^{-1}$. Replacing y by y/c, we may assume that $^g y = y$ for every $g \in \mathfrak{G}$. So, $y \in A = (A \otimes_k k')^{\mathfrak{G}}$. Now each element $a \in A$ is expressed in $A \otimes_k k' = k'[x_1, \ldots, x_n, y, y^{-1}]$ as a Laurent polynomial

$$a = \sum_{i \in \mathbb{Z}} a'_i y^i, \quad a'_i \in k'[x_1, \ldots, x_n].$$

Since ${}^g a = a$, we have ${}^g(a'_i) = a'_i$ for every $i \in \mathbb{Z}$. Hence $a'_i \in A$. Let B be the set of elements $a \in A$ such that ${}^g a = a$ for $g \in \mathfrak{G}$ and a is constant as a Laurent polynomial in y. Then B is a k-algebra such that $(B[y, y^{-1}]) \otimes_k k' = A \otimes_k k'$. Then it follows that $A = B[y, y^{-1}]$. Since

$$(B \otimes_k k')[y, y^{-1}] = k'[x_1, \ldots, x_n][y, y^{-1}],$$

comparison of the constant rings of Laurent polynomial rings on both sides implies that $B \otimes_k k' = k'[x_1, \ldots, x_n]$. Thus $Y := \operatorname{Spec} B$ is a k-form of $\mathbb{A}^n$ trivialized by the field extension k'/k. □

Corollary 1 *Assume that $N = \mathfrak{G}$ and $n = 1$ or 2. Then X is k-isomorphic to $\mathbb{A}^n \times \mathbb{A}^1_*$.*

Proof Since there are no non-trivial k-forms of $\mathbb{A}^n$ if $n = 1, 2$, the assertion follows from Lemma 30. □

Assume that $N \neq \mathfrak{G}$ and $k'' = (k')^N$. Then k'' is a quadratic extension with Galois group $\mathfrak{G}/N \cong \mathbb{Z}/2\mathbb{Z}$. Let σ be a generator of the group $\mathfrak{G}/N$. Write $k'' = k(\alpha)$ with $a = \alpha^2 \in k$. By Lemma 30, there exists a k''-algebra A'' such that $A \otimes_k k'' = A''[y, y^{-1}]$ and $A'' \otimes_{k''} k' \cong k'[x_1, \ldots, x_n]$. Since ${}^\sigma y = cy^{-1}$ with $c \in (k'')^*$ and $\sigma^2 = e$, it follows that ${}^\sigma c = c$, whence $c \in k$. Let $C = \operatorname{Spec} R$ be an affine plane curve defined by

$$R = k[Y, Y']/(Y^2 - aY'^2 = 4c).$$

Note that C is k-isomorphic to $\mathbb{A}^1_*$ if and only if $\alpha \in k$. Then we have the following result.

Lemma 31 *With the above notations and assumptions, the following assertions hold.*

(1) The k-algebra R is identified with a subalgebra of A. Hence there exists a morphism $f : X \to C$.
(2) Each fiber of f is a k-form of $\mathbb{A}^n$.
(3) If $n = 1, 2$, the morphism f defines an $\mathbb{A}^n$-bundle over C.

Proof (1) Set

$$Y = y + \frac{c}{y}, \quad Y' = \frac{1}{\alpha}\left(y - \frac{c}{y}\right).$$

Then ${}^\sigma Y = Y$ and ${}^\sigma Y' = Y'$. Hence $Y, Y' \in A$ and there is a relation $Y^2 - aY'^2 = 4c$. Further, since we have

$$y = \frac{1}{2}(Y + \alpha Y'), \quad y^{-1} = \frac{1}{2c}(Y - \alpha Y'),$$

it follows that $R := k[Y, Y']/(Y^2 - aY'^2 = 4c)$ is a k-subalgebra of A which is a k-form of $k[y, y^{-1}]$.

(2) Let P be a closed (or generic) point of C and let $k'(P) = k' \otimes_k k(P)$, where $k(P)$ is the residue field of C at P. Then we have

$$\begin{aligned} A \otimes_R k'(P) &= (A \otimes_k k') \otimes_{R \otimes_k k'} k'(P) \\ &= k'[x_1, \ldots, x_n, y, y^{-1}] \otimes_{k'[y, y^{-1}]} k'(P) \\ &= k'[x_1, \ldots, x_n] \otimes_{k'} k'(P). \end{aligned}$$

This implies that the fiber $f^{-1}(P)$, which is defined over $k(P)$, is a $k(P)$-form of $\mathbb{A}^n$.

(3) If $n = 1, 2$, the generic fiber as well as general fibers of f is a form of $\mathbb{A}^n$, which is trivial. Hence f is an $\mathbb{A}^n$-bundle. □

5 Cancellation Problem in Dimension Three

The base field k is as before an algebraically closed field of characteristic zero. We can apply the arguments in the previous Subsect. 4.4 to the cancellation problem in dimension 3. Namely we prove the following result.

Theorem 10 *Let X be an affine variety of dimension 3 such that $X \times \mathbb{A}^{n-3} \cong \mathbb{A}^n$ with $n \geq 4$. Suppose that X has a proper and q-tight G_a-action. Then X is isomorphic to $\mathbb{A}^3$.*

The proof will be given as a consequence of lemmas below. Since the projection $\mathbb{A}^n \to X$ is faithfully flat, X is smooth, contractible and factorial. Let A be the coordinate ring of X. We have further $A^* = k^*$. Since the graph morphism $\Psi := (\sigma, \mathrm{id}_X) : G_a \times X \to X \times X$ is proper, the action is nontrivial. Let $q : X \to Y$ be the quotient morphism, where $Y = \operatorname{Spec} B$ with $B = A^{G_a}$. Note that B is a factorial affine domain of dimension two such that $B^* = k^*$ and q is an $\mathbb{A}^1$-fibration. As in the proof of Lemma 26, the action σ is fixed-point free, and every non-empty fiber of q is a G_a-orbit. We set $S = Y \setminus q(X)$.

Lemma 32 *The following assertions hold.*

(1) Let $g = q \circ p : \mathbb{A}^n \to Y$ with the canonical projection $p : \mathbb{A}^n \to X$. Then g has no fibers which contains an irreducible component of dimension $n - 1$.
(2) Let P_i $(1 \leq i \leq r)$ be a finite set of closed points of Y. Then there exists a linear plane L of $\mathbb{A}^n$ such that $g|_L : L \to Y$ is a dominant morphism and $\dim g^{-1}(P_i) \cap L \leq 0$ for every $1 \leq i \leq r$.
(3) Let $Y' = Y - \operatorname{Sing}(Y)$. Then $\overline{\kappa}(Y') = -\infty$.
(4) Y is isomorphic to $\mathbb{A}^2$.

Proof (1) There is a $G_a^{n-2} = G_a^{n-3} \times G_a$-action on $\mathbb{A}^n$ such that one factor G_a^{n-3} acts along the fibers of the trivial $\mathbb{A}^{n-3}$-bundle $\mathbb{A}^n \to X$ and another G_a is the given G_a-action on X. Suppose that g has an irreducible component F of dimension $n - 1$.

Then F is defined by an equation $b = 0$ for $b \in \Gamma(\mathbb{A}^n, \mathcal{O}_{\mathbb{A}^n})$. Since $g(F)$ is a point, F is stable by the G_a^{n-2}-action on $\mathbb{A}^n$. Hence $b \in B$. This gives a contradiction because the subvariety of Y defined by $b = 0$ has dimension one and $g(F)$ is a point.

(2) This follows from [39, Lemma 4].

(3) Note that Y is factorial because Y is the algebraic quotient X/G_a. Hence $\mathrm{Sing}(Y)$ is either empty or a finite set. Let $Y' = Y - \mathrm{Sing}(Y)$. Then $(g|_L)^{-1}(\mathrm{Sing}Y)$ is an empty set or a finite set by the assertion (2). Hence $L' := L - (g|_L)^{-1}(\mathrm{Sing}Y)$ has log Kodaira dimension $-\infty$ as the complement of a finite set of closed points in L. Then $\overline{\kappa}(Y') = -\infty$.

(4) The proof is the same as for Lemma 25. □

We need the following result of Dutta [6].

Lemma 33 *Let $q : X \to Y$ be a faithfully flat affine morphism of finite type of locally noetherian schemes. Assume that Y is normal and that the following conditions are satisfied.*

(1) The fiber of q over the generic point of Y is $\mathbb{A}^1$.
(2) The fiber of q over the generic point of each irreducible reduced closed subscheme of Y of codimension one is geometrically integral.

Then X is an $\mathbb{A}^1$-bundle over Y. In particular, if Y is an affine scheme then X is a line bundle over Y.

In order to use the result of Dutta, we have to show the following result.

Lemma 34 *In our setting of $q : X \to Y$, the condition (1) and (2) in Lemma 33 are satisfied. Hence X is an $\mathbb{A}^1$-bundle over the open set $q(X)$ of Y, where $S := X \setminus q(X)$ is a finite set.*

Proof We apply Lemma 33 to the morphism $q : X \to q(Y)$. Since X and Y are smooth and since q is equi-dimensional, it follows that q is a flat morphism. The local slice theorem (see [12]) implies the condition (1). Since every non-empty fiber of q is a G_a-orbit, the fiber of q over the generic point of an irreducible curve C of Y is geometrically irreducible (see the proof of Lemma 27). If it is not geometrically reduced, a general fiber X_y with $y \in C$ is a multiple fiber of q. The argument of Lemma 28 implies that q has no multiple fibers. Hence the fiber of q over the generic point of C is geometrically integral, and the condition (2) is satisfied.

Since $q : X \to Y$ is a flat morphism, the set S is a closed set. If it contains an irreducible component C of dimension one, C is defined by a prime element p. Since the empty set $q^{-1}(C)$ is defined by $p = 0$ in X, it follows that $p \in A^*$. Since $A^* = k^*$, this is impossible. So, S is a finite set. □

Remark 5 Lemma 34 is not true without the properness assumption of the G_a-action. In fact, let δ be an LND on $A = \mathbb{C}[x, y, z]$ defined by $\delta(x) = 0$, $\delta(y) = 2z$ and $\delta(z)=x$. Then $h = xy - z^2 \in \mathrm{Ker}\,\delta$, where $\mathrm{Ker}\,\delta$ is eventually a polynomial ring $B = \mathbb{C}[x, h]$. In B, x is a prime element and $T := f^{-1}(L) = \mathbb{A}^2 = \mathrm{Spec}\,\mathbb{C}[y, z]$, where $L = \{x = 0\}$ on $\mathbb{A}^2 = \mathrm{Spec}\, B$. But the restriction $f|_T : T \to L$ is not a fibration. For a point $h = c \in \mathbb{C}$ on the line L, the inverse image of this point in T is two lines $z = \pm\sqrt{c}$ parametrized by y if $c \neq 0$ and the double line if $c = 0$.

We need to show the following result.

Lemma 35 *The morphism $q : X \to Y$ is surjective.*

Proof This is proved by the same argument as in Theorem 9 if one notes that X is contractible and $q : X \to q(X)$ is a locally trivial $\mathbb{A}^1$-bundle. □

By combining the above lemmas together, we know that X is an $\mathbb{A}^1$-bundle over $Y \cong \mathbb{A}^2$. Hence $X \cong \mathbb{A}^3$.

Acknowledgements The present article is based on the materials prepared for a research to be conducted as a Rip program 2018 in the period August 5 to September 1, 2018 at the Mathematisches Forschungsinstitute Oberwolfach (MFO). We are very much grateful to MFO for the hospitality and the nice research environment. The first author would like to thank the Department of Atomic Energy of the Government of India for Dr. Raja Ramanna Fellowship during this work. The second author is supported by Grant-in-Aid for Scientific Research (C), No. 15K04831, JSPS, and the third author is supported by Grant-in-Aid for Scientific Research (C), No. 16K05115, JSPS.

References

1. Bass, H.: Big projective modules are free. Illinois J. Math. **7**, 24–31 (1963). https://projecteuclid.org/euclid.ijm/1255637479
2. Bass, H., Connell, E., Wright, D.: Locally polynomial algebras are symmetric algebras, Invent. Math. **38**(3), 279–299 (1976/77). https://doi.org/10.1007/BF01403135
3. Bonnet, P.: Surjectivity of quotient maps for algebraic $(\mathbb{C}, +)$-actions. Transform. Groups **7**, 3–14 (2002). https://doi.org/10.1007/BF01253462
4. Daigle, D.: Triangular derivations of $k[X, Y, Z]$. J. Pure Appl. Algebra **214**(7), 1173–1180 (2010). https://doi.org/10.1016/j.jpaa.2009.10.004
5. Daigle, D., Kaliman, Sh: A note on locally nilpotent derivations and variables of $k[X, Y, Z]$. Canad. Math. Bull. **52**(4), 535–543 (2009). https://doi.org/10.4153/CMB-2009-054-5
6. Dutta, A.K.: On $\mathbb{A}^1$-bundles of affine morphisms. J. Math. Kyoto Univ. **35**(3), 377–385 (1995). https://projecteuclid.org/euclid.kjm/1250518702
7. Dutta, A.K., Gupta, N., Lahiri, A.: A note on separable $\mathbb{A}^2$ and $\mathbb{A}^3$-forms. preprint
8. Dubouloz, A.: Hedén, I., Kishimoto, T.: Equivariant extensions of G_a-torsors over punctured surfaces. arXiv: 1707.08768v1
9. Dubouloz, A.: The cylinder over the Koras-Russell cubic threefold has a trivial Makar-Limanov invariant. Transform. Groups **14**(3), 531–539 (2009). https://doi.org/10.1007/S00031-009-9051-3
10. Dolgachev, I.V., Ju, B.: Weisfeiler, Unipotent group schemes over integral rings. Izv. Akad. Nauk SSSR Ser. Mat. **38**, 751–799 (1974)
11. Fieseler, K.-H.: On complex affine surfaces with $\mathbb{C}^+$-action. Comment. Math. Helv. **69**(1), 5–27 (1994). https://doi.org/10.1007/BF02564471
12. Freudenburg, G.: Invariant Theory and Algebraic Transformation Groups, VII. Algebraic theory of locally nilpotent derivations. Encyclopaedia of Mathematical Sciences, 2nd edn, p. xii+319. Springer, Berlin (2017). https://doi.org/10.1007/978-3-662-55350-3

13. Gurjar, R.V., Koras, M., Masuda, K., Miyanishi, M., Russell, P.: $\mathbb{A}^1_*$-fibrations on affine threefolds. The proceedings of the AAG conference, Osaka, vol. 2013, pp. 62–102. World Scientific (2011). https://doi.org/10.1142/9789814436700_0004
14. Gurjar, R.V., Masuda, K., Miyanishi, M.: $\mathbb{A}^1$-fibrations on affine threefolds. J. Pure Applied Algebra **216**, 296–313 (2012). https://doi.org/10.1016/j.jpaa.2011.06.013
15. Gurjar, R.V., Masuda, K., Miyanishi, M.: Deformations of $\mathbb{A}^1$-fibrations. Proceedings of "Groups of Automorphisms in Birational and Affine Geometry", Springer, Levico Terme (Trento), Italy (2012). https://doi.org/10.1007/978-3-319-05681-4_19
16. Gurjar, R.V., Koras, M., Masuda, K., Miyanishi, M., Russell, P.: Affine threefolds admitting G_a-actions. Math. Ann. Online first (2018). https://doi.org/10.1007/s00208-017-1622-3
17. Gurjar, R.V., Koras, M., Masuda, K., Miyanishi, M., Russell, P.: Triviality of affine space fibrations. preprint
18. Hartshorne, R.: Algebraic Geometry. Graduate Texts in Mathematics, vol. 52, pp. xvi+496. Springer, New York (1977)
19. Hironaka, H.: Flattening theorem in complex-analytic geometry. Am. J. Math. **97**, 503–547 (1975). https://doi.org/10.2307/2373721
20. Humphreys, J.E.: Linear algebraic groups, Graduate Texts in Mathematics, vol. 21, pp. xiv+247. Springer, New York (1975)
21. Kaliman, Sh: Polynomials with general $\mathbb{C}^2$-fibers are variables. Pacific J. Math. **203**(1), 161–190 (2002). https://doi.org/10.2140/pjm.2002.203.161
22. Kaliman, Sh: Free $\mathbb{C}^+$-actions on $\mathbb{C}^3$ are translations. Invent. Math. **156**(1), 163–173 (2004). https://doi.org/10.1007/s00222-003-0336-1
23. Kaliman, Sh: Proper G_a-actions on $\mathbb{C}^4$ preserving a coordinate. Algebra Number Theory **12**(2), 227–258 (2018). https://doi.org/10.2140/ant.2018.12.227
24. Kaliman, Sh, Saveliev, N.: $\mathbb{C}^+$-actions on contractible threefolds. Michigan Math. J. **52**(3), 619–625 (2004). https://doi.org/10.1307/mmj/1100623416
25. Kaliman, Sh, Vénéreau, S., Zaidenberg, M.: Simple birational extensions of the polynomial ring $\mathbb{C}^{[3]}$. Trans. Am. Math. Soc. **35**(2), 509–555 (2004). https://doi.org/10.1090/S0002-9947-03-03398-1
26. Kaliman, Sh, Zaidenberg, M.: Vénéreau polynomials and related fiber bundles. J. Pure Appl. Algebra **192**(1–3), 275–286 (2004). https://doi.org/10.1016/j.jpaa.2004.01.009
27. Kaliman, Sh., Zaidenberg, M.: Families of affine planes: the existence of a cylinder. Michigan Math. J. **49**(2), 353–367 (2001). https://doi.org/10.1307/mmj/1008719778
28. Kambayashi, T.: On the absence of nontrivial separable forms of the affine plane. J. Algebra **35**, 449–456 (1975). https://doi.org/10.1016/0021-8693(75)90058-7
29. Kambayashi, T., Miyanishi, M.: On flat fibrations by the affine line. Illinois J. Math. **22**, 662–671 (1978). https://projecteuclid.org/euclid.ijm/1256048473
30. Kambayashi, T., Wright, D.: Flat families of affine lines are affine-line bundles. Illinois J. Math. **29**(4), 672–681 (1985). https://projecteuclid.org/euclid.ijm/1256045502
31. Kollár, J.: Rational Curves on Algebraic Varieties. Ergebnisse der Mathematik und ihrer Grenzgebiete, 3 Folge, vol. 32. Springer (1996). https://doi.org/10.1007/978-3-662-03276-3
32. Koras, M., Russell, P.: Separable forms of G_m-actions on $\mathbb{A}^3$. Transform. Groups **18**(4), 1155–1163 (2013). https://doi.org/10.1007/s00031-013-9242-9
33. Kraft, H., Russell, P.: Families of group actions, generic isotriviality, and linearization. Transform. Groups **19**(3), 779–792 (2014). https://doi.org/10.1007/s00031-014-9274-9
34. Lazard, D.: Sur les modules plats. C. R. Acad. Sci. Paris **258**, 6313–6316 (1964)
35. Miyanishi, M.: Curves on rational and unirational surfaces, Tata Institute of Fundamental Research Lectures on Mathematics and Physics. Tata Institute of Fundamental Research, Bombay, vol. 60, pp. ii+302. Narosa Publishing House, New Delhi (1978)
36. Miyanishi, M.: Singularities of normal affine surfaces containing cylinderlike open sets. J. Algebra **68**(2), 268–275 (1981). https://doi.org/10.1016/0021-8693(81)90264-7
37. Miyanishi, M.: G_a-actions and completions. J. Algebra **319**, 2845–2854 (2008). https://doi.org/10.1016/j.jalgebra.2007.11.024

38. Miyanishi, M.: Open Algebraic Surfaces, CRM Monograph series, vol. 12. American Mathematical Soc (2001)
39. Miyanishi, M.: Normal Affine Subalgebras of a Polynomial Ring, Algebraic and Topological Theories (Kinosaki, 1984), 37–51. Kinokuniya, Tokyo (1986)
40. Nagata, M.: A treatise on the fourteenth problem of Hilbert. Mem. Sci. Kyoto Univ. **30**, 57–70 (1956). https://projecteuclid.org/euclid.kjm/1250777137
41. Popov, V.L.: Around the Abhyankar-Sathaye conjecture, Doc. Math. 2015, Extra volume: Alexander S. Merkurjev's sixtieth birthday, 513–528
42. Sakai, F.: Classification of normal surfaces, Algebraic geometry, Bowdoin, 1985 Part 1. Proceedings of Symposia in Pure Mathematics, vol. 46, pp. 451–465. American Mathematical Society, Providence, RI (1987)
43. Sathaye, A.: Polynomial ring in two variables over a DVR: a criterion. Invent. Math. **74**(1), 159–168 (1983). https://doi.org/10.1007/BF01388536
44. Sathaye, A.: On linear planes. Proc. Am. Math. Soc. **56**, 1–7 (1976). https://doi.org/10.1090/S0002-9939-1976-0409472-6
45. Seidenberg, A.: The hyperplane sections of normal varieties. Trans. Am. Math. Soc. **69**, 357–386 (1950). https://doi.org/10.2307/1990364
46. Seidenberg, A.: Derivations and integral closure. Pacific J. Math. **16**, 167–171 (1966). https://projecteuclid.org/euclid.pjm/1102995095
47. Seshadri, C.S.: Quotient spaces modulo reductive algebraic groups. Ann. Math. **95**, 511–556 (1972). https://doi.org/10.2307/1970870
48. Suzuki, M.: Propriétés topologiques des polynômes de deux variables complexes, et automorphismes algébriques de l'espace $\mathbb{C}^2$. J. Math. Soc. Japan **26**, 241–257 (1974). https://projecteuclid.org/euclid.jmsj/1240435278
49. Zaidenberg, M.: Isotrivial families of curves on affine surfaces and characterization of the affine plane. Izv. Akad. Nauk SSSR Ser. Mat. **51**(3), 534–567, 688 (1987); Translation in Math. USSR-Izv. **30**(3), 503–532 (1988). https://doi.org/10.1070/IM1988v030n03ABEH001027
50. Zariski, O.: Interprétations algébrico-géométriques du quatorzième problème de Hilbert. Bull. Sci. Math. **78**(2), 155–168 (1954)
51. Xiao, G.: π_1 of elliptic and hyperelliptic surfaces. Internat. J. Math. **2**, 599–615 (1991). https://doi.org/10.1142/S0129167X91000338

A Graded Domain Is Determined at Its Vertex. Applications to Invariant Theory

R. V. Gurjar

Abstract We will prove that a positively graded domain/$\mathbf{C}$ is uniquely determined by its completion at the irrelevant maximal ideal. As an application we will prove that the logarithmic Kodaira dimension of the smooth locus of a quotient of an affine space modulo a reductive algebraic group is $-\infty$.

Keywords Graded ring · Logarithmic Kodaira dimension · Invariant theory

Mathematics Subject Classification (2000) 14L24 · 14L30 · 13A02

1 Introduction

The aim of this paper is to prove two results which were conjectured by the author twenty five years ago. At that time the author was not aware of the results of Hauser–Müller [4], [6] which are very useful in this paper. The proof depends crucially on ideas of Olivier Mathieu.

Theorem 1 *Let G be a complex reductive algebraic group acting linearly and rationally on an affine space* $\mathbf{C}^n$. *Let* $V := \mathbf{C}^n//G$ *be the quotient corresponding to the ring of invariants* $\mathbf{C}[X_1, X_2, \ldots, X_n]^G$. *Then the logarithmic Kodaira dimension of the smooth locus of* V, $\overline{\kappa}(V - Sing\ V)$, *is* $-\infty$.

The next result is a very general result about affine graded domains.

Theorem 2 *Let* R, S *be positively graded affine domains* $/\mathbf{C}$. *Let* $(V, p), (W, q)$ *be the corresponding affine varieties with good* $\mathbf{C}^*$*-actions with vertices* p, q *respectively. If the complex analytic germs of* V, W *at* p, q *are isomorphic then* V, W *are isomorphic as affine varieties (not necessarily as* $\mathbf{C}^*$*-varieties).*

R. V. Gurjar (✉)
Department of Mathematics, Indian Institute of Technology Bombay, Mumbai 400076, India
e-mail: gurjar@math.iitb.ac.in

S. Kuroda et al. (eds.), *Polynomial Rings and Affine Algebraic Geometry*,
Springer Proceedings in Mathematics & Statistics 319,
https://doi.org/10.1007/978-3-030-42136-6_7

Theorem 2 is an easy consequence of a result of Hauser and Müller [4].

The proof of Theorem 1 uses a result proved by the author which proved a stronger form of a conjecture of C.T.C. Wall [1].

The crucial idea in the proof of Theorem 2 is due to O. Mathieu. In fact, around 2003 he showed a proof of this result to the author which was somewhat more complicated than the simplified proof here.

Mathieu's proof used some results about proalgebraic groups, proLevi subgroups, etc. After all these years Mathieu's proof has still not appeared in print (he has also not replied to many emails). The author feels that this result will be useful to other mathematicians, and Theorem 1 is a very general result about quotients of an affine space modulo a reductive algebraic group which will also be of interest to many mathematicians. So he has dared to publish a more accessible proof using the results of Hauser-Müller.

It was observed by the author in 1990 that Theorem 1 follows from Theorem 2. In fact, this was the motivation to conjecture Theorem 2. In [2] Theorem 1 was proved when G is either a semisimple group or a torus.

It is clear that Theorem 2 makes the well-known étale slice theorem of Luna a global result is some sense.

2 Preliminaries

We will only deal with complex algebraic or analytic varieties. Using Lefschetz Principle Theorems 1 and 2 are both valid for any algebraically closed field of characteristic 0. In Theorem 2 we have to replace complex analytic germs by completions of the local rings of V, W at their irrelevant maximal ideals.

For any smooth irreducible algebraic variety Z the logarithmic Kodaira dimension, $\overline{\kappa}(Z)$, was introduced by Iitaka [5]. We will only use the easy result that if $f : Y \to Z$ is a dominant morphism between smooth irreducible algebraic varieties of the same dimension then $\overline{\kappa}(Y) \geq \overline{\kappa}(Z)$.

We will first recall the work of Hauser–Müller in [4], [6].

Let X be a complex analytic germ corresponding to a local analytic ring $R := \mathbf{C}\{Z_1, Z_2, \ldots, Z_n\}/I$, where I is a reduced ideal in the ring of convergent power series $\mathbf{C}\{Z_1, Z_2, \ldots, Z_n\}$. Let M be the maximal ideal of R and let $A := Aut\ X$ be the group of complex analytic automorphisms of X. An algebraic group G which is a subgroup A is called an *algebraic subgroup* of A if the natural action of G on the finite dimensional vector spaces M^l/M^{l+1} is rational for $l > 0$. Let σ be the representation of A on M/M^2 and let $A_1 := \sigma(A)$. Then A_1 is an algebraic subgroup of $GL(M/M^2)$. By a standard result [3, Chapter VIII, Theorem 4.3], A_1 has a maximal reductive algebraic subgroup L, called a *Levi subgroup*. Any reductive subgroup of A_1 is conjugate (by an algebraic automorphism) to a subgroup of L. A reductive algebraic subgroup G of A is called a Levi subgroup of A if $\sigma(G)$ is a Levi subgroup of A_1. It is proved in [6] that a rational action of a reductive subgroup on R can be lifted to an action on $\mathbf{C}\{Z_1, Z_2, \ldots, Z_n\}$, linear in suitable generators of

$(Z_1, Z_2, \ldots, Z_n)$. It follows that the action of any reductive subgroup of $Aut\ X$ is linearizable.

The next result is one of the main results in [4] (stated in a slightly weaker form).

Proposition 1 *With the above notation if I is generated by polynomials in $Z_1, \ldots, Z_n$ (i.e. if R is algebraic) then $Aut\ X$ has a Levi subgroup.*

In particular, if I is weighted homogeneous w.r.t. weights of $Z_i > 0$ for all i then $Aut\ X$ contains a Levi subgroup.

Next we will recall the result in [1] which will be a crucial ingredient in the proof of Theorem 1.

Proposition 2 *Let G be a reductive algebraic group acting rationally on a smooth affine variety Y. Let $p \in Y$ be a point with closed orbit. Let $V := Y//G$, and let $\pi : Y \to V$ be the quotient morphism. Then there is a closed reductive subgroup $H \subset G$ acting linearly and rationally on an affine space $\mathbf{C}^m$ such that if $W := \mathbf{C}^m//H$ then the analytic germs of V at $\pi(p)$ and of W at the image of $O \in \mathbf{C}^m$ are isomorphic. Further, the inverse image of the singular locus of W in $\mathbf{C}^m$ has codimension > 1 in $\mathbf{C}^m$.*

The first part of this result is just Luna's étale slice theorem. The second part is the author's main contribution in [1].

3 Proof of Theorem 2

We must prove that $R \cong S$ as affine domains $/\mathbf{C}$.

Let X, Y be the analytic germs of V, W at p, q respectively. By assumption, $X \cong Y$ as analytic germs. Using this isomorphism we will identify X, Y, but keep track of the two actions of $\mathbf{C}^*$ on X. Let L be a Levi subgroup of $Aut\ X$, as guaranteed by Proposition 1. Now R, S are positively graded rings which are subrings of the analytic local ring of X at p. There exist rational actions of $\mathbf{C}^*$ on X such that p is the only closed orbit for both the actions. For notational clarity we denote the two $\mathbf{C}^*$s by groups G_1, G_2. Let M be the maximal ideal of X at p. We have a short exact sequence of L-modules

$$(0) \to M^2 \to M \to M/M^2 \to (0).$$

The action of L on M/M^2 is linear. By a result in [6, Hilfssatz2], there is an L-splitting of the above sequence. This means that there is a basis $x_1, x_2, \ldots, x_n$ of M w.r.t. which the action of L is $\mathbf{C}$-linear. By Proposition 1 we can assume that G_1, G_2 are both subgroups of L. We can assume without loss of generality that for the action of G_1 we have $\rho_{1,\lambda} x_i = \lambda^{e_i} x_i$ with $e_i > 0$ for all i.

Let M_1, M_2 be the irrelevant maximal ideals of R, S resp. There is a basis of M_1, say $y_1, \ldots, y_n$, such that y_i is a semi-invariant for G_1. Similarly M_2 has a basis $z_1, \ldots, z_n$

such that z_j is a semi-invariant for G_2. Each y_i is a power series in $x_1, x_2, ..., x_n$. Since y_j and each x_l is a semiinvariant for the action of G_1 it is easy to see that y_j is actually a polynomial in $x_1, ..., x_n$, i.e. $R \subset T := \mathbf{C}[x_1, ..., x_n]$. Similarly, $S \subset T$. Denote by M_0 the maximal ideal $(x_1, ..., x_n)$ of T. The G_1-action on T makes T a graded domain and R is a graded subring for this grading. Similarly, the G_2-action on T gives another grading on T for which S is a grading subring of T. For the action of G_1 on the $\mathbf{C}$-isomorphic vector spaces M_1/M_1^2 and M_0/M_0^2 the weight spaces for each weight are of same dimension. It follows that $R = T$. Similarly $S = T$, showing that $R \cong S$ as affine domains.

This completes the proof of Theorem 2.

4 Proof of Theorem 1

We use Proposition 2. Let $H, W, \varphi : \mathbf{C}^m \to W$ be as in Proposition 2. Then φ^{-1} $(Sing\ W)$ has codimension > 1 in $\mathbf{C}^m$. Let Z be a general linear subspace of $\mathbf{C}^m$ of same dimension as W. We denote the restriction of φ to Z by the same symbol. Then $Z \cap \varphi^{-1}(Sing\ W)$ has codimension > 1 in Z. It follows easily that $\overline{\kappa}(Z \setminus \varphi^{-1}(Sing\ W)) = -\infty$. Since $Z \setminus \varphi^{-1}(Sing\ W) \to W \setminus Sing\ W$ is a dominant morphism we see that $\overline{\kappa}(W \setminus Sing\ W) = -\infty$.

By Theorem 2, $\mathbf{C}^n//G$ and $\mathbf{C}^m//H$ are isomorphic as affine varieties. Hence $\overline{\kappa}(V \setminus Sing\ V) = -\infty$.

This completes the proof of Theorem 1.

References

1. Gurjar, R.V.: On a conjecture of C.T.C. Wall. J. Kyoto Univ. **31**, 1121–1124 (1991)
2. Gurjar, R.V., Simha, R.R.: Some results on the topology of varieties dominated by C^n. Math. Zeit. **211**, 333–340 (1992)
3. Hochschild, G.: Basic Theory of Algebraic Groups and Lie Algebras. Graduate Texts in Mathematics. Springer, Berlin (1981)
4. Hauser, H., Müller, G.: Algebraic singularities have maximal reductive automorphism groups. Nagoya Math. J. **113**, 181–186 (1989)
5. Iitaka, S.: Algebraic Geometry. An Introduction to Birational Geometry of Algebraic Varieties. Graduate Texts in Mathematics, vol. 76. Springer, Berlin (1982)
6. Müller, G.: Reduktive automorphismengruppen von analytischer C-algebren. J. Reine Angew. Math. **364**, 26–34 (1986)

Singularities of Normal Log Canonical del Pezzo Surfaces of Rank One

Hideo Kojima

Abstract Let X be a normal del Pezzo surface of rank one with only rational log canonical singular points. In this paper, we prove that X can have at most one non klt singular point.

Keywords Normal del Pezzo surface · Log canonical singular point · klt singular point · Logarithmic Kodaira dimension

2010 Mathematics Subject Classification Primary 14J26 · Secondary 14J17 · 14R25

1 Introduction

We work over an algebraically closed field k. Let X be a normal complete algebraic surface. We say that X is a *normal del Pezzo surface* if $(-K_X)^2 > 0$ and $(-K_X)C > 0$ for every curve C on X (with Mumford's rational intersection number (see [18, 20])). Note that a normal del Pezzo surface is projective (see [21, Corollary 4.4]). A normal del Pezzo surface is said to have rank one if its Picard number equals one.

We recall some of the results on normal del Pezzo surfaces. Normal del Pezzo surfaces with only Gorenstein singularities were studied by Hidaka–Watanabe [11], Furushima [10], Ye [24], etc. Normal del Pezzo surfaces of rank one containing at least one non-rational singular point were classified by Cheltsov [4] in the case of $\text{char}(k) = 0$ and by Schröer [22] in any characteristic. Normal del Pezzo surfaces with only klt (Kawamata log terminal) singular points (usually called log del Pezzo surfaces) have been studied from many points of view. There are partial classification results of log del Pezzo surfaces, mainly in the case $\text{char}(k) = 0$. See, e.g., Alexeev–Nikulin [2], Nakayama [19], Fujita [7, 8] and Fujita–Yasutake [9].

H. Kojima (✉)
Department of Mathematics, Faculty of Science, Niigata University, 8050 Ikarashininocho, Nishi-ku, Niigata 950-2181, Japan
e-mail: kojima@math.sc.niigata-u.ac.jp

S. Kuroda et al. (eds.), *Polynomial Rings and Affine Algebraic Geometry*, Springer Proceedings in Mathematics & Statistics 319,
https://doi.org/10.1007/978-3-030-42136-6_8

In this paper, we study normal del Pezzo surfaces of rank one with only log canonical singular points. Such surfaces form a class of end results in the minimal model theory of normal projective surfaces with only log canonical singular points, which was developed by Fujino and Tanaka (see [5, 6, 23]). The main result of this paper is the following theorem.

Theorem 1 *Every normal del Pezzo surface of rank one with only rational log canonical singular points can have at most one non klt singular point.*

The author thinks that Theorem 1 is well-known for experts when $\mathrm{char}(k) = 0$. We prove Theorem 1 by using the minimal model theory of open algebraic surfaces due to Miyanishi and Tsunoda (see [16, Chap. 2] and [17]), which works also in the case $\mathrm{char}(k) > 0$. So our proof of Theorem 1 works also in the case $\mathrm{char}(k) > 0$.

2 Preliminaries

We employ the following notations.

K_V: the canonical divisor on V.
$\rho(V)$: the Picard number of V.
$\overline{\kappa}(S)$: the logarithmic Kodaira dimension of S. (See [16] for its definition.)
$\#D$: the number of irreducible components of a reduced effective divisor D.
$\lfloor Q \rfloor$: the integral part of a $\mathbb{Q}$-divisor Q.
$D_1 \equiv D_2$: D_1 and D_2 are numerically equivalent.
$\mu^*(D)$: the total transform of D by μ.
$\mu'(D)$: the proper transform of D by μ.

We recall some basic notions in the theory of peeling. For more details, see [16, Chap. 2] and [17, Chap. 1]. A reduced effective divisor on a smooth algebraic variety is called an *SNC-divisor* if it has only simple normal crossings. Let X be a smooth projective surface and B an SNC-divisor on X. We call such a pair (X, B) an *SNC-pair*. B is said to be *minimal* if, for any (-1)-curve $E \subset \mathrm{Supp}\, B$, either $E(B - E) \geq 3$ or there exists an irreducible component E' of $B - E$ such that $EE' = E(B - E) = 2$. A connected curve consisting only of irreducible components of B is called a connected curve in B for shortness. A connected curve T in B is said to be *admissible* (resp. *rational*) if there are no (-1)-curves in $\mathrm{Supp}\, T$ and the intersection matrix of T is negative definite (resp. if it consists only of rational curves).

A connected curve T in B is called a *twig* if its dual graph is a linear chain and T meets $D - T$ in a single point at one of the end components of T. An admissible rational twig in B is said to be *maximal* if it is not extended to an admissible rational twig with more irreducible components of B. A connected curve in B is called a *rod* (resp. a *fork*) if it is a connected component of B and its dual graph is a linear chain (resp. its dual graph is that of the exceptional curves of a minimal resolution of a klt singular point and is not a linear chain). By a (-2)-rod (resp. a (-2)-fork), we mean a rod (resp. a fork) consisting only of (-2)-curves.

Let $\{T_\lambda\}$ (resp. $\{R_\mu\}$, $\{F_\nu\}$) be the set of all admissible rational maximal twigs (resp. all admissible rational rods, all admissible rational forks), where no irreducible components of T_λ's belong to R_μ's or F_ν's. Then there exists a unique decomposition of B as a sum of effective $\mathbb{Q}$-divisors $B = B^\# + \mathrm{Bk}(B)$ such that the following two conditions (a) and (b) are satisfied:

(a) $\mathrm{Supp}(\mathrm{Bk}(B)) = (\cup_\lambda T_\lambda) \cup (\cup_\mu R_\mu) \cup (\cup_\nu F_\nu)$.
(b) $(K_V + B^\#)Z = 0$ for every irreducible component Z of $\mathrm{Supp}(\mathrm{Bk}(B))$.

Definition 1 An SNC-pair (X, B) is said to be *almost minimal* if, for every irreducible curve C on X, either $C(K_X + B^\#) \geq 0$ or $C(K_X + B^\#) < 0$ and the intersection matrix of $C + \mathrm{Bk}(B)$ is not negative definite.

Let (X, B) be an SNC-pair. It then follows from [16, Theorem 2.3.11.1 (p. 107)], which is the same as [17, Theorem 1.11], that there exists a birational morphism $\mu : X \to V$ onto a smooth projective surface V such that the following conditions (1)–(4) are satisfied:

(1) $D := \mu_*(B)$ is a minimal SNC-divisor. So $D^\#$ and $\mathrm{Bk}(D)$ are defined.
(2) $\mu_*(\mathrm{Bk}(B)) \leq \mathrm{Bk}(D)$ and $\mu_*(K_X + B^\#) \geq K_V + D^\#$.
(3) $\overline{\kappa}(V - D) = \overline{\kappa}(X - B)$.
(4) (V, D) is almost minimal.

We call the pair (V, D) an *almost minimal model* of (V, D).

We recall the following lemma on the almost minimal SNC-pairs of $\overline{\kappa} = -\infty$.

Lemma 1 *Let (V, D) be an almost minimal SNC-pair of $\overline{\kappa}(V - D) = -\infty$. Let $\pi : V \to \overline{V}$ be the contraction of* $\mathrm{Supp}(\mathrm{Bk}(D))$ *to normal points and set $\overline{D} := \pi_*(D)$. (Here we note that $\overline{V}$ has only klt singular points.) Then one of the following cases takes place.*

(A) There exists a $\mathbb{P}^1$-fibration $h : \overline{V} \to C$ onto a smooth projective curve C such that every fiber of h is irreducible and $\overline{D}F \leq 1$ for a fiber F of h.
(B) $\rho(\overline{V}) = 1$ and $-(K_{\overline{V}} + \overline{D})$ is an ample $\mathbb{Q}$-Cartier divisor.

Proof See [16, Lemmas 2.3.14.3 and 2.3.14.4 (pp. 113–114)]. □

3 Proof of Theorem 1

In this section, we prove Theorem 1. Let X be a normal del Pezzo surface of rank one with only rational log canonical singular points and let $\pi : V \to X$ be the minimal resolution of singularities on X, here we assume that $\mathrm{Sing}\, X \neq \emptyset$. Since X has only rational singularities, $D = \pi^{-1}(\mathrm{Sing}\, X)$ becomes an SNC-divisor and consists only of smooth rational curves (cf. [3]). It is clear that X is a birationally ruled surface. Moreover, X is a rational surface since it has only rational singular points (see [12, Theorem 1]).

Let $D = \sum_i D_i$ be the irreducible decomposition of D. We have an effective $\mathbb{Q}$-divisor $\Delta = \sum_i \alpha_i D_i$ such that $K_V + \Delta \equiv \pi^* K_X$. Since X has only log canonical singular points, we know that $\Delta = D^\#$, where $D^\# = D - \mathrm{Bk}(D)$ is the one defined in Sect. 2. We set $D' = \lfloor D^\# \rfloor$.

Let $P_1, \ldots, P_t$ ($t = \#\,\mathrm{Sing}\, X$) be the singular points on X and let $D = \sum_{i=1}^t D^{(i)}$ be the decomposition of D into connected components such that $D^{(i)} = \pi^{-1}(P_i)$ for $i = 1, \ldots, t$. We assume that P_j ($j = 1, \ldots, s$) is a non klt singular point and $P_{j'}$ ($j' = s+1, \ldots, t$) is a klt singular point. Note that Supp D' consists of s connected components. In order to prove Theorem 1, we may assume that $s \geq 1$.

Lemma 2 *With the same notations and assumptions as above, we have* $\overline{\kappa}(V - D) = -\infty$.

Proof We take a positive integer n_0 such that $n_0 D^\#$ is a Cartier divisor. Suppose to the contrary that $\overline{\kappa}(V - D) \geq 0$. Take a positive integer n such that $n_0 \mid n$ and $|n(K_V + D)| \neq \emptyset$. Then $\lfloor n(K_V + D^\#) \rfloor = n(K_V + D^\#)$. By [16, Lemma 2.3.10.1 (p. 106)], we have

$$h^0(V, n(K_V + D)) = h^0(V, \lfloor n(K_V + D^\#) \rfloor) = h^0(V, n(K_V + D^\#)) > 0.$$

Take an ample divisor H on V. Since $|n(K_V + D^\#)| \neq \emptyset$, $H(K_V + D^\#) \geq 0$. On the other hand, since $-(K_V + D^\#) \equiv -\pi^* K_X$ is nef and big, $-H(K_V + D^\#) = -H\pi^* K_X > 0$, i.e., $H(K_V + D^\#) < 0$. This is a contradiction. □

Lemma 3 *Assume that* (V, D) *is almost minimal. Then the following assertions hold true*:

(1) D' *is irreducible. In particular,* $s = 1$.
(2) $V - D$ *is affine ruled.*

Proof By Lemma 2, one of the cases (A) and (B) in Lemma 1 takes place. Here we use the same notations as in Lemma 1. Since $\rho(\overline{V}) = \rho(V) - \#(D - D') = \#D' + 1 \geq 2$, the case (B) does not take place. So $\rho(\overline{V}) = 2$ and hence $\#D' = 1$, which proves the assertion (1).

Set $\Phi := h \circ \pi : V \to C$, which is a $\mathbb{P}^1$-fibration on V. Here, $C \cong \mathbb{P}^1$ since V is a rational surface. Let F be a fiber of Φ. Then $F(K_V + D^\#) < 0$ and $FD^\# = FD'$ since $D - D'$ is contained in fibers of Φ. So $FD = FD' \leq 1$. In particular, $V - D$ is affine ruled. □

Theorem 1 is thus verified when the SNC-pair (V, D) is almost minimal.

From now on, we assume that (V, D) is not almost minimal. Then there exists an irreducible curve E on V such that $E(K_V + D^\#) < 0$ and the intersection matrix of $E + (D - D')$ is negative definite. Since $E \not\subset \mathrm{Supp}\, D$, E is a (-1)-curve. If $P_1, \ldots, P_s \notin \pi(E)$, i.e., $ED^{(i)} = 0$ for $i = 1, \ldots, s$, then the intersection matrix of $E + D$ is negative definite. This contradicts $\rho(V) = 1 + \#D$. Hence we may assume that $ED^{(1)} \geq 1$. Since $ED^\# < -EK_V = 1$, $E\lfloor D^\# \rfloor = 0$. In particular, $E\lfloor (D^{(1)})^\# \rfloor = 0$.

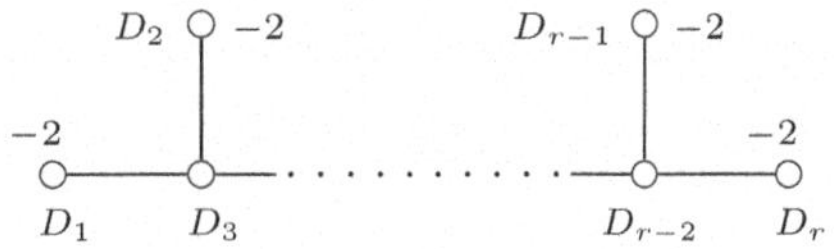

Fig. 1 The weighted dual graph of $D^{(1)}$ in Case 1

By [16, Lemmas 2.3.6.3 and 2.3.8.4 (p. 96, p. 102)], which is the same as [17, Lemmas 1.6.2 and 1.8.3], we know that:

(1) $ED \leq 2$.
(2) $ED^{(1)} = 1$. In particular, E meets an admissible rational maximal twig, say $T^{(1)}$, in $D^{(1)}$.
(3) If $ED = 2$, then the connected component of $D - D^{(1)}$ meeting E is an admissible rational rod, say R, in D and $E + T^{(1)} + R$ can be contracted to either an admissible rational rod or a smooth point. In particular, E meets a terminal component of $T^{(1)}$ or a terminal component of R.

Set $S := X - \pi(E)$. Since $\rho(X) = 1$ and X is $\mathbb{Q}$-factorial, S is a normal affine surface.

Lemma 4 *With the same notations and assumptions as above, assume further that $ED = ED^{(1)} = 1$. Then there exists a birational morphism $\mu : W \to V$ from a smooth projective surface W such that the following conditions are satisfied*:

(1) μ is a composite of blowing ups at a point on $\mathrm{Supp}(E + D^{(1)})$ *and its infinitely near points.*
(2) There exists a $\mathbb{P}^1$-fibration $\Phi : W \to \mathbb{P}^1$ such that $F\mu^*(E + D)_{\mathrm{red}} = 1$ for a fiber F of Φ.

Proof We note that $E + D^{(1)}$ supports a big divisor. In particular, the intersection matrix of $E + D^{(1)}$ is not negative definite. Since P_1 is not a klt singular point, the weighted dual graph of $D^{(1)}$ is one of the dual graphs (6)–(8) in [1, p. 58]. We consider the following cases separately.

Case 1. The dual graph of D is given as one of (7) and (8) in [1, p. 58]. Let $D^{(1)} = \sum_{i=1}^{r} D_i$ be the decomposition of $D^{(1)}$ into irreducible components and set $a_i = -(D_i)^2$ for $i = 1, \ldots, r$. In this case, $r \geq 5$ and the dual graph of $D^{(1)}$ is given as in Fig. 1.

In this case, we know that $(D^{(1)})^{\#} = \frac{1}{2}(D_1 + D_2 + D_{r-1} + D_r) + \sum_{i=3}^{r-2} D_i$ and some of $a_3, \ldots, a_{r-2} \geq 3$. Since $ED^{(1)} = 1$ and $E(D^{(1)})^{\#} < -EK_V = 1$, we may assume that $ED^{(1)} = ED_1 = 1$. Since the intersection matrix of $E + D^{(1)}$ (resp. $D^{(1)}$) is not negative definite (resp. negative definite), we know that $a_3 = 2$ and $r \geq 6$. Moreover, $a_4 = 2$ and $r \geq 7$. Then the divisor $F := 2(E + D_1 + D_3) + D_2 + D_4$ defines a $\mathbb{P}^1$-fibration $\Phi_{|F|} : V \to \mathbb{P}^1$, D_5 becomes a section of Φ and $D - D_5$ is contained in fibers of $\Phi_{|F|}$. In this case, Lemma 4 is verified by setting $\mu := \mathrm{id}_V$ and $\Phi := \Phi_{|F|}$.

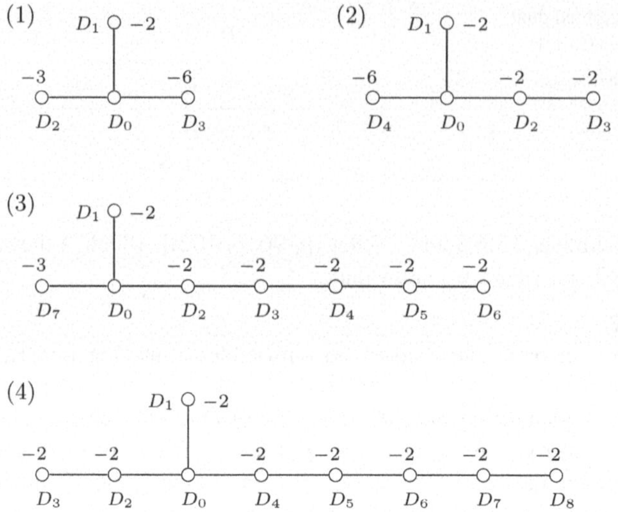

Fig. 2 The possible weighted dual graphs of $D^{(1)}$ with $(\Delta_1, \Delta_2, \Delta_3) = (2, 3, 6)$

Case 2. The dual graph of $D^{(1)}$ is given as (6) in [1, p. 58]. We treat only the case where $(\Delta_1, \Delta_2, \Delta_3) = (2, 3, 6)$ with the notations in [1, p. 58]; the other cases can be treated easier.

From now on, we assume that $(\Delta_1, \Delta_2, \Delta_3) = (2, 3, 6)$. Then the dual graph of $D^{(1)}$ is given as one of (1)–(4) in Fig. 2.

We consider the following cases separately.

Case 2-1. The dual graph of $D^{(1)}$ is given as (1) in Fig. 2, where $D^{(1)} = D_0 + D_1 + D_2 + D_3$ is the decomposition of D into irreducible components. Then, since $ED^{(1)} = 1$, we know that the intersection matrix of $E + D^{(1)}$ is negative definite. This is a contradiction.

Case 2-2. The dual graph of $D^{(1)}$ is given as (2) in Fig. 2, where $D^{(1)} = D_0 + D_1 + \cdots + D_4$ is the decomposition of $D^{(1)}$ into irreducible components. Then $(D^{(1)})^\# = D_0 + \frac{1}{2}D_1 + \frac{2}{3}D_2 + \frac{1}{3}D_3 + \frac{5}{6}D_4$. Since $ED^{(1)} = 1$, $E(D^{(1)})^\# < -EK_V = 1$, $a_0 = -(D_0)^2 \geq 2$ and the intersection matrix of $E + D^{(1)}$ is not negative definite, we know that $ED^{(1)} = ED_2 = 1$ and $a_0 = 2$. Then the divisor $G := 2(E + D_2) + D_0 + D_3$ defines a $\mathbb{P}^1$-fibration $\Phi_{|G|} : V \to \mathbb{P}^1$, D_1 and D_4 become sections of $\Phi_{|G|}$ and $D - (D_1 + D_4)$ is contained in fibers of $\Phi_{|G|}$. By [16, Corollary 2.2.11.1 (p. 82)], $V - D$ is affine ruled. Moreover, as seen from the proof of [16, Corollary 2.2.11.1 (p. 82)], we know that there exists a birational morphism $\mu : W \to V$ from a smooth projective surface W such that:

(1) μ is a composite of blowing ups at the point $D_0 \cap D_4$ and its infinitely near points.

(2) There exists a $\mathbb{P}^1$-fibration $\Phi : W \to \mathbb{P}^1$ such that $F\mu^*(E+D)_{\text{red}} = 1$ for a fiber F of Φ.

Of course, we can give the birational morphism μ explicitly; we omit the description. In this case, Lemma 4 is thus verified.

Case 2-3. The dual graph of $D^{(1)}$ is given as (3) in Fig. 2, where $D^{(1)} = D_0 + D_1 + \cdots + D_7$ is the decomposition of $D^{(1)}$ into irreducible components. Then $(D^{(1)})^\# = D_0 + \frac{1}{2}D_1 + \frac{5}{6}D_2 + \frac{2}{3}D_3 + \frac{1}{2}D_4 + \frac{1}{3}D_5 + \frac{1}{6}D_6 + \frac{2}{3}D_7$. Since $ED^{(1)} = 1$, $E(D^{(1)})^\# < -EK_V = 1$, $a_0 = -(D_0)^2 \geq 2$ and the intersection matrix of $E + D^{(1)}$ is not negative definite, we know that $ED^{(1)} = ED_i = 1$ for some i, $1 \leq i \leq 5$. We consider the following subcases separately.

Subcase 1. $i = 1$. Then $a_0 = 2$ because the intersection matrix of $E + D^{(1)}$ is not negative definite. So the divisor $F := 3(E + D_1 + D_0) + 2D_2 + D_3 + D_7$ defines a $\mathbb{P}^1$-fibration $\Phi_{|F|} : V \to \mathbb{P}^1$, D_4 becomes a section of Φ and $D - D_4$ is contained in fibers of $\Phi_{|F|}$. In this subcase, Lemma 4 is verified by setting $\mu := \mathrm{id}_V$ and $\Phi := \Phi_{|F|}$.

Subcase 2. $3 \leq i \leq 5$. Then the divisor $G := 2(E + D_i) + D_{i-1} + D_{i+1}$ defines a $\mathbb{P}^1$-fibration $\Phi_{|G|} : V \to \mathbb{P}^1$, D_{i-2} and D_{i+2} become sections of Φ and $D - (D_{i-2} + D_{i+2})$ is contained in fibers of Φ. By using the argument as in Case 2-2, we can prove Lemma 4 in this subcase.

Subcase 3. $i = 2$. Since the intersection matrix of $E + D^{(1)}$ is not negative definite, $2 \leq a_0 \leq 5$. If we contract $E, D_2, \ldots, D_6$, then the image of D_0 becomes a smooth rational curve with non-negative self-intersection number. By [16, Corollary 2.2.11.1 (p. 82)], $V - D$ is affine ruled. Moreover, as seen from the proof of [16, Corollary 2.2.11.1 (p. 82)], we know that there exists a birational morphism $\mu : W \to V$ from a smooth projective surface W such that:

(1) μ is a composite of blowing ups at the point $D_0 \cap D_7$ and its infinitely near points.
(2) There exists a $\mathbb{P}^1$-fibration $\Phi : W \to \mathbb{P}^1$ such tht $F\mu^*(E+D)_{\text{red}} = 1$ for a fiber F of Φ.

Of course, we can give the birational morphism μ explicitly; we omit the description. In this subcase, Lemma 4 is thus verified.

Case 2-4. The dual graph of $D^{(1)}$ is given as (4) in Fig. 2, where $D^{(1)} = D_0 + D_1 + \cdots + D_8$ is the decomposition of $D^{(1)}$ into irreducible components. Then $(D^{(1)})^\# = D_0 + \frac{1}{2}D_1 + \frac{2}{3}D_2 + \frac{1}{3}D_3 + \frac{5}{6}D_4 + \frac{2}{3}D_5 + \frac{1}{2}D_6 + \frac{1}{3}D_7 + \frac{1}{6}D_8$. Since $ED^{(1)} = 1$, $ED^\# < 1, a_0 \geq 3$ and the intersection matrix of $E + D^{(1)}$ is not negative definite, we know that $ED^{(1)} = ED_i = 1$ for some $i \in \{2, 4, 5, 6, 7\}$. We consider the following subcases separately.

Subcase 1. $i = 2$. Then $a_0 = 3$. So the divisor $F := 4(E + D_2) + 2(D_0 + D_3) + D_1 + D_4$ defines a $\mathbb{P}^1$-fibration $\Phi_{|F|} : V \to \mathbb{P}^1$, D_5 becomes a section of $\Phi_{|F|}$ and $D - D_5$ is contained in fibers of $\Phi_{|F|}$. In this subcase, Lemma 4 is verified by setting $\mu := \mathrm{id}_V$ and $\Phi := \Phi_{|F|}$.

Subcase 2. $5 \leq i \leq 7$. By using the argument as in Case 2-2, we can prove Lemma 4 in this subcase.

Subcase 3. $i = 4$. Since the intersection matrix of $E + D^{(1)}$ is not negative definite, $3 \leq a_0 \leq 6$. When $3 \leq a_0 \leq 5$, we can prove Lemma 4 by using the same argument as in Subcase 3 of Case 2–3. Suppose that $a_0 = 6$. Then $(10(E + D_4) + 8D_5 + 6D_6 + 4D_7 + 2(D_0 + D_8) + D_1 + D_2)^2 = 0$. So we can prove Lemma 4 by using the same argument as in Subcase 1.

Lemma 4 is thus verified. □

Lemma 5 *With the same notations and assumptions as in Lemma 4, $s = 1$ and $D^{(2)}, \ldots, D^{(t)}$ are admissible rational rods.*

Proof Let $\mu : W \to V$ and $\Phi : W \to \mathbb{P}^1$ be the same as in Lemma 4. Then $D^{(2)}, \ldots, D^{(t)}$ are not affected by the birational morphism μ. Φ is a $\mathbb{P}^1$-fibration and $\Phi|_{W-\mu^*(E+D)_{\mathrm{red}}}$ gives rise to an $\mathbb{A}^1$-fibration. By virtue of [15, Lemmas 1 and 2], where we note that the proofs of these lemmas use elementary properties on singular fibers of a $\mathbb{P}^1$-fibration only and work in any characteristic, $\mu'(D^{(2)}), \ldots, \mu'(D^{(t)})$ are linear chains of $\mathbb{P}^1$'s. This proves Lemma 5. □

From now on, we assume that $ED = 2$. Then we may assume that $R = D^{(s+1)}$ is an admissible rational rod. Let $f : V \to V_1$ be the contraction of E and the contractions of all subsequently contractible curves in Supp Bk(D). In fact, f contracts curves in Supp$(E + T^{(1)} + R)$ only. By [16, Lemma 2.3.7.1 (3) (p. 97)] (or [17, Lemma 1.7 (3)]), we know that, if $f_*(E + T^{(1)} + R) \neq 0$, then $f_*(E + T^{(1)} + R)$ is an admissible rational twig in $f_*(E + D)$. By [16, Lemma 2.3.7.1 (2) (p. 97)], which is the same as [17, Lemma 1.7 (2)], $f_*(E + D)$ is an SNC-divisor.

Since $\rho(V) = 1 + \#D = \#(E + D)$, $\rho(V_1) = \#f_*(E + D)$. Since $S = X - \pi(E)$ is a normal affine surface, Supp $f_*(E + D^{(1)} + R)$ supports a big divisor. Let $g : V \to \tilde{V}$ be a successive contractions of (-1)-curves in Supp $f_*(E + D^{(1)} + R)$ such that $g_*(f_*(E + D^{(1)} + R))$ becomes a minimal SNC-divisor. We set $\mu := g \circ f$ and $\tilde{D} := \mu_*(E + D)$. Note that $D^{(2)}, \ldots, D^{(s)}, D^{(s+2)}, \ldots, D^{(t)}$ are not affected by the birational morphism μ.

Lemma 6 *Assume that the SNC-divisor $(\tilde{V}, \tilde{D})$ is not almost minimal. Then $s = 1$ and $D^{(2)}, \ldots, D^{(t)}$ are admissible rational rods.*

Proof We note that $\overline{\kappa}(\tilde{V} - \tilde{D}) = \overline{\kappa}(V - D) = -\infty$, $\#\tilde{D} = \rho(\tilde{V})$ and $\tilde{D} = \mu_*(E + D^{(1)} + R) + \mu_*(D^{(2)}) + \cdots + \mu_*(D^{(s)}) + \mu_*(D^{(s+2)}) + \cdots + \mu_*(D^{(t)})$ gives the decomposition of $\tilde{D}$ into connected components. Since $(\tilde{V}, \tilde{D})$ is not almost minimal and $\tilde{D}$ is a minimal SNC-divisor, there exists a (-1)-curve $\tilde{E} \not\subset \mathrm{Supp}\, \tilde{D}$ such that $\overline{\kappa}(\tilde{V} - (\tilde{E} + \tilde{D})) = \overline{\kappa}(\tilde{V} - \tilde{D}) = -\infty$, $\tilde{E}(K_{\tilde{V}} + \tilde{D}^{\#}) < 0$ and the intersection matrix of $\tilde{E} + \mathrm{Bk}(\tilde{D})$ is negative definite. Then $\tilde{E}\tilde{D} \leq 2$. If $\tilde{E}\mu_*(E + D^{(1)} + R) = 0$, then $\pi(\mu'(\tilde{E}))$ is a complete curve contained in $S = X - \pi(E)$. This is a contradiction because S is affine. Therefore, $\tilde{E}\mu_*(E + D^{(1)} + R) > 0$.

Since $\rho(\tilde{V}) - \#(\tilde{E} + \tilde{D}) = -1 < 0$, we infer from [14, Lemma 2.8] (see the proof of [13, Proposition 2.1]) that there exists a birational morphism $\nu : \tilde{W} \to \tilde{V}$ from a smooth projective surface $\tilde{W}$ such that the following conditions are satisfied:

(1) ν is a composite of blowing ups at a point $\tilde{P}$ on $\mathrm{Supp}(\tilde{E} + \tilde{D})$ and its infinitely near points.
(2) There exists a $\mathbb{P}^1$-fibration $\tilde{\Phi} : \tilde{W} \to \mathbb{P}^1$ such that $\tilde{F}\nu^*(\tilde{E} + \tilde{D})_{\mathrm{red}} = 1$ for a fiber $\tilde{F}$ of $\tilde{\Phi}$.

If $\tilde{P} \notin \mathrm{Supp}\,\mu_*(E + D^{(1)} + R)$, then $\mathrm{Supp}\,\nu^*(\mu_*(E + D^{(1)} + R))$ is contained in a fiber of $\tilde{\Phi}$, which is a contradiction. Hence $\tilde{P} \in \mathrm{Supp}\,\mu_*(E + D^{(1)} + R)$. By the same argument as in the proof of Lemma 5, we know that every connected component of $\tilde{E} + \tilde{D} - \mu_*(E + D^{(1)} + R)$ is a linear chain of $\mathbb{P}^1$'s. Therefore, $s = 1$ and $D^{(2)}, \ldots, D^{(t)}$ are admissible rational rods. □

Lemma 7 *With the same notations and assumptions as above, assume further that the SNC-pair $(\tilde{V}, \tilde{D})$ is almost minimal. Then $s = 1$.*

Proof Set $\tilde{D}' := \lfloor \tilde{D}^{\#} \rfloor$. Since the intersection matrix of $\mu_*(E + D^{(1)} + R)$ is not negative definite, we have $\mu_*(E + D^{(1)} + R)^{\#} \neq 0$. So $\mathrm{Supp}\lfloor \mu_*(E + D^{(1)} + R)^{\#} \rfloor \subseteq \mathrm{Supp}\,\tilde{D}' = \mathrm{Supp}\lfloor \tilde{D}^{\#} \rfloor$.

Let $\tilde{\pi} : \tilde{V} \to V'$ be the contraction of $\mathrm{Supp}\,\mathrm{Bk}(\tilde{D}) = \mathrm{Supp}(\tilde{D} - \tilde{D}')$. Then V' has only klt singular points. By Lemma 1, one of the following cases take place:

(A') There exists a $\mathbb{P}^1$-fibration $h' : V' \to \mathbb{P}^1$ such that every fiber of h' is irreducible and $\tilde{\pi}_*(\tilde{D}')F' = 1$ for a fiber F' of h'. Here $\tilde{\pi}_*(\tilde{D}')F' = 1$ follows since the intersection matrix of $\tilde{D}$ is not negative semi-definite.
(B') $\rho(V') = 1$ and $-(K_{V'} + \tilde{\pi}_*(\tilde{D}'))$ is an ample $\mathbb{Q}$-Cartier divisor.

Suppose that the case (B') takes place. Then $\rho(\tilde{V}) = 1 + \#(\tilde{D} - \tilde{D}')$. Since $\rho(\tilde{V}) = \#\tilde{D}$, $\tilde{D}' = \lfloor \tilde{D}^{\#} \rfloor$ is irreducible. So $\tilde{D}' = \lfloor \mu_*(E + D^{(1)} + R)^{\#} \rfloor$. It follows that each connected component of $\mathrm{Supp}(\tilde{D} - \mu_*(E + D^{(1)} + R))$ is either an admissible rational rod or an admissible rational fork. Therefore, $s = 1$.

Suppose that the case (A') takes place. Then $\tilde{\Phi} := h' \circ \pi' : \tilde{V} \to \mathbb{P}^1$ is a $\mathbb{P}^1$-fibration and $\tilde{F}\tilde{D} = \tilde{F}\mu_*(E + D^{(1)} + R) = 1$ for a fiber $\tilde{F}$ of $\tilde{\Phi}$. By using the argument as in the proof of Lemma 5, we know that each connected component of $\mathrm{Supp}(\tilde{D} - \mu_*(E + D^{(1)} + R))$ is an admissible rational rod. Therefore, $s = 1$. □

The Proof of Theorem 1 is thus completed.

References

1. Alexeev, V.: Classification of log canonical surface singularities: arithmetical proof. Flips and abundance for algebraic threefolds. Kollár, J. et al.: Astérisque **211**, 47–58 (1992). (Société Mathématique de France)
2. Alexeev, V., Nikulin, V.V.: Del Pezzo and K3 Surfaces. MSJ Memoirs, vol. 15. Mathematical Society of Japan (2006)
3. Artin, M.: Some numerical criteria for contractability of curves on algebraic surfaces. Am. J. Math. **84**, 485–496 (1962)
4. Cheltsov, I.: Del Pezzo surfaces with nonrational singularities. Math. Notes **62**, 377–389 (1997)

5. Fujino, O.: Minimal model theory for log surfaces. Publ. Res. Inst. Math. Sci. **48**, 339–371 (2012)
6. Fujino, O., Tanaka, H.: On log surfaces. Proc. Jpn. Acad. Ser. A Math. Sci. **88**, 109–114 (2012)
7. Fujita, K.: Log del Pezzo surfaces with large volumes. Kyushu J. Math. **70**, 131–147 (2016)
8. Fujita, K.: Log del Pezzo surfaces with not small fractional indices. Math. Nachr. **289**, 34–59 (2016)
9. Fujita, K., Yasutake, K.: Classification of log del Pezzo surfaces of index three. J. Math. Soc. Jpn. **69**, 163–225 (2017)
10. Furushima, M.: Singular del Pezzo surfaces and analytic compactifications of 3-dimensional complex affine space $\mathbb{C}^3$. Nagoya Math. J. **104**, 1–28 (1986)
11. Hidaka, F., Watanabe, K.: Normal Gorenstein surfaces with ample anti-canonical divisor. Tokyo J. Math. **4**, 319–330 (1981)
12. Kojima, H.: On normal surfaces with strictly nef anticanonical divisors. Arch. Math. **77**, 517–521 (2001)
13. Kojima, H.: Rational unicuspidal curves on $\mathbb{Q}$-homology planes whose complements have logarithmic Kodaira dimension $-\infty$. Nihonkai Math. J. **29**, 29–43 (2018)
14. Kojima, H.: Some results on open algebraic surfaces of logarithmic Kodaira dimension zero. J. Algebra **5**47, 238–261 (2020)
15. Miyanishi, M.: Singularities of normal affine surfaces containing cylinderlike open sets. J. Algebra **68**, 268–275 (1981)
16. Miyanishi, M.: Open Algebraic Surfaces. CRM Monograph Series, vol. 12. American Mathematical Society, Providence (2001)
17. Miyanishi, M., Tsunoda, S.: Non-complete algebraic surfaces with logarithmic Kodaira dimension $-\infty$ and with non-connected boundaries at infinity. Jpn. J. Math. **10**, 195–242 (1984)
18. Mumford, D.: The topology of normal surface singularities of an algebraic surface and a criterion for simplicity. Publ. IHES **9**, 5–22 (1961)
19. Nakayama, N.: Classification of log del Pezzo surfaces of index two. J. Math. Sci. Univ. Tokyo **14**, 293–498 (2007)
20. Sakai, F.: Weil divisors on normal surfaces. Duke Math. J. **51**, 877–887 (1984)
21. Schröer, S.: On contractible curves on normal surfaces. J. Reine Angew. Math. **524**, 1–15 (2000)
22. Schröer, S.: Normal del Pezzo surfaces containing a nonrational singularity. Manuscripta Math. **104**, 257–274 (2001)
23. Tanaka, H.: Minimal models and abundance for positive characteristic log surfaces. Nagoya Math. J. **216**, 1–70 (2014)
24. Ye, Q.: On Gorenstein log del Pezzo surfaces. Jpn. J. Math. **28**, 87–136 (2002)

$O_2(\mathbb{C})$-Vector Bundles and Equivariant Real Circle Actions

L. Moser-Jauslin

Abstract The main goal of this article is to give an expository overview of some new results on real circle actions on affine four-space and their relation to previous results on $O_2(\mathbb{C})$-equivariant vector bundles. In Moser-Jauslin (Infinite families of inequivalent real circle actions on affine four-space, 2019, [13]), we described infinite families of equivariant real circle actions on affine four-space. In the present note, we will describe how these examples were constructed, and some consequences of these results.

Keywords Circle group action · Affine space · G-vector bundle

1 Introduction

In recent years, the study of equivariant real structures on complex varieties has had many new developments. There are several different approaches to this theory. Suppose G is a complex algebraic group with a fixed real structure, and X is a complex G-variety. Then one can ask if there exist real structures on X which are compatible with the group action, and, if so, how many inequivalent real structures exist. When the group of G-equivariant automorphisms is small, for example when X is a homogeneous space, or when it is a spherical variety, the question of existence is of central interest. The case of toric varieties has been studied in [7, 16]. Different cases of spherical varieties have been studied in [1–3, 14]. On the other hand, in this article, we consider a special case where the group of equivariant automorphisms is large. More precisely, we fix the group $G = \mathbb{C}^*$ with the real structure whose fixed point set defines the unit circle, and we choose X to be affine four-space, endowed with a

This work was supported by the French "Investissements d'Avenir" program, project ISITE-BFC (contract ANR-lS-IDEX-OOOB).

L. Moser-Jauslin (✉)
Institut de Mathématiques de Bourgogne, UMR 5584 du CNRS, Université de Bourgogne Franche-Comté, 9 av. Alain Savary, BP 47870, 21078 Dijon, France
e-mail: moser@u-bourgogne.fr

S. Kuroda et al. (eds.), *Polynomial Rings and Affine Algebraic Geometry*, Springer Proceedings in Mathematics & Statistics 319,
https://doi.org/10.1007/978-3-030-42136-6_9

particular linear $\mathbb{C}^*$-action, such that a special equivariant real structure exists. The question then is to study the different equivariant real structures. In [6], one example of a non-linearizable circle action was given. In [13], this example was generalized to give explicit examples where there are uncountably many inequivalent equivariant real structures. These examples are the main topic of this note.

In Sect. 2 we give a brief review of the definitions of real structures in several different settings. Then in Sect. 3, we describe $O_2(\mathbb{C})$-vector bundles. These objects are the key to the examples described in Sects. 4 and 5. In the final section, we give some remarks about the nature of the examples.

These results concern only $\mathbb{C}^*$-varieties which are isomorphic to affine space endowed with a linear action. For a systematic approach of the study of affine equivariant real structures of $\mathbb{C}^*$-varieties, see [4, 5], where they consider the Altmann–Hausen representation of $\mathbb{C}^*$-varieties and the DPD representation for surfaces with a hyperbolic $\mathbb{C}^*$-action to describe real structures.

2 Real Structures

To define an equivariant real structure on a complex G-variety, we first define a real structure on the group G.

A real structure on a complex algebraic group G is an anti-regular involution $\sigma : G \to G$ which is a group automorphism. Two real structures σ_1 and σ_2 are equivalent if there exists a group automorphism $\varphi \in Aut_{\mathbb{C}}(G)$ such that $\sigma_2 = \varphi\sigma_1\varphi^{-1}$.

Given a fixed real structure σ of G, a real (G, σ)-structure on a complex G-variety X is an anti-regular involution μ on X such that $\mu(g \cdot x) = \sigma(g)\mu(x)$ for all $g \in G$ and $x \in X$. Two (G, σ)-structures μ_1 and μ_2 are equivalent if there exists a G-equivariant automorphism $\psi \in Aut_G(X)$ such $\mu_2 = \psi \circ \mu_1 \circ \psi^{-1}$.

For the cases described here, we consider real structures on total spaces of trivial G-vector bundles. Therefore, we will define an intermediate notion of real equivariant vector bundle structures. Suppose $E = B \times F$ is the total space of a trivial G-vector bundle $E \to B$, where F is a G-module, and B is a G-variety. Let μ_B be a (G, σ)-real structure on B. Then a (G, σ, μ_B)-real vector bundle structure on E is a (G, σ)-real structure μ_E on E such that μ_E is also a real vector-bundle automorphism. Two such real structures μ_E and μ'_E are equivalent if there exists a G-equivariant complex vector-bundle automorphism ψ such that $\psi \circ \mu_E \circ \psi^{-1} = \mu'_E$. Clearly, if two equivariant vector bundle real structures are equivalent, then they are equivalent as equivariant real structures of the total space. However, the converse does not hold in general.

For the case discussed here, consider the group $G = \mathbb{C}^*$, the multiplicative group of $\mathbb{C}$, and σ the involution given by $\sigma(t) = \bar{t}^{-1}$. The fixed point set of σ is the unit circle, and therefore we say that a $(\mathbb{C}^*, \sigma)$-structure on a $\mathbb{C}^*$-variety X defines a *circle action* on X. In particular, if μ is such a structure, then the fixed point set of σ acts on the fixed point set of μ on X.

Let X be a linear $\mathbb{C}^*$-module with weights $(k_1, -k_1, k_2, -k_2)$. There is an obvious real $(\mathbb{C}^*, \sigma)$-real structure μ_0 given by $\mu_0(a, b, x, y) = (\overline{b}, \overline{a}, \overline{y}, \overline{x})$. Note that μ_0 is an $\mathbb{R}$-linear map and that the fixed point set is a real four-dimensional vector space on which the fixed point set of σ acts linearly. For this reason, we say that μ_0 defines a real linear circle action on affine four-space. Now consider X as the total space of the trivial $\mathbb{C}^*$-vector bundle where the base B is a two-dimensional vector space with weights $(k_1, -k_1)$ and fiber F with weights $(k_2, -k_2)$. Consider the $(\mathbb{C}^*, \sigma)$-real structure μ_B on B defined by $\mu_B(a, b) = (\overline{b}, \overline{a})$. Then μ_0 defines a $(\mathbb{C}^*, \sigma, \mu_B)$-vector bundle real structure. We will define other $(\mathbb{C}^*, \sigma, \mu_B)$-vector bundle real structures on $X = B \times F$. The main result of [13] is to construct infinite families of inequivalent equivariant real structures on the total space of these $\mathbb{C}^*$-vector bundles.

This is done in two steps. First one distinguishes the equivariant real vector bundle structures. As stated above, this does not directly imply that one has inequivalent real structures on the total space. To achieve this second part, one uses an argument of Masuda and Petrie [10] that was originally used to distinguish $O_2(\mathbb{C})$-bundles. This will be briefly explained in Sect. 5.

Note that the real structures discussed here are all vector bundle structures μ on $B \times F$ on which the induced real structure on the base are real linear. Thus the fixed point set on the base is a real vector space of dimension two, and since all real vector bundles on $\mathbb{R}^2$ are trivial, the fixed point set of μ is diffeomorphic to $\mathbb{R}^4$. More precisely, by construction, all the equivariant real structures discussed here are equivalent if one forgets the $\mathbb{C}^*$-group action. In other words, all vector bundle equivariant structures μ of $B \times F$ are equivalent to μ_0 as a real structure on $\mathbb{A}^4_{\mathbb{C}}$. We are thus describing inequivalent circle actions on affine four-space.

3 $O_2(\mathbb{C})$-Vector Bundles

The construction of the real structures given here comes from examples of non-trivial $O_2(\mathbb{C})$-bundles given by Schwarz [15] in 1989. In this section, we will recall these results and give an explicit description of these equivariant vector bundles, based on the works of [8–11]. We then describe how we can use them to find explicit examples of real equivariant circle actions.

From now on, set $G = O_2(\mathbb{C}) = \mathbb{C}^* \rtimes \mathbb{Z}/2\mathbb{Z}$. Let $s \in G$ be the involution generating the second factor. For $k \in \mathbb{N} \setminus \{0\}$, let V_k be the irreducible $O_2(\mathbb{C})$-module of dimension two where $\lambda(x, y) = (\lambda^k x, \lambda^{-k} y)$ for $\lambda \in \mathbb{C}^*$, and $s(x, y) = (y, x)$. Recall that an equivariant G-vector bundle $\pi : E \to B$ is a vector bundle where the total space E and the base B are G-varieties, the projection π is G-equivariant, and for each $g \in G$ and each $b \in B$, the induced map of g from $\pi^{-1}(b)$ to $\pi^{-1}(gb)$ is linear. In the case considered here, B will be a G-module and one has in particular that the zero fiber is another G-module, denoted by F. Given two G-modules B and F, $VEC(B; F)$ denotes the set of equivalence classes of G-vector bundles with base B and zero fiber F. Two G-vector bundles over B are equivalent if they are isomorphic as equivariant vector bundles by an isomorphism which is G-equivariant. A G-vector bundle is trivial if it is equivalent to $B \times F$.

In his article, G. Schwarz, described the sets $VEC(V_1; V_k)$ with base V_1 and zero fiber V_k for $k \geq 2$, and $VEC(V_2; V_{2m+1})$ for $m \geq 1$. It was shown by Schwarz that any G-vector bundle with base V_1 and zero-fiber V_k can be stabilized by the trivial vector bundle of rank 1, $V_1 \times \mathbb{C}$, where G acts trivially on $\mathbb{C}$. That is, the direct sum of any vector bundle $E \to V_1$ with zero fiber V_k and the trivial vector bundle of rank 1, $V_1 \times \mathbb{C} \to V_2$ is isomorphic to $(V_k \oplus \mathbb{C}) \times V_1 \to V_1$. Also, any G-vector bundle with base V_2 and zero fiber V_{2m+1} can be stabilized by the trivial G-vector bundle $V_2 \times V_1$. That is, the direct sum of any vector bundle $E \to V_2$ with zero fiber V_{2m+1} and the trivial vector bundle, $V_2 \times V_1 \to V_2$ is isomorphic to $(V_{2m+1} \oplus V_1) \times V_2 \to V_2$. Also, all of these G-vector bundles are trivial when considered as $\mathbb{C}^*$-vector bundles for the subgroup $\mathbb{C}^*$ of $O_2(\mathbb{C})$. This fact could be easily seen for this case by Schwarz. It is also a consequence of a more general result that for any complex abelian reductive group H, all H-vector bundles having an H-module as a base are trivial (see [12]).

The original construction by Schwarz of all G-vector bundle classes in $VEC(V_1; V_k)$ and in $VEC(V_2; V_{2m+1})$ used local trivializations and gluing functions. We will use here another approach, developed by Masuda and Petrie (see [9]) using stable trivialization. Either method can be used to find explicit descriptions of the G-vector bundles.

Case 1: $B = V_1$ and $F = V_k, k \geq 2$.
Here, the result of Schwarz is that $VEC(V_1; V_k) \cong \mathbb{C}^{k-1}$. One way to describe the different G-vector bundles is to use the fact that any such G-vector bundle is $\mathbb{C}^*$-equivalent to the trivial vector bundle $V_1 \times V_k \to V_1$, and thus the involution defined by the action of s determines the G-vector bundle structure. Note that by the comments above, any G-vector bundle with base V_1 and zero fiber V_k can be realized as the kernel of a surjective G-vector bundle map between trivial G-vector bundles of the form $f : V_1 \times (V_k \oplus \mathbb{C}) \to V_1 \times \mathbb{C}$.

Also the kernels of two surjections f_1 and f_2 are equivalent if and only if there exists a G-vector bundle automorphism Ψ of $V_1 \times (V_k \oplus \mathbb{C})$ such that $f_2 = f_1 \circ \Psi$.

To describe the surjective G-vector bundle maps, choose coordinates (a, b) of the base V_1, coordinates (x, y) of V_k and z of $\mathbb{C}$ where $\mathbb{C}^*$ acts on $V_1 \times (V_k \oplus \mathbb{C})$ by $\lambda((a, b), (x, y, z)) = ((\lambda a, \lambda^{-1} b), (\lambda^k x, \lambda^{-k} y, z))$ for all $\lambda \in \mathbb{C}^*$, and $s((a, b), (x, y, z)) = ((b, a), (y, x, z))$. Note that the invariant ring of the regular functions on the base is generated by $T = ab$.

Let $h \in \mathbb{C}[T]$. Let f_h be the surjective vector bundle morphism $V_1 \times (V_k \oplus \mathbb{C}) \to V_1 \times \mathbb{C} : ((a, b), (x, y, z)) \to ((a, b), (b^k x + a^k y + (1 - Th)z)$. In other words, f_h restricts to a linear map on the fiber over (a, b), defined by $\begin{pmatrix} x \\ y \\ z \end{pmatrix} \to L_h \begin{pmatrix} x \\ y \\ z \end{pmatrix}$, where

$$L_h = \begin{pmatrix} b^k & a^k & 1 - Th \end{pmatrix}.$$

In fact, it can be shown that all surjective G-vector bundle maps of $V_1 \times (V_k \oplus \mathbb{C})$ are equivalent to one of the form f_h for some choice of h. Thus we are left to find the equivalence classes of the kernels of these surjections. This can be done as follows. First of all, as stated above, the kernel is trivial as a $\mathbb{C}^*$-vector bundle. Therefore, one can choose a $\mathbb{C}^*$-vector bundle automorphism which trivializes the $\mathbb{C}^*$-action on the kernel. For example, one can choose

$$A_h = \begin{pmatrix} 1 & 0 & 0 \\ 0 & \sum_{j=0}^{k-1}(Th)^j & -b^k h^k \\ b^k & a^k & 1-Th \end{pmatrix},$$

and

$$A_h^{-1} = \begin{pmatrix} 1 & 0 & 0 \\ -b^{2k}h^k & 1-Th & b^k h^k \\ -b^k \sum_{j=0}^{k-1}(Th)^j & -a^k & \sum_{j=0}^{k-1} Th \end{pmatrix}.$$

By using A_h to change the coordinates (x, y, z) on $V_1 \times (V_k \oplus \mathbb{C})$, we determine a coordinate system of $\ker f_h$ on which $\mathbb{C}^*$ acts linearly. More precisely, the coordinate system on the fibers given by $x' = x$ and $y' = \sum_{j=0}^{k-1}(Th)^k y - b^k h^k z$ generate the kernel of f_h, and $\mathbb{C}^*$ acts linearly on the fiber with weights $\pm k$. It remains to determine the action of $s \in G$, which is given by an involution which we call τ_h. In the new coordinates,

$$\tau_h\left(\begin{pmatrix} a \\ b \end{pmatrix}, \begin{pmatrix} x' \\ y' \end{pmatrix}\right) = \left(\begin{pmatrix} b \\ a \end{pmatrix}, \begin{pmatrix} 1-Th & a^{2k}h^k \\ -b^{2k}h^k & \sum_{j=0}^{2k-1}(Th)^j \end{pmatrix}\begin{pmatrix} y' \\ x' \end{pmatrix}\right),$$

where $\begin{pmatrix} a \\ b \end{pmatrix} \in V_1$ and $\begin{pmatrix} x' \\ y' \end{pmatrix} \in V_k$.

Let

$$M_{1,h} = \begin{pmatrix} 1-Th & a^{2k}h^k \\ -b^{2k}h^k & \sum_{j=0}^{2k-1}(Th)^j \end{pmatrix} \in GL(\mathbb{C}[a,b]).$$

From what we have seen, for each $h \in \mathbb{C}[T]$, $M_{1,h}$ determines an involution τ_h which therefore defines a G-equivariant vector bundle structure whose class is an element of $VEC(V_1; V_k)$. Also, all G-equivariant vector bundles are equivalent to one of this form. Finally, in this setting, the result of Schwarz can be deduced from the fact that M_{1,h_1} and M_{1,h_2} define equivalent G-vector bundles if and only if $h_2 \equiv h_1 \mod (T^{k-1})$. This gives a natural bijection between $VEC(V_1, V_k)$ and $\mathbb{C}^{k-1}$ (see also [9]).

I will now discuss another point of view of these objects which we will use later to compare the situation of G-vector bundles to that of the structure of real equivariant vector bundle structures for the group $\mathbb{C}^*$ with real structure σ.

We will show that, since all G-vector bundles with base V_1 are trivial as $\mathbb{C}^*$-vector bundles, there is a natural bijection between the set $VEC(V_1, V_k)$ and a certain cohomology pointed set. Let $\mathcal{A}_{2k}$ be the group of $\mathbb{C}^*$-vector bundle automorphisms of the trivial bundle $V_1 \times V_k$ over V_1, viewed as a $\mathbb{C}^*$-vector bundle. This group can be identified with

$$\mathcal{A}_{2k} = \{\begin{pmatrix} P & a^{2k}Q \\ b^{2k}S & R \end{pmatrix} \in GL_2(\mathbb{C}[a,b]\},$$

where, $P, Q, R, S \in \mathbb{C}[T]$ are polynomials in the variable $T = ab$. The group $C_2 = \{1, s\}$ of order two acts on $\mathcal{A}_{2k}$ where the action of s is given by conjugation by τ_0. If $\varphi \in \mathcal{A}_{2k}$, we denote by ${}^s\varphi$ the automorphism $\tau_0\varphi\tau_0$. This action defines a cohomology pointed set. More precisely, the set of 1-cycles $Z^1(C_2; \mathcal{A}_{2k})$ is given by $\{\varphi \in \mathcal{A}_{2k} | \varphi \circ ({}^s\varphi) = id\}$. The cohomology set $H^1(C_2; \mathcal{A}_{2k}) = Z^1(C_2; \mathcal{A}_{2k})/\sim$, where the 1-cycles φ_1 and φ_2 are equivalent if and only if there exists $\psi \in \mathcal{A}_{2k}$ such that $\psi\varphi_1({}^s\psi^{-1}) = \varphi_2$.

Let $\Phi_1 : Z^1(C_2; \mathcal{A}_{2k}) \to VEC(V_1; V_k)$ be the map which sends $\varphi \in Z^1(C_2; \mathcal{A}_{2k})$ to the class of the vector bundle in $VEC(V_1; V_k)$ which is trivial as a $\mathbb{C}^*$-bundle, with weights $\pm k$, and on which the involution s acts by the involution $\varphi \circ \tau_0$.

Proposition 3.1 *The map Φ_1 induces a bijection from the pointed set $H^1(C_2; \mathcal{A}_{2k})$ to $VEC(V_1; V_k)$.*

Proof First of all, note that Φ_1 is surjective, since every class of $VEC(V_1; V_k)$ has a representative which is trivial as a $\mathbb{C}^*$-vector bundle. It remains to show that, given $\varphi_1, \varphi_2 \in Z_1(C_2; \mathcal{A}_{2k})$, the vector bundle classes $\Phi_1(\varphi_1)$ and $\Phi_1(\varphi_2)$ are equal if and only if $\varphi_1 \sim \varphi_2$ in $H^1(C_2; \mathcal{A}_{2k})$. Note that φ_1 and φ_2 are equivalent if and only if there exists $\psi \in \mathcal{A}_{2k}$ such that $\psi\varphi_1(\tau_0\psi^{-1}) = \varphi_2\tau_0$. This is exactly the condition that the two vector bundles $\Phi_1(\varphi_1)$ and $\Phi_1(\varphi_2)$ are equivalent. □

This point of view can be used to give another proof of the result of Schwarz. I will not describe the details here, since it does not give any additional information. However, I will show how one can interpret the action of C_2 and look at a special key case of the result to compare it later to the case of real equivariant vector bundle structures.

If $N = \begin{pmatrix} P & a^{2k}Q \\ b^{2k}S & R \end{pmatrix} \in \mathcal{A}_{2k}$, then ${}^sN = \begin{pmatrix} R & a^{2k}S \\ b^{2k}Q & P \end{pmatrix}$. Note that, as expected, $M_{1,h}^{-1} = {}^sM_{1,h}$, thus it is a cocycle in $Z^1(C_2; \mathcal{A}_{2k})$.

Proposition 3.2 *Suppose that $h \in \mathbb{C}[T]$ is a non-zero polynomial and $\deg(h) \leq k-2$. Then the class of M_h in $H^1(C_2; \mathcal{A}_{2k})$ is not equivalent to the identity matrix I_2 In other words, $\ker(f_h)$ is not equivalent to the trivial G-vector bundle $V_1 \times V_k$.*

Proof Suppose the degree of h is $\leq k-2$. If $M_{1,h}$ is equivalent to I_2 then there exists $K \in \mathcal{A}_{2k}$ such that $K \cdot {}^sK^{-1} = M_{1,h}$. Suppose that $K = \begin{pmatrix} P & a^{2k}Q \\ b^{2k}S & S \end{pmatrix}$. Let $\Delta = \det(K) \in \mathbb{C}^*$. Then the first coefficient of $K \cdot {}^sK^{-1}$ is $(P^2 - T^{2k}Q^2)/\Delta$. In particular, if $K \cdot {}^sK^{-1} = M_{1,h}$, then the degree of $(P + T^kQ)(P - T^kQ) = \Delta(1 - Th)$ is at most $k-1$. This means that each factor has degree bounded by $k-1$, and thus the degrees of P and of T^kQ are at most $k-1$. In other words, $Q = 0$, and, since Δ is constant, P is constant, and thus $h = 0$. □

Case 2: $B = V_2$ and $F = V_{2m+1}$, $m \geq 1$.
To simplify notation, let $n = 2m + 1$. This case can be treated similarly to the previous case. Any such G-vector bundle is the kernel of a surjective G-vector bundle map of $V_2 \times (V_n \oplus V_1) \to V_2 \times V_1$. Consider the following special case. Let $h \in \mathbb{C}[T]$ where, as before, $T = ab$, and let $f_h : V_2 \times (V_n \oplus V_1) \to V_2 \times V_1$ be defined by:

$$f_h(\begin{pmatrix} a \\ b \end{pmatrix}, (\begin{pmatrix} x \\ y \end{pmatrix}, \begin{pmatrix} z \\ w \end{pmatrix})) = (\begin{pmatrix} a \\ b \end{pmatrix}, L_h \begin{pmatrix} x \\ y \\ z \\ w \end{pmatrix})$$

where

$$L_h = \begin{pmatrix} b^m & 0 & 1 & ah \\ 0 & a^m & bh & 1 \end{pmatrix}.$$

As before, one can show that every G-vector bundle with base V_2 and zero fiber V_n is equivalent to a kernel of a surjection of this form, and ker f_{h_1} is equivalent to ker f_{h_2} if and only if $h_1 \equiv h_2 (\mod T^m)$. This gives the result of Schwarz that $VEC(V_2; V_n) \cong \mathbb{C}^m$.

By using the automorphism

$$A_h = \begin{pmatrix} 1 & 0 & 0 & 0 \\ 0 & \sum_{j=0}^{m-1}(Th^2)^j & 0 & -b^m h^{2m} \\ b^m & 0 & 1 & ah \\ 0 & a^m & bh & 1 \end{pmatrix},$$

one can choose coordinates $x' = x$ and $y' = \sum_{j=0}^{m-1}(Th^2)^j y - b^m h^{2m} z$ for the kernel $\ker(f_h)$. One then finds that the element s acts on the kernel by the involution τ_h where

$$\tau_h(\begin{pmatrix} a \\ b \end{pmatrix}, \begin{pmatrix} x' \\ y' \end{pmatrix}) = (\begin{pmatrix} b \\ a \end{pmatrix}, M_{2,h} \begin{pmatrix} y' \\ x' \end{pmatrix})$$

with

$$M_{2,h} = \begin{pmatrix} 1 - Th^2 & a^n h^n \\ -b^n h^n & \sum_{j=0}^{n-1}(Th^2)^j \end{pmatrix}.$$

As in the previous case, one can identify the set $VEC(V_2; V_n)$ with a cohomology set. Let $\mathcal{A}_n$ be the group of $\mathbb{C}^*$-vector bundle automorphisms of the trivial bundle $V_2 \times V_n \to V_2$, viewed as a $\mathbb{C}^*$-vector bundle. This group can be identified with

$$\mathcal{A}_n = \{\begin{pmatrix} P & a^n Q \\ b^n S & R \end{pmatrix} \in GL_2(\mathbb{C}[a, b]\},$$

where, P, Q, R and S are polynomials in $T = ab$. The group $C_2 = \{1, s\}$ of order two acts on $\mathcal{A}_n$ where s acts by conjugation by τ_0. If $\varphi \in \mathcal{A}_n$, we denote by ${}^s\varphi$ the automorphism $\tau_0\varphi\tau_0$. This action defines a cohomology set. More precisely, the set of 1-cycles $Z^1(C_2; \mathcal{A}_n)$ is given by the set $\{\varphi \in \mathcal{A}_n | \varphi \circ ({}^s\varphi) = id\}$, and the cohomology set is $H^1(C_2; \mathcal{A}_n) = Z^1(C_2; \mathcal{A}_n)/\sim$, where the 1-cycles φ_1 and φ_2 are equivalent if and only if there exists $\psi \in \mathcal{A}_n$ such that $\psi\varphi_1({}^s\psi^{-1}) = \varphi_2$.

Now, as in the previous case, let $\Phi_2 : Z^1(C_2; \mathcal{A}_n) \to VEC(V_2; V_n)$ be the map which sends $\varphi \in Z^1(C_2; \mathcal{A}_n)$ to the class of the vector bundle in $VEC(V_2; V_n)$ which is trivial as a $\mathbb{C}^*$-bundle, with weights $\pm n$, and on which the involution s acts by the automorphism $\varphi \circ \tau_0$.

Proposition 3.3 *The map Φ_2 induces a bijection from the pointed set $H^1(C_2; \mathcal{A}_n)$ to $VEC(V_2; V_n)$.*

The proof is identical to the one in the previous case.

Also, analogously to the previous case, we will give the proof of one key special case to compare with the real vector bundle structures in the next section.

Proposition 3.4 *Let $h \in \mathbb{C}[T]$. Then the class of $M_{2,h}$ in $H^1(C_2; \mathcal{A}_n)$ is equivalent to the identity matrix I_2 if and only if T^m divides h. In other words,* $\ker(f_h)$ *is the trivial G-vector bundle $V_2 \times V_n$ if and only if T^m divides h.*

Proof $M_{2,h}$ is equivalent to I_2 if and only if there exists $K_h \in \mathcal{A}_n$ such that $K_h \cdot {}^sK_h^{-1} = M_{2,h}$. If $h = T^m\widehat{h}$, then one can choose $K_h = \begin{pmatrix} 1 & a^n\widehat{h} \\ b^n\widehat{h}\sum_{j=0}^{m-1}(Th^2)^j & \sum_{j=0}^{m}(Th^2)^j \end{pmatrix}$. One can verify that $\det(K_h) = 1$ and that $K_h \cdot {}^sK_h^{-1} = M_{2,h}$.

On the other hand, if T^m does not divide h, then by using the previous case, we may assume that $\deg(h) \le m - 1$. Suppose that $K \cdot {}^sK^{-1} = M_{2,h}$ with $K = \begin{pmatrix} P & a^nQ \\ b^nS & R \end{pmatrix}$. The determinant Δ is a non-zero constant, and the first coefficient of $K \cdot {}^sK^{-1}$ is $(P^2 - T^nQ^2)/\Delta = 1 - Th^2$. Since n is odd, the degree of T^nQ^2 is odd, and the degree of P^2 is even. In particular, if $Q \neq 0$, then this polynomial is of degree at least $n = 2m + 1$. This is not the case, since $\deg(h) \le m - 1$. Therefore $Q = 0$, P is constant, and therefore the only possibility is $h = 0$. □

4 Real Vector Bundle Structures

Now we will relate these results on G-vector bundles to the construction of real $(\mathbb{C}^*, \sigma, \mu_B)$-real vector bundle structures.

We consider two cases analogous to the two cases of G-bundles from the last section. The base B is a two-dimensional $\mathbb{C}^*$-module with weights ± 1 (resp. ± 2), and $E = B \times F$ where F is a two-dimensional $\mathbb{C}^*$-module with weights $\pm k$ where $k \ge 2$ (resp. $\pm n$, with $n = 2m + 1$ and $m \ge 1$). In both cases, the real structure of

B is given by μ_B with $\mu_B(a, b) = (\overline{b}, \overline{a})$. Then the real structure μ_0 on $B \times F$ with $\mu_0(a, b, x, y) = (\overline{b}, \overline{a}, \overline{y}, \overline{x})$ is a $(\mathbb{C}^*, \sigma, \mu_B)$-real structure. Note that μ_0 is the composition of the complex involution τ_0 with conjugation, and these two involutions commute. Now if τ is any complex involution which is a $\mathbb{C}^*$-vector bundle automorphism, and if τ commutes with conjugation, then the composition of τ with conjugation defines an anti-regular involution, and thus a $(\mathbb{C}^*, \sigma, \mu_B)$-real vector bundle structure. In other words, if $h \in \mathbb{R}[T]$, then $\mu_h = \tau_h \circ conj$ defines a real equivariant vector bundle structure.

Let $\Gamma = \{1, \gamma\}$ be the Galois group of $\mathbb{C}/\mathbb{R}$. We use μ_0 to define an action of Γ on $\mathcal{A}_{2k}$, (resp. $\mathcal{A}_n$). More precisely, $\gamma \cdot \varphi = {}^{\gamma}\varphi = \mu_0 \varphi \mu_0$. Then, as for the G-vector bundles, one can define the set of all equivalence classes of $(\mathbb{C}^*, \sigma, \mu_B)$-real vector bundle structures by a cohomology set.

Case 1: $B = B_1$ has weights ± 1 and $F = F_k$ has weights $\pm k$, $k \geq 2$.
In this case, the set of equivalence classes of real $(\mathbb{C}^*, \sigma, \mu_B)$ real vector bundle structures is in bijection with $H^1(\Gamma; \mathcal{A}_{2k})$. This set looks very similar to the set $H^1(C_2; \mathcal{A}_{2k})$, however the action of the group of two element on $\mathcal{A}_{2k}$ differs in the two settings. First of all, the action of γ, the non-trivial element of Γ is given by conjugation by μ_0 instead of τ_0. Thus if

$$N = \begin{pmatrix} P & a^{2k} Q \\ b^{2k} S & R \end{pmatrix} \in \mathcal{A}_{2k},$$

then

$${}^{\gamma}N = \begin{pmatrix} \overline{R} & a^{2k} \overline{S} \\ b^{2k} \overline{Q} & \overline{P} \end{pmatrix}.$$

Here, if $f = \sum a_j T^j \in \mathbb{C}[T]$, then $\overline{f} = \sum \overline{a_j} T^j$.

This action gives a different result for the equivalence classes. One can find common cocycles for the two settings for which equivalences are not the same. For example, in the case that $k = 2$, where $VEC_G(V_1, V_2) \cong \mathbb{C}$, we find for the real equivariant vector bundles, the following result.

Proposition 4.1 *Let $k = 2$. For any $h \in \mathbb{R}[T]$ the real (G, σ, μ_B)-vector bundle structure $\mu_h = \tau_h \circ conj$, is equivalent to μ_0.*

Proof Set

$$K = \begin{pmatrix} 1 - \frac{(1-i)}{2} Th - \frac{(1+i)}{4} (Th)^2 & \frac{a^4 h^2}{4} (1 - i) \\ \frac{-b^4 h^2}{4} (3 - i + (1 + i) Th) & 1 + \frac{1-i}{4} (2Th + (Th)^2 + (Th)^3) \end{pmatrix}.$$

It suffices to note that $K \cdot {}^{\gamma}K^{-1} = M_{1,h}$. □

For the cases of $k \geq 3$, I have not checked the details, but I would expect that the same result is true.

In particular, in this case, we do not know of a way to find any non-trivial equivariant real structures. However it is possible that they exist, using another construction of cocycles, not coming from $O_2(\mathbb{C})$-vector bundles.

Case 2: $B = B_2$ has weights ± 2 and $F = F_n$ has weights $\pm n$ (As before, $n = 2m + 1$ is odd, and $m \geq 1$)
In this case, something different occurs. One finds that, up to a sign, the equivalences of the cocycles determined by $M_{2,h}$ for all $h \in \mathbb{R}[T]$ are the same for $H^1(C_2; \mathcal{A}_n)$ and $H^1(\Gamma_2; \mathcal{A}_n)$. More precisely, the classes of M_{2,h_1} and M_{2,h_2} in $H^1(\Gamma; \mathcal{A}_n)$ are the same if and only if the class of M_{2,h_1} equals either the class of M_{2,h_2} or of $M_{2,-h_2}$ in $H^1(C_2; \mathcal{A}_n)$. Of course, this does not imply that there is a relation between the complete cohomology sets. It simply means that restricted to the common cocycles defined by $M_{2,h}$, $h \in \mathbb{R}[T]$, the equivalences are strongly related.

Here we give the proof of a special case. Note that the proof is almost identical to that of the analogous G-vector bundle case.

Proposition 4.2 *Suppose $h \in \mathbb{R}[T]$. Then μ_h is equivalent to μ_0 as real $(\mathbb{C}^*, \sigma, \mu_B)$-real vector bundle structures if and only if T^m divides h.*

Proof The real structures μ_h and μ_0 are equivalent if and only if $M_{2,h}$ is equivalent to I_2. In other words, they are equivalent if and only if there exists $K_h \in \mathcal{A}_n$ such that $K_h \cdot {}^{\gamma}K_h^{-1} = M_{2,h}$. If $h = T^m\widehat{h}$, then the same choice of K_h as the one given in Proposition 3.4 holds, since the coefficients of K_h are all real.

If, on the other hand, T^m does not divide h, then we may assume that $\deg(h) \leq m - 1$. Suppose that $K \cdot {}^{\gamma}K^{-1} = M_{2,h}$ with $K = \begin{pmatrix} P & a^{2k}Q \\ b^{2k}S & R \end{pmatrix}$. The determinant Δ is a non-zero constant, and the first coefficient of $K \cdot {}^{s}K^{-1}$ is $(P\overline{P} - T^n Q\overline{Q})/\Delta = 1 - Th^2$. Since n is odd, the degree of $T^n Q\overline{Q}$ is odd, and the degree of $P\overline{P}$ is even. In particular, if $Q \neq 0$, then this polynomial is of degree at least $n = 2m + 1$. This is not the case. Therefore $Q = 0$, P is constant, and therefore the only possibility is $h = 0$. □

This proposition can be generalized to the following result.

Proposition 4.3 *μ_{h_1} is equivalent to μ_{h_2} as real $(\mathbb{C}^*, \sigma, \mu_B)$-real vector bundle structures on $B_2 \times F_n$ if and only if $h_1 \equiv \pm h_2 (\operatorname{mod} T^m)$.*

For the proof of this result, we refer to [13], where this result is a step in the proof of Theorem 3.1. A main ingredient of the proof is to show that the real equivariant vector bundle structures are all equivalent over the open subset U_0 of the base B where $T \neq 0$. This can be done using equivalences coming from the matrices used in Propositions 3.4 and 4.2. There, we supposed that T^m divided h, and the matrices therefore have coefficients in $\mathbb{C}[a, b]$. In general, the coefficients are in $\mathbb{C}[a, b, T^{-1}]$, and define an equivalence on U_0. In fact, this idea also comes from the study of G-vector bundles, where an analogous result holds.

Remark 4.4 There is a minor difference between the G-vector bundle case and the real vector bundle structures. If T^m does not divide h, we still have that $M_{2,h}$ is equivalent to $M_{2,-h}$ in $H^1(\Gamma; \mathcal{A}_n)$, but not in $H^1(C_2; \mathcal{A}_n)$. The equivalence with the group action from Γ comes from the fact that $DM_h{}^{\gamma}D^{-1} = M_h$ where $D = diag(i, -i)$.

5 Families of Real Structures on Affine Four-Space

In the previous section, it was shown how to distinguish real vector bundle structures in the case where $B = B_2$. These examples lead to the main result proven in [13] on real circle actions:

Theorem 5.1 ([13]) *Let $h_1, h_2 \in \mathbb{R}[T]$ be two real polynomials. Then the real $(\mathbb{C}^*, \sigma)$-structures defined by μ_{h_1} and μ_{h_2} on the total space of the $\mathbb{C}^*$-vector bundle $B_2 \times F_n$ are equivalent if and only if there exists $r \in \mathbb{R}^*$ such that $h_1(T) \equiv r h_2(r^2 T)$* mod (T^m).

As explained in the introduction, this is proven by using a result of Masuda and Petrie to compare G-actions on the total space of G-vector bundles. The idea is to use the fact that the fixed point set of the action of the subgroup $\{\pm 1\}$ of $\mathbb{C}^*$ is exactly the base of the vector bundle. Then, by considering the normal bundle to the base, it is shown that any equivalence as equivariant real structures on the total space would yield an equivalence as equvariant real vector bundle structures, up to an equivariant automorphism of the base.

6 Final Comments

Note first of all that, unlike the case of $O_2(\mathbb{C})$-bundles, we do not claim to find all real vector bundle structures. We only treat those that arise from the special case of involutions defining an $O_2(\mathbb{C})$-bundle which commute with conjugation. It is interesting to note the difference in the two cases where the base of the vector bundle is V_1 or V_2. In the first case, one finds families of inequivalent $O_2(\mathbb{C})$-vector bundles, but the corresponding involutions can yield trivial real structures. On the other hand, for the case where the base is V_2, inequivalent $O_2(\mathbb{C})$-vector bundles where the involution commutes with conjugation yield many inequivalent real circle actions. In both cases, there could be many other equivariant real circle actions that do not arise from the vector bundle structure.

One can also remark that for all the real circle actions described here the actions on the fixed point sets are all diffeomorphic. More precisely, there is an open algebraic neighborhood of the fixed point set on which the induced real structures are all trivial. This is a consequence of the following result.

Proposition 6.1 *For any $h \in \mathbb{R}[T]$, let μ_h be the real $(\mathbb{C}^*, \sigma, \mu_{B_2})$-vector bundle structure defined on $\pi : B_2 \times F_n \to B_2$ determined by $M_{2,h}$. Then there exists a Zariski open neighborhood U of $B_2^{\mu_{B_2}} = \{(a, \overline{a} | a \in \mathbb{C}\} \subset B_2$ satisfying the following properties:*

(i) U is $\mathbb{C}^*$-stable;
(ii) U is μ_B-stable;
(iii) The restriction of μ_h on $\pi^{-1}(U)$ is equivalent to the restriction of μ_0 on $\pi^{-1}(U)$.

Proof Remember that $M_{2,h} = \begin{pmatrix} 1 - Th^2 & a^n h^n \\ -b^n h^n & R \end{pmatrix}$ where $R = \sum_{j=0}^{n-1}(Th^2)$. Note that $R \in \mathbb{R}[T]$ is a real polynomial with no real roots. Indeed, none of the roots of the polynomial in $\sum_{j=0}^{n-1} X^j$ in $\mathbb{C}[X]$ are real. Since $h \in \mathbb{R}[T]$, for any real value T_0 of T, $h(T_0)$ is real, and thus $R(T_0) \neq 0$. Now let

$$K = \begin{pmatrix} 1 & a^n h^n \\ 0 & R \end{pmatrix} \in GL_n(\mathbb{C}[a, b, R^{-1}].$$

Then

$${}^{\gamma}K^{-1} = \begin{pmatrix} R^{-1} & 0 \\ -b^n h^n R^{-1} & 1 \end{pmatrix},$$

and $K \cdot {}^{\gamma}K^{-1} = M_{2,h}$. Thus if $U \subset B_2$ is the open set defined by $R(T) \neq 0$, then the restriction of μ_h on $U \times F_n \to U$ is equivalent to the restriction of μ_0, and U satisfies the conditions of the proposition. □

References

1. Akhiezer, D., Cupit-Foutou, S.: On the canonical real structure on wonderful varieties. Journal für die Reine und Angewandte Mathematik **693**, 231–244 (2014)
2. Borovoi, M., Gagliardi, G.: Existence of equivariant models of spherical homogeneous spaces and other G-varieties. arxiv:1810.08960 (2018)
3. Cupit-Foutou, S.: Anti-holomorphic involutions and spherical subgroups of reductive groups. Transform. Groups **20**(4), 969–984 (2015)
4. Dubouloz, A., Liendo, A.: Normal real affine varieties with circle actions. arxiv:1810.11712 (2018)
5. Dubouloz, A., Petitjean, C.: Rational real algebraic models of compact differential surfaces with circle actions. In this volume
6. Freudenburg, G., Moser-Jauslin, L.: Real and rational forms of certain $O_2(\mathbb{C})$-actions, and a solution to the weak complexification problem. Transform. Groups **9**(3), 257–272 (2004)
7. Huruguen, M.: Toric varieties and spherical embeddings over an arbitrary field. J. Algebra **342**, 212–234 (2011)
8. Kraft, H., Schwarz, G.: Reductive group actions with one-dimensional quotient. Inst. Hautes Études Sci. Publ. Math. **76**, 1–97 (1992)
9. Masuda, M., Petrie, T.: Algebraic families of O(2)-actions on affine space $\mathbf{C}^4$. In: Algebraic Groups and Their Generalizations: Classical Methods (University Park, PA, 1991), Proceedings of Symposia in Pure Mathematics, vol. 56, pp. 347–354. American Mathematical Society, Providence, RI (1994)

10. Masuda, M., Petrie, T.: Stably trivial equivariant algebraic vector bundles. J. Am. Math. Soc. **8**(3), 687–714 (1995)
11. Masuda, M., Moser-Jauslin, L., Petrie, T.: Equivariant algebraic vector bundles over cones with smooth one-dimensional quotient. J. Math. Soc. Japan **50**(2), 379–414 (1995)
12. Masuda, M., Moser-Jauslin, L., Petrie, T.: The equivariant Serre problem for abelian groups. Topology **35**(2), 329–334 (1996)
13. Moser-Jauslin, L.: Infinite families of inequivalent real circle actions on affine four-space. Épijournal de Géométrie Algébrique, vol. 3. https://epiga.episciences.org (2019)
14. Moser-Jauslin, L., Terpereau, R.: Real structures on horospherical varietie. arXiv:1808.10793 (2018)
15. Schwarz, G.: Exotic algebraic group actions. C. R. Acad. Sci. Paris Sér. I Math. **309**(2), 89–94 (1989)
16. Wedhorn, T.: Spherical spaces. Ann. Inst. Fourier **68**(1), 229–256 (2018)

On Some Sufficient Conditions for Polynomials to Be Closed Polynomials over Domains

Takanori Nagamine

Abstract In this paper, we study closed polynomials of the polynomial ring in n variables over an integral domain. By using the techniques on $\mathbb{Z}$-gradings on the polynomial ring, we give some sufficient conditions for a polynomial f to be a closed polynomial. We also give a correspondence between closed polynomials and derivations in the polynomial ring $R[x, y]$ in two variables over a UFD R containing $\mathbb{Q}$.

Keywords Polynomial ring · Closed polynomial · Derivation

1 Introduction

Let k be a field of characteristic $p \geq 0$ and let B be the polynomial ring in n variables over k. A polynomial $f \in B$ is said to be a *closed polynomial* if $f \notin k$ and the ring $k[f]$ is integrally closed in B.

Closed polynomials in B are studied by several mathematicians. See e.g., Nowicki [10], Nowicki and Nagata [11], Ayad [2], Arzhantsev and Petravchuk [1], Kato and Kojima [5], etc. Historically, in 1988, Nowicki and Nagata [11] introduced the notation of closed polynomials for understanding derivations and their kernels. Around the same time, in 1989, Stein [12] announced the concepts about the total reducibility order of a polynomial. This concept is essentially the same as closed polynomials. But his approach is differ from Nowicki–Nagata's one. After that, in 2007, Arzhantsev and Petravchuk [1] improved these results as below.

Theorem 1 (cf. [1, Theorem 1]) *Let k be a field and let B be the polynomial ring in n variables over k. $\bar{k}$ denotes an algebraically closed field containing k. For a non-constant polynomial $f \in B \setminus k$, the following conditions are equivalent:*

T. Nagamine (✉)
Graduate School of Science and Technology, Niigata University, Niigata, Japan
e-mail: t.nagamine14@m.sc.niigata-u.ac.jp

S. Kuroda et al. (eds.), *Polynomial Rings and Affine Algebraic Geometry*, Springer Proceedings in Mathematics & Statistics 319,
https://doi.org/10.1007/978-3-030-42136-6_10

(i) f *is a closed polynomial over* k.
(ii) $k[f] \subset k[g]$ *implies* $k[f] = k[g]$ *for* $g \in B$.

If the characteristic of k *equals zero, then the condition (i) is equivalent to the following condition (iii):*

(iii) There exists a k*-derivation* d *on* B *such that* $\ker d = k[f]$.

If k *is a perfect field, then the condition (i) is equivalent to the following conditions (iv) and (v):*

(iv) $f - \lambda$ *is irreducible over* $\bar{k}$ *for all but finitely many* $\lambda \in \bar{k}$.
(v) $\#\{\lambda \in \bar{k} \mid f - \lambda \text{ is reducible }\} < \deg f$, *where* deg *is the standard degree function on* B.

The condition (iv) in Theorem 1 gives a geometrical meaning for closed polynomials over an algebraically closed field k. Let $f \in B \setminus k$ and let

$$\Phi_f : \mathbb{A}_k^n \cong_k \operatorname{Spec} B \to \operatorname{Spec} k[f] \cong_k \mathbb{A}_k^1$$

be the surjective morphism defined by the natural inclusion $k[f] \to B$. Then the condition (iv) in Theorem 1 implies that f is a closed polynomial if and only if general fibers of Φ_f are irreducible and reduced.

This paper is an announcement of our results and the detailed versions are [6–8]. The aim of this paper is to give some kinds of sufficient conditions for polynomials to be closed over an integral domain. The main criterion is Theorem 3 (cf. [8, Propositions 3.10 and 3.11]).

In Sect. 2, we give a proof of Theorem 3 based on [8, Sect. 3]. In Sect. 3, we give two kinds of criteria (Proposition 2 and Theorem 4). Proposition 2 says that a polynomial satisfying the Jacobian condition is closed. Theorem 4 gives a necessary and sufficient condition for polynomials of prime degree to be closed. Finally, we describe a correspondence between equivalence classes of closed polynomials and equivalence classes of derivations, in the polynomial ring $R[x, y]$ in two variables over a UFD R containing $\mathbb{Q}$.

2 Criteria of Closed Polynomials over Domains

Let R be an integral domain. $Q(R)$ denotes the field of fractions of R and $R^{[n]}$ denotes the polynomial ring in n variables over R. Let $B = R[x_1, \dots, x_n] \cong_R R^{[n]}$. A polynomial $f \in B$ is a *closed polynomial* over R if $f \notin R$ and the ring $R[f]$ is integrally closed in B.

The following theorem is a generalization of Theorem 1 in the case where the coefficient ring of the polynomial ring is an arbitrary integral domain. For more detail, refer to [7, Theorem 3.1].

Theorem 2 *Let R be an integral domain, let $B \cong_R R^{[n]}$ be the polynomial ring in n variables over R and let $K = Q(R)$. Let $f \in B \setminus R$ be a non-constant polynomial such that $K[f] \cap B = R[f]$. Then the following conditions (i), (ii) and (iii) are equivalent:*

(i) f is a closed polynomial over R.
(ii) f is a closed polynomial over K.
(iii) $K[f] \subset K[g]$ implies $K[f] = K[g]$ for $g \in B \otimes_K K$.

Moreover, if the characteristic of R equals zero (resp. if the field extension $Q(B)/Q(R[f])$ is separable), then the condition (i) is equivalent to the following condition (iv) (resp. (v)):

(iv) There exists an R-derivation d on B such that $B^d = R[f]$.
(v) There exists a rational higher R-derivation D on B such that $B^D = R[f]$ and $Q(R[f]) = Q(B)^{\overline{D}}$, where $\overline{D}$ denotes the extension of D to $Q(B)$.

A polynomial $f \in B$ is called a *variable* (or *coordinate*) over R if there exist $g_2, \dots, g_n \in B$ such that $R[f, g_2, \dots, g_n] = B$, or equivalently, there exists an R-automorphism φ of B such that $\varphi(x_1) = f$.

Proposition 1 *If $f \in B$ is a variable over R, then it is closed over R.*

Proof Take $g_2, \dots, g_n \in B$ such that $R[f, g_2, \dots, g_n] = B$. Define the $\mathbb{Z}$-grading $\mathfrak{g}$ on B by $\deg_{\mathfrak{g}} f = 0$ and $\deg_{\mathfrak{g}} g_i = 1$. Using this grading, we see that $R[f]$ is factorially closed in B, that is, for $b_1, b_2 \in B \setminus \{0\}$, $b_1 b_2 \in R[f]$ implies $b_i \in R[f]$. In particular, $R[f]$ is integrally closed in B. □

For polynomials $f, g \in B$, we write $f \in Rg$ if there exists $r \in R$ such that $f = rg$. Let $\mathbf{w} = (w_1, \dots, w_n) \in \mathbb{N}^n$. We consider the $\mathbb{Z}$-grading $\mathfrak{g}_{\mathbf{w}}$ on B with respect to $\mathbf{w}$ by $\deg_{\mathfrak{g}_{\mathbf{w}}} x_i = w_i$ for $1 \le i \le n$, and set $\deg_{\mathbf{w}} := \deg_{\mathfrak{g}_{\mathbf{w}}}$. We write simply deg by the standard degree function on B, where the standard degree function is the degree function of $\mathbf{w} = (1, \dots, 1)$. A homogeneous polynomial $f \in B$ for $\mathfrak{g}_{\mathbf{w}}$ is defined to be *decomposable* with respect to $\mathfrak{g}_{\mathbf{w}}$ if there exists a homogeneous polynomial $g \in B$ for $\mathfrak{g}_{\mathbf{w}}$ such that $f \in Rg^m$ for some $m \ge 2$. Also we say that f is *primitive* with respect to $\mathfrak{g}_{\mathbf{w}}$ if it is not decomposable with respect to $\mathfrak{g}_{\mathbf{w}}$.

Set $B_K := K \otimes_R B \cong_K K^{[n]}$. For a polynomial $f \in B$, we define $\hat{f} \in B_K$ by

$$\hat{f} := \gcd(f_{x_1}, \dots, f_{x_n}),$$

where f_{x_i} is the partial derivative of f with respect to x_i and we take the greatest common divisor of $f_{x_1}, \dots, f_{x_n}$ as polynomials in B_K, hence $\hat{f} \in B_K$.

Definition 1 Let $f \in B$ and $\mathbf{w} \in \mathbb{N}^n$. Assume that $\deg_{\mathbf{w}} f \ge 2$. Then $\mathrm{LD}_{\mathbf{w}}(f)$ denotes the smallest positive prime integer dividing $\deg_{\mathbf{w}} f$.

The number $\mathrm{LD}_{\mathbf{w}}(f)$ is the most important thing for Theorem 3. We give some examples as below.

Example 1 For $f = x^9 + x^6y^2 + x^3y^4 \in \mathbb{Z}[x, y] \cong_{\mathbb{Z}} \mathbb{Z}^{[2]}$, we can easily see that:

(a) for $\mathbf{u} = (1, 1)$, $\deg_{\mathbf{u}} f = \deg f = 9$ and $\mathrm{LD}_{\mathbf{u}}(f) = 3$,
(b) for $\mathbf{v} = (0, 1)$, $\deg_{\mathbf{v}} f = 4$ and $\mathrm{LD}_{\mathbf{v}}(f) = 2$,
(c) for $\mathbf{w} = (1, 2)$, $\deg_{\mathbf{w}} f = 11$ and $\mathrm{LD}_{\mathbf{w}}(f) = 11$.

As seen in the above example, if $\deg_{\mathbf{w}} f$ is a prime number, then $\deg_{\mathbf{w}} f = \mathrm{LD}_{\mathbf{w}}(f)$. The aim of this section is to prove the following theorem based on [8, Sect. 3].

Theorem 3 *Let $B \cong_R R^{[n]}$ be the polynomial ring in n variables over an integral domain R. Let $f \in B \setminus R$ be a non-constant polynomial such that $K[f] \cap B = R[f]$. Then the following assertions hold true.*

(a) Suppose that f is homogeneous for some $\mathbb{Z}$-grading $\mathfrak{g}_{\mathbf{w}}$, $\mathbf{w} \in \mathbb{N}^n$. Then f is a closed polynomial over R if and only if it is primitive in B_K with respect to $\mathfrak{g}_{\mathbf{w}}$.
(b) Suppose that the characteristic of R is zero. If there exists $\mathbf{w} \in \mathbb{N}^n$ such that $\deg_{\mathbf{w}} f = 1$ or, $\deg_{\mathbf{w}} f \geq 2$ and

$$\deg_{\mathbf{w}} \hat{f} < \frac{\mathrm{LD}_{\mathbf{w}}(f) - 1}{\mathrm{LD}_{\mathbf{w}}(f)} \deg_{\mathbf{w}} f,$$

then f is a closed polynomial over R.

First of all, we prove the following two lemmas needed later.

Lemma 1 *Let $f \in B \setminus R$. Assume that f is homogeneous for some $\mathfrak{g}_{\mathbf{w}}$ of $\deg_{\mathbf{w}} f > 0$, where $\mathbf{w} \in \mathbb{N}^n$. For $g \in B$ such that $f \in R[g]$ and $g(0, \ldots, 0) = 0$, the following assertions hold true.*

(a) g is homogeneous for $\mathfrak{g}_{\mathbf{w}}$.
(b) $f \in Rg^m$ for some positive integer $m \geq 1$.

Proof Since f is homogeneous for $\mathfrak{g}_{\mathbf{w}}$ of $\deg_{\mathbf{w}} f > 0$, we see that $f(0, \ldots, 0) = 0$. Hence we can write f as

$$f = u(g) = \sum_{i=\ell}^{m} u_i g^i,$$

where $u_i \in R$, $u_\ell \neq 0$, $u_m \neq 0$ and $m \geq \ell \geq 1$. Then $\deg_{\mathbf{w}} f = m \deg_{\mathbf{w}} g$, so $\deg_{\mathbf{w}} g > 0$. Set $h = \sum_{i=\ell}^{m} u_i g^{i-\ell}$. Then $f = g^\ell h$ and g does not divide h. Since f is homogeneous, g and h are also homogeneous for $\mathfrak{g}_{\mathbf{w}}$. This completes the proof for the part (a).

For the part (b), we suppose that $m > \ell$. Since $\deg_{\mathbf{w}} h = (m - \ell) \deg_{\mathbf{w}} g > 0$ and h is homogeneous, $h(0, \ldots, 0) = 0$. But $h(0, \ldots, 0) = u_\ell \neq 0$, which is a contradiction. Therefore we have $m = \ell$, which implies that $f = u_m g^m$. □

Lemma 2 *Let $\mathbf{w} \in \mathbb{N}^n$ and let $f, g \in B \setminus R$ with $f \in R[g]$. Assume that $\deg_{\mathbf{w}} f > 0$ and $f = u(g)$ for $u(t) \in R[t] \cong_R R^{[1]}$ of t-degree $m \geq 1$. Then the following assertions hold true.*

(a) $\deg_{\mathbf{w}} f = m \deg_{\mathbf{w}} g$, *hence* m *divides* $\deg_{\mathbf{w}} f$.
(b) If the characteristic of R *equals zero, then*

$$\deg_{\mathbf{w}} \hat{f} \geq \frac{m-1}{m} \deg_{\mathbf{w}} f.$$

Proof **(a)** We write $f = u(g)$ as follows:

$$f = u(g) = u_0 g^m + u_1 g^{m-1} + \cdots + u_{m-1} g + u_m,$$

where $u_0 \in R \setminus \{0\}$ and $u_1, \ldots, u_m \in R$. Since $\deg_{\mathbf{w}} f > 0, \deg_{\mathbf{w}} g > 0$. This implies that $\deg_{\mathbf{w}} g^i \geq \deg_{\mathbf{w}} g^j$ if $i \geq j$. So,

$$\deg_{\mathbf{w}} f = \deg_{\mathbf{w}}(u(g)) = \deg_{\mathbf{w}}(u_0 g^m) = m \deg_{\mathbf{w}} g.$$

(b) Since $f = u(g)$, $f_{x_i} = u'(g) g_{x_i}$ for $1 \leq i \leq n$, where $u'(t) = du/dt$. This implies that each f_{x_i} is divisible by $u'(g)$, so $u'(g)$ divides $\hat{f}$ in B_K. Therefore $\deg_{\mathbf{w}} \hat{f} \geq \deg_{\mathbf{w}}(u'(g))$. On the other hand, since the characteristic of R equals zero, $m u_0 \neq 0$. Therefore $\deg_{\mathbf{w}} u'(g) = (m-1) \deg_{\mathbf{w}} g$, so we have

$$\deg_{\mathbf{w}} \hat{f} \geq \deg_{\mathbf{w}}(u'(g)) = (m-1) \deg_{\mathbf{w}} g = \frac{m-1}{m} \deg_{\mathbf{w}} f.$$

□

We give the proof of Theorem 3 below.

Proof of Theorem 3 **(a)** Suppose that f is homogeneous for some $\mathbb{Z}$-grading $\mathfrak{g}_{\mathbf{w}}$, $\mathbf{w} \in \mathbb{N}^n$. It is clear that, if f is not primitive for $\mathfrak{g}_{\mathbf{w}}$, then f is not closed over R.

Conversely, we suppose that f is primitive in B_K for $\mathfrak{g}_{\mathbf{w}}$. According to Theorem 2, it suffices to prove the maximality of the ring $K[f]$. Let $g \in B_K \setminus K$ be any polynomial such that $f \in K[g]$. Without loss of generality we may assume that $g(0, \ldots, 0) = 0$. By Lemma 1 (a) and (b), g is homogeneous for $\mathfrak{g}_{\mathbf{w}}$ and $f \in K g^m$ for some positive integer $m \geq 1$. Since f is primitive in B_K, we have $m = 1$ and hence $K[f] = K[g]$.

(b) Let $g \in B_K \setminus K$ with $f \in K[g]$. Since $f \in K[g]$, there exists $u(t) \in K[t]$ of degree m such that $f = u(g)$. We write $u(t)$ as

$$u(t) = u_0 t^m + u_1 t^{m-1} + \cdots + u_{m-1} t + u_m,$$

for some $u_i \in K$ and $u_0 \neq 0$. By Lemma 2 (a), $\deg_{\mathbf{w}} f = m \deg_{\mathbf{w}}(g)$. It is enough to show that $m = 1$. Indeed, if $m = 1$, then $f = u_0 g + u_1$. This implies $g \in K[f]$, so $K[f] = K[g]$.

If $\deg_{\mathbf{w}} f = 1$, then obviously $m = 1$. On the other hand, we suppose that $\mathbf{w} \in \mathbb{N}^n$ satisfies $\deg_{\mathbf{w}} f \geq 2$ and

$$\deg_{\mathbf{w}} \hat{f} < \frac{\mathrm{LD}_{\mathbf{w}}(f) - 1}{\mathrm{LD}_{\mathbf{w}}(f)} \deg_{\mathbf{w}} f.$$

Since the characteristic of R equals zero, by Lemma 2 (b),

$$\deg_{\mathbf{w}} \hat{f} \geq \frac{m-1}{m} \deg_{\mathbf{w}} f.$$

By comparing the above two inequalities, we have $\mathrm{LD}_{\mathbf{w}}(f) > m$. By using Lemma 2 (a), we see that m divides $\deg_{\mathbf{w}} f$. But the number $\mathrm{LD}_{\mathbf{w}}(f)$ is the smallest positive prime dividing $\deg_{\mathbf{w}} f$, hence $m = 1$. Therefore it follows from Theorem 2 that f is a closed polynomial over R. □

By using Theorem 3 (b), we have the following. This is a generalization of [2, Proposition 14].

Corollary 1 *Let $f \in B \setminus R$ be a non-constant polynomial such that $K[f] \cap B = R[f]$. Suppose that the characteristic of R is zero. If there exists $1 \leq i < j \leq n$ such that $f_{x_i} \neq 0$, $f_{x_j} \neq 0$ and $\hat{f} \in R[x_i]$, then f is a closed polynomial over R.*

Proof If $\hat{f} \in R$, then $\deg_{\mathbf{w}} \hat{f} = 0$ for any $\mathbf{w} \in \mathbb{N}^n$. Hence the inequality in Theorem 3 (b) is satisfied for any $\mathbf{w} \in \mathbb{N}^n$.

Suppose that $\hat{f} \in R[x_i] \setminus R$. Set $\mathbf{w}_{i,j} = (w_1, \ldots, w_n) \in \mathbb{N}^n$, where $w_i = 0$, $w_j = 2$ and $w_\ell = 1$ for $1 \leq \ell \leq n$ with $\ell \neq i, j$. Then $\deg_{\mathbf{w}_{i,j}} \hat{f} = 0$ and $\deg_{\mathbf{w}_{i,j}} f \geq 2$. Then the inequality in Theorem 3 (b) is satisfied for $\mathbf{w}_{i,j}$. Therefore f is a closed polynomial over R. □

We give some remarks on Theorem 3 (b). In the case where R is a field, the assumption "$K[f] \cap B = R[f]$" satisfies for any $f \in B$ automatically. Furthermore, we give later (Lemma 3) a necessary and sufficient condition for a polynomial f to satisfy "$K[f] \cap B = R[f]$" when R is a UFD. For this reason, we can confirm that almost all a given polynomial is to be a closed polynomial. However, there are closed polynomials which do not satisfy the assumption on Theorem 3 (b) for any $\mathbf{w} \in \mathbb{N}^n$ as below:

Example 2 Let $f = x^6y^4 + x^4y^6 \in \mathbb{Q}[x, y] \cong_{\mathbb{Q}} \mathbb{Q}^{[2]}$. Then f is a closed polynomial over $\mathbb{Q}$, but it does not satisfy the assumption on Theorem 3 (b) for any $\mathbf{w} \in \mathbb{N}^2$.

Indeed, it is clear that f is homogeneous and primitive for the standard $\mathbb{Z}$-grading. By Theorem 3 (a), f is a closed polynomial.

On the other hand, for any $\mathbf{w} = (w_1, w_2) \in \mathbb{N}^2$, we have

$$\deg_{\mathbf{w}} f = \max\{6w_1 + 4w_2, 4w_1 + 6w_2\}.$$

Hence $\deg_{\mathbf{w}} f$ is divisible by 2, which implies that $\deg_{\mathbf{w}} f \geq 2$ and $\mathrm{LD}_{\mathbf{w}}(f) = 2$. We may assume that $\deg_{\mathbf{w}} f = 6w_1 + 4w_2$. Also, we can see easily that $\hat{f} = x^3y^3$, hence $\deg_{\mathbf{w}} \hat{f} = 3(w_1 + w_2)$. Then

$$\frac{\mathrm{LD}_{\mathbf{w}}(f)-1}{\mathrm{LD}_{\mathbf{w}}(f)}\deg_{\mathbf{w}} f = 3w_1 + 2w_2 \leq 3w_1 + 3w_2 = \deg_{\mathbf{w}} \hat{f}.$$

Therefore f does not satisfy the assumption on Theorem 3 (b) for any $\mathbf{w} \in \mathbb{N}^2$.

Also, in the case where the characteristic of R is positive, Theorem 3 (b) and Corollary 1 do not hold true in general. We give two examples below.

Example 3 Let k be a field of characteristic $p > 0$ and let $k[x, y] \cong_k k^{[2]}$. The following two polynomials are not closed.

(a) $f = x^p + y^p + (x + y)^{p-1}$.
(b) $g = x^p + y^p + x + y$.

Indeed, for (a), we can see easily that $\hat{f} = (x + y)^{p-2}$. Hence $\deg_{\mathbf{w}} f$ satisfies the assumption on Theorem 3 (b) for $\mathbf{w} = (1, 1)$. However, $k[f] \subsetneq k[x + y]$, which implies that f is not a closed polynomial.

Also, for (b), we see that $\hat{g} = 1$. Hence $\deg_{\mathbf{w}} f$ satisfies the assumption on Corollary 1. However, $k[g] \subsetneq k[x + y]$, which implies that g is not a closed polynomial.

Finally, we give a necessary and sufficient condition for a polynomial f to be satisfied "$K[f] \cap B = R[f]$" when R is a UFD.

Lemma 3 (cf. [6, Lemma 1.3]) *Suppose that R is a UFD. For a non-constant polynomial $f \in B \setminus R$. $c(f) \in R$ denotes the greatest common divisor of the coefficients of f. Then the following two conditions are equivalent:*

(i) $c(f - f(0, \ldots, 0)) \in R^*$.
(ii) $K[f] \cap B = R[f]$.

3 Closed Polynomials in Special Cases

Let $B = R[x_1, \ldots, x_n] \cong_R R^{[n]}$ be the polynomial ring in n variables over an integral domain R and set $K = Q(R)$, the quotient field of R.

For polynomials $f_1, \ldots, f_n \in B$, let $F := (f_1, \ldots, f_n)$. $J(F)$ denotes the Jacobian matrix of F with respect to variables $x_1, \ldots, x_n$, namely,

$$J(F) = \left(\frac{\partial f_i}{\partial x_j}\right)_{1 \leq i, j \leq n}.$$

Proposition 2 (cf. [6, Proposition 2.2]) *Suppose that the characteristic of R is zero. Let $F := (f_1, \ldots, f_n)$ for polynomials $f_1, \ldots, f_n \in B$. Assume that* $\det J(F) \in R \setminus \{0\}$ *and $K[f_i] \cap B = R[f_i]$ for $1 \leq i \leq n$. Then these polynomials $f_1, \ldots, f_n$ are closed polynomials. In particular, for $g \in B \setminus R$ satisfying $K[g] \cap B = R[g]$, if $\hat{g} = \gcd(g_{x_1}, \ldots, g_{x_n}) \in R \setminus \{0\}$, then it is a closed polynomial over R.*

In the case where R is a field of characteristic zero, the above proposition gives a relation between the Jacobian conjecture and closed polynomials. Let k be a field of characteristic zero and let $k[\mathbf{X}] = k[x_1, \dots, x_n]$ be the polynomial ring in n variables over k. For $f_1, \dots, f_n \in k[\mathbf{X}]$, let $F = (f_1, \dots, f_n)$. We can regard F as a k-endomorphism on $k[\mathbf{X}]$. Here, we consider the following two conditions (a) and (b).

(a) F is a k-automorphism on $k[\mathbf{X}]$.
(b) $\det J(F) \in k^*$.

It is easy to show that the implication "(a) $\Longrightarrow$ (b)" holds true. The Jacobian conjecture says that the converse implication "(b) $\Longrightarrow$ (a)" holds true, namely, the above two conditions (a) and (b) are equivalent. Let $f \in k[\mathbf{X}]$ be a polynomial. We say that f *satisfies the Jacobian condition* if there exist $f_2, \dots, f_n \in k[\mathbf{X}]$ such that $F = (f, f_2, \dots, f_n)$ satisfies the condition (b). $\mathrm{JC}_k(k[\mathbf{X}])$ denotes the set of polynomials satisfying the Jacobian condition. Then it is well known that the Jacobian conjecture is equivalent to the following assertion (c) (see e.g., [3]):

(c) Every $f \in \mathrm{JC}_k(k[\mathbf{X}])$ is a variable.

In fact, Proposition 2 implies that every $f \in \mathrm{JC}_k(k[\mathbf{X}])$ is at least a closed polynomial.

The following result gives a necessary and sufficient condition for a polynomial of prime degree to be closed.

Theorem 4 (cf. [6, Theorem 2.5]) *Let $B \cong_R R^{[n]}$ be the polynomial ring in n variables over an integral domain R of characteristic zero. Consider the standard degree function* deg *on B. For a non-constant polynomial $f \in B \setminus R$ of prime degree such that $K[f] \cap B = R[f]$, the following conditions are equivalent:*

(i) f *is a closed polynomial.*
(ii) $\deg \hat{f} < \deg f - 1$.

In the rest of this section, we assume further that R contains $\mathbb{Q}$. For $d, \delta \in \mathrm{Der}_R B$, we write $d \sim \delta$ when $fd = g\delta$ for some $f, g \in B \setminus \{0\}$.

Theorem 5 (cf. [8, Theorem 2.5]) *Let R be a UFD containing $\mathbb{Q}$ and let d and δ be R-derivations of $R[x, y] \cong_R R^{[2]}$ such that $R[x, y]^d \neq R$ and $R[x, y]^\delta \neq R$. Then $R[x, y]^d = R[x, y]^\delta$ if and only if $d \sim \delta$.*

Let $R[x, y] \cong_R R^{[2]}$ and let $d \in \mathrm{Der}_R R[x, y] \setminus \{0\}$. Then there exists $f \in R[x, y]$ such that $R[x, y]^d = R[f]$. If $R[f] \neq R$, then f is a closed polynomial. For $f, g \in R[x, y]$, we write $f \equiv g$ if $f = ag + b$ for some $a \in R^*$ and $b \in R$. Clearly, $f \equiv g$ if and only if $R[f] = R[g]$. Here, we define the following:

$\mathfrak{Der}_R R[x, y] :=$ the set of equivalence classes of $\mathrm{Der}_R R[x, y] \setminus \{0\}$,
$\mathfrak{CL}_R R[x, y] :=$ the set of equivalence classes of closed polynomials,
$\mathrm{KDer}_R R[x, y] := \{R[x, y]^d \mid d \in \mathrm{Der}_R R[x, y] \setminus \{0\}\}$.

Here, $[f]$ denotes the equivalence class of $f \in R[x, y]$ and $[d]$ denotes the equivalence class of $d \in \mathrm{Der}_R R[x, y]$. By Theorem 5, we have the following result:

Proposition 3 *Define the map* $\alpha : \mathfrak{Der}_R R[x, y] \to \mathfrak{CL}_R R[x, y] \cup \{1\}$ *by, for* $[d] \in \mathfrak{Der}_R R[x, y]$,

$$\alpha([d]) = \begin{cases} [f] & \text{if } R[x, y]^d = R[f] \neq R, \\ 1 & \text{if } R[x, y]^d = R. \end{cases}$$

Then α *is a well-defined bijective mapping.*

Proof The above map α factors through $\mathrm{KDer}_R R[x, y]$ as below:

$$\beta : \mathrm{KDer}_R R[x, y] \to \mathfrak{CL}_R R[x, y] \cup \{1\},$$

where β is defined by $\beta(R[x, y]^d) = [f]$ if $R \neq R[x, y]^d = R[f]$, otherwise $\beta(R[x, y]^d) = 1$. It is clear that β is injective. By Theorem 2, β is surjective, hence β is bijective. Furthermore, by Theorem 5, the natural mapping $[d] \mapsto R[x, y]^d$ is bijective. Therefore, α is bijective. □

References

1. Arzhantsev, I.V., Petravchuk, A.P.: Closed polynomials and saturated subalgebras of polynomial algebras. Ukrainian Math. J. **59**, 1783–1790 (2007)
2. Ayad, M.: Sur les plynômes $f(X, Y)$ tels que $K[f]$ est intégralement fermé dans $K[X, Y]$. Acta Arith. **105**, 9–28 (2002)
3. Bass, H., Connell, E., Wright, D.: The Jacobian conjecture: reduction of degree and formal expansion of the inverse. Bull. Am. Math. Soc. **7**, 287–330 (1982)
4. Jędrzejewicz, P.: Positive characteristic analogue of closed polynomials. Cent. Eur. J. Math. **9**, 50–56 (2011)
5. Kato, M., Kojima, H.: Closed polynomials in polynomial rings over unique factorization domains. Commun. Algebra **43**, 1935–1938 (2015)
6. Kitazawa, C., Kojima, H., Nagamine, T.: Closed polynomials and their applications for computations of kernels of monomial derivations. J. Algebra **553**, 266–282 (2019)
7. Kojima, H., Nagamine, T.: Closed polynomials in polynomial rings over integral domains. J. Pure Appl. Algebra **219**, 5493–5499 (2015)
8. Nagamine, T.: Derivations having divergence zero and closed polynomials over domains. J. Algebra **462**, 67–76 (2016)
9. Nagamine, T.: On some properties of coordinates in polynomial rings. Commun. Algebra **46**, 4265–4272 (2018)
10. Nowicki, A.: On the equation $J(f, g) = 0$ for polynomials in $k[x, y]$. Nagoya Math. J. **109**, 151–157 (1988)
11. Nowicki, A., Nagata, M.: Rings of constants for k-derivations in $k[x_1, \ldots, x_n]$. J. Math. Kyoto Univ. **28**, 111–118 (1988)
12. Stein, Y.: The total reducibility order of a polynomial in two variables. Israel J. Math. **68**, 109–122 (1989)

Variations on the Theme of Zariski's Cancellation Problem

Vladimir L. Popov

Abstract This is an expanded version of the talk by the author at the conference *Polynomial Rings and Affine Algebraic Geometry*, February 12–16, 2018, Tokyo Metropolitan University, Tokyo, Japan. Considering a local version of the Zariski Cancellation Problem naturally leads to exploration of some classes of varieties of special kind and their equivariant versions. We discuss several topics inspired by this exploration, including the problem of classifying a class of affine algebraic groups that are naturally singled out in studying the conjugacy problem for algebraic subgroups of the Cremona groups.

Keywords Zariski Cancellation Problem · Flattenable variety · Algebraic group · Action

1 Introduction

This is an expanded version of the talk by the author at the conference *Polynomial Rings and Affine Algebraic Geometry*, February 12–16, 2018, Tokyo Metropolitan University, Tokyo, Japan.

Our starting point is a local version of the Zariski Cancellation Problem (LZCP). Its consideration naturally leads to distinguishing a class of varieties of a special kind, called here flattenable, and a more general class of locally flattenable varieties. We discuss the relevant examples, including flattenability of affine algebraic groups and the related varieties, in particular, we prove that all smooth spherical varieties are locally flattenable. This is completed by answering (LZCP). We then consider the equivariant versions of flattenability and obtain a series of results on equivariant flattenability of affine algebraic groups. In particular, we prove that a reductive algebraic

V. L. Popov (✉)
Steklov Mathematical Institute, Russian Academy of Sciences, Gubkina 8,
Moscow 119991, Russia
e-mail: popovvl@mi-ras.ru
URL: http://www.mathnet.ru/php/person.phtml?&personid=8935&option_lang=eng

S. Kuroda et al. (eds.), *Polynomial Rings and Affine Algebraic Geometry*,
Springer Proceedings in Mathematics & Statistics 319,
https://doi.org/10.1007/978-3-030-42136-6_11

group is equivariantly flattenable if and only if it is linearly equivariantly flattenable, and we prove that equivariant flattenability of a Levi subgroup of a connected affine algebraic group G implies that of G. The latter yields that every connected solvable affine algebraic group is equivariantly flattenable. As an application, we briefly survey a special role of equivariantly flattenable subgroups in the rational linearization problem. Then we dwell on the classification problem of equivariantly flattenable affine algebraic groups G. We prove that every such G is special in the sense of Serre, which implies that if G is reductive equivariantly flattenable, then its derived group if a product of the groups of types SL and Sp. We complete this discussion with the unexpected recent examples of reductive equivariantly flattenable groups, whose derived groups do contain factors of type Sp. In the last section, the local version of equivariantly flattenable varieties and the relevant version of the Gromov problem are briefly considered.

1.1 Notation, Conventions, and Terminology

We fix an algebraically closed field k of characteristic zero. In what follows, as in [7, 26], variety means algebraic variety over k in the sense of Serre (so algebraic group means algebraic group over k). Unless otherwise specified, all topological terms refer to the Zariski topology. We use freely the standard notation and conventions of loc. cit., where the proofs of the unreferenced claims and/or the relevant references can be found. Action of an algebraic group on an algebraic variety means algebraic (morphic) left action; homomorphism of an algebraic group means algebraic homomorphism.

Recall that a *Levi subgroup* of a connected affine algebraic group G is its (necessarily connected reductive) subgroup L such that G is the semi-direct product of L and the unipotent radical $\mathcal{R}_u G$ of G; since $\operatorname{char}(k) = 0$, by a result of Mostow, Levi subgroups exist and are conjugate in G (cf. [7, 11.22]).

We also use the following notation:

- A^* is the group of units of an associative k-algebra A with identity.
- $\mathrm{Mat}_{n\times m}$ is the k-vector space of all $n \times m$-matrices with entries in k; for $n = m$, it is naturally endowed with the k-algebra structure.
- $A^\top$ is the transpose of a matrix $A \in \mathrm{Mat}_{n\times m}$.

2 The Zariski Cancellation Problem

So is called the following question:

$$\begin{aligned}&\textit{Are there affine varieties } X \textit{ and } Y \textit{ such that } Y \textit{ and}\\ &X \times Y \textit{ are isomorphic respectively to } \mathbf{A}^s \textit{ and } \mathbf{A}^{s+d},\\ &\textit{but } X \textit{ is not isomorphic to } \mathbf{A}^d?\end{aligned} \qquad \text{(ZCP)}$$

At this writing (January 2019), for $d > 2$, it is still open. A historical survey about the Zariski Cancellation Problem is given in [16].

Our starting point is a local version of this problem. Making precise its formulation (see (LZCP) in Sect. 5) leads to distinguishing the following class of varieties:

Definition 1 An irreducible variety X is called

(a) *flattenable* if X isomorphic to an open subset of an affine space;
(b) *locally flattenable* if for every point $x \in X$ there is a flattenable open subset of X containing x.

3 Terminology

Under other names, locally flattenable varieties appeared in the literature long ago. The earliest reference known to the author is [11, p. 2-09] where Chevalley calls them *special* varieties. In [20] Chevalley terminology is used for the definition of R-equivalence. In [1] these varieties appear as *algebraic spaces*, in [3] as *plain varieties*, and in [4, 21] as *uniformly rational varieties*. The term *flattenable variety* is coined in [24], where special properties of linearly equivariantly flattenable algebraic subgroups (see below Definition 4) of the Cremona groups have been revealed (this topic is briefly surveyed in Sect. 9 below).

By Definition 1, every locally flattenable variety is rational. Whether the converse is true is open at this writing (January 2019):

Is every irreducible smooth rational variety locally flattenable? (Gr)

This problem was raised by Gromov in [14, 3.5.E$'''$] (for projective varieties).

4 Examples of Locally Flattenable Varieties

4.1 Homogeneous Spaces

Theorem 1 *Let X be an irreducible variety. If the natural action of* Aut(X) *on X is transitive, then the following properties are equivalent:*

(a) X is a rational variety;
(b) X is a locally flattenable variety.

Proof If (a) holds, then X contains an open flattenable subset U. Since U and gU for any $g \in \mathrm{Aut}(X)$ are isomorphic, (b) follows from the equality $X = \bigcup_{g\in G} gU$ (the latter holds because of the transitivity condition). Definition 1 implies (b)⇒(a). □

Corollary 1 *Let G be a connected affine algebraic group and let H be a closed subgroup of G. Then the following properties are equivalent:*

(a) G/H is a rational variety;
(b) G/H is a locally flattenable variety.

Remark 1 Maintain the notation of Corollary 1. There are nonrational (and even not stably rational) varieties G/H, where G is

$$\mathrm{SL}_{n_1} \times \cdots \times \mathrm{SL}_{n_r} \times \mathrm{Sp}_{2m_1} \times \cdots \times \mathrm{Sp}_{2m_s}$$

and H is finite; see [25, Theorem 2]. It is an old problem, still open at this writing (October 2019), whether there are nonrational homogeneous spaces G/H with connected H. For connected H of various special types, rationality of G/H is known; see [12] and Remark 5 below. In particular, G is rational as a variety [10] (cf. [25, Lemma 2]).

4.2 Vector Bundles and Homogeneous Fiber Spaces

Definition 2 Given three varieties X, Z, F, we say that a surjective morphism $\varphi\colon X \to Z$ is a *locally trivial fibration over Z with fiber F* if for every point $z \in Z$, there are a neighbourhood U of z in Z (called *trivializing neighborhood*) and an isomorphism $\tau_U\colon \varphi^{-1}(U) \to U \times F$ over U. If F has a structure of an algebraic vector space over k and, for any pair of trivializing neighborhoods U and V, the automorphism $\tau_V \circ \tau_U^{-1}\colon (U \cap V) \times F \to (U \cap V) \times F$ over $U \cap V$ is linear over every point of $U \cap V$, then we say that φ is *a vector bundle over Z with fiber F*.

The claim of Theorem 2 below for vector bundles is mentioned in [4, Example 2.1]:

Theorem 2 *Let $X \to Z$ be a locally trivial fibration over an irreducible variety Z with fiber $\mathbf{A}^n$. If Z is locally flattenable, then X is locally flattenable as well.*

Proof This follows from Definitions 1 and 2. □

We recall (see [29, 3.2], [26, 4.8]) a construction used several times below. Let G be a connected algebraic group, H its closed subgroup, and F a variety endowed with an action of H. Then H acts on $G \times F$ by the formula

$$h \cdot (g, f) \mapsto (gh^{-1}, h \cdot f).$$

By [29, Proposition 4], a mild restriction on F ensures the existence of the quotient of this action (in the sense of [5, 6.3]): namely, it exists if every finite subset of F lies in affine open subset of F (for instance, any quasiprojective F shares this property). The corresponding quotient variety is denoted by $G \times^H F$. The natural projection $G \times F \to G$ is H-equivariant and therefore induces the surjective morphism of the quotients $\pi_{G,H,F}\colon G \times^H F \to G/H$; its fibers are isomorphic to F. The action of G on $G \times F$ by left multiplications on the first factor commutes with the action of

H and therefore descends to the action of G on $G \times^H F$; the morphism $\pi_{G,H,F}$ is equivariant with respect to this action and the natural action of G on G/H. Given the aforesaid, $G \times^H F$ is called the *(algebraic) homogeneous fiber space over G/H with fiber F*. In general, $\pi_{G,H,F}$ is not a locally trivial fibration over G/H with fiber F (in the sense of Definition 2). However, if F is a vector space V over k and the action of H on V is linear, then [29, Theorem 2] implies that $\pi_{G,H,F}$ is a vector bundle over G/H with fiber V (in the sense of Definition 2).

Combining Corollary 1 and Theorem 2 yields

Corollary 2 *Maintain the above notation. If G/H is rational, then $G \times^H V$ is locally flattenable.*

Theorem 3 *Let G be a connected reductive algebraic group and let X be a smooth affine variety endowed with an action of G. Assume that*

(a) $k[X]^G = k$;
(b) the (*unique*, see, e.g., [26, Corollary of Theorem 4.7]) *closed G-orbit $\mathcal{O}$ in X is rational.*

Then X is locally flattenable.

Proof By [19, p. 98, Corollary 2] (see also [26, Theorem 6.7]), (a) and smoothness of X imply that X is G-equivariantly isomorphic to $G \times^H V$, where H is the G-stabilizer of a point of $\mathcal{O}$, and V is a finite-dimensional H-module. The claim then follows from Corollary 2. □

4.3 Spherical Varieties

Let G be a connected reductive algebraic group and let B be a Borel subgroup of G. Recall that a variety X endowed with an action of G is called *spherical variety of G* if there is a dense open B-orbit in X.

Theorem 4 *Every smooth spherical variety is locally flattenable.*

Proof Let X be a smooth spherical variety of a connected reductive group G.

First, X is rational because every B-orbit is rational (the latter is isomorphic to the complement of a union of several coordinate hyperplanes in some affine space [15, Cop. p. 5-02]).

Secondly, every G-orbit in X is spherical (see, e.g., [32, Proposition 15.14]), hence rational. Therefore, by Theorem 3, if X is affine, then X is locally flattenable.

Thirdly, arbitrary X is covered by open subsets, each of which is isomorphic to a variety of the form $P \times^L Z$, where P and L are respectively a parabolic subgroups of G and a Levi subgroup of P, and Z is an affine spherical variety of L; see, e.g., [32, Theorem 15.17]. Since X is smooth, Z is smooth as well. Therefore, as explained above, Z is locally flattenable. The variety $P \times^L Z$ is isomorphic to the product of Z and the underlying variety of the unipotent radical of P. Since this underlying variety is isomorphic to an affine space [15, Corollary p. 5-02], we infer that $P \times^L Z$ is locally flattenable. Therefore, X is locally flattenable, too. □

Since every toric variety is spherical, Theorem 4 implies

Corollary 3 ([4, Example 2.2]) *Every smooth toric variety is locally flattenable.*

4.4 Blow-Ups with Nonsingular Centers

Theorem 5 ([14, p. 885, Proposition], [3, Theorem 4.4], [4, Proposition 2.6]) *The blow-up of a locally flattenable variety along a smooth subvariety is locally flattenable.*

4.5 Curves and Surfaces

For varieties of dimension $\leqslant 2$, the answer to (Gr) is affirmative:

Theorem 6 ([3, Proposition 3.2], [4, Proposition 2.6]) *Every irreducible rational smooth algebraic curve or surface X is locally flattenable.*

Proof If X is a curve, it admits an open embedding in $\mathbf{P}^1$. If X is a surface, it admits an open embedding in a projective smooth surface, which, being rational, is obtained by repeated point blow-ups of a minimal model, i.e., either $\mathbf{P}^2$ or a Hirzebruch surface F_n, $n \neq 1$. Since $\mathbf{P}^1$, $\mathbf{P}^2$, and F_n are toric varieties, the claim follows from Corollary 3 and Theorem 5. □

5 Local Version of (ZCP)

Given Definition 1, the local version of the Zariski Cancellation Problem mentioned in Sect. 2 is formulated as follows:

Are there affine varieties X and Y such that Y and $X \times Y$ are flattenable, but X is not flattenable? (LZCP)

In Sect. 7 we show that the answer to (LZCP) is affirmative.

6 Flattenable Varieties Versus Locally Flattenable Varieties

Flattenable varieties have special properties:

Lemma 1 *Let X be an affine flattenable variety and let $\varphi\colon X \hookrightarrow \mathbf{A}^n$ be an open embedding. If $k[X]^* = k^*$, then $\varphi(X) = \mathbf{A}^n$.*

Proof Assume that the closed set $\mathbf{A}^n \setminus \varphi(X)$ is nonempty. Then, since X is affine, the dimension of every irreducible component of this set is $n-1$. Therefore, $\mathrm{Pic}(\mathbf{A}^n) = 0$ implies that $\mathbf{A}^n \setminus \varphi(X)$ is the set of zeros of a certain function $f \in k[\mathbf{A}^n]$. Then $f \circ \varphi$ is a nonconstant element of $k[X]^*$, a contradiction. Hence $\varphi(X) = \mathbf{A}^n$. □

Lemma 2 *For a connected affine algebraic group G, the following properties are equivalent:*

(a) *as a variety, G is isomorphic to an affine space;*
(b) *as a group, G is unipotent.*

Proof Assume that (a) holds. If G is not unipotent, there exists a nontrivial torus T among the closed subgroups of G. The action of T on G by left multiplication then gives a fixed point free action of T on an affine space, which is impossible by [2, Theorem 1]. This contradiction proves (a)⇒(b).

Conversely, (a) follows from (b) by [15, Corollary p. 5-02]. □

Theorem 7 *Let G be a connected affine algebraic group, and let $\mathcal{R}G$ be its radical.*

(a) *If G is solvable, then G is flattenable.*
(b) *If G is flattenable and nonsolvable, then $\mathcal{R}G$ is not unipotent.*

Proof Let G be solvable. Then G, as a variety, is isomorphic to the complement of a union of several coordinate hyperplanes in some affine space [15, Corollary p. 5-02]; whence (a).

Assume that G is flattenable and nonsolvable. The latter implies that G is not unipotent, hence, by Lemma 2, as a variety, G is not isomorphic to an affine space. Lemma 1 then implies that there is a nonconstant invertible function $f \in k[G]$. By [28, Theorem 3], the map $G \to \mathrm{GL}_1$, $g \mapsto f(g)/f(e)$, is then a nontrivial character. According to [23, Lemma 1.1], the existence of such a character is equivalent to the property that $\mathcal{R}G$ is not unipotent; whence (b). □

Corollary 4 *Let G be a nontrivial connected reductive algebraic group. If G is flattenable, then the dimension of its center is positive. In particular, every semisimple G is not flattenable.*

7 Answering (LZCP)

Theorem 8 *There are affine varieties X and Y such that*

(a) *X is not flattenable;*
(b) *Y and $X \times Y$ are flattenable.*

Proof As a variety, any SL_n for $n > 1$ is not flattenable by Corollary 4. On the other hand, being open in the affine space $\mathrm{Mat}_{m\times m}$, any GL_m is flattenable. The morphism

$$\mathrm{SL}_n \times \mathrm{GL}_1 \to \mathrm{GL}_n, \quad (s, a) \mapsto s\,\mathrm{diag}(a, 1, \dots, 1), \tag{1}$$

is an isomorphism of varieties: its inverse is

$$\mathrm{GL}_n \to \mathrm{SL}_n \times \mathrm{GL}_1, \quad g \mapsto \big(g\,\mathrm{diag}(1/\det(g), 1, \dots, 1), \det(g)\big).$$

Hence we can take $X = \mathrm{SL}_n$ for $n > 1$, and $Y = \mathrm{GL}_1$. □

In Remark 6 below one can find other examples.

8 Equivariantly Flattenable Varieties

Definition 3 A variety X endowed with an action of an algebraic group G is called *equivariantly* (respectively, *linearly equivariantly*) *flattenable* if there are

- an action (respectively, a linear action) of G on some $\mathbf{A}^n$;
- a G-equivariant open embedding $X \hookrightarrow \mathbf{A}^n$.

Definition 4 An algebraic group G is called *equivariantly* (respectively, *linearly equivariantly*) *flattenable* if G, as a variety endowed with the G-action by left multiplication, is *equivariantly* (respectively, *linearly equivariantly*) *flattenable*.

Example 1 Every GL_n is linearly equivariantly flattenable since GL_n is an invariant open set of $\mathrm{Mat}_{n\times n}$ endowed with the GL_n-action by left multiplication.

Example 2 Every (connected) unipotent affine algebraic group G is, as a variety, isomorphic to an affine space; hence G is equivariantly flattenable. In fact, a more general statement, Theorem 10 below, holds. It is easily seen that G is linearly equivariantly flattenable only if it is trivial. On the other hand, the example of a Borel subgroup of SL_2 naturally acting on k^2 shows that there are nontrivial solvable linearly equivariantly flattenable groups.

Example 3 If $G_1, \dots, G_m$ are equivariantly (respectively, linearly equivariantly) flattenable groups, then, clearly, $G_1 \times \cdots \times G_m$ is equivariantly (respectively, linearly equivariantly) flattenable as well. In particular, the group

$$\mathrm{GL}_{n_1} \times \cdots \times \mathrm{GL}_{n_s} \tag{2}$$

is linearly equivariantly flattenable for any $n_1, \dots, n_s$. Taking $n_1 = \ldots = n_s = 1$ yields that every affine algebraic torus is linearly equivariantly flattenable.

Example 4 Generalizing Example 1, let A be a finite-dimensional associative k-algebra with identity. The group A^* is a connected affine algebraic group. It is open in A and invariant with respect to the action of A^* on A by left multiplication, cf. [7, I, 1.6(9)]. Hence A^* is a linearly equivariantly flattenable group. For $A = \mathrm{Mat}_{n\times n}$, we obtain $A^* = \mathrm{GL}_n$. More generally, if A is semisimple, then A^* is a group of type (2), and all groups of type (2) are obtained in this way.

Example 5 Every $G = \mathrm{SL}_n \times \mathrm{GL}_1$ is equivariantly flattenable. Indeed, consider the G-module structure on $V = \mathrm{Mat}_{n\times n}$ defined by the formula

$$G \times V \to V, \quad ((s, a), x) \mapsto s\, x\, \mathrm{diag}(a, 1, \dots, 1).$$

For $x = \mathrm{diag}(1, \dots, 1)$, the orbit map $G \to V$, $g \mapsto g \cdot x$ is then the G-equivariant open embedding (1).

Lemma 3 *Let X be an algebraic variety endowed with an action of an algebraic group G.*

(a) If X^G is reducible, then X is not linearly equivariantly flattenable.
(b) If X is not linearly equivariantly flattenable, but equivariantly flattenable, then the action α of G on $\mathbf{A}^n$, extending that on X, is nonlinearizable.

Proof (a) Arguing on the contrary, assume that X is linearly equivariantly flattenable. Then there are a linear action of G on $\mathbf{A}^n$ and a G-equivariant open embedding $X \hookrightarrow \mathbf{A}^n$. We identify X with its image. Then $X^G = X \cap (\mathbf{A}^n)^G$. Since $(\mathbf{A}^n)^G$ is a linear subspace of $\mathbf{A}^n$, and X is open in $\mathbf{A}^n$, this implies that X^G is irreducible, contrary to the assumption. This proves (a).

(b) Arguing on the contrary, assume that α is linearizable, i.e., there is a linear action β of G on $\mathbf{A}^n$ and a G-equivariant automorphism $\varphi\colon \mathbf{A}^n \to \mathbf{A}^n$, where the left (resp. the right) $\mathbf{A}^n$ is endowed with the action α (resp., β). Then the equivariant open embedding of X in the left $\mathbf{A}^n$ composed with φ is an equivariant open embedding of X in the right $\mathbf{A}^n$. Hence X is linearly equivariantly flattenable, contrary to the assumption. This proves (b). □

Theorem 9 *The following properties of a connected reductive algebraic group G are equivalent:*

(a) G is equivariantly flattenable;
(b) G is linearly equivariantly flattenable.

Proof Let G be equivariantly flattenable. By Definitions 3, 4, we may (and shall) identify G with an open orbit in some $\mathbf{A}^n$ endowed with a regular action of G. Openness of this orbit implies $k[\mathbf{A}^n]^G = k$. Hence, by [19, p. 98, Corollary 2] (see also [26, Theorem 6.7]), there are a closed reductive subgroup L of G and a finite-dimensional algebraic L-module V such that $\mathbf{A}^n$ and $G \times^L V$ are G-equivariantly isomorphic. We claim that this implies

$$L = G. \tag{3}$$

If (3) is proved, then $\mathbf{A}^n$ and V are G-equivariantly isomorphic, which proves (a)⇒(b).

So it remains to prove (3). In view of connectedness of G, to this end it suffices to prove the equality $\dim(L) = \dim(G)$.

Since $\mathrm{char}(k) = 0$, by the Lefschetz principle we may (and shall) assume that $k = \mathbf{C}$; in the remainder of the proof topological terms are related to the Hausdorff $\mathbf{C}$-topology. Since $\mathbf{A}^n$ is simply connected, G/L is simply connected as well, hence L is connected.

We now note that the dimension any connected complex reductive algebraic group R is equal to the maximum m_R of i such that $H_i(R) \neq 0$ (singular homology with complex coefficients). Indeed, if K is a maximal compact subgroup of R, then the Iwasawa decomposition of R shows that R, as a manifold, is a product of K and a Euclidean space. Hence R and K have the same homology. Since K is a compact oriented manifold, this shows that m_R is equal to the dimension of the Lie group K. As R is the complexification of K, the statement follows.

So to prove (3) is the same as to prove $m_G = m_L$. In fact, since L is a subgroup of G, the above equality $m_R = \dim(R)$ yields $m_G \geqslant m_L$, so to prove $m_G = m_L$ we only need to prove the inequality

$$m_G \leqslant m_L. \tag{4}$$

As is known (see, e.g., [17, Chap. IX, Theorem 11.1]), the spectral sequence of the natural fiber bundle $G \to G/L$ yields the following inequality for the Betti numbers

$$\dim_{\mathbf{C}}(H_{m_G}(G)) \leqslant \sum_{i+j=m_G} \dim_{\mathbf{C}}(H_i(G/L)) \dim_{\mathbf{C}}(H_j(L)). \tag{5}$$

On the other hand, since $G \times^L V$ is a vector bundle over G/L and $G \times^L V$ is isomorphic to $\mathbf{A}^n$, we have

$$\dim_{\mathbf{C}}(H_i(G/L)) = \dim_{\mathbf{C}}(H_i(\mathbf{A}^n)) = \begin{cases} 1 & \text{for } i = 0, \\ 0 & \text{for } i > 0. \end{cases} \tag{6}$$

From (5), (6) we infer that

$$0 < \dim_{\mathbf{C}}(H_{m_G}(G)) \leqslant \dim_{\mathbf{C}}(H_{m_G}(L)). \tag{7}$$

The definition of m_L and (7) then yield (4). This completes the proof. □

Remark 2 Using the same argument, but (in the spirit of [6]) étale cohomology in place of singular homology, one can avoid applying the Lefschetz principle and adapt the above proof to the case of base field of arbitrary characteristic.

Theorem 10 *Let G be a connected affine algebraic group and let L be a Levi subgroup of G. If L is equivariantly flattenable, then so is G.*

Proof Let L be equivariantly flattenable. Then by Theorem 9, we may (and shall) assume that L, as a variety with the L-action by left multiplication, is an open orbit $\mathcal{O}$ of an algebraic L-module V. This implies that $\dim(L) = \dim(V)$, and therefore,

$$\dim(G \times^L V) = \dim(G) - \dim(L) + \dim(V) = \dim(G). \tag{8}$$

We identify V with the fiber of $G \times^L V \to G/L$ over the point of G/L corresponding to L. Since the L-stabilizer of any point $v \in \mathcal{O}$ is trivial, the G-stabilizer of v is trivial as well. From this and (8) we infer that $G \to G \times^L V$, $g \mapsto g \cdot v$, is a G-equivariant (with respect to the action of G on itself by left multiplication) open embedding. Now we note that G/L is isomorphic to the underlying variety of $\mathcal{R}_u G$, therefore, by Lemma 2, to an affine space. Since, by Quillen–Suslin [27, 31], algebraic vector bundles over affine spaces are trivial, we conclude that the variety $G \times^L V$ is isomorphic to an affine space. This completes the proof. □

Corollary 5 *Every connected solvable affine algebraic group is equivariantly flattenable.*

Proof Levi subgroups of connected solvable affine algebraic groups are tori. As the latter are equivariantly flattenable, the claim follows from Theorem 10. □

9 Equivariant Flattenability Versus Linear Equivariant Flattenability

The following shows that there are affine equivariantly flattenable varieties which are not linearly equivariantly flattenable.

Example 6 As is known (see, e.g., references in survey [18]), there are affine spaces endowed with nonlinearizable actions of reductive algebraic groups. So they are equivariantly flattenable, but, by Lemma 1, not linearly equivariantly flattenable.

Question 1 Are there flattenable reductive algebraic groups, which are not equivariantly flattenable?

10 Equivariant Flattenability Versus Linearizability

In this section, we construct a series of affine flattenable varieties endowed with actions of finite groups, which are not linearly equivariantly flattenable.

The construction is as follows.

Take a pair S, G, where

(c_1) S is a connected semisimple algebraic group;
(c_2) G is a finite subgroup of S such that $\mathcal{Z}_S(G)$ (the centralizer of G in S) is finite and nontrivial.

Note that such pairs S, G do exist.

Example 7 Let F be a nontrivial finite group, satisfying the conditions:

(i) there are no nontrivial characters $F \to \mathrm{GL}_1$;
(ii) there is a faithful irreducible representation $\varphi\colon F \to \mathrm{GL}_n$ for some n.

For instance, (i) and (ii) hold for any nontrivial simple F. By (i), we have $\varphi(F) \subset \mathrm{SL}_n$, and by (ii) we may (and shall) identify F with $\varphi(F)$; thus F is a nontrivial subgroup of SL_n whose natural linear action on k^n is irreducible. By Schur's lemma, $\mathcal{Z}_{\mathrm{SL}_n}(F)$ is the cyclic group $\{\varepsilon I_n \mid \varepsilon \in k^*, \varepsilon^n = 1\}$ of order n. Since F is nontrivial, $n \geqslant 2$. Hence $S = \mathrm{SL}_n$, $G = F$ is the pair of interest.

Example 8 (*See* [5, 13, 30]) Let S be a connected semisimple algebraic group. Let T and $N_S(T)$ be resp. a maximal torus of S and its normalizer in S. Assume that the Weyl group $N_S(T)/T$ contains -1; if S is simple, this is equivalent to the condition that S is *not* of type A_ℓ for $\ell \geqslant 2$, D_ℓ for odd ℓ, or E_6 (see [8, Tables I–IX]). Let n be an element of $N_S(T)$ representing -1 of $N_S(T)/T$. Then either $n^2 = 1$ (this is so if S is adjoint) or n^2 is an element of order 2 of the center of S. Assume that $n^2 = 1$ and let G be the subgroup of S generated by n and all the elements of T whose order is 2. Then G is an elementary Abelian group of order $2^{\mathrm{rk}(S)+1}$, for which condition (c_2) holds.

We return back to the description of the construction. By (c_2), there is a nonidentity element $z \in \mathcal{Z}_S(G)$, and its order is finite. In view of char $(k) = 0$, the finiteness of the order entails that z is semisimple. Hence (see [7, 11.10]) the element z lies is a maximal torus of S. Since, in turn, any torus of S lies in a Borel subgroup of S (cf. [7, 11.3]), we infer that there is a Borel subgroup B of S containing two different elements of $\mathcal{Z}_S(G)$:

$$1, z \in B \cap \mathcal{Z}_S(G), \quad z \neq 1. \tag{9}$$

Let B^- be the Borel subgroup of S opposite to B. Then the "big cell" $\Theta := B^-B$ is an open subset of S isomorphic to the complement of a union of several coordinate hyperplanes in $L := \mathbf{A}^{\dim(S)}$; in particular, Θ is an affine flattenable variety. In view of (9), we have

$$1, z \in \Theta \cap \mathcal{Z}_S(G), \quad z \neq 1. \tag{10}$$

Now we consider the conjugating action of G on S. Its fixed point set S^G is $\mathcal{Z}_S(G)$. The variety

$$X := \bigcap_{g \in G} g\Theta g^{-1} \tag{11}$$

is a G-stable open subset of S. Since Θ is an affine flattenable variety, X is such a variety, too. From (10), (11), and the finiteness of $\mathcal{Z}_S(G)$ we conclude that

$$X^G = X \cap S^G \text{ is a finite set containing two different points 1 and } z. \tag{12}$$

Proposition 1 *The affine flattenable G-variety X defined by formula* (11) *shares the following properties:*

(i) X is not linearly equivariantly flattenable.
(ii) The alternative holds:

- *either X is not equivariantly flattenable,*
- *or X is equivariantly flattenable, but the action of G on an affine space, extending that on X, is nonlinearizable.*

Proof In view of (12), the variety X^G is a reducible. By Lemma 3(a), this implies (i). In turn, (i) and Lemma 3(b) imply (ii). □

Thus there are two a priori possibilities: our construction yields either an example of an action of a finite group on a flattenable variety, which is not equivariantly flattenable, or an example of a nonlinearizable action of a finite group on an affine space. The linearization problem for reductive group actions on affine spaces has received much attention in the literature, but, to the best of my knowledge, at this writing (October 2019) the existence of nonlinearizable actions of finite Abelian groups on affine spaces remains an unsolved problem. Being linked with this subject, our construction leads to the intriguing questions:

Question 2 Is each of these two a priori possibilities realizable for an appropriate pair S, G?

Question 3 Take $S = \mathrm{PGL}_2$, and let G be the group from Example 8. This G is the Klein four-group, it is isomorphic to $\mathbb{Z}/2 \oplus \mathbb{Z}/2$. The variety X is isomorphic to an affine open subset of $\mathbf{A}^3$. Which of the two a priori possibilities indicated above for the action of G on X is realized in this case?

11 Equivariantly Flattenable Subgroups of the Cremona Groups

As an application, below is briefly surveyed a special role of equivariantly flattenable subgroups in the conjugacy problem for algebraic subgroups of the Cremona groups Cr_n (i.e., in the rational linearization problem). We refer to [24, 26] and references therein regarding the basic definitions and properties of rational algebraic group actions and, in particular, the definition of embeddings

$$\mathrm{Cr}_1 \subset \mathrm{Cr}_2 \subset \cdots \subset \mathrm{Cr}_n \subset \mathrm{Cr}_{n+1} \subset \cdots . \tag{13}$$

Theorem 11 ([24, Theorem 2.1]) *Let G be a connected algebraic subgroup of the Cremona group Cr_n. Assume that*

(i) G is linearly equivariantly flattenable;
(ii) the natural rational action of G on $\mathbf{A}^n$ is locally free.

If the field extension $k(\mathbf{A}^n)^G/k$ is purely transcendental, then G is conjugate in Cr_n to a subgroup of GL_n (i.e., the natural rational action of G on $\mathbf{A}^n$ is rationally linearizable).

For tori, the sufficient condition of Theorem 11 for rational linearization is also necessary:

Theorem 12 ([24, Theorem 2.4]) *Let T be an affine algebraic torus in the Cremona group* Cr_n. *The following properties are equivalent:*

(a) *T is conjugate in* Cr_n *to a subgroup of* GL_n;
(b) *the field extension* $k(\mathbf{A}^n)^T/k$ *is purely transcendental.*

The sufficient condition of Theorem 11 for rational linearization always holds true in the stable range:

Theorem 13 ([24, Lemma 2.2 and Theorem 2.2]) *Let G be a connected algebraic subgroup of the Cremona group* Cr_n *such that assumptions* (i), (ii) *of Theorem* 11 *hold. Then there is an integer* $s \geqslant 0$ *such that, for the rational action of G on* $\mathbf{A}^{n+s}$ *determined by embedding* (13), *the field extension* $k(\mathbf{A}^{n+s})^G/k$ *is purely transcendental.*

Remark 3 By [24, Theorem 2.6], if G in Theorem 13 is a torus, then one can take $s = \dim(G)$.

Remark 4 In general, s in Theorem 13 is strictly positive. For example, by [24, Corollary 2.5], the Cremona group Cr_n for $n \geqslant 5$ contains an $(n-3)$-dimensional affine algebraic torus, which is not conjugate in Cr_n to a subgroup of GL_n.

12 Equivariantly Flattenable Groups and Special Groups in the Sense of Serre

Recall from [29, 4.1] that an algebraic group G is called *special* if every principal G-bundle (which, by definition [29, 2.2], is locally trivial in étale topology) is locally trivial in the Zariski topology. By [29, Sect. 4.1, Theorem 1] special group is automatically connected and affine. Special groups are classified:

Theorem 14 *The following properties of a connected affine algebraic group G are equivalent:*

(a) *G is special;*
(b) *maximal connected semisimple subgroups of G are isomorphic to a group of type*

$$\mathrm{SL}_{n_1} \times \cdots \times \mathrm{SL}_{n_r} \times \mathrm{Sp}_{2m_1} \times \cdots \times \mathrm{Sp}_{2m_s}.$$

Proof The implications (a)⇒(b) and (b)⇒(a) are proved respectively in [15] and [29]. □

In [24, Lemma 2.2] is sketched a reduction of the following claim to [22, Theorem 1.4.3]. Below is given the complete self-contained argument.

Theorem 15 *Every linearly equivariantly flattenable affine algebraic group G is special.*

Proof By Definitions 3, 4, there is a finite-dimensional algebraic G-module V with a G-orbit $\mathcal{O}$ such that

(i) $\mathcal{O}$ is open in V;
(ii) the points of $\mathcal{O}$ have trivial G-stabilizers.

By (ii), V is faithful; hence we may (and shall) identify G with a closed subgroup of $\mathrm{GL}(V)$.

Next, consider the ith direct summand of

$$L := V \oplus \cdots \oplus V \quad (n := \dim(V) \text{ copies}),$$

as a linear subspace V_i of L, denote by π_i the natural projection $L \to V_i$, and identify V with V_1. The $\mathrm{GL}(V)$-module L contains a $\mathrm{GL}(V)$-orbit $\widetilde{\mathcal{O}}$ such that

(iii) $\widetilde{\mathcal{O}}$ is open in L;
(iv) the points of $\widetilde{\mathcal{O}}$ have trivial $\mathrm{GL}(V)$-stabilizers.

From (i), (iii) we infer that $\mathcal{O} \cap \pi_1(\widetilde{\mathcal{O}}) \neq \varnothing$ and $\pi_1^{-1}(\mathcal{O}) \cap \widetilde{\mathcal{O}}$ is an nonempty open subset of $\widetilde{\mathcal{O}}$. Take a point

$$a \in \mathcal{O} \cap \pi(\widetilde{\mathcal{O}}) \tag{14}$$

and consider in L the affine subspace

$$A := \{a + v_2 + \cdots + v_n \in L \mid v_i \in V_i \text{ for all } i\}. \tag{15}$$

From (iii), (14), (15) we deduce that $A \cap \widetilde{\mathcal{O}}$ is a nonempty open subset of A, and from (ii) that the G-orbit of every point of $\pi_1^{-1}(\mathcal{O}) \cap \widetilde{\mathcal{O}}$ intersects $A \cap \widetilde{\mathcal{O}}$ at a single point. This means that the natural action of G on $\widetilde{\mathcal{O}}$ admits a rational section. In view of (iv), this, in turn, means that the natural map $\mathrm{GL}(V) \to \mathrm{GL}(V)/G$ admits a rational section. Since the group $\mathrm{GL}(V)$ is special, this implies, according to [29, Sect. 4.3, Theorem 2], that the group G is special as well. □

Remark 5 Since A is rational and $A \cap \widetilde{\mathcal{O}}$ is open in A, the above proof of Theorem 15 shows that $\mathrm{GL}(V)/G$ is a rational variety.

13 Classifying Equivariantly Flattenable Groups

Theorem 10 naturally leads to the following

Problem 1 Obtain a classification of equivariantly flattenable reductive algebraic groups.

Theorem 16 *For every nontrivial equivariantly flattenable reductive algebraic group G, the following properties hold:*

(a) the derived group L of G is a semisimple group of type

$$\mathrm{SL}_{n_1} \times \cdots \times \mathrm{SL}_{n_r} \times \mathrm{Sp}_{2m_1} \times \cdots \times \mathrm{Sp}_{2m_s};$$

(b) the radical of G is a central torus C of positive dimension;
(c) $G = L \cdot C$ and $C \cap L$ is finite.

Proof Since L is a maximal connected semisimple subgroup of G, combining Theorems 15 and 14 yields (a). Combining Corollary 4 and [7, 14.2, Proposition] implies (b) and (c). □

Problem 1 looks manageable. Initially, being influenced by Example 1.4, the author was even overoptimistic and put forward the conjecture that all equivariantly flattenable reductive algebraic groups are that of type (2) (see [24, p. 221]); this overoptimism was shared by some of the participants of the 2013 Oberwolfach meeting on algebraic groups who even sketched a plan of possible proof. So it came as a surprise, when in [9] the examples of equivariantly flattenable reductive groups of a type distinct from (2) have been revealed, making Problem 1 even more intriguing. Below we describe them.

Theorem 17 ([9, 2.1]) *For every positive integer n, the group*

$$G := \mathrm{Sp}_{2n} \times \mathrm{GL}_{2n-1} \times \mathrm{GL}_{2n-2} \times \cdots \times \mathrm{GL}_1 \tag{16}$$

is equivariantly flattenable.

Proof *(Sketch of proof)* Consider the vector space

$$V := \mathrm{Mat}_{2n\times 1} \oplus \mathrm{Mat}_{2n\times(2n-1)} \oplus \mathrm{Mat}_{(2n-1)\times(2n-2)} \oplus \cdots \oplus \mathrm{Mat}_{2\times 1} \tag{17}$$

and the linear action of G on V defined for the elements

$$\begin{aligned} g &:= (A, B_{2n-1}, B_{2n-2}, \ldots, B_2, B_1) \in G,\ A \in \mathrm{Sp}_n,\ B_d \in \mathrm{GL}_d,\\ v &:= (X, Y_{2n-1}, Y_{2n-2}, \ldots, Y_1) \in V,\ X \in \mathrm{Mat}_{2n\times 1},\ Y_d \in \mathrm{Mat}_{(d+1)\times d} \end{aligned}$$

by the formula

$$g \cdot v := (AX, AY_{2n-1}B_{2n-1}^{\top}, B_{2n-1}Y_{2n-2}B_{2n-2}^{\top}, \ldots, B_2Y_1B_1^{\top}).$$

From (16) and (17) we deduce

$$\dim(G) \overset{(16)}{=} 2n^2 + n + \sum_{i=1}^{2n-1} i^2 = 2n + \sum_{i=1}^{2n-1} i(i+1) \overset{(17)}{=} \dim(V). \tag{18}$$

Next, one shows the existence of a point $v_0 \in V$ whose G-stabilizer is trivial. In view of (18), the orbit map $G \to V$, $g \mapsto g \cdot v_0$ is then a G-equivariant open embedding. □

Remark 6 Since the group Sp_{2n} is not flattenable by Corollary 4, but $\mathrm{GL}_{2n-1} \times \mathrm{GL}_{2n-2} \times \cdots \times \mathrm{GL}_1$ is flattenable, Theorem 17 provides other (than that in the proof of Theorem 8) examples, which yield the affirmative answer to (LZCP).

14 Locally Equivariantly Flattenable Varieties

Similarly to flattenability, equivariant flattenability admits an evident local version:

Definition 5 ([21, Definition 4(iii)], *up to change of terminology*) A variety X endowed with an action of an algebraic group G is called *equivariantly* (respectively, *linearly equivariantly*) *locally flattenable* if for every point $x \in X$ there is an equivariantly (respectively, linearly equivariantly) flattenable G-stable open subset of X containing x.

Definition 5 leads to the following equivariant version of Gromov's question (Gr):

Is every irreducible smooth rational G-variety equivariantly locally flattenable? (EqGr)

The examples in [21, Sect. 4] show that the answer to (EqGr) is negative.

Acknowledgements I am grateful to the referee for thoughtful reading and suggestions.

References

1. Akbulut, S.: Lectures on algebraic spaces. In: 1992 Proceedings of KAIST Mathematics Workshop on Algebra and Topology, pp.1–15. KAIST, Daejeon, Republic of Korea (1993)
2. Białynicki-Birula, A.: Remarks on the action of an algebraic torus on k^n. Bull. Acad. Polon. Sci, Sér. sci. math., astr., phys. **XIV**(4), 177–181 (1966)
3. Bodnár, G., Hauser, H., Schicho, J., Villamayor U.O.: Plain varieties. Bull. Lond. Math. Soc. **40**(6), 965–971 (2008)
4. Bogomolov, F., Böhning, C.: On uniformly rational varieties. In: Topology, Geometry, Integrable Systems, and Mathematical Physics. American Mathematical Society Translations: Series 2, vol. 234, pp. 33–48. Advances in the Mathematical Sciences 67. American Mathematical Society, Providence, RI (2014)
5. Borel, A.: Sous-groupes commutatifs et torsion des groupes de Lie compacts connexes. Tôhoku Math. J. **13**, 216–240 (1961)
6. Borel, A.: On affine algebraic homogeneous spaces. Arch. Math. **45**, 74–78 (1985)
7. Borel, A.: Linear Algebraic Groups. Graduate Text in Mathematics, vol. 126, 2nd edn. Springer, New York (1991)

8. Bourbaki, N.: Groupes et Algèbres de Lie. Chapter IV, V, VI. Hermann (1968)
9. Burde, D., Globke, W., Minchenko, A.: Étale representations for reductive algebraic groups with factors Sp_n or SO_n. Transform. Groups **24**, 769–780 (2019). https://doi.org/10.1007/s00031-018-9483-8
10. Chevalley, C.: On algebraic group varieties. J. Math. Soc. Japan **6**(3–4), 303–324 (1954)
11. Chevalley, C.: Les classes d'équivalence rationnelle, I. In: Anneaux de Chow et Applications. Séminaire Claude Chevalley, Exp. no. 2, pp. 1–14 (1958)
12. Chin, C., Zhang, D.-Q.: Rationality of homogeneous varieties. Trans. Am. Math. Soc. **369**, 2651–2673 (2017)
13. Cohen, A.M., Seitz, G.M.: The r-rank of the groups of exceptional Lie type. Indag. Math. **49**, 251–259 (1987)
14. Gromov, M.: Oka's principle for holomorphic sections of elliptic bundles. J. Am. Math. Soc. **2**(4), 851–897 (1989)
15. Grothendieck, A.: Torsion homologique et sections rationnelle. In: Anneaux de Chow et Applications, Séminaire Claude Chevalley, vol. 3, Exp. no. 5, pp. 1–29. Secrétariat math., Paris (1958)
16. Gupta, N.: A survey on Zariski cancellation problem. Indian J. Pure Appl. Math. **46**(6), 865–877 (2015)
17. Hu, S.-T.: Homotopy Theory. Academic Press, New York (1959)
18. Kraft, H.: Challenging problems in affine n-space. Astérisque **47**(802), 295–317 (1996)
19. Luna, D.: Slices étales. Mém. SMF **33**, 81–105 (1973)
20. Manin, Y.: Cubic Forms. North-Holland, Amsterdam (1974)
21. Petitjean, C.: Equivariantly uniformly rational varieties. Michugan Math. J. **66**(2), 245–268 (2017)
22. Popov, V.L.: Sections in invariant theory. In: Proceedings of the Sophus Lie Memorial Conference, Oslo 1992, pp. 315–362. Scandinavian University Press, Oslo (1994)
23. Popov, V.L.: On the Makar-Limanov, Derksen invariants, and finite automorphism groups of algebraic varieties. In: Affine Algebraic Geometry: The Russell Festschrift, CRM Proceedings and Lecture Notes, vol. 54, pp. 289–311. American Mathematical Society (2011)
24. Popov, V.L.: Some subgroups of the Cremona groups. In: Affine Algebraic Geometry, Proceedings (Osaka, Japan, 3–6 March 2011), pp. 213–242. World Scientific, Singapore (2013)
25. Popov, V.L.: Rationality and the FML invariant. J. Ramanujan Math. Soc. **28A**, 409–415 (2013)
26. Popov, V.L., Vinberg, E.B.: Invariant theory. In: Algebraic Geometry, IV, Encyclopaedia of Mathematical Sciences, vol. 55, pp. 123–284. Springer, Berlin (1994)
27. Quillen, D.: Projective modules over polynomial rings. Invent. Math. **36**(1), 167–171 (1976)
28. Rosenlicht, M.: Toroidal algebraic groups. Proc. Am. Math. Soc. **12**, 984–988 (1961)
29. Serre, J.-P.: Espaces fibrés algébriques. In: Anneaux de Chow et Applications, Séminaire Claude Chevalley, vol. 3, Exp. no. 1, pp. 1–37. Secrétariat mathématique, Paris (1958)
30. Serre, J.-P.: Sous-groupes finis des groupes de Lie. Séminaire Bourbaki, Exp. no. 864 (1998–99). In: Serre, J.-P. (ed.) Exposés de Séminaires 1950–1999, pp. 233–248. Documents Mathématiques, Soc. Math. de France, Paris (2001)
31. Suslin, A.A.: Projective modules over polynomial rings are free. Soviet Math. **17**(4), 1160–1164 (1976)
32. Timashev, D.A.: Homogeneous Spaces and Equivariant Embeddings. Encyclopaedia of Mathematikcal Sciences, vol. 138. Subseries Invariant Theory and Algebraic Transformation Groups, vol. VIII. Springer, Berlin (2011)

Tango Structures on Curves in Characteristic 2

Yoshifumi Takeda

Abstract The (pre-)Tango structure is a certain ample invertible sheaf of exact differential 1-forms on a projective algebraic variety and it implies some typical pathological phenomena in positive characteristic. Moreover, by using the notion of (pre-)Tango structure, we can construct another variety accompanied by similar pathological phenomena. In this article, we explicitly show several interesting and mysterious phenomena on the induced uniruled surfaces from (pre-)Tango structures on curves in characteristic 2.

Keywords Characteristic 2 · Pathological phenomenon · Tango structure · Non-closed differential form · Non-reduced automorphism group scheme

1 Introduction

Let X be a projective algebraic variety over an algebraically closed field k of characteristic $p > 0$ and let $F_X : \widetilde{X} \to X$ be the relative Frobenius morphism over k. We then have a short exact sequence

$$0 \to \mathcal{O}_X \to F_{X*}\mathcal{O}_{\widetilde{X}} \to F_{X*}\mathcal{B}^1_{\widetilde{X}} \to 0,$$

where $\mathcal{B}^1_{\widetilde{X}}$ is the sheaf of first coboundaries of the de Rham complex $\Omega^\bullet_{\widetilde{X}}$ of $\widetilde{X}$. Suppose that there exists an ample invertible subsheaf $\mathcal{L}$ of $F_{X*}\mathcal{B}^1_{\widetilde{X}}$ provided that $F_{X*}\mathcal{B}^1_{\widetilde{X}}$ is regarded as an $\mathcal{O}_X$-module. We call $\mathcal{L}$ a *pre-Tango structure* (see the next section and see also Tango [18], Mukai [6], Takeda-Yokogawa [17]). Let us consider the exact sequence

$$0 \to \mathcal{L}^{-1} \to F_{X*}\mathcal{O}_{\widetilde{X}} \otimes_{\mathcal{O}_X} \mathcal{L}^{-1} \to F_{X*}\mathcal{B}^1_{\widetilde{X}} \otimes_{\mathcal{O}_X} \mathcal{L}^{-1} \to 0.$$

Y. Takeda (✉)
Department of Mathematics and Statistics, Wakayama Medical University,
Wakayama 641-8509, Japan
e-mail: ytakeda@wakayama-med.ac.jp

S. Kuroda et al. (eds.), *Polynomial Rings and Affine Algebraic Geometry*,
Springer Proceedings in Mathematics & Statistics 319,
https://doi.org/10.1007/978-3-030-42136-6_12

Taking their cohomologies, we have

$$0 \to H^0(X, \mathcal{L}^{-1}) \to H^0(X, F_{X*}\mathcal{O}_{\tilde{X}} \otimes_{\mathcal{O}_X} \mathcal{L}^{-1}) \to H^0(X, F_{X*}\mathcal{B}^1_{\tilde{X}} \otimes_{\mathcal{O}_X} \mathcal{L}^{-1})$$
$$\to H^1(X, \mathcal{L}^{-1}) \to \cdots$$

Since $H^0(X, F_{X*}\mathcal{O}_{\tilde{X}} \otimes_{\mathcal{O}_X} \mathcal{L}^{-1}) = 0$ and $H^0(X, F_{X*}\mathcal{B}^1_{\tilde{X}} \otimes_{\mathcal{O}_X} \mathcal{L}^{-1}) \neq 0$, we know that $H^1(X, \mathcal{L}^{-1}) \neq 0$. Hence, if X is a smooth variety of dimension greater than one, then the pair $(X, \mathcal{L})$ is a counter-example to the Kodaira vanishing theorem in positive characteristic. It is, however, hard to find such a pair in dimension greater than one.

On the other hand, regarding in dimension one, we know that almost all non-singular projective curves have pre-Tango structures (see [17]). In fact, Raynaud's famous counter-example in [8] is an uniruled surface constructed by using a Tango structure (i.e. very suitable pre-Tango structure) on a curve (see also Kurkr [3], Lang [4, 5], Russell [10], Takeda [11]).

2 Tango Structures and Pre-Tango Structures on Curves

From now on, suppose that k is an algebraically closed field of characteristic $p = 2$, C is a non-singular projective algebraic curve over k of genus $g > 1$, and $F : \tilde{C} \to C$ is the relative Frobenius morphism over k. We then have a short exact sequence

$$0 \to \mathcal{O}_C \to F_*\mathcal{O}_{\tilde{C}} \to F_*\mathcal{B}^1_{\tilde{C}} \to 0,$$

where $\mathcal{B}^1_{\tilde{C}}$ is the sheaf of first coboundaries of the de Rham complex $\Omega^\bullet_{\tilde{C}}$ of $\tilde{C}$. Moreover, we know that $F_*\mathcal{O}_{\tilde{C}}$ is a locally free $\mathcal{O}_C$-module of rank 2 and $F_*\mathcal{B}^1_{\tilde{C}}$ is an invertible sheaf of degree $g - 1$ since $\chi(F_*\mathcal{B}^1_{\tilde{C}}) = \chi(F_*\mathcal{O}_{\tilde{C}}) - \chi(\mathcal{O}_C) = 0$.

Definition We say an ample invertible subsheaf $\mathcal{L} \subset F_*\mathcal{B}^1_{\tilde{C}}$ (regarded as $\mathcal{O}_C$-modules) a *pre-Tango structure*. Furthermore, if $\mathcal{L} = F_*\mathcal{B}^1_{\tilde{C}}$, then we say $\mathcal{L}$ is a *Tango structure*.

Suppose that $\mathcal{L} \subset F_*\mathcal{B}^1_{\tilde{C}}$ is a pre-Tango structure. We then have its adjoint inclusion $F^*\mathcal{L} \subset \mathcal{B}^1_{\tilde{C}}$ and thus $F^*\mathcal{L} \subset \Omega^1_{\tilde{C}}$. Hence, if $\mathcal{L}$ is a Tango structure, i.e. $\mathcal{L} = F_*\mathcal{B}^1_{\tilde{C}}$, then we know that $F^*\mathcal{L} = \Omega^1_{\tilde{C}}$ by comparing their degrees, and $\mathcal{L}^2 \cong \Omega^1_C \cong \omega_C$.

Suppose that $\mathcal{L} \subset F_*\mathcal{B}^1_{\tilde{C}}$ is a pre-Tango structure, once again. We then have

$$\begin{array}{ccccccccc} 0 \to & \mathcal{O}_C & \to & F_*\mathcal{O}_{\tilde{C}} & \to & F_*\mathcal{B}^1_{\tilde{C}} & \to & 0 \\ & \parallel & & \cup & & \cup & & \\ 0 \to & \mathcal{O}_C & \to & \mathcal{E} & \to & \mathcal{L} & \to & 0, \end{array}$$

where $\mathcal{E}$ is the inverse image of $\mathcal{L}$ in $F_*\mathcal{O}_{\tilde{C}}$. We know that rank $\mathcal{E} = 2$, $\deg \mathcal{E} = \deg \mathcal{L} > 0$ and $\mathcal{E}$ is *stable*. Indeed, if there exists a positive invertible subsheaf $\mathcal{M} \subset \mathcal{E}$, then $\mathcal{M} \subset F_*\mathcal{O}_{\tilde{C}}$ and thus $\mathcal{M}^p \subset \mathcal{O}_C$, this is a contradiction. However, by considering that the pull-back $F^*\mathcal{E}$ splits into $\mathcal{O}_{\tilde{C}} \oplus F^*\mathcal{L}$, we know that it is *unstable*.

Suppose that $\mathcal{L} = F_*\mathcal{B}^1_{\tilde{C}}$ is a Tango structure. In addition, if there exists an invertible sheaf $\mathcal{N}$ such that $\mathcal{N}^n \cong \mathcal{L}$ with an odd number $n > 2$, then we obtain a *uniruled surface* accompanied by a *counter-example* to the Kodaira Vanishing Theorem, by a certain construction (mentioned in the next section) similar to the method of Raynaud.

3 The Induced Uniruled Surfaces

Let $\mathcal{L}$ be a pre-Tango structures on C and suppose that there exists an invertible sheaf $\mathcal{N}$ such that $\mathcal{N}^n \cong \mathcal{L}$ with an odd number $n > 2$.[1] Recall the diagram

$$\begin{array}{ccccccc} 0 \to \mathcal{O}_C \to & F_*\mathcal{O}_{\tilde{C}} & \xrightarrow{F_*d} & F_*\mathcal{B}^1_{\tilde{C}} & \to 0 \\ \| & \cup & & \cup & \\ 0 \to \mathcal{O}_C \to & \mathcal{E} & \longrightarrow & \mathcal{L} & \to 0. \end{array}$$

Take an affine open covering $\{U_i\}_{i \in I}$ of C and local functions $\{q_i\}$ on $\tilde{C}$ such that $\{dq_i\}$ generate $\mathcal{L}$, i.e.

$$\begin{array}{ccl} \mathcal{O}_{\tilde{C}} & \xrightarrow{d} & \mathcal{B}^1_{\tilde{C}} \\ q_i & \mapsto & dq_i \text{ generating locally } \mathcal{L} \text{ on } C. \end{array}$$

Then there exist local functions $\{b_{ij}\}$ on C and transition functions $\{d_{ij}\}$ of $\mathcal{N}$ such that

$$q_i = d_{ij}^n q_j + b_{ij}.$$

Set $a_i = q_i^2$. We then have

$$a_i = d_{ij}^{2n} a_j + b_{ij}^2$$

and a_i is a local function on C.

Let us consider schemes

$$\begin{aligned} &\operatorname{Spec} \mathcal{O}_C(U_i)[x_i, y_i]/(y_i^2 + x_i + a_i x_i^{n+1}), \\ &\operatorname{Spec} \mathcal{O}_C(U_i)[u_i, v_i]/(v_i^2 + u_i^n + a_i). \end{aligned}$$

By glueing these schemes together under the relations

---

[1] Here, note the *divisibility* of the Picard variety.

$$u_i = d_{ij}^2 u_j, \quad v_i = d_{ij}^n v_j + b_{ij}, \quad x_i = u_i^{-1}, \quad y_i = v_i x_i^{\frac{n+1}{2}},$$

we obtain a normal uniruled surface X with a fibration $\varphi : X \to C$ such that

The fibration φ has a section Γ defined locally by $y_i = 0$;

The general fibre is a rational curve with one cusp of type $u^n + v^2 = 0$;

The locus Σ of moving singularities is defined locally by $u_i = 0$.

(For details, see Takeda [15].)

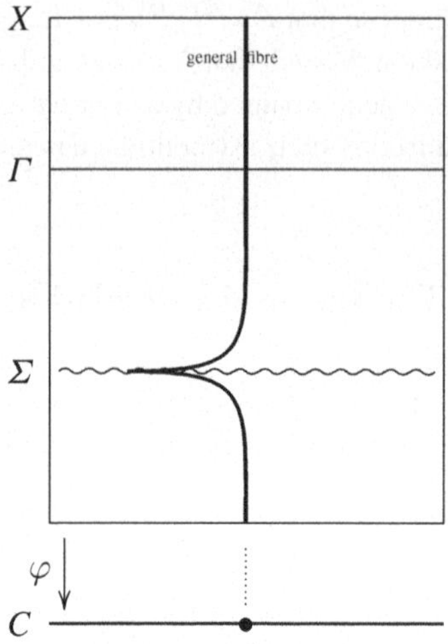

It is easy to verify that the singular points of X are lying on Σ, and over the support of the cokernel of the injection $\mathcal{L} \to F_*\mathcal{B}^1_{\tilde{C}}$. Therefore, if $\mathcal{L}$ is a Tango structure, then the surface X is smooth. Furthermore, considering an invertible sheaf $\mathcal{O}_X((n-1)\Gamma) \otimes_{\mathcal{O}_X} \varphi^*\mathcal{N}$ on X, we can verify that it is a Tango structure (see [6, 8]. See also [15]). Therefore, it gives a *counter-example* to the Kodaira Vanishing Theorem. In addition, X is *not liftable* over the ring of Witt-vectors of length 2 and its Hodge-de Rham spectral sequence

$$E_1^{ij} = H^j(X, \Omega^i_{X/k}) \implies H^{i+j}_{DR}(X/k) = \mathbb{H}^{i+j}(X, \Omega^\bullet_{X/k})$$

does not degenerate at E_1(cf. Deligne-Illusie [1]).

4 Differential Forms

Let X be the previous uniruled surface induced from a pre-Tango structure $\mathcal{N}^n$ on C. The local equation $v_i^2 + u_i^n + a_i = 0$ implies an equation of differential forms $u_i^{n-1} du_i = da_i$, and thus

$$du_i = x_i^{n-1} da_i.$$

Note that du_i is regular near the locus of moving singularities Σ and that $x_i^{n-1} da_i$ is regular near the section Γ. Hence we know that the previous differential form is regular on $X|_{U_i}$.

Similarly, we know that the differential form

$$v_i du_i = y_i x_i^{n-1-\frac{n+1}{2}} da_i$$

is regular on $X|_{U_i}$. Moreover, we know that the differential forms

$$\begin{array}{lll} du_i, & \ldots, & u_i^{\frac{n+1}{2}-2} du_i, \\ v_i du_i, & \ldots, & v_i u_i^{\frac{n+1}{2}-2} du_i \end{array}$$

are regular on $X|_{U_i}$.

Now consider those regular forms in the direct image $\varphi_* \Omega^1_{X/k}$ on C. We have an invertible sheaf in $\varphi_* \Omega^1_{X/k}$, which is generated locally by the images of du_i's, and which is isomorphic to $\mathcal{N}^2$ since $du_i = d_{ij}^2 du_j$. Furthermore, we have a locally free $\mathcal{O}_C$-module of rank 2 in $\varphi_* \Omega^1_{X/k}$, which is generated locally by the images of du_i's and $v_i du_i$'s, and which is isomorphic to $\mathcal{E} \otimes \mathcal{N}^2$ since $v_i = d_{ij}^n v_j + b_{ij}$.[2] Therefore, we obtain an extension on C, which is isomorphic to

$$0 \to \mathcal{N}^2 \to \mathcal{E} \otimes \mathcal{N}^2 \to \mathcal{N}^{n+2} \to 0.$$

In the same fashion, we have extensions, which are isomorphic respectively to

$$\begin{gathered} 0 \to \mathcal{N}^2 \to \mathcal{E} \otimes \mathcal{N}^2 \to \mathcal{N}^{n+2} \to 0, \\ 0 \to \mathcal{N}^4 \to \mathcal{E} \otimes \mathcal{N}^4 \to \mathcal{N}^{n+4} \to 0, \\ \vdots \\ 0 \to \mathcal{N}^{2(\frac{n+1}{2}-1)} \to \mathcal{E} \otimes \mathcal{N}^{2(\frac{n+1}{2}-1)} \to \mathcal{N}^{n+2(\frac{n+1}{2}-1)} \to 0. \\ \left(\text{i.e. } 0 \to \mathcal{N}^{n-1} \to \mathcal{E} \otimes \mathcal{N}^{n-1} \to \mathcal{N}^{2n-1} \to 0.\right) \end{gathered}$$

Note that $v_i du_i, \ldots, v_i u_i^{\frac{n+1}{2}-2} du_i$ are non-closed differential 1-forms on X. Meanwhile, if $\sigma : \widehat{X} \to X$ is a desingularization, then we have an injection $\sigma^* \Omega^1_{X/k} \to \Omega^1_{\widehat{X}/k}$ of sheaves. Hence we conclude that

Lemma *If one of the injections*

$$\begin{gathered} H^0(C, \mathcal{N}^2) \to H^0(C, \mathcal{E} \otimes \mathcal{N}^2), \\ H^0(C, \mathcal{N}^4) \to H^0(C, \mathcal{E} \otimes \mathcal{N}^4), \\ \vdots \\ H^0(C, \mathcal{N}^{n-1}) \to H^0(C, \mathcal{E} \otimes \mathcal{N}^{n-1}) \end{gathered}$$

is not surjective, then there exist global non-closed differential 1-forms on any desingularization of X.

---

[2] Recall the extension $0 \to \mathcal{O}_C \to \mathcal{E} \to \mathcal{N}^n \to 0$, where $\mathcal{E}$ is generated locally by 1 and q_i's subjected to $q_i = d_{ij}^n q_j + b_{ij}$ (see Sect. 3).

5 Case of Genus >11

We have already known

Theorem 1 (Takeda [14]) *On any curve of genus greater than* 11, *there exists a pre-Tango structure involving*[3] *a projective smooth uniruled surface with global* non-closed *differential 1-forms.*

Here we shall describe a brief proof for reader's convenience. Take an ample invertible sheaf $\mathcal{L}$ such that

$$\begin{cases} \mathcal{L} = F_*\mathcal{B}^1_{\tilde{C}} \text{ (then } \deg \mathcal{L} = g-1) \text{ if } g \text{ is even;} \\ \mathcal{L} \subset F_*\mathcal{B}^1_{\tilde{C}} \text{ with } \deg \mathcal{L} = g-2 \text{ if } g \text{ is odd,} \end{cases}$$

and take an invertible sheaf $\mathcal{N}$ such that $\mathcal{N}^n \cong \mathcal{L}$ with $n = \deg \mathcal{L}$, $\deg \mathcal{N} = 1$. It follows that n is an odd number and $g-2 \leq n \leq g-1$. Recall the extension mentioned in the previous section, which is isomorphic to

$$0 \to \mathcal{N}^{n-1} \to \mathcal{E} \otimes \mathcal{N}^{n-1} \to \mathcal{N}^{2n-1} \to 0\,.$$

Considering the Euler characteristic $\chi(\mathcal{N}^{n-1}) = n-1+1-g = n-g$, we have

$$\dim H^1(\mathcal{N}^{n-1}) = \dim H^0(\mathcal{N}^{n-1}) + g - n \geq 0 + 1.$$

Thus $H^1(\mathcal{N}^{n-1}) \neq 0$, which implies that $\mathcal{N}^{n-1}$ is special. By virtue of Clifford's theorem, we know

$$\dim H^0(\mathcal{N}^{n-1}) \leq \frac{n-1}{2} + 1.$$

Therefore, we estimate

$$\begin{aligned} \dim H^1(\mathcal{N}^{n-1}) &= \dim H^0(\mathcal{N}^{n-1}) + g - n \\ &\leq \frac{n-1}{2} + 1 + g - n = g - \frac{n-1}{2} \\ &\leq g - \frac{g-2-1}{2} = \frac{g+3}{2}. \end{aligned}$$

On the other hand, since $\chi(\mathcal{N}^{2n-1}) = 2n-1+1-g = 2n-g$, we have

$$\dim H^0(\mathcal{N}^{2n-1}) = \dim H^1(\mathcal{N}^{2n-1}) + 2n - g.$$

Hence $\dim H^0(\mathcal{N}^{2n-1}) \geq 2n - g \geq 2(g-2) - g = g-4$.

[3]Here 'involving' means 'inducing' or 'inducing a normal uniruled surface whose desingularization is'.

Take the long exact sequence of the above-mentioned extension:

$$\begin{aligned} 0 \to H^0(\mathcal{N}^{n-1}) \to H^0(\mathcal{E} \otimes \mathcal{N}^{n-1}) &\to H^0(\mathcal{N}^{2n-1}) \\ &\quad (\dim \geq g-4) \\ \to H^1(\mathcal{N}^{n-1}) \to H^1(\mathcal{E} \otimes \mathcal{N}^{n-1}) &\to H^1(\mathcal{N}^{2n-1}) \to 0. \\ (\dim \leq \frac{g+3}{2}) \quad & \end{aligned}$$

Since

$$g - 4 - \frac{g+3}{2} = \frac{2g - 8 - g - 3}{2} = \frac{g - 11}{2},$$

we know $\mathrm{Ker}\,[H^0(\mathcal{N}^{2n-1}) \to H^1(\mathcal{N}^{n-1})] \neq 0$ provided that the genus $g > 11$. Therefore, the injection $H^0(\mathcal{N}^{n-1}) \to H^0(\mathcal{E} \otimes \mathcal{N}^{n-1})$ is *not* surjective. Hence, by virtue of Lemma in Sect. 4, we obtain the assertion.

6 Case of Small Genera

In the case of small genera, it is difficult to know whether any curve has a suitable pre-Tango structure or not. However, provided that $g - 1$ is divided by an odd number $n \geq 3$, there necessarily exists a Tango structure which induces a smooth uniruled surface with several pathological phenomena, e.g. not liftable, counter-example to the Kodaira Vanishing Theorem, ... Here we shall explicitly exhibit the differential forms on the induced uniruled surface.

6.1 Case of Genus 11

Suppose that the genus g is 11. We then have $\deg F_*\mathcal{B}^1_{\tilde{C}} = 10$. Thus there exists an invertible sheaf $\mathcal{N}$ such that $\mathcal{N}^5 \cong F_*\mathcal{B}^1_{\tilde{C}}$ and $\deg \mathcal{N} = 2$.

Recall the extensions

$$0 \to \mathcal{N}^2 \to \mathcal{E} \otimes \mathcal{N}^2 \to \mathcal{N}^{5+2} \to 0,$$
$$0 \to \mathcal{N}^4 \to \mathcal{E} \otimes \mathcal{N}^4 \to \mathcal{N}^{5+4} \to 0$$

and the injections

$$H^0(\mathcal{N}^2) \to H^0(\mathcal{E} \otimes \mathcal{N}^2), \quad H^0(\mathcal{N}^4) \to H^0(\mathcal{E} \otimes \mathcal{N}^4).$$

From $\mathcal{E} = F_*\mathcal{O}_{\tilde{C}}$, it follows that $\mathcal{E} \otimes \mathcal{N}^2 \approx \mathcal{N}^4$ and $\mathcal{E} \otimes \mathcal{N}^4 \approx \mathcal{N}^8$, where the isomorphisms are not of $\mathcal{O}_C$-modules but of sheaves of additive groups. Therefore, we have

$$H^0(\mathcal{N}^2) \to H^0(\mathcal{E} \otimes \mathcal{N}^2) \approx H^0(\mathcal{N}^4) \to H^0(\mathcal{E} \otimes \mathcal{N}^4) \approx H^0(\mathcal{N}^8).$$

Since $\chi(\mathcal{N}^2) = 2 \cdot 2 + 1 - 11 = -6$, we know

$$\dim H^1(\mathcal{N}^2) = \dim H^0(\mathcal{N}^2) + 6.$$

Considering the dualizing sheaf $\omega_C \cong \mathcal{N}^{10}$, we have

$$\dim H^0(\mathcal{N}^8) = \dim H^1(\mathcal{N}^2) = \dim H^0(\mathcal{N}^2) + 6.$$

Therefore, at least one of the above-mentioned two injections is not surjective. Hence the induced uniruled surface has global non-closed 1-forms.

6.2 *Case of Genus* 10

Suppose that the genus g is 10. We then have $\deg F_*\mathcal{B}^1_{\tilde{C}} = 9$. Thus there exists an invertible sheaf $\mathcal{N}$ such that $\mathcal{N}^9 \cong F_*\mathcal{B}^1_{\tilde{C}}$ and $\deg \mathcal{N} = 1$.

Recall the extensions

$$0 \to \mathcal{N}^2 \to \mathcal{E} \otimes \mathcal{N}^2 \to \mathcal{N}^{9+2} \to 0,$$
$$0 \to \mathcal{N}^4 \to \mathcal{E} \otimes \mathcal{N}^4 \to \mathcal{N}^{9+4} \to 0,$$
$$0 \to \mathcal{N}^8 \to \mathcal{E} \otimes \mathcal{N}^8 \to \mathcal{N}^{9+8} \to 0$$

and the injections

$$H^0(\mathcal{N}^2) \to H^0(\mathcal{E} \otimes \mathcal{N}^2),\ H^0(\mathcal{N}^4) \to H^0(\mathcal{E} \otimes \mathcal{N}^4),\ H^0(\mathcal{N}^8) \to H^0(\mathcal{E} \otimes \mathcal{N}^8).$$

We know that $\mathcal{E} \otimes \mathcal{N}^2 \approx \mathcal{N}^4$, $\mathcal{E} \otimes \mathcal{N}^4 \approx \mathcal{N}^8$ and $\mathcal{E} \otimes \mathcal{N}^8 \approx \mathcal{N}^{16}$. Therefore, we have

$$\begin{array}{llll} H^0(\mathcal{N}^2) \to H^0(\mathcal{E} \otimes \mathcal{N}^2) & & & \\ \qquad\qquad \wr\wr & & & \\ \qquad\qquad H^0(\mathcal{N}^4) \to H^0(\mathcal{E} \otimes \mathcal{N}^4) & & & \\ \qquad\qquad\qquad\qquad \wr\wr & & & \\ \qquad\qquad\qquad\qquad H^0(\mathcal{N}^8) \to H^0(\mathcal{E} \otimes \mathcal{N}^8) & & & \\ \qquad\qquad\qquad\qquad\qquad\qquad \wr\wr & & & \\ \qquad\qquad\qquad\qquad\qquad\qquad H^0(\mathcal{N}^{16}). & & & \end{array}$$

Since $\chi(\mathcal{N}^2) = 2 + 1 - 10 = -7$, we know

$$\dim H^1(\mathcal{N}^2) = \dim H^0(\mathcal{N}^2) + 7.$$

Considering the dualizing sheaf $\omega_C \cong \mathcal{N}^{18}$, we have

$$\dim H^0(\mathcal{N}^{16}) = \dim H^1(\mathcal{N}^2) = \dim H^0(\mathcal{N}^2) + 7.$$

Therefore, at least one of the above-mentioned three injections is not surjective, which is the required result.

6.3 *Case of Genus* 8

Suppose that the genus g is 8. We then have $\deg F_*\mathcal{B}^1_{\tilde{C}} = 7$. Thus there exists an invertible sheaf $\mathcal{N}$ such that $\mathcal{N}^7 \cong F_*\mathcal{B}^1_{\tilde{C}}$ and $\deg \mathcal{N} = 1$.

Recall the extensions

$$0 \to \mathcal{N}^2 \to \mathcal{E} \otimes \mathcal{N}^2 \to \mathcal{N}^{7+2} \to 0,$$
$$0 \to \mathcal{N}^4 \to \mathcal{E} \otimes \mathcal{N}^4 \to \mathcal{N}^{7+4} \to 0,$$
$$0 \to \mathcal{N}^6 \to \mathcal{E} \otimes \mathcal{N}^6 \to \mathcal{N}^{7+6} \to 0$$

and the injections

$$H^0(\mathcal{N}^2) \to H^0(\mathcal{E} \otimes \mathcal{N}^2),\ H^0(\mathcal{N}^4) \to H^0(\mathcal{E} \otimes \mathcal{N}^4),\ H^0(\mathcal{N}^6) \to H^0(\mathcal{E} \otimes \mathcal{N}^6).$$

We know that $\mathcal{E} \otimes \mathcal{N}^2 \approx \mathcal{N}^4$, $\mathcal{E} \otimes \mathcal{N}^4 \approx \mathcal{N}^8$ and $\mathcal{E} \otimes \mathcal{N}^6 \approx \mathcal{N}^{12}$.

Suppose that the first and second injections

$$H^0(\mathcal{N}^2) \to H^0(\mathcal{E} \otimes \mathcal{N}^2), \quad H^0(\mathcal{N}^4) \to H^0(\mathcal{E} \otimes \mathcal{N}^4),$$

are surjective. We then have

$$\dim H^0(\mathcal{N}^2) = \dim H^0(\mathcal{N}^4) = \dim H^0(\mathcal{N}^8).$$

On the other hand, since $\chi(\mathcal{N}^6) = 6 + 1 - 8 = -1$, we know

$$\dim H^1(\mathcal{N}^6) = \dim H^0(\mathcal{N}^6) + 1.$$

Besides, since $\chi(\mathcal{N}^2) = 2 + 1 - 8 = -5$, we know

$$\dim H^1(\mathcal{N}^2) = \dim H^0(\mathcal{N}^2) + 5.$$

Considering the dualizing sheaf $\omega_C \cong \mathcal{N}^{14}$, we know

$$\dim H^0(\mathcal{N}^6) = \dim H^1(\mathcal{N}^6) - 1 = \dim H^0(\mathcal{N}^8) - 1 = \dim H^0(\mathcal{N}^2) - 1,$$
$$\dim H^0(\mathcal{N}^{12}) = \dim H^1(\mathcal{N}^2) = \dim H^0(\mathcal{N}^2) + 5.$$

Therefore, we conclude that the injection $H^0(\mathcal{N}^6) \to H^0(\mathcal{E} \otimes \mathcal{N}^6) \approx H^0(\mathcal{N}^{12})$ cannot be surjective. Hence at least one of the above-mentioned three injections is not surjective.

6.4 Case of Genus 7

Suppose that the genus g is 7. We then have $\deg F_*\mathcal{B}^1_{\tilde{C}} = 6$. Thus there exists an invertible sheaf $\mathcal{N}$ such that $\mathcal{N}^3 \cong F_*\mathcal{B}^1_{\tilde{C}}$ and $\deg \mathcal{N} = 2$.[4]

Recall the extensions

$$0 \to \mathcal{N}^2 \to \mathcal{E} \otimes \mathcal{N}^2 \to \mathcal{N}^{3+2} \to 0$$

and the injection

$$H^0(\mathcal{N}^2) \to H^0(\mathcal{E} \otimes \mathcal{N}^2) \approx H^0(\mathcal{N}^4).$$

Since $\chi(\mathcal{N}^2) = 2 \cdot 2 + 1 - 7 = -2$, we know

$$\dim H^1(\mathcal{N}^2) = \dim H^0(\mathcal{N}^2) + 2.$$

Considering the dualizing sheaf $\omega_C \cong \mathcal{N}^6$, we have

$$\dim H^0(\mathcal{N}^4) = \dim H^1(\mathcal{N}^2) = \dim H^0(\mathcal{N}^2) + 2.$$

Therefore, the above-mentioned injection is not surjective.

6.5 Case of Genus 6

Suppose that the genus g is 6. We then have $\deg F_*\mathcal{B}^1_{\tilde{C}} = 5$. Thus there exists an invertible sheaf $\mathcal{N}$ such that $\mathcal{N}^5 \cong F_*\mathcal{B}^1_{\tilde{C}}$ and $\deg \mathcal{N} = 1$.

[4] The induced uniruled surface is a quasi-elliptic surface of $\kappa = 1$.

Recall the extensions

$$0 \to \mathcal{N}^2 \to \mathcal{E} \otimes \mathcal{N}^2 \to \mathcal{N}^{5+2} \to 0,$$
$$0 \to \mathcal{N}^4 \to \mathcal{E} \otimes \mathcal{N}^4 \to \mathcal{N}^{5+4} \to 0$$

and the injections

$$H^0(\mathcal{N}^2) \to H^0(\mathcal{E} \otimes \mathcal{N}^2) \approx H^0(\mathcal{N}^4) \to H^0(\mathcal{E} \otimes \mathcal{N}^4) \approx H^0(\mathcal{N}^8).$$

Since $\chi(\mathcal{N}^2) = 2 + 1 - 6 = -3$, we know

$$\dim H^1(\mathcal{N}^2) = \dim H^0(\mathcal{N}^2) + 3.$$

Considering the dualizing sheaf $\omega_C \cong \mathcal{N}^{10}$, we have

$$\dim H^0(\mathcal{N}^8) = \dim H^1(\mathcal{N}^2) = \dim H^0(\mathcal{N}^2) + 3.$$

Therefore, at least one of the above-mentioned two injections is not surjective.

6.6 *Case of Genus* 4

Suppose that the genus g is 4. We then have $\deg F_*\mathcal{B}^1_{\tilde{C}} = 3$. Thus there exists an invertible sheaf $\mathcal{N}$ such that $\mathcal{N}^3 \cong F_*\mathcal{B}^1_{\tilde{C}}$ and $\deg \mathcal{N} = 1$.[5]

Recall the extensions

$$0 \to \mathcal{N}^2 \to \mathcal{E} \otimes \mathcal{N}^2 \to \mathcal{N}^{3+2} \to 0$$

and the injection

$$H^0(\mathcal{N}^2) \to H^0(\mathcal{E} \otimes \mathcal{N}^2) \approx H^0(\mathcal{N}^4).$$

Since $\chi(\mathcal{N}^2) = 2 + 1 - 4 = -1$, we know

$$\dim H^1(\mathcal{N}^2) = \dim H^0(\mathcal{N}^2) + 1.$$

Considering the dualizing sheaf $\omega_C \cong \mathcal{N}^6$, we have

$$\dim H^0(\mathcal{N}^4) = \dim H^1(\mathcal{N}^2) = \dim H^0(\mathcal{N}^2) + 1.$$

Therefore, the above-mentioned injection is not surjective.

Remark 1 Regrettably, for genera

[5]The induced uniruled surface is also a quasi-elliptic surface of $\kappa = 1$.

$$9 = 2^3 + 1, \quad 5 = 2^2 + 1, \quad 3 = 2 + 1, \quad 2 = 1 + 1,$$

there are no good ideas at present.

7 Automorphisms of the Induced Uniruled Surfaces

Let X be the *smooth* uniruled surface induced from a *Tango* structure $\mathcal{N}^n$ on C with $n \geq 3$. Recall that X is obtained by glueing the schemes

$$\operatorname{Spec} \mathcal{O}_C(U_i)[x_i, y_i]/(y_i^2 + x_i + a_i x_i^{n+1}) \text{ (near } \Gamma),$$
$$\operatorname{Spec} \mathcal{O}_C(U_i)[u_i, v_i]/(v_i^2 + u_i^n + a_i) \text{ (near } \Sigma)$$

together under the relations

$$u_i = d_{ij}^2 u_j, \; v_i = d_{ij}^n v_j + b_{ij}, \; x_i = u_i^{-1}, \; y_i = v_i x_i^{\frac{n+1}{2}}.$$

We have the followings:

Theorem 2 (cf. Takeda [13, 15, 16] and also Russell [9]) *In the same notation and under the same assumption as above, we have an injection*

$$H^0(C, \mathcal{N}) \to \operatorname{Aut}_k(X \otimes k[\varepsilon]),$$

where $k[\varepsilon]$ is the ring of dual numbers with $\varepsilon^2 = 0$.

Proof Suppose there exists a global section $\gamma \in H^0(C, \mathcal{N})$ which is represented by local function $\{\gamma_i\}$ such that $d_{ij}\gamma_i = \gamma_j$.

Consider the U_i-automorphism of $X|_{U_i} \otimes k[\varepsilon]$ such that

$$y_i \mapsto y_i + \varepsilon\gamma_i, \quad x_i \mapsto x_i, \quad v_i \mapsto v_i + \varepsilon\gamma_i u_i^{\frac{n+1}{2}}, \quad u_i \mapsto u_i.$$

Note that these are compatible with the relations

$$y_i^2 = x_i + a_i x_i^{n+1}, \; v_i^2 = u_i^n + a_i, \; x_i = u_i^{-1}, \; y_i = v_i x_i^{\frac{n+1}{2}}.$$

We can glue those automorphisms together so that we obtain an automorphism of $X \otimes k[\varepsilon]$. Indeed, we have

$$
\begin{array}{ccc}
v_i & \overset{\gamma_i}{\mapsto} & v_i + \varepsilon\gamma_i u_i^{\frac{n+1}{2}} \\
\| & & \\
d_{ij}^n v_j + b_{ij} & \overset{\gamma_j}{\mapsto} & d_{ij}^n (v_j + \varepsilon\gamma_j u_j^{\frac{n+1}{2}}) + b_{ij}
\end{array}
$$

$$
\begin{aligned}
&= d_{ij}^n v_j + b_{ij} + d_{ij}^n \varepsilon\gamma_j u_j^{\frac{n+1}{2}} \\
&= v_i + \varepsilon\gamma_i d_{ij}^{n+1} u_j^{\frac{n+1}{2}} \\
&= v_i + \varepsilon\gamma_i u_i^{\frac{n+1}{2}}.
\end{aligned}
$$

Therefore, we conclude that there exists an injection

$$H^0(C, \mathcal{N}) \to \mathrm{Aut}_k(X \otimes k[\varepsilon]).$$

Remark 2 In fact, by the same fashion as Ganong-Russell [2] and Takeda [12], it can be verified that

$$H^0(C, \mathcal{N}) \cong H^0(X, \Theta_X),$$

where Θ_X is the tangent bundle of X.[6] Since $H^0(X, \Theta_X)$ is identified with $\mathrm{Aut}_k(X \otimes k[\varepsilon])$, we know that the above-mentioned injection is an isomorphism.

Remark 3 In the same notation and under the same assumption as above, we know that X is a smooth minimal surface since X has a fibration onto the base curve C of genus > 1 such that every fibre is a rational curve with one cusp. Meanwhile it is also verified that

$$\omega_X \cong \mathcal{O}_X((n-3)\Gamma) \otimes_{\mathcal{O}_X} \varphi^* \mathcal{N}^{2+n},$$

which is ample if $n > 3$, and

$$(\Gamma^2) = \deg \mathcal{N} > 0$$

(see [15]). Hence, provided $n > 3$, we obtain that X is a surface of *general type* and that the automorphism group is a *finite* group (see e.g. Matsumura [7]). In other words, the k-rational points of the automorphism group scheme $\mathbf{Aut}_{X/k}$ are finite. On the other hand, we know that the tangent space at the identity of $\mathbf{Aut}_{X/k}$ is identified with $\mathrm{Aut}_k(X \otimes k[\varepsilon])$. Therefore, it is concluded that the scheme $\mathbf{Aut}_{X/k}$ is *not reduced* provided that $H^0(C, \mathcal{N}) \neq 0$.

Example Let C be the plane curve defined by

$$y^{2n} + y = x^{2n-1} + c_{2n-3}x^{2n-3} + c_{2n-5}x^{2n-5} + \cdots + c_1 x$$

[6] For details, see the author's forthcoming paper.

with an odd number $n > 3$ and $c_i \in k$. We have

$$dy = (x^{2n-2} + c_{2n-3}x^{2n-4} + c_{2n-5}x^{2n-6} + \cdots + c_1)dx.$$

Hence we know that x is a local parameter near the origin. Let $u = x^{-1}$ and $v = yu$. We then have

$$v^{2n} + vu^{2n-1} = u + c_{2n-3}u^3 + c_{2n-5}u^5 + \cdots + c_1u^{2n-1}$$

and

$$u^{2n-1}dv + vu^{2n-2}du = (1 + c_{2n-3}u^2 + c_{2n-5}u^4 + \cdots + c_1u^{2n-2})du,$$

that is

$$u^{2n-1}dv = (1 + c_{2n-3}u^2 + c_{2n-5}u^4 + \cdots + c_1u^{2n-2} + vu^{2n-2})du.$$

Hence we know that v is a local parameter near the point at infinity P_∞.

Meanwhile, we can calculate as follows:

$$\begin{aligned} v^{2n} &= u + c_{2n-3}u^3 + c_{2n-5}u^5 + \cdots + c_1u^{2n-1} + vu^{2n-1} \\ v^{2n} &= u(1 + c_{2n-3}u^2 + c_{2n-5}u^4 + \cdots + c_1u^{2n-2} + vu^{2n-2}). \end{aligned}$$

Therefore, we know $u = v^{2n}(1 + c_{2n-3}u^2 + c_{2n-5}u^4 + \cdots + c_1u^{2n-2} + vu^{2n-2})^{-1}$. Furthermore,

$$\begin{aligned} u^{-1} &= v^{-2n}(1 + c_{2n-3}u^2 + c_{2n-5}u^4 + \cdots + c_1u^{2n-2} + vu^{2n-2}) \\ x &= v^{-2n}(1 + c_{2n-3}u^2 + c_{2n-5}u^4 + \cdots + c_1u^{2n-2}) + v\,v^{-2n}u^{2n-2} \\ x &= v\,v^{-2n}u^{2n-2} + v^{-2n}(1 + \sqrt{c_{2n-3}}\,u + \sqrt{c_{2n-5}}\,u^2 + \cdots + \sqrt{c_1}\,u^{n-1})^2 \\ x &= v\,v^{-2n}(v^{2n})^{2n-2}(1 + c_{2n-3}u^2 + c_{2n-5}u^4 + \cdots + c_1u^{2n-2} + vu^{2n-2})^{-(2n-2)} \\ &\quad + v^{-2n}(1 + \sqrt{c_{2n-3}}\,u + \sqrt{c_{2n-5}}\,u^2 + \cdots + \sqrt{c_1}\,u^{n-1})^2 \\ x &= v\,v^{2n(2n-3)}(1 + c_{2n-3}u^2 + c_{2n-5}u^4 + \cdots + c_1u^{2n-2} + vu^{2n-2})^{-(2n-2)} \\ &\quad + v^{-2n}(1 + \sqrt{c_{2n-3}}\,u + \sqrt{c_{2n-5}}\,u^2 + \cdots + \sqrt{c_1}\,u^{n-1})^2. \end{aligned}$$

Here, note that the last equation is corresponding to the relation $a_i = d_{ij}^{2n}a_j + b_{ij}^2$ in Sect. 3. We obtain an equation of differential forms on C:

$$dx = v^{2n(2n-3)}(1 + c_{2n-3}u^2 + c_{2n-5}u^4 + \cdots + c_1u^{2n-2} + vu^{2n-2})^{-(2n-2)}dv.$$

This implies that the divisor of the differential form dx is

$$(dx) = 2n(2n - 3)P_\infty.$$

Set $\mathcal{N} = \mathcal{O}_C((2n-3)P_\infty)$. We then have a Tango structure $\mathcal{N}^n \cong F_*\mathcal{B}^1_{\tilde{C}}$. Thus, we obtain a smooth surface X of general type such that

$$H^0(X, \Theta_X) \supset H^0(C, \mathcal{N}) \neq 0.$$

Hence X has non-zero regular vector fields and its automorphism group scheme is not reduced.

References

1. Deligne, P., Illusie, L.: Relèvements modulo p^2 et décomposition du complexe de de Rham. Invent. Math. **89**, 247–270 (1987)
2. Ganong, R., Russell, P.: The tangent bundle of a ruled surface. Math. Ann. **271**, 527–548 (1985)
3. Kurke, H.: Example of false ruled surfaces. In: Proceedings of Symposium on Algebraic Geometry Kinosaki, pp. 203–223 (1981)
4. Lang, W.E.: Quasi-elliptic surfaces in characteristic three. Ann. Sci. Ecole Norm. Sup. **12**(4), 473–500 (1979)
5. Lang, W.E.: Examples of surfaces of general type with vector fields. In: Artin, M., Tate, J. (eds.) Arithmetic and Geometry, vol. II, pp. 167–173. Birkhäuser, Boston (1983)
6. Mukai, S.: Counterexamples to Kodaira's vanishing and Yau's inequality in positive characteristics. Kyoto J. Math. **53**(2), 515–532 (2013). Japanese version: In: Proceedings of the Symposium on Algebraic Geometry, Kinosaki, pp. 9–31 (1979)
7. Matsumura, H.: On algebraic groups of birational transformations. Atti Accad. Naz. dei Lincei **34**, 151–155 (1963)
8. Raynaud, M.: Contre-exemple au "Vanishing Theorem" em caractéristique $p > 0$. In: Ramanujan C.P. (ed.) A Tribute. Tata Institute of Fundamental Research Studies in Mathematics, vol. 8, pp. 273–278. Springer, Berlin (1978)
9. Russell, P.: Forms of the affine line and its additive group. Pacific J. Math. **32**, 527–539 (1970)
10. Russell, P.: Factoring the Frobenius morphism of an algebraic surface. In: Algebraic Geometry, Bucharest 1982, Lecture Notes in Math. 1056, vol. 1984, pp. 366–380. Springer, Berlin (1982)
11. Takeda, Y.: Fibrations with moving cuspidal singularities. Nagoya Math. J. **122**, 161–179 (1991)
12. Takeda, Y.: Vector fields and differential forms on generalized Raynaud surfaces. Tôhoku Math. J. **44**, 359–364 (1992)
13. Takeda, Y.: Groups of Russell type over a curve. J. Pure Appl. Algebra **128**, 93–108 (1998); Corrigendum **148**, 317–318 (2000)
14. Takeda, Y.: Pre-Tango structures in characteristic two. Japanese J. Math. **28**, 81–86 (2002)
15. Takeda, Y.: Pre-Tango structures and uniruled varieties. Colloq. Math. **108**(2), 193–216 (2007)
16. Takeda, Y.: Groups of Russell Type and Tango Structures. In: Daigle, D., Ganong, R., Koras, M. (eds.) Affine Algebraic Geometry, The Russell Festschrift, CRM Proceedings and Lecture Notes, vol. 54, pp. 327–334. Centre de Recherches Mathématiques, Montéal, AMS (2011)
17. Takeda, Y., Yokogawa, K.: Pre-Tango structures on curves. Tôhoku Math. J. **54**, 227–237 (2002); Errata and addenda **55**, 611–614 (2003)
18. Tango, H.: On the behavior of extensions of vector bundles under the Frobenius map. Nagoya Math. J. **48**, 73–89 (1972)

Exponential Matrices of Size Five-By-Five

Ryuji Tanimoto

Abstract In the article, we supply examples of exponential matrices in positive characteristic, and then give an overlapping classification of exponential matrices of size five-by-five in positive characteristic. In characteristic zero, we can easily classify exponential matrices up to equivalence. But, in positive characteristic, we meet difficulties of classifying exponential matrices up to equivalence. At the conference *Polynomial Rings and Affine Algebraic Geometry*, February 12–16, 2018, Tokyo Metropolitan University, we gave a talk about classifying exponential matrices of size four-by-four in positive characteristic, up to equivalence. So, the article can be regarded as a continuation of the talk.

Keywords Matrix theory · Modular representation theory

1 Introduction

In the article, we supply examples of exponential matrices in positive characteristic, and then give an overlapping classification of exponential matrices of size five-by-five. This study is a continuation of the articles [4, 5, 7], in which we classify exponential matrices of size at most four-by-four in positive characteristic.

A motivation for studying exponential matrices comes from invariant theory of the additive group $\mathbb{G}_a$ of an algebraically closed field. The Weitzenböck problem asks whether or not any linear action of $\mathbb{G}_a$ on a polynomial ring gives rise to a finitely generated $\mathbb{G}_a$-invariant ring. In characteristic zero, Weitzenböck [9] solved the problem (see [3, 8]). On the other hand, in positive characteristic, we have partial positive answers to the problem (see [1, 4, 6, 10]). So, we seek to understand linear actions of $\mathbb{G}_a$ on polynomial rings in positive characteristic. Such linear $\mathbb{G}_a$-actions are afforded by exponential matrices.

R. Tanimoto (✉)
Shizuoka University, 836 Ohya, Suruga-ku, Shizuoka 422-8529, Japan
e-mail: tanimoto.ryuji@shizuoka.ac.jp

S. Kuroda et al. (eds.), *Polynomial Rings and Affine Algebraic Geometry*,
Springer Proceedings in Mathematics & Statistics 319,
https://doi.org/10.1007/978-3-030-42136-6_13

Let k be a field of characteristic $p \geq 0$. For any polynomial matrix $A(T) \in GL(n, k[T])$, we say that $A(T)$ is an *exponential matrix* if $A(T)$ satisfies the following conditions (1) and (2):

(1) $A(0) = I_n$.
(2) $A(T)A(T') = A(T + T')$, where T, T' are indeterminates over k.

Any exponential matrix $A(T)$ of $GL(n, k[T])$ is an element of $SL(n, k[T])$ (see [7]). Let $\mathcal{E}(n, k[T])$ be the set of all exponential matrices of size $n \times n$. For two exponential matrices $A(T), B(T) \in \mathcal{E}(n, k[T])$, we say that $A(T)$ and $B(T)$ are *equivalent* if there exists a regular matrix $P \in GL(n, k)$ such that $P^{-1}A(T)P = B(T)$.

We are concerned with classifying exponential matrices, up to equivalence. In characteristic zero, we can easily classify exponential matrices, up to equivalence. But, in positive characteristic, we meet difficulties of classifying exponential matrices up to equivalence. Even so, in positive characteristic, we can classify exponential matrices of size at most four-by-four, up to equivalence (see [7]).

In this article, we classify exponential matrices of $\mathfrak{U}_{[n,2]}$ in positive characteristic, where $n \geq 3$ (see Theorem 1), and give an overlapping classification of exponential matrices of size five-by-five in positive characteristic (see Theorem 2 and Corollary 2). As a consequence of Theorem 2, we obtain an overlapping classification of five-dimensional modular representations of elementary abelian p-groups (see Theorem 3).

Notations and Definitions

Let $\mathfrak{R}$ be a ring with unity, where $\mathfrak{R}$ is not necessarily commutative. Let R be a commutative ring with unity. Let k be a field of characteristic $p \geq 0$, and let $k[T]$ be a polynomial ring in one variable over k. Let $\mathbb{Z}$ denote the ring of all integers.

We make the following index for promptly finding the definitions of notations:

Matrices

$\mathcal{E}(n, k[T])$, §1	$GL(n, R)$, 1.2	
$GLC^{(n+1,\ldots,n+s)}(n+s, k)$, 1.2		$GLS(n, R)$, 1.2
$\mathrm{Mat}_{m,n}(\mathfrak{R})$, 1.1	$\mathrm{Mat}(n, \mathfrak{R})$, 1.1	$SL(n, R)$, 1.2
$\mathrm{Sym}(n, R)$, 1.2		

Miscellaneous notations

$A_{(i,j)}$, 1.1	$\bigoplus_{i=1}^{r} A_i$, 1.1	e_i, 1.1
$\mathrm{Exp}(A(T))$, 1.9	$J(f_0, \ldots, f_{n-1})$, 1.7	k^n, 1.1
ν_n, 1.7	$\mathfrak{P}$, 1.8	Φ_n, 1.9
Ψ_{n-1}, 1.9	S^E, 1.3	$S \rightsquigarrow S'$, 1.4
$\sigma_{i,j}$, 1.6	${}^tA$, 1.1	${}^TA$, 1.5
${}^T\mathfrak{S}$, 1.5		

Sets of polynomial matrices

$\mathfrak{A}(i_1, i_2, i_3)$, 1.14	$\mathfrak{H}_{m+2}$, 1.15	$\mathfrak{H}_{m+2}{}^{\circ}$, 1.15
$\mathfrak{H}_4{}^{\varepsilon}$, 6.7	$\mathfrak{H}\mathfrak{H}_5$, 6.2	$\mathfrak{H}\mathfrak{H}_5{}^{\circ}$, 6.2

$\mathfrak{J}_{[n]}$, 1.11	$\mathfrak{J}_{[n,1]}$, 1.12	$\mathfrak{J}^0_{[n,1]}$, 1.12
$\mathfrak{J}^1_{[n,1]}$, 1.12	$\mathfrak{J}_{[1,n]}$, 1.13	$\mathfrak{J}^0_{[1,n]}$, 1.13
$\mathfrak{J}^1_{[1,n]}$, 1.13	$\mathfrak{J}^{0,0}_{[n,2]}$, 3.1	$\mathfrak{J}^{0,1}_{[n,2]}$, 3.2
$\mathfrak{J}^{1,0}_{[n,2]}$, 3.3	$\mathfrak{J}^{1,1}_{[n,2]}$, 3.4	$\mathfrak{J}^{0,0}_{[2,n]}$, §4
$\mathfrak{J}^{0,1}_{[2,n]}$, §4	$\mathfrak{J}^{1,0}_{[2,n]}$, §4	$\mathfrak{J}^{1,1}_{[2,n]}$, §4
$\mathfrak{U}_{[d_1,\ldots,d_t]}$, 1.10	$\mathfrak{U}_n$, 1.10	

Finite dimensional algebras

$\mathfrak{a}(i_1, i_2, i_3)$, 7.7	$\mathfrak{h}_{m+2}(S)$, 7.8	$\mathfrak{hh}_{m+3}(S_1, S_2)$, 7.9
$\mathfrak{j}_{[n]}$, 7.2	$\mathfrak{j}^{1,0}_{[n,2]}(S)$, 7.3	$\mathfrak{j}^{1,1}_{[n,2]}(\lambda)$, 7.4
$\mathfrak{j}^{0,1}_{[2,n]}(S)$, 7.5	$\mathfrak{j}^{1,1}_{[2,n]}(\lambda)$, 7.6	$\mathfrak{u}_n$, 7.1

Notations for constructing exponential matrices and exponential algebras

$A(T) \Lsh \alpha(T)$, §5	$A \Lsh' \alpha$, 7.1	$\exp(\mathfrak{e})$, §8
$\mathrm{Hk}_{r,s}$, 6.7	$\mathrm{hk}_{r,s}$, §9	$\mathrm{Hook}_{r,s}(\mathfrak{R})$, §5
$[\![I_r,\ \mathfrak{S},\ I_s]\!]$, §5	$\mathfrak{S} \Lsh \mathfrak{T}$, §5	$\mathfrak{S} \Lsh' \mathfrak{T}$, 7.1

1.1 $\mathbf{Mat}_{m,n}(\mathfrak{R})$, $\mathbf{Mat}(n, \mathfrak{R})$, ${}^tA$, $A_{(i,j)}$, $\bigoplus_{i=1}^r A_i$, k^n, e_i

We denote by $\mathrm{Mat}_{m,n}(\mathfrak{R})$ the set of all $m \times n$ matrices whose entries belong to $\mathfrak{R}$. We simply write $\mathrm{Mat}(n, \mathfrak{R})$ instead of $\mathrm{Mat}_{n,n}(\mathfrak{R})$. We denote by I_n the identity matrix of $\mathrm{Mat}(n, \mathfrak{R})$. The zero matrix $O_{m,n}$ of $\mathrm{Mat}_{m,n}(\mathfrak{R})$ is frequently referred as O. For any matrix $A \in \mathrm{Mat}_{m,n}(\mathfrak{R})$, we denote by ${}^tA$ the transpose of A. For any matrix $A \in \mathrm{Mat}(n, \mathfrak{R})$ and for integers $1 \leq i, j \leq n$, we denote by $A_{(i,j)}$ the submatrix of A formed by deleting the i-th row and the j-th column of A.

For matrices $A_i \in \mathrm{Mat}(n_i, \mathfrak{R})$ $(1 \leq i \leq r)$, we denote by $\bigoplus_{i=1}^r A_i$ denote the direct sum of the matrices A_i $(1 \leq i \leq r)$. In particular if the A_i are the same matrix A, we also use the notation $A^{\oplus r}$ in place of $\bigoplus_{i=1}^r A$.

Let k^n be the column space of dimension n. We denote by e_i the column vector of k^n whose i-th entry is 1 and the other entries are zero.

1.2 $GL(n, R)$, $SL(n, R)$, $\mathbf{Sym}(n, R)$*, $GLS(n, R)$, $GLC^{(n+1,\ldots,n+s)}(n+s, k)$, Polynomial Matrices and Their Equivalence*

We say that a matrix $A \in \mathrm{Mat}(n, R)$ is *regular* if there exists a matrix $X \in \mathrm{Mat}(n, R)$ such that $AX = XA = I_n$. We denote by $GL(n, R)$ the group of all regular matrices of $\mathrm{Mat}(n, R)$. We denote by $SL(n, R)$ the subgroup of all regular matrices of $GL(n, R)$ whose determinants are 1. We denote by $\mathrm{Sym}(n, R)$ the set of all symmetric matrices of $\mathrm{Mat}(n, R)$. We denote by $GLS(n, R)$ the set of all symmetric matrices of $GL(n, R)$, i.e., $GLS(n, R) := GL(n, R) \cap \mathrm{Sym}(n, R)$.

For integers $n \geq 1$ and $s \geq 1$, we denote by $GLC^{(n+1,\dots,n+s)}(n+s, k)$ the set of all matrices $P \in GL(n+s, k)$ with the form

$$P = \left(\begin{array}{c|c} I_n & P_{1,2} \\ \hline O & P_{2,2} \end{array} \right),$$

where the sizes of $P_{1,2}$, $P_{2,2}$ are $n \times s$, $s \times s$, respectively. Clearly, $GLC^{(n+1,\dots,n+s)}(n+s, k)$ is a subgroup of $GL(n+s, k)$.

Any element $A(T)$ of $\mathrm{Mat}(n, k[T])$ is said to be a *polynomial matrix* of size $n \times n$. For two polynomial matrices $A(T), B(T) \in \mathrm{Mat}(n, k[T])$, we say that $A(T)$ and $B(T)$ are *equivalent* if there exists a regular matrix $P \in GL(n, k)$ such that $P^{-1}A(T)P = B(T)$.

1.3 S^E

For any subset S of $\mathrm{Mat}(n, k[T])$, we denote by S^E the subset of S consisting of all exponential matrices belonging to S. Clearly, $S^E = S \cap \mathcal{E}(n, k[T])$.

1.4 $S \rightsquigarrow S'$

For two subsets S, S' of $\mathrm{Mat}(n, k[T])$, we write $S \rightsquigarrow S'$ if for any $A(T) \in S$ there exists a regular matrix $P \in GL(n, k)$ such that $P^{-1}A(T)P \in S'$.

1.5 ${}^{\tau}A$, ${}^{\tau}\mathfrak{S}$

Let $A = (a_{i,j})$ be a matrix of $\mathrm{Mat}(n, \mathfrak{R})$. We can draw the other diagonal line l of the matrix A. We denote by ${}^{\tau}A$ the transposed matrix of A with respect to the other diagonal line l, i.e., the (i, j)-th entry of ${}^{\tau}A$ is defined by $a_{n-j+1,n-i+1}$ for all $1 \leq i, j \leq n$. We say that ${}^{\tau}A$ is *the other transpose* of A. For any subset $\mathfrak{S}$ of $\mathrm{Mat}(n, \mathfrak{R})$, we define a subset ${}^{\tau}\mathfrak{S}$ of $\mathrm{Mat}(n, \mathfrak{R})$ as ${}^{\tau}\mathfrak{S} := \{{}^{\tau}A \in \mathrm{Mat}(n, \mathfrak{R}) \mid A \in \mathfrak{S}\}$.

1.6 $\sigma_{i,j}$

We denote by S_n the symmetric group on the set $\{i \in \mathbb{Z} \mid 1 \le i \le n\}$. Let $\rho : S_n \to GL(n,k)$ be the permutation representation of S_n. For any transposition $(i,j) \in S_n$, we simply write $\sigma_{i,j}$ for $\rho((i,j))$.

1.7 ν_n, $J(f_0, \dots, f_{n-1})$

Let $\nu_n = (\varepsilon_{i,j})$ be the nilpotent matrix of $\mathrm{Mat}(n,k)$ defined by

$$\varepsilon_{i,j} = \begin{cases} 1 & \text{if } 1 \le i \le n-1 \text{ and } j = i+1, \\ 0 & \text{otherwise.} \end{cases}$$

Clearly, $\nu_n^n = O_n$.

For $f_0, \dots, f_{n-1} \in k[T]$ $(n \ge 1)$, we let

$$J(f_0, \dots, f_{n-1}) := \sum_{i=0}^{n-1} f_i\, \nu_n^i \in \mathrm{Mat}(n, k[T]).$$

1.8 $\mathfrak{P}$

Assume that the characteristic p of k is positive. We say that a polynomial $f(T) \in k[T]$ is a *p-polynomial* if $f(T)$ can be written in the form

$$f(T) = \sum_{i \ge 0} a_i T^{p^i},$$

where $a_i \in k$ for all $i \ge 0$. Let $\mathfrak{P}$ be the set of all p-polynomials. This $\mathfrak{P}$ becomes a ring with multiplication by the composition of functions, i.e., $f(T) \circ g(T) := f(g(T))$ for all $f(T), g(T) \in \mathfrak{P}$. Note that T is the unity of $\mathfrak{P}$, i.e., $1_{\mathfrak{P}} = T$.

1.9 $\mathrm{Exp}(A(T))$, Φ_n, Ψ_{n-1}

Assume that the characteristic p of k is positive. For any $A(T) \in \mathrm{Mat}(n, k[T])$, we define the *truncated exponential* $\mathrm{Exp}(A(T))$ of $A(T)$ as

$$\mathrm{Exp}(A(T)) := \sum_{i=0}^{p-1} \frac{1}{i!} A(T)^i.$$

For $2 \leq n \leq p$, we let

$$\begin{cases} \Phi_n := \{\text{the first row of } \mathrm{Exp}\big(J(0, f_1, \ldots, f_{n-1})\big) \mid f_1, \ldots, f_{n-1} \in \mathfrak{P}\}, \\ \Psi_{n-1} := \{(a_1, \ldots, a_{n-1}) \in k[T]^{n-1} \mid (1, a_1, \ldots, a_{n-1}) \in \Phi_n\}. \end{cases}$$

1.10 $\mathfrak{U}_n$, $\mathfrak{U}_{[d_1,\ldots,d_t]}$

Let $\mathfrak{U}_n$ be the set of all upper triangular polynomial matrices $A(T) = (a_{i,j}(T))$ of $SL(n, k[T])$ whose all diagonal entries are 1, i.e., $a_{i,i}(T) = 1$ for all $1 \leq i \leq n$. Clearly, $\mathfrak{U}_n$ is a subgroup of $SL(n, k[T])$.

We define a partition of a polynomial matrix of $\mathfrak{U}_n$. For any positive integer n, an ordered sequence $[d_1, \ldots, d_t]$ of positive integers d_i $(1 \leq i \leq t)$ is said to be an *ordered partition* of n if $[d_1, \ldots, d_t]$ satisfies $\sum_{i=1}^{t} d_i = n$. For any $A(T) = (a_{i,j}(T)) \in \mathfrak{U}_n$, we say that $A(T)$ *has an ordered partition* $[d_1, d_2, \ldots, d_t]$ if $A(T)$ satisfies

$$\{i \in \{1, \ldots, n\} \mid a_{i,i+1}(T) = 0\} = \{d_1,\ d_1 + d_2,\ \ldots,\ d_1 + \cdots + d_t\},$$

where we let $a_{n,n+1}(T) = 0$. We denote by $\mathfrak{U}_{[d_1,\ldots,d_t]}$ the set of all polynomial matrices of $\mathfrak{U}_n$ having the partition $[d_1, \ldots, d_t]$. We have

$$\mathfrak{U}_n = \bigcup_{[d_1,\ldots,d_t] \text{ is an ordered partition of } n} \mathfrak{U}_{[d_1,\ldots,d_t]}.$$

1.11 $\mathfrak{J}_{[n]}$

We denote by $\mathfrak{J}_{[n]}$ the set of all matrices $A(T)$ of $SL(n, k[T])$ with the form $A(T) = J(1, a_1, \ldots, a_{n-1})$, where $a_1, \ldots, a_{n-1} \in k[T]$ and $a_1 \neq 0$ (see Sect. 1.7).

1.12 $\mathfrak{J}_{[n,1]}$, $\mathfrak{J}^0_{[n,1]}$, $\mathfrak{J}^1_{[n,1]}$

We denote by $\mathfrak{J}_{[n,1]}$ the set of all matrices $A(T)$ of $SL(n+1, k[T])$ with the following form:

$$A(T) = \left(\begin{array}{c|c} J(1, a_1, \ldots, a_{n-1}) & a_n\, e_1 \\ \hline 0 & 1 \end{array}\right) \qquad (a_1, \ldots, a_{n-1}, a_n \in k[T],\quad a_1 \neq 0).$$

We denote by $\mathfrak{J}^0_{[n,1]}$ the set of all matrices $A(T)$ of $\mathfrak{J}_{[n,1]}$ with $a_n = 0$, and by $\mathfrak{J}^1_{[n,1]}$ the set of all matrices $A(T)$ of $\mathfrak{J}_{[n,1]}$ with a_1, a_n being linearly independent over k.

1.13 $\mathfrak{J}_{[1,n]}$, $\mathfrak{J}^0_{[1,n]}$, $\mathfrak{J}^1_{[1,n]}$

Let $\mathfrak{J}_{[1,n]} := {}^{\tau}(\mathfrak{J}_{[n,1]})$, $\mathfrak{J}^0_{[1,n]} := {}^{\tau}(\mathfrak{J}^0_{[n,1]})$ and $\mathfrak{J}^1_{[1,n]} := {}^{\tau}(\mathfrak{J}^1_{[n,1]})$.

1.14 $\mathfrak{A}(i_1, i_2, i_3)$

Let $i_1 \geq 1$, $i_2 \geq 0$, $i_3 \geq 1$ be integers. For any matrix $\alpha(T) \in \mathrm{Mat}_{i_1,i_3}(k[T])$, we define a matrix $\Lambda(i_1, i_2, i_3; \alpha(T))$ of $\mathrm{Mat}(i_1 + i_2 + i_3, k[T])$ as

$$\Lambda(i_1, i_2, i_3; \alpha(T)) = \left(\begin{array}{c|c|c} I_{i_1} & O & \alpha(T) \\ \hline O & I_{i_2} & O \\ \hline O & O & I_{i_3} \end{array}\right).$$

We denote by $\mathfrak{A}(i_1, i_2, i_3)$ the set of all polynomial matrices $A(T)$ of $\mathrm{Mat}(i_1 + i_2 + i_3, k[T])$ with the form $A(T) = \Lambda(i_1, i_2, i_3; \alpha(T))$, where $\alpha(T) \in \mathrm{Mat}_{i_1,i_3}(k[T])$. Clearly, $\mathfrak{A}(i_1, i_2, i_3)$ becomes a commutative subgroup of $GL(i_1 + i_2 + i_3, k[T])$. We frequently use the notation $\Lambda(i_1, i_3; \alpha(T))$ in place of $\Lambda(i_1, i_2, i_3; \alpha(T))$ if we can understand the value of $i_1 + i_2 + i_3$ from the context, and also use the notation $\mathfrak{A}(i_1, i_3)$ in place of $\mathfrak{A}(i_1, i_2, i_3)$.

1.15 $\mathfrak{H}_{m+2}$, $\mathfrak{H}_{m+2}{}^{\circ}$

Let $m \geq 1$ be an integer. For $x_1, \ldots, x_m, y_1, \ldots, y_m, z \in R$, we define an upper triangular matrix $\eta(x_1, \ldots, x_m, y_1, \ldots, y_m, z)$ of $SL(m + 2, R)$ as

$$\eta(x_1, \ldots, x_m, y_1, \ldots, y_m, z) := \begin{pmatrix} 1 & x_1 & \cdots & \cdots & x_m & z \\ & 1 & 0 & \cdots & 0 & y_1 \\ & & 1 & \ddots & \vdots & \vdots \\ & & & \ddots & 0 & \vdots \\ & & & & 1 & y_m \\ & & & & & 1 \end{pmatrix}.$$

Let $\mathfrak{H}(m+2, R)$ be the set of all matrices of $SL(m+2, R)$ with the form $\eta(x_1, \ldots, x_m, y_1, \ldots, y_m, z)$. Clearly, $\mathfrak{H}(m+2, R)$ is a subgroup of $SL(m+2, R)$. We call the subgroup $\mathfrak{H}(m+2, R)$ the *Heisenberg group* over R of *degree* $m+2$.

For simplicity, we write $\mathfrak{H}_{m+2}$ for $\mathfrak{H}(m+2, k[T])$. We denote by $\mathfrak{H}_{m+2}{}^{\circ}$ the set of all polynomial matrices $A(T) = \eta(a_1, \ldots, a_m, a_{m+1}, \ldots, a_{2m}, a_{2m+1})$ of $\mathfrak{H}_{m+2}$ satisfying $a_1, \ldots, a_m$ are linearly independent over k, and $a_{m+1}, \ldots, a_{2m}$ are also linearly independent over k.

2 Preliminaries

From now on, we assume that the characteristic p of k is positive.

In this section, we recall several results obtained in [7].

2.1 $\mathfrak{U}_n^E$

We have

$$\mathfrak{U}_n^E = \bigcup_{[d_1,\ldots,d_t] \text{ is an ordered partition of } n} \mathfrak{U}_{[d_1,\ldots,d_t]}^E.$$

2.2 $\mathfrak{U}_{[n]}^E$

We know the following lemma (see [7, Corollary 1.15, Lemma 1.18 and Lemma 1.20]).

Lemma 1 *Let $n \geq 2$ be an integer. Then the following assertions (1), (2), (3), (4) hold true:*

(1) We have $\mathfrak{U}_{[n]}^E \rightsquigarrow \mathfrak{J}_{[n]}^E$.
(2) $\mathfrak{U}_{[n]}^E \neq \emptyset$ if and only if $2 \leq n \leq p$.
(3) Let $A(T) = J(1, a_1, \ldots, a_{n-1}) \in \mathfrak{J}_{[n]}$. Then $A(T) \in \mathfrak{J}_{[n]}^E$ if and only if $(a_1, \ldots, a_{n-1}) \in \Psi_{n-1}$.
(4) Let $A(T), B(T) \in \mathfrak{U}_{[n]}$. If $P \in GL(n, k)$ satisfies $P^{-1}A(T)P = B(T)$, then P is an upper triangular matrix.

2.3 $\mathfrak{U}_{[n,1]}^E$

We know the following lemma (see [7, Corollary 1.23 and Lemma 1.24]).

Lemma 2 *Let $n \geq 2$ be an integer. Then the following assertions (1), (2), (3) hold true:*

(1) *We have $\mathfrak{U}^E_{[n,1]} \rightsquigarrow \mathfrak{J}^E_{[n,1]}$.*
(2) *$\mathfrak{U}^E_{[n,1]} \neq \emptyset$ if and only if $2 \leq n \leq p$.*
(3) *Let*

$$A(T) = \left(\begin{array}{c|c} J(1, a_1, \ldots, a_{n-1}) & a_n\, e_1 \\ \hline 0 & 1 \end{array}\right) \in \mathfrak{J}_{[n,1]}.$$

Then $A(T) \in \mathfrak{J}^E_{[n,1]}$ if and only if $(a_1, \ldots, a_{n-1}) \in \Psi_{n-1}$ and $a_n \in \mathfrak{P}$.

2.4 $\mathfrak{U}^E_{[1,n]}$

We know the following lemma (see [7, Corollary 1.27 and Lemma 1.28]).

Lemma 3 *Let $n \geq 2$ be an integer. Then the following assertions (1), (2), (3) hold true:*

(1) *We have $\mathfrak{U}^E_{[1,n]} \rightsquigarrow \mathfrak{J}^E_{[1,n]}$.*
(2) *$\mathfrak{U}^E_{[1,n]} \neq \emptyset$ if and only if $2 \leq n \leq p$.*
(3) *Let*

$$A(T) = \left(\begin{array}{c|c} 1 & a_n\, {}^{\tau}e_1 \\ \hline 0 & J(1, a_1, \ldots, a_{n-1}) \end{array}\right) \in \mathfrak{J}_{[1,n]}.$$

Then $A(T) \in \mathfrak{J}^E_{[1,n]}$ if and only if $(a_1, \ldots, a_{n-1}) \in \Psi_{n-1}$ and $a_n \in \mathfrak{P}$.

2.5 $\mathfrak{A}(i_1, i_2, i_3)^E$

We know the following lemma (see [7, Lemma 1.29]).

Lemma 4 *Let $i_1 \geq 1$, $i_2 \geq 0$, $i_3 \geq 1$ be integers. Let $A(T) = \Lambda(i_1, i_2, i_3; \alpha(T)) \in \mathfrak{A}(i_1, i_2, i_3)$. Then the following conditions (1), (2), (3) are equivalent:*

(1) *$A(T) \in \mathfrak{A}(i_1, i_2, i_3)^E$.*
(2) *$\alpha(T) + \alpha(T') = \alpha(T + T')$, where T, T' are indeterminates over k.*
(3) *All entries of $\alpha(T)$ are p-polynomials.*

2.6 $(\mathfrak{H}_{m+2}{}^{\circ})^E$

We know the following lemma (see [7, Theorem 3.3]).

Lemma 5 *Let $A(T)=\eta(a_1, \ldots, a_m, a_{m+1}, \ldots, a_{2m}, a_{2m+1}) \in \mathfrak{H}_{m+2}{}^{\circ}$. Then $A(T) \in (\mathfrak{H}_{m+2}{}^{\circ})^E$ if and only if the following conditions (1), (2), (3) hold true:*

(1) The polynomials a_i $(1 \le i \le 2m)$ are p-polynomials.
(2) There exists a unique regular symmetric matrix $S = (s_{i,j})_{1\le i,j\le m} \in GL(m,k)$ such that

$$(a_{m+1}, \ldots, a_{2m}) = (a_1, \ldots, a_m)\, S,$$

and furthermore if $p = 2$, the diagonal entires $s_{i,i}$ $(1 \le i \le m)$ of S are zero.
(3) There exists a unique p-polynomial $\alpha(T) \in k[T]$ such that

$$a_{2m+1}(T) = \begin{cases} \alpha(T) + \sum\limits_{1\le i<j\le m} s_{i,j} \cdot a_i(T) \cdot a_j(T) & \text{if } p = 2, \\ \alpha(T) + \dfrac{1}{2} \cdot a(T) \cdot S \cdot {}^t a(T) & \text{if } p \ge 3, \end{cases}$$

where $a(T) := (a_1(T), \ldots, a_m(T)) \in k[T]^{\oplus m}$.

3 Polynomial Matrices of $\mathfrak{U}_{[n,2]}$

In this section, we define four subsets $\mathfrak{J}^{0,0}_{[n,2]}, \mathfrak{J}^{0,1}_{[n,2]}, \mathfrak{J}^{1,0}_{[n,2]}, \mathfrak{J}^{1,1}_{[n,2]}$ of $\mathfrak{U}_{[n,2]}$, give necessary and sufficient conditions for a polynomial matrix of any one of the four subsets to be exponential (see Lemmas 7, 9, 10, 11), and show that the four subsets $\mathfrak{J}^{0,0}_{[n,2]}, \mathfrak{J}^{0,1}_{[n,2]}, \mathfrak{J}^{1,0}_{[n,2]}, \mathfrak{J}^{1,1}_{[n,2]}$ are mutually $GL(n+2,k)$-disjoint (see Lemma 13).

3.1 $\mathfrak{J}^{0,0}_{[n,2]}$

For $n \ge 3$, we denote by $\mathfrak{J}^{0,0}_{[n,2]}$ the set of all polynomial matrices $A^{0,0}(T)$ of $SL(n+2, k[T])$ with the form

$$A^{0,0}(T) = \left(\begin{array}{c|c|c} J(1, a_1, \ldots, a_{n-1}) & 0 & {}^t(c, 0, \ldots, 0) \\ \hline 0 & 1 & b \\ \hline 0 & 0 & 1 \end{array}\right)$$

$$\big(a_1, \ldots, a_{n-1}, b, c \in k[T], \quad a_1 \ne 0, \quad b \ne 0\big)$$

Clearly, $\mathfrak{J}^{0,0}_{[n,2]} \subset \mathfrak{U}_{[n,2]}$ and $(\mathfrak{J}^{0,0}_{[n,2]})^E \subset (\mathfrak{U}_{[n,2]})^E$.

Lemma 6 *Let $P := \sigma_{n,n+1}\sigma_{n-1,n} \cdots \sigma_{1,2} \in GL(n+2,k)$. Then*

$$P^{-1} \cdot A^{0,0}(T) \cdot P = \left(\begin{array}{c|c|c} 1 & 0 & b \\ \hline 0 & J(1, a_1, \ldots, a_{n-1}) & c\, e_1 \\ \hline 0 & 0 & 1 \end{array}\right).$$

In particular, we have $\mathfrak{J}^{0,0}_{[n,2]} \rightsquigarrow \mathfrak{U}_{[1,n,1]}$.

Proof The proof is straightforward. □

Lemma 7 *Assume* $3 \leq n \leq p$. *Let* $A^{0,0}(T) \in \mathfrak{J}^{0,0}_{[n,2]}$. *Then* $A^{0,0}(T) \in (\mathfrak{J}^{0,0}_{[n,2]})^E$ *if and only if the following conditions (1) and (2) are satisfied:*

(1) $(a_1, \ldots, a_{n-1}) \in \Psi_{n-1}$.
(2) b, c *are p-polynomials.*

Proof $A^{0,0}(T)$ is an exponential matrix if and only if $J(1, a_1, \ldots, a_{n-1})$ is an exponential matrix and $b, c \in \mathfrak{P}$. Using assertion (3) of Lemma 1, we have the desired equivalence. □

3.2 $\mathfrak{J}^{0,1}_{[n,2]}$

For $n \geq 3$, we denote by $\mathfrak{J}^{0,1}_{[n,2]}$ the set of all polynomial matrices $A^{0,1}(T)$ of $SL(n+2, k[T])$ with the form

$$A^{0,1}(T) = \left(\begin{array}{c|c|c} J(1, a_1, \ldots, a_{n-1}) & 0 & {}^t(c, a_{n-1}, \ldots, a_1) \\ \hline 0 & 1 & b \\ \hline 0 & 0 & 1 \end{array}\right)$$

$$\left(\begin{array}{l} a_1, \ldots, a_{n-1}, b, c \in k[T], \\ a_1, b \text{ are linearly independent over } k \end{array}\right).$$

Clearly, $\mathfrak{J}^{0,1}_{[n,2]} \subset \mathfrak{U}_{[n,2]}$ and $(\mathfrak{J}^{0,1}_{[n,2]})^E \subset (\mathfrak{U}_{[n,2]})^E$.

Lemma 8 *Let* $P := \sigma_{n,n+1}\sigma_{n-1,n} \cdots \sigma_{1,2} \in S_{n+2}$. *Then*

$$P^{-1} \cdot A^{0,1}(T) \cdot P = \left(\begin{array}{c|c} 1 & b\, {}^t e_{n+1} \\ \hline 0 & J(1, a_1, \ldots, a_{n-1}, c) \end{array}\right).$$

In particular, we have both $\mathfrak{J}^{0,1}_{[n,2]} \rightsquigarrow \mathfrak{J}^{1}_{[1,n+1]}$ *and* $\mathfrak{J}^{1}_{[1,n+1]} \rightsquigarrow \mathfrak{J}^{0,1}_{[n,2]}$.

Proof The proof is straightforward. □

Lemma 9 *Assume* $3 \leq n \leq p$. *Let* $A^{0,1}(T) \in \mathfrak{J}^{0,1}_{[n,2]}$. *Then* $A^{0,1}(T) \in (\mathfrak{J}^{0,1}_{[n,2]})^E$ *if and only if the following conditions (1) and (2) are satisfied:*

(1) $(a_1, \ldots, a_{n-1}, c) \in \Psi_n$.

(2) *b is a p-polynomial.*

Proof We know from Lemma 8 that $A^{0,1}(T)$ is an exponential matrix if and only if $J(1, a_1, \ldots, a_{n-1}, c)$ is an exponential matrix and $b \in \mathfrak{P}$. Using assertion (3) of Lemma 1, we have the desired equivalence. □

3.3 $\mathfrak{J}^{1,0}_{[n,2]}$

For $n \geq 3$, we denote by $\mathfrak{J}^{1,0}_{[n,2]}$ the set of all polynomial matrices $A^{1,0}(T)$ of $SL(n+2, k[T])$ with the form

$$A^{1,0}(T) = \left(\begin{array}{c|c|c} J(1, a_1, \ldots, a_{n-1}) & a_n\, e_1 & {}^t(c_1, c_2, 0, \ldots, 0) \\ \hline 0 & 1 & b \\ \hline 0 & 0 & 1 \end{array}\right)$$

$$\left(\begin{array}{l} a_1, \ldots, a_{n-1}, a_n, b, c_1, c_2 \in k[T], \\ a_1, a_n \text{ are linearly independent over } k, \\ \text{and } b \neq 0 \end{array}\right).$$

Clearly, $\mathfrak{J}^{1,0}_{[n,2]} \subset \mathfrak{U}_{[n,2]}$ and $(\mathfrak{J}^{1,0}_{[n,2]})^E \subset (\mathfrak{U}_{[n,2]})^E$.

Lemma 10 *Assume $3 \leq n \leq p$. Let $A^{1,0}(T) \in \mathfrak{J}^{1,0}_{[n,2]}$. Then $A^{1,0}(T) \in (\mathfrak{J}^{1,0}_{[n,2]})^E$ if and only if the following conditions (1), (2), (3), (4) are satisfied:*

(1) $(a_1, \ldots, a_{n-1}) \in \Psi_{n-1}$.
(2) *a_n, c_2, b are p-polynomials.*
(3) $(c_2, b) = (\lambda\, a_1 + \mu\, a_n,\ \mu\, a_1 + \nu\, a_n)$ *for some $\lambda, \mu, \nu \in k$.*
(4) $c_1 - \big((\lambda/2)\, a_1^2 + \mu\, a_1\, a_n + (\nu/2)\, a_n^2\big)$ *is a p-polynomial.*

Proof $A^{1,0}(T)$ is an exponential matrix if and only if

$$(*) \quad \begin{cases} (a_1, \ldots, a_{n-1}) \in \Psi_{n-1}, \\ a_n, c_2, b \text{ are } p\text{-polynomials}, \\ c_1(T') + a_1(T)\, c_2(T') + a_n(T)\, b(T') + c_1(T) = c_1(T + T'). \end{cases}$$

We shall prove the implication $(*) \Longrightarrow$ (1), (2), (3), (4). Since a_1, a_n are linearly independent over k, the third condition of $(*)$ implies that

$$(c_2, b) = (\lambda\, a_1 + \mu\, a_n,\ \mu\, a_1 + \nu\, a_n) \text{ for some } \lambda, \mu, \nu \in k$$

(see [7, Theorem 3.4]). So, (3) is obtained. Letting

$$f(T) := c_1 - \left(\frac{\lambda}{2}\, a_1^2 + \mu\, a_1\, a_n + \frac{\nu}{2}\, a_n^2\right),$$

we have

$$\begin{aligned}&f(T+T')-f(T)-f(T')\\&\quad=c_1(T+T')-c_1(T)-c_1(T')-(a_1(T)\,c_2(T')+a_n(T)\,b(T'))=0,\end{aligned}$$

which implies $f\in\mathfrak{P}$. Thus (4) is obtained. The proof of the implication (1), (2), (3), (4) $\Longrightarrow$ (∗) is clear. □

3.4 $\mathfrak{J}^{1,1}_{[n,2]}$

For $n\geq 3$, we denote by $\mathfrak{J}^{1,1}_{[n,2]}$ the set of all polynomial matrices $A^{1,1}(T)$ of $SL(n+2,k[T])$ with the form

$$A^{1,1}(T)=\left(\begin{array}{c|c|c}J(1,a_1,\ldots,a_{n-1}) & a_n\,e_1 & {}^t(c,a_{n-1},\ldots,a_1)\\\hline 0 & 1 & b\\\hline 0 & 0 & 1\end{array}\right)$$

$$\left(\begin{array}{l}a_1,\ldots,a_{n-1},a_n,b,c\in k[T],\\a_1,a_n\text{ are linearly independent over }\;k,\\a_n=\lambda\,b\text{ for some }\;\lambda\in k\setminus\{0\}\end{array}\right).$$

Clearly, a_1,b are linearly independent over k, $\mathfrak{J}^{1,1}_{[n,2]}\subset\mathfrak{U}_{[n,2]}$ and $(\mathfrak{J}^{1,1}_{[n,2]})^E\subset(\mathfrak{U}_{[n,2]})^E$.

Lemma 11 *Assume $3\leq n\leq p$. Let $A^{1,1}(T)\in\mathfrak{J}^{1,1}_{[n,2]}$. Then $A^{1,1}(T)\in(\mathfrak{J}^{1,1}_{[n,2]})^E$ if and only if the following conditions (1) and (2) are satisfied:*

(1) $(a_1,\ldots,a_{n-1},c-(\lambda b^2)/2)\in\Psi_n$.
(2) a_n,b are p-polynomials.

Proof $A^{1,1}(T)$ is an exponential matrix if and only if

$$(*)\quad\begin{cases}(a_1,\ldots,a_{n-1})\in\Psi_{n-1},\\a_n,b\text{ are }p\text{-polynomials},\\c(T')+\left(\displaystyle\sum_{i=1}^{n-1}a_i(T)\,a_{n-i}(T')\right)+a_n(T)\,b(T')+c(T)=c(T+T').\end{cases}$$

We shall prove the implication (∗) $\Longrightarrow$ (1), (2). Let $\widetilde{c}:=c-(\lambda/2)\,b^2$. We have

$$\widetilde{c}(T+T')-\widetilde{c}(T)-\widetilde{c}(T')=c(T+T')-c(T)-c(T')-a_n(T)\,b(T').$$

So, we know from the third condition of (∗) that $J(1,a_1,\ldots,a_{n-1},\widetilde{c})$ is an exponential matrix, which implies $(a_1,\ldots,a_{n-1},\widetilde{c})\in\Psi_n$. The proof of the implication (1), (2) $\Longrightarrow$ (∗) is clear. □

3.5 $\mathfrak{J}^{0,0}_{[n,2]}, \mathfrak{J}^{0,1}_{[n,2]}, \mathfrak{J}^{1,0}_{[n,2]}, \mathfrak{J}^{1,1}_{[n,2]}$ Are Mutually $GL(n+2,k)$-Disjoint

Let $A(T)$ be a polynomial matrix of $\mathrm{Mat}(n, k[T])$. We denote by V_n the column space of dimension n over k, and by V_n^* the row space of dimension n over k. We can regard the polynomial matrix $A(T)$ as a k-linear map from V_n to $k[T] \otimes_k V_n$ and also a k-linear map from V_n^* to $k[T] \otimes_k V_n^*$. We naturally regard V_n as a subspace of $k[T] \otimes_k V_n$, and V_n^* as a subspace of $k[T] \otimes_k V_n^*$. We define the subspaces ${V_n}^{A(T)}$ and $(V_n^*)^{A(T)}$ as

$$\begin{cases} {V_n}^{A(T)} := \{v \in V_n \mid A(T)v = v\}, \\ (V_n^*)^{A(T)} := \{v^* \in V_n^* \mid v^* A(T) = v^*\}. \end{cases}$$

If two polynomial matrices $A(T)$ and $B(T)$ are equivalent, we have $\dim_k {V_n}^{A(T)} = \dim_k {V_n}^{B(T)}$ and $\dim_k (V_n^*)^{A(T)} = \dim_k (V_n^*)^{B(T)}$.

Lemma 12 *Let $n \geq 3$ be an integer. Then the following assertions (1), (2), (3), (4) hold true:*

(1) For any $A(T) \in \mathfrak{J}^{0,0}_{[n,2]}$, we have

$$\begin{cases} {V_{n+2}}^{A(T)} = \{ {}^t(v_1, 0, \ldots, 0, v_{n+1}, 0) \mid v_1, v_{n+1} \in k \}, \\ (V_{n+2}^*)^{A(T)} = \{ (0, \ldots, 0, v_n^*, 0, v_{n+2}^*) \mid v_n^*, v_{n+2}^* \in k \}. \end{cases}$$

In particular, we have $\dim_k {V_{n+2}}^{A(T)} = \dim_k (V_{n+2}^*)^{A(T)} = 2$.

(2) For any $A(T) \in \mathfrak{J}^{0,1}_{[n,2]}$, we have

$$\begin{cases} {V_{n+2}}^{A(T)} = \{ {}^t(v_1, 0, \ldots, 0, v_{n+1}, 0) \mid v_1, v_{n+1} \in k \}, \\ (V_{n+2}^*)^{A(T)} = \{ (0, \ldots, 0, v_{n+2}^*) \mid v_{n+2}^* \in k \}. \end{cases}$$

In particular, we have $\dim_k {V_{n+2}}^{A(T)} = 2$ and $\dim_k (V_{n+2}^*)^{A(T)} = 1$.

(3) For any $A(T) \in \mathfrak{J}^{1,0}_{[n,2]}$, we have

$$\begin{cases} {V_{n+2}}^{A(T)} = \{ {}^t(v_1, 0, \ldots, 0) \mid v_1 \in k \}, \\ (V_{n+2}^*)^{A(T)} = \{ (0, \ldots, 0, v_n^*, 0, v_{n+2}^*) \mid v_n^*, v_{n+2}^* \in k \}. \end{cases}$$

In particular, we have $\dim_k {V_{n+2}}^{A(T)} = 1$ and $\dim_k (V_{n+2}^*)^{A(T)} = 2$.

(4) For any $A(T) \in \mathfrak{J}^{1,1}_{[n,2]}$, we have

$$\begin{cases} {V_{n+2}}^{A(T)} = \{ {}^t(v_1, 0, \ldots, 0) \mid v_1 \in k \}, \\ (V_{n+2}^*)^{A(T)} = \{ (0, \ldots, 0, v_{n+2}^*) \mid v_{n+2}^* \in k \}. \end{cases}$$

In particular, we have $\dim_k V_{n+2}{}^{A(T)} = \dim_k (V^*_{n+2})^{A(T)} = 1$.

Proof The proofs of assertions (1), (2), (3), (4) are straightforward. □

Let $S_1, \dots, S_N$ be subsets of $GL(n,k)$. We say that $S_1, \dots, S_N$ are *mutually $GL(n,k)$-disjoint* if $(PS_iP^{-1}) \cap (P'S_jP'^{-1}) = \emptyset$ for all $1 \le i, j \le N$ with $i \ne j$ and for all $P, P' \in GL(n,k)$.

Lemma 13 *Let (i,j), (i',j') be distinct two elements of the set $\{(0,0), (0,1), (1,0), (1,1)\}$. Then $\mathfrak{J}^{i,j}_{[n,2]}$ and $\mathfrak{J}^{i',j'}_{[n,2]}$ are mutually $GL(n+2,k)$-disjoint.*

Proof The proof follows from Lemma 12. □

3.6 Exponential Matrices of $\mathfrak{U}_{[n,2]}$

Lemma 14 *Assume $2 \le n \le p$. Let $A(T) \in SL(n+1, k[T])$ be an exponential matrix with the form*

$$A(T) = \left(\begin{array}{c|c} J(1, a_1, \dots, a_{n-1}) & {}^t(b_n, b_{n-1}, \dots, b_1) \\ \hline 0 & 1 \end{array}\right) \quad \left(\begin{array}{l} a_1, \dots, a_{n-1}, b_1, \dots, b_n \in k[T], \\ a_1 \ne 0 \end{array}\right)$$

Then there exists an upper triangular regular matrix P of $GLC^{(n+1)}(n+1,k)$ such that

$$P^{-1}A(T)P \in \begin{cases} (\mathfrak{J}^0_{[n,1]})^E \cup (\mathfrak{J}^1_{[n,1]})^E & \text{if } b_1 = 0, \\ \mathfrak{J}^E_{[n+1]} & \text{if } b_1 \ne 0. \end{cases}$$

Proof See the proofs of Theorem 1.22 and Lemma 1.14 in [7] (see Sect. 1.2 for the definition of $GLC^{(n+1)}(n+1,k)$). □

Theorem 1 *Let $n \ge 3$. Then the following assertions (1), (2), (3) hold true:*

(1) We have

$$\mathfrak{U}^E_{[n,2]} \rightsquigarrow (\mathfrak{J}^{0,0}_{[n,2]})^E \cup (\mathfrak{J}^{0,1}_{[n,2]})^E \cup (\mathfrak{J}^{1,0}_{[n,2]})^E \cup (\mathfrak{J}^{1,1}_{[n,2]})^E.$$

(2) $\mathfrak{U}^E_{[n,2]} \ne \emptyset$ if and only if $3 \le n \le p$.

(3) The four subsets $(\mathfrak{J}^{0,0}_{[n,2]})^E$, $(\mathfrak{J}^{0,1}_{[n,2]})^E$, $(\mathfrak{J}^{1,0}_{[n,2]})^E$, $(\mathfrak{J}^{1,1}_{[n,2]})^E$ of $SL(n+2, k[T])$ are $GL(n+2,k)$-mutually disjoint.

Proof (1) Let $A(T) \in \mathfrak{U}^E_{[n,2]}$. Then $A(T)$ has the form

$$A(T) = \left(\begin{array}{c|c|c} \alpha_{1,1} & \alpha_{1,2} & \alpha_{1,3} \\ \hline 0 & 1 & \alpha_{2,3} \\ \hline 0 & 0 & 1 \end{array}\right) \quad \left(\begin{array}{l} \alpha_{1,1} \in \mathfrak{U}_n, \quad \alpha_{1,2}, \alpha_{1,3} \in \mathrm{Mat}_{n,1}(k[T]), \\ \text{the } (n,1)\text{-th entry of } \alpha_{1,2} \text{ is zero}, \\ \alpha_{2,3} \in k[T] \setminus \{0\} \end{array}\right).$$

By assertions (1) and (4) of Lemma 1, there exists an upper triangular regular matrix P of $GL(n,k)$ such that $P^{-1}\alpha_{1,1}P$ has the form $P^{-1}\alpha_{1,1}P = J(1, a_1, \ldots, a_{n-1})$, where $a_i \in k[T]$ $(1 \le i \le n-1)$ and $a_1 \ne 0$. Thus, $A(T)$ is equivalent to a matrix $A_1(T)$ with the form

$$A_1(T) = \left(\begin{array}{c|c|c} J(1, a_1, \ldots, a_{n-1}) & \beta_{1,2} & \beta_{1,3} \\ \hline 0 & 1 & \beta_{2,3} \\ \hline 0 & 0 & 1 \end{array}\right)$$

$$\left(\begin{array}{l} \beta_{1,2},\ \beta_{1,3} \in \mathrm{Mat}_{n,1}(k[T]), \\ \text{the } (n,1)\text{-th entry of } \beta_{1,2} \text{ is zero}, \\ \beta_{2,3} \in k[T] \setminus \{0\} \end{array}\right).$$

Apply Lemma 14 to the exponential matrix $A_1(T)_{(n+2,n+2)}$. Then $A_1(T)$ is equivalent to a matrix $A_2(T)$ with the following form:

$$A_2(T) = \left(\begin{array}{c|c|c} J(1, a_1, \ldots, a_{n-1}) & a\, e_1 & \gamma_{1,3} \\ \hline 0 & 1 & \gamma_{2,3} \\ \hline 0 & 0 & 1 \end{array}\right)$$

$$\left(\begin{array}{l} a_1 \ne 0, \\ a_1, a \text{ are linearly independent over } k \text{ if } a \ne 0, \\ \gamma_{1,3} \in \mathrm{Mat}_{n,1}(k[T]), \quad \gamma_{2,3} \in k[T] \setminus \{0\} \end{array}\right).$$

Let $A_3(T) := {\sigma_{n+1,n+2}}^{-1} \cdot A_2(T) \cdot \sigma_{n+1,n+2}$. So,

$$A_3(T) := \left(\begin{array}{c|c|c} J(1, a_1, \ldots, a_{n-1}) & \gamma_{1,3} & a\, e_1 \\ \hline 0 & 1 & 0 \\ \hline 0 & \gamma_{2,3} & 1 \end{array}\right).$$

Note that the matrix $(A_3(T)_{(1,1)})_{(n+1,n+1)}$ is an exponential matrix. So, we can apply Lemma 14 to the exponential matrix $(A_3(T)_{(1,1)})_{(n+1,n+1)}$. Write $\gamma_{1,3} = {}^t(c_n, \ldots, c_1)$. Then $A_3(T)$ is equivalent to a matrix $A_4(T)$ with the form

$$A_4(T) = \begin{cases} \left(\begin{array}{c|c|c} J(1, a_1, \ldots, a_{n-1}) & {}^t(d_n, d_{n-1}, 0, \ldots, 0) & \mu\, a\, e_1 \\ \hline 0 & 1 & 0 \\ \hline 0 & \gamma & 1 \end{array}\right) & \text{if} \quad c_1 = 0, \\ \left(\begin{array}{c|c|c} J(1, a_1, \ldots, a_{n-1}) & {}^t(a_n, a_{n-1}, \ldots, a_1) & \mu\, a\, e_1 \\ \hline 0 & 1 & 0 \\ \hline 0 & \gamma & 1 \end{array}\right) & \text{if} \quad c_1 \ne 0, \end{cases}$$

where $\mu \in k \setminus \{0\}$. Clearly, $\gamma \ne 0$. In the case where $c_1 = 0$ and $a = 0$, we can apply Lemma 14 to $A_4(T)_{(n+2,n+2)}$, and thereby $A_4(T)$ is equivalent to a matrix with the form $A^{0,0}(T)$. In the case where $c_1 = 0$ and $a \ne 0$, then $A_4(T)$ is equivalent to a matrix ${\sigma_{n+1,n+2}}^{-1} \cdot A_4(T) \cdot \sigma_{n+1,n+2}$, which has the form $A^{1,0}(T)$. In the case where $c_1 \neq 0$ and $a = 0$, then $A_4(T)$ is equivalent to a matrix ${\sigma_{n+1,n+2}}^{-1} \cdot A_4(T) \cdot \sigma_{n+1,n+2}$, which has the form $A^{0,1}(T)$. In the case where $c_1 \neq 0$ and $a \neq 0$, we have

$$a_n(T') + \left(\sum_{i=1}^{n-1} a_i(T) \cdot a_{n-i}(T') \right) + a_n(T) + \mu\, a(T) \cdot \gamma(T') = a_n(T + T')$$

by calculating the $(1, n+1)$-th entries of the both sides of the equality $A_4(T) \cdot A_4(T') = A_4(T + T')$. So, $\mu\, a(T) = \lambda\, \gamma(T)$ for some $\lambda \in k \setminus \{0\}$. Thus $A_4(T)$ is equivalent to a matrix with the form $A^{1,1}(T)$. Hence we have

$$\mathfrak{U}^E_{[n,2]} \rightsquigarrow (\mathfrak{J}^{0,0}_{[n,2]})^E \cup (\mathfrak{J}^{0,1}_{[n,2]})^E \cup (\mathfrak{J}^{1,0}_{[n,2]})^E \cup (\mathfrak{J}^{1,1}_{[n,2]})^E .$$

(2) The proof follows from assertion (2) of Lemma 1.

(3) The proof follows from Lemma 13. □

4 Polynomial Matrices of $\mathfrak{U}_{[2,n]}$

For any $n \geq 3$, let

$$\mathfrak{J}^{0,0}_{[2,n]} := \{ {}^\tau A(T) \mid A(T) \in \mathfrak{J}^{0,0}_{[n,2]} \}, \qquad \mathfrak{J}^{0,1}_{[2,n]} := \{ {}^\tau A(T) \mid A(T) \in \mathfrak{J}^{1,0}_{[n,2]} \},$$
$$\mathfrak{J}^{1,0}_{[2,n]} := \{ {}^\tau A(T) \mid A(T) \in \mathfrak{J}^{0,1}_{[n,2]} \}, \qquad \mathfrak{J}^{1,1}_{[2,n]} := \{ {}^\tau A(T) \mid A(T) \in \mathfrak{J}^{1,1}_{[n,2]} \}.$$

Clearly, $\mathfrak{J}^{0,0}_{[2,n]}$, $\mathfrak{J}^{0,1}_{[2,n]}$, $\mathfrak{J}^{1,0}_{[2,n]}$ and $\mathfrak{J}^{1,1}_{[2,n]}$ are subsets of $\mathfrak{U}_{[2,n]}$.

As a Corollary of Theorem 1, we have the following:

Corollary 1 *Let $n \geq 3$ be an integer. Then the following assertions (1), (2), (3) hold true:*

(1) We have

$$\mathfrak{U}^E_{[2,n]} \rightsquigarrow (\mathfrak{J}^{0,0}_{[2,n]})^E \cup (\mathfrak{J}^{0,1}_{[2,n]})^E \cup (\mathfrak{J}^{1,0}_{[2,n]})^E \cup (\mathfrak{J}^{1,1}_{[2,n]})^E .$$

(2) $\mathfrak{U}^E_{[2,n]} \neq \emptyset$ if and only if $3 \leq n \leq p$.

(3) The four subsets $(\mathfrak{J}^{0,0}_{[2,n]})^E$, $(\mathfrak{J}^{0,1}_{[2,n]})^E$, $(\mathfrak{J}^{1,0}_{[2,n]})^E$, $(\mathfrak{J}^{1,1}_{[2,n]})^E$ of $SL(n+2, k[T])$ are $GL(n+2, k)$-mutually disjoint.

5 A Construction of Exponential Matrices

Let r, s be integers satisfying $r, s \geq 0$. Let $\mathfrak{R}$ be a ring with unity, where $\mathfrak{R}$ is not necessarily commutative. We denote by $\mathrm{Hook}_{r,s}(\mathfrak{R})$ the set of all matrices α of

$\mathrm{Mat}_{r+1,s+1}(\mathfrak{R})$ with the form

$$\alpha = \begin{pmatrix} \alpha_{1,1} & \alpha_{1,2} \\ 0 & \alpha_{2,2} \end{pmatrix},$$

where the sizes of the submatrices $\alpha_{1,1}$, $\alpha_{1,2}$, $\alpha_{2,2}$ of α are $r \times 1$, $r \times s$, $1 \times s$, respectively.

Let $n \geq 2$ be an integer and let $A(T) \in \mathrm{Mat}(n, k[T])$ be a polynomial matrix with the following form:

$$A(T) = \begin{pmatrix} A_{1,1}(T) & A_{1,2}(T) \\ A_{2,1}(T) & A_{2,2}(T) \end{pmatrix},$$

where the sizes of the submatrices $A_{1,1}(T)$, $A_{1,2}(T)$, $A_{2,1}(T)$, $A_{2,2}(T)$ of $A(T)$ are $1 \times (n-1)$, 1×1, $(n-1) \times (n-1)$, $(n-1) \times 1$, respectively. For any $\alpha(T) \in \mathrm{Hook}_{r,s}(k[T])$, we can define a polynomial matrix $A(T) \urcorner \alpha(T)$ of $\mathrm{Mat}(r+n+s, k[T])$ as follows:

$$A(T) \urcorner \alpha(T) := \left(\begin{array}{|c|c|c|c|} \hline I_r & O & \alpha_{1,1}(T) & \alpha_{1,2}(T) \\ \hline O & A_{1,1}(T) & A_{1,2}(T) & \alpha_{2,2}(T) \\ \hline O & A_{2,1}(T) & A_{2,2}(T) & O \\ \hline O & O & O & I_s \\ \hline \end{array}\right).$$

Let $\mathfrak{S}$ be a subset of $\mathrm{Mat}(n, k[T])$ and let $\mathfrak{T}$ be a subset of $\mathrm{Hook}_{r,s}(k[T])$. We define a subset $\mathfrak{S} \urcorner \mathfrak{T}$ of $\mathrm{Mat}(r+n+s, k[T])$ as

$$\mathfrak{S} \urcorner \mathfrak{T} := \{A(T) \urcorner \alpha(T) \in \mathrm{Mat}(r+n+s, k[T]) \mid A(T) \in \mathfrak{S} \text{ and } \alpha(T) \in \mathfrak{T}\}.$$

Clearly, we have $\mathfrak{J}_{[n,1]} = \mathfrak{J}_{[n]} \urcorner \mathrm{Hook}_{0,1}(k[T])$ and $\mathfrak{J}_{[1,n]} = \mathfrak{J}_{[n]} \urcorner \mathrm{Hook}_{1,0}(k[T])$.

Lemma 15 *Let $A(T) \in \mathfrak{U}_n$ and let $\alpha(T) \in \mathrm{Hook}_{r,s}(k[T])$. Then the following conditions (1) and (2) are equivalent:*

(1) $A(T) \urcorner \alpha(T)$ is an exponential matrix.
(2) $A(T)$ is an exponential matrix, and all entries of $\alpha(T)$ are p-polynomials.

Proof The proof is straightforward. □

Let n, r, s be integers satisfying $n \geq 2$ and $r, s \geq 0$. Let $\mathfrak{S}$ be a subset of $\mathrm{Mat}(n, k[T])$. We define a subset $[\![I_r, \mathfrak{S}, I_s]\!]$ of $\mathrm{Mat}(r+n+s, k[T])$ as

$$[\![I_r, \mathfrak{S}, I_s]\!] := \left\{ \left(\begin{array}{c|c|c} I_r & B_{1,2} & B_{1,3} \\ \hline O & A & B_{2,3} \\ \hline O & O & I_s \end{array}\right) \;\middle|\; \begin{array}{l} A \in \mathfrak{S}, \\ \text{the first column of } B_{1,2} \text{ is a zero vector,} \\ \text{and the } n\text{-th row of } B_{2,3} \text{ is a zero vector} \end{array} \right\}.$$

Clearly, $[\![I_1, \mathfrak{U}_n, I_1]\!] = \mathfrak{U}_{[1,n,1]}$ and ${}^{T}[\![I_r, \mathfrak{S}, I_s]\!] = [\![I_s, {}^{T}\mathfrak{S}, I_r]\!]$.

Lemma 16 *For all integers r, n, s satisfying $r \geq 0$, $2 \leq n \leq p$ and $s \geq 0$, we have*

$$[\![I_r,\ \mathfrak{U}_n,\ I_s]\!]^E \rightsquigarrow \mathfrak{J}^E_{[n]} \,\Lsh\, \mathrm{Hook}_{r,s}(\mathfrak{P}).$$

Proof We know from Lemma 1 that $[\![I_r,\ \mathfrak{U}_n,\ I_s]\!]^E \rightsquigarrow [\![I_r,\ \mathfrak{J}^E_{[n]},\ I_s]\!]$. We shall prove the following statement $(*)$:

$(*)$ For any integer $s \geq 1$ and for any $A(T) \in [\![\mathfrak{J}^E_{[n]},\ I_s]\!]$, there exists a regular matrix $P \in GLC^{(n+1,\dots,n+s)}(n+s, k)$ such that $P^{-1}A(T)P \in \mathfrak{J}^E_{[n]} \,\Lsh\, \mathrm{Hook}_{0,s}(\mathfrak{P})$.

We shall prove $(*)$ by induction on s. If $s = 1$, the proof follows from Lemma 14. So, assume the statement $(*)$ holds true for $s = i$ (≥ 1). Then take any matrix $A(T)$ of $[\![\mathfrak{J}^E_{[n]},\ I_{i+1}]\!]$. We can write $A(T)$ as

$$A(T) = \left(\begin{array}{c|c} J(1, a_1, \dots, a_{n-1}) & A_{1,2} \\ \hline O & I_{i+1} \end{array} \right),$$

where $a_1, \dots, a_{n-1} \in k[T]$, $a_1 \neq 0$, $A_{1,2} \in \mathrm{Mat}_{n,i+1}(k[T])$, and the n-th row of $A_{1,2}$ is a zero vector. There exists a regular matrix $P_1 \in GLC^{(n+1,\dots,n+i)}(n+i, k)$ such that

$$P_1^{-1} A(T)_{(n+i+1,\, n+i+1)} P_1 \in \mathfrak{J}^E_{[n]} \,\Lsh\, \mathrm{Hook}_{0,i}(\mathfrak{P}).$$

So, let

$$P_2 := \left(\begin{array}{c|c} P_1 & 0 \\ \hline 0 & 1 \end{array} \right) \cdot \sigma_{n+1,n+i+1}.$$

and consider the matrix $P_2^{-1}A(T)P_2 \in [\![\mathfrak{J}^E_{[n]},\ I_{i+1}]\!]$. We can apply the induction hypothesis to the matrix $(P_2^{-1}A(T)P_2)_{(n+i+1,\, n+i+1)}$. Then there exists a matrix $P_3 \in GLC^{(n+1,\dots,n+i)}(n+i, k)$ such that $P_3^{-1}(P_2^{-1}A(T)P_2)_{(n+i+1,\, n+i+1)}P_3 \in \mathfrak{J}^E_{[n]} \,\Lsh$ $\mathrm{Hook}_{0,i}(\mathfrak{P})$. So, let

$$P_4 := P_2 \cdot \left(\begin{array}{c|c} P_3 & 0 \\ \hline 0 & 1 \end{array} \right).$$

Then $P_4^{-1}A(T)P_4 \in \mathfrak{J}^E_{[n]} \,\Lsh\, \mathrm{Hook}_{0,i+1}(\mathfrak{P})$.

We can prove the following statement $(**)$:

$(**)$ For any integer $r \geq 1$ and for any $A(T) \in [\![I_r,\ \mathfrak{J}^E_{[n]}]\!]$, there exists a regular matrix $Q \in GLR^{(1,\dots,r)}(n+r, k)$ such that $Q^{-1}A(T)Q \in \mathfrak{J}^E_{[n]} \,\Lsh\, \mathrm{Hook}_{0,s}(\mathfrak{P})$, where $GLR^{(1,\dots,r)}(n+r, k) := {}^{\tau}GLC^{(n+1,\dots,n+r)}(n+r, k)$.

Now, take any matrix $A(T) \in [\![I_r,\ \mathfrak{J}^E_{[n]},\ I_s]\!]$. By considering submatrices of $A(T)$, we let P be as in $(*)$ and Q be as in $(**)$. Let

$$R := (I_r \oplus P) \cdot (Q \oplus I_s).$$

Then $R^{-1}A(T)R \in \mathfrak{J}^E_{[n]} \urcorner \mathrm{Hook}_{r,s}(\mathfrak{P})$. □

6 Polynomial Matrices of $\mathfrak{U}_5$

6.1 $\mathfrak{H}_5$

Lemma 17 *We have*

$$\begin{aligned}\mathfrak{H}_5 \rightsquigarrow & \left(\mathfrak{U}_{[3]} \urcorner \mathrm{Hook}_{0,2}(k[T])\right) \cup \mathfrak{U}_{[1,3,1]} \cup \left(\mathfrak{U}_{[3]} \urcorner \mathrm{Hook}_{2,0}(k[T])\right) \\ & \cup \mathfrak{A}(1,4) \cup \mathfrak{A}(2,3) \cup \mathfrak{A}(3,2) \cup \mathfrak{A}(4,1) \\ & \cup \mathfrak{H}_5{}^{\circ} \\ & \cup \left(\mathfrak{H}_4{}^{\circ} \urcorner \mathrm{Hook}_{0,1}(k[T])\right) \cup \left(\mathfrak{H}_4{}^{\circ} \urcorner \mathrm{Hook}_{1,0}(k[T])\right).\end{aligned}$$

Proof Let $A(T) \in \mathfrak{H}_5$. Write

$$A(T) = \left(\begin{array}{c|ccc|c} 1 & x_1 & x_2 & x_3 & z \\ \hline 0 & 1 & 0 & 0 & y_1 \\ 0 & 0 & 1 & 0 & y_2 \\ 0 & 0 & 0 & 1 & y_3 \\ \hline 0 & 0 & 0 & 0 & 1 \end{array}\right).$$

Let $\ell := \dim_k \mathrm{Span}_k\{x_1, x_2, x_3\}$. If $\ell = 0$, then $A(T) \in \mathfrak{A}(4,1)$. If $\ell = 1$, there exists a matrix P of $GL(5,k)$ such that $P^{-1}A(T)P \in \mathfrak{A}(3,2) \cup (\mathfrak{U}_{[3]} \urcorner \mathrm{Hook}_{2,0}(k[T]))$. If $\ell = 2$, then there exists a matrix P of $GL(5,k)$ such that

$$P^{-1}A(T)P = \begin{pmatrix} 1 & 0 & 0 & 0 & z^* \\ 0 & 1 & x_1^* & x_2^* & y_1^* \\ 0 & 0 & 1 & 0 & y_2^* \\ 0 & 0 & 0 & 1 & y_3^* \\ 0 & 0 & 0 & 0 & 1 \end{pmatrix} \quad (x_1^*, x_2^* \text{ are linearly independent over } k).$$

Let $r^* := \dim_k \mathrm{Span}_k\{y_2^*, y_3^*\}$. If $r^* = 0$, then $P^{-1}A(T)P \in \mathfrak{A}(2,3)$. If $r^* = 1$, then there exists a matrix P' of $GL(5,k)$ such that $P'^{-1}P^{-1}A(T)PP' \in \mathfrak{U}_{[1,3,1]}$. If $r^* = 2$, then $P^{-1}A(T)P \in \mathfrak{H}_4{}^{\circ} \urcorner \mathrm{Hook}_{1,0}(k[T])$. Thus if $\ell = 0, 1, 2$, then there exists a matrix P of $GL(5,k)$ such that

$$P^{-1}A(T)P \in \mathfrak{U}_{[1,3,1]} \cup \big(\mathfrak{U}_{[3]} \,\Lsh\, \mathrm{Hook}_{2,0}(k[T])\big) \\ \cup\, \mathfrak{A}(2,3) \cup \mathfrak{A}(3,2) \cup \mathfrak{A}(4,1) \cup \big(\mathfrak{H}_4{}^{\circ} \,\Lsh\, \mathrm{Hook}_{1,0}(k[T])\big).$$

If $\ell = 3$ and $\dim_k \mathrm{Span}_k\{y_1, y_2, y_3\} \le 2$, we can argue as above, and have

$$P^{-1}A(T)P \in \big(\mathfrak{U}_{[3]} \,\Lsh\, \mathrm{Hook}_{0,2}(k[T])\big) \cup \mathfrak{A}(1,4) \cup \big(\mathfrak{H}_4{}^{\circ} \,\Lsh\, \mathrm{Hook}_{0,1}(k[T])\big)$$

for some P of $GL(5,k)$. If $\ell = 3$ and $\dim_k \mathrm{Span}_k\{y_1, y_2, y_3\} = 3$, then $A(T) \in \mathfrak{H}_5{}^{\circ}$. □

6.2 $\mathfrak{H}\mathfrak{H}_5$, $\mathfrak{H}\mathfrak{H}_5{}^{\circ}$

We denote by $\mathfrak{H}\mathfrak{H}_5$ the set of all matrices $A(T)$ of $\mathfrak{U}_5$ with the following form:

$$A(T) = \left(\begin{array}{c|cc|c|c} 1 & x_1 & x_2 & z & w \\ \hline 0 & 1 & 0 & u_1 & v_1 \\ 0 & 0 & 1 & u_2 & v_2 \\ \hline 0 & 0 & 0 & 1 & 0 \\ \hline 0 & 0 & 0 & 0 & 1 \end{array}\right) \qquad (x_1, x_2, u_1, u_2, v_1, v_2, z, w \in k[T]).$$

We denote by $\mathfrak{H}\mathfrak{H}_5{}^{\circ}$ the set of all matrices $A(T)$ of $\mathfrak{H}\mathfrak{H}_5$ satisfying the following conditions (1), (2), (3), (4):

(1) x_1, x_2 are linearly independent over k.
(2) u_1, u_2 are linearly independent over k.
(3) v_1, v_2 are linearly independent over k.
(4) The two vectors (u_1, u_2), (v_1, v_2) of $k[T]^{\oplus 2}$ are linearly independent over k.

6.3 $\mathfrak{U}_{[3,1,1]}$, $\mathfrak{U}_{[1,3,1]}$, $\mathfrak{U}_{[1,1,3]}$

Lemma 18 *The following assertions (1), (2), (3) hold true:*

(1) $\mathfrak{U}_{[3,1,1]} \rightsquigarrow \mathfrak{U}_{[4,1]} \cup [\![\mathfrak{U}_{[3]},\ I_2]\!]$.
(2) $\mathfrak{U}_{[1,3,1]} = [\![I_1,\ \mathfrak{U}_{[3]},\ I_1]\!]$.
(3) $\mathfrak{U}_{[1,1,3]} \rightsquigarrow \mathfrak{U}_{[1,4]} \cup [\![I_2,\ \mathfrak{U}_{[3]}]\!]$.

Proof (1) Let

$$A(T) = \begin{pmatrix} 1 & a_1 & a_2 & a_3 & a_4 \\ 0 & 1 & a_5 & a_6 & a_7 \\ 0 & 0 & 1 & 0 & a_8 \\ 0 & 0 & 0 & 1 & 0 \\ 0 & 0 & 0 & 0 & 1 \end{pmatrix} \in \mathfrak{U}_{[3,1,1]}.$$

If $a_8 = 0$, then $A(T) \in [\![\mathfrak{U}_{[3]},\ I_2]\!]$. If $a_8 \neq 0$, then $\sigma_{4,5}^{-1} \cdot A(T) \cdot \sigma_{4,5} \in \mathfrak{U}_{[4,1]}$.

(2) is clear.

(3) is clear by taking the other transpose of the formula appearing in the above (1). □

6.4 $\mathfrak{U}_{[2,2,1]}$, $\mathfrak{U}_{[2,1,2]}$, $\mathfrak{U}_{[1,2,2]}$

Lemma 19 *Let*

$$A(T) = \begin{pmatrix} 1 & x_1 & x_2 & z & w \\ 0 & 1 & 0 & u_1 & v_1 \\ 0 & 0 & 1 & u_2 & v_2 \\ 0 & 0 & 0 & 1 & 0 \\ 0 & 0 & 0 & 0 & 1 \end{pmatrix} \in SL(5, k[T]).$$

Let $\ell := \dim_k \mathrm{Span}\{x_1, x_2\}$. *Then the following assertions (1), (2), (3) hold true:*

(1) If $\ell = 0$, *then* $A(T) \in \mathfrak{A}(3, 2)$.

(2) If $\ell = 1$, *then there exists a regular matrix* $P \in GL(5, k)$ *such that*

$$P^{-1}A(T)P \in [\![I_1,\ \mathfrak{U}_{[3]},\ I_1]\!] \cup \mathfrak{A}(2, 3).$$

(3) *If* $\ell = 2$, *then there exists a regular matrix* $P \in GL(5, k)$ *such that*

$$P^{-1}A(T)P \in \mathfrak{U}_{[3,2]} \cup [\![\mathfrak{U}_{[3]},\ I_2]\!] \cup \mathfrak{H}_5 \cup \mathfrak{H}\mathfrak{H}_5{}^{\circ}.$$

Proof (1) is clear.

(2) If $x_1 = 0$, $u_2 = 0$ and $v_2 = 0$, then $A(T) \in \mathfrak{A}(2, 3)$. If $x_1 = 0$, $u_2 = 0$ and $v_2 \neq 0$, then $\sigma_{4,5}^{-1} \cdot \sigma_{1,2}^{-1} \cdot A(T) \cdot \sigma_{1,2} \cdot \sigma_{4,5} \in \mathfrak{U}_{[1,3,1]}$. If $x_1 = 0$ and $u_2 \neq 0$, then $\sigma_{1,2}^{-1} \cdot A(T) \cdot \sigma_{1,2} \in \mathfrak{U}_{[1,3,1]}$.

If $x_1 \neq 0$, we have $x_2 = \lambda x_1$ for some $\lambda \in k$. Let

$$P := \big(I_1 \oplus J(1, -\lambda) \oplus I_2\big) \cdot \sigma_{2,3} \cdot \sigma_{1,2}.$$

If $u_1 + \lambda u_2 = 0$ and $v_1 + \lambda v_2 = 0$, then $P^{-1}A(T)P \in \mathfrak{A}(2, 3)$. If $u_1 + \lambda u_2 = 0$ and $v_1 + \lambda v_2 \neq 0$, then $\sigma_{4,5}^{-1} \cdot P^{-1}A(T)P \cdot \sigma_{4,5} \in \mathfrak{U}_{[1,3,1]}$. If $u_1 + \lambda u_2 \neq 0$, then $P^{-1}A(T)P \in \mathfrak{U}_{[1,3,1]}$.

(3) Let $u := {}^t(u_1, u_2) \in k[T]^{\oplus 2}$ and $v := {}^t(v_1, v_2) \in k[T]^{\oplus 2}$. Let d be the dimension of $\mathrm{Span}_k\{u, v\}$. If $0 \leq d \leq 1$, then there exists a regular matrix P of $GL(5, k)$ such that $P^{-1}A(T)P \in \mathfrak{H}_5$. So, assume $d = 2$. Let $\delta_1 := \dim_k \mathrm{Span}_k\{u_1, u_2\}$ and $\delta_2 := \dim_k \mathrm{Span}_k\{v_1, v_2\}$. Clearly, $\delta_1 \geq 1$ and $\delta_2 \geq 1$. Now, one of the following cases (a), (b), (c), (d), (e) occurs:

(a) $(\delta_1, \delta_2) = (2, 2)$.

(b) $(\delta_1, \delta_2) = (2, 1)$.
(c) $(\delta_1, \delta_2) = (1, 2)$.
(d) $(\delta_1, \delta_2) = (1, 1)$, and (u_1, v_1), (u_2, v_2) are linearly independent over k.
(e) $(\delta_1, \delta_2) = (1, 1)$, and (u_1, v_1), (u_2, v_2) are linearly dependent over k.

In the case (a), $A(T) \in \mathfrak{H}\mathfrak{H}_5{}^{\circ}$. In the case (b), the polynomial matrix $\sigma_{4,5}{}^{-1} \cdot A(T) \cdot \sigma_{4,5}$ arrives at the case (c). If the case (c) occurs, there exists a matrix P of $GL(5, k)$ such that $P^{-1}A(T)P \in \mathfrak{U}_{[3,2]}$. In fact, let

$$P := \begin{cases} \sigma_{2,3} \cdot \sigma_{3,4} & \text{if} \quad u_1 = 0, \\ \left(I_1 \oplus {}^t J(1, \lambda) \oplus I_2 \right) \cdot \sigma_{3,4} & \text{if} \quad u_1 \neq 0 \textit{ and } u_2 = \lambda u_1 (\lambda \in k). \end{cases}$$

In the case (d), we can similarly argue as in the case (c), and have $P^{-1}A(T)P \in \mathfrak{U}_{[3,2]}$ for some $P \in GL(5, k)$. The left case is (e). In this case, there exists a matrix P of $GL(5, k)$ such that $P^{-1}A(T)P \in [\![\mathfrak{U}_{[3]},\ I_2]\!]$. In fact, let

$$P := \begin{cases} \sigma_{2,3} \cdot \sigma_{3,4} & \text{if} \quad (u_1, v_1) = (0, 0), \\ \left(I_1 \oplus {}^t J(1, \lambda) \oplus I_2 \right) \cdot \sigma_{3,4} & \text{if} \quad (u_1, v_1) \neq (0, 0) \text{ and} \\ & \qquad (u_2, v_2) = \lambda(u_1, v_1) \ (\lambda \in k). \end{cases}$$

□

Lemma 20 *Let*

$$A(T) = \begin{pmatrix} 1 & x_1 & x_2 & z & w \\ 0 & 1 & 0 & u_1 & v_1 \\ 0 & 0 & 1 & 0 & v_2 \\ 0 & 0 & 0 & 1 & v_3 \\ 0 & 0 & 0 & 0 & 1 \end{pmatrix} \in SL(5, k[T]) \qquad (x_1 \neq 0).$$

Then there exists a regular matrix $P \in GL(5, k)$ *such that*

$$P^{-1}A(T)P \in \begin{cases} \mathfrak{H}_5 & \textit{if} \quad u_1 = 0, \\ \mathfrak{U}_{[3,1,1]} & \textit{if} \quad u_1 \neq 0 \textit{ and } v_2 = 0, \\ \mathfrak{U}_{[3,2]} & \textit{if} \quad u_1 \neq 0 \textit{ and } v_2 \neq 0. \end{cases}$$

Proof If $u_1 = 0$, the proof is clear. If $u_1 \neq 0$ and $v_2 = 0$, we have $\sigma_{3,4}^{-1} \cdot A(T) \cdot \sigma_{3,4} \in \mathfrak{U}_{[3,1,1]}$. If $u_1 \neq 0$ and $v_2 \neq 0$, we have $\sigma_{3,4}^{-1} \cdot A(T) \cdot \sigma_{3,4} \in \mathfrak{U}_{[3,2]}$. □

Lemma 21 *The following assertions (1), (2), (3) hold true:*

(1) $\mathfrak{U}_{[2,2,1]} \rightsquigarrow \mathfrak{U}_{[3,2]} \cup [\![\mathfrak{U}_{[3]},\ I_2]\!] \cup [\![I_1,\ \mathfrak{U}_{[3]},\ I_1]\!] \cup \mathfrak{A}(2, 3) \cup \mathfrak{H}_5 \cup \mathfrak{H}\mathfrak{H}_5{}^{\circ}$.
(2) $\mathfrak{U}_{[2,1,2]} \rightsquigarrow \mathfrak{U}_{[4,1]} \cup \mathfrak{U}_{[3,2]} \cup [\![\mathfrak{U}_{[3]},\ I_2]\!] \cup \mathfrak{H}_5$.
(3) $\mathfrak{U}_{[1,2,2]} \rightsquigarrow \mathfrak{U}_{[2,3]} \cup [\![I_1,\ \mathfrak{U}_{[3]},\ I_1]\!] \cup [\![I_2,\ \mathfrak{U}_{[3]}]\!] \cup \mathfrak{A}(3, 2) \cup \mathfrak{H}_5 \cup {}^{\tau}(\mathfrak{H}\mathfrak{H}_5{}^{\circ})$.

Proof (1) Use assertions (2) and (3) of Lemma 19.
(2) Use Lemmas 20 and 18.
(3) Use the above assertion (1). □

6.5 $\mathfrak{U}_{[2,1,1,1]}$, $\mathfrak{U}_{[1,2,1,1]}$, $\mathfrak{U}_{[1,1,2,1]}$, $\mathfrak{U}_{[1,1,1,2]}$, $\mathfrak{U}_{[1,1,1,1,1]}$

Lemma 22 *Let*

$$A(T) = \begin{pmatrix} 1 & 0 & x_1 & z_1 & z_2 \\ 0 & 1 & x_2 & z_3 & z_4 \\ 0 & 0 & 1 & y_1 & y_2 \\ 0 & 0 & 0 & 1 & 0 \\ 0 & 0 & 0 & 0 & 1 \end{pmatrix} \in SL(5, k[T]).$$

Then the following assertions (1), (2), (3) hold true:

(1) If $(x_1, x_2) = (0, 0)$, then $A(T) \in \mathfrak{A}(3, 2)$.
(2) If $(y_1, y_2) = (0, 0)$, then $A(T) \in \mathfrak{A}(2, 3)$.
(3) If $(x_1, x_2) \neq (0, 0)$ and $(y_1, y_2) \neq (0, 0)$, then there exists a regular matrix $P \in GL(5, k)$ such that $P^{-1}A(T)P \in [\![I_1, \mathfrak{U}_{[3]}, I_1]\!]$.

Proof (1) is clear.
(2) is also clear.
(3) Let

$$P := \begin{cases} \sigma_{1,2} \cdot \sigma_{4,5} & \text{if } x_2 = 0, \quad y_1 = 0, \\ \sigma_{1,2} & \text{if } x_2 = 0, \quad y_1 \neq 0, \\ \sigma_{4,5} & \text{if } x_2 \neq 0, \quad y_1 = 0, \\ I_5 & \text{if } x_2 \neq 0, \quad y_1 \neq 0. \end{cases}$$

Then we have $P^{-1}A(T)P \in [\![I_1, \mathfrak{U}_{[3]}, I_1]\!]$. □

Lemma 23 *The following assertions (1), (2), (3), (4), (5) hold true:*

(1) $\mathfrak{U}_{[2,1,1,1]} \rightsquigarrow \mathfrak{U}_{[4,1]} \cup \mathfrak{U}_{[3,2]} \cup [\![\mathfrak{U}_{[3]}, I_2]\!] \cup \mathfrak{H}_5$.
(2) $\mathfrak{U}_{[1,2,1,1]} \rightsquigarrow [\![I_1, \mathfrak{U}_{[3]}, I_1]\!] \cup \mathfrak{A}(2, 3)$.
(3) $\mathfrak{U}_{[1,1,2,1]} \rightsquigarrow [\![I_1, \mathfrak{U}_{[3]}, I_1]\!] \cup \mathfrak{A}(3, 2)$.
(4) $\mathfrak{U}_{[1,1,1,2]} \rightsquigarrow \mathfrak{U}_{[2,3]} \cup \mathfrak{U}_{[1,4]} \cup [\![I_2, \mathfrak{U}_{[3]}]\!] \cup \mathfrak{H}_5$.
(5) $\mathfrak{U}_{[1,1,1,1,1]} \rightsquigarrow [\![I_1, \mathfrak{U}_{[3]}, I_1]\!] \cup \mathfrak{A}(2, 3) \cup \mathfrak{A}(3, 2)$.

Proof (1) Use Lemmas 20 and 18.
(2) Use Lemma 22.
(3) Use the above assertion (2).
(4) Use the above assertion (1).
(5) Use Lemma 22. □

6.6 $\mathfrak{U}_5$

We have

$$\mathfrak{U}_5 = \mathfrak{U}_{[5]} \cup \mathfrak{U}_{[4,1]} \cup \mathfrak{U}_{[3,2]} \cup \mathfrak{U}_{[2,3]} \cup \mathfrak{U}_{[1,4]}$$
$$\cup\, \mathfrak{U}_{[3,1,1]} \cup \mathfrak{U}_{[1,3,1]} \cup \mathfrak{U}_{[1,1,3]} \cup \mathfrak{U}_{[2,2,1]} \cup \mathfrak{U}_{[2,1,2]} \cup \mathfrak{U}_{[1,2,2]}$$
$$\cup\, \mathfrak{U}_{[2,1,1,1]} \cup \mathfrak{U}_{[1,2,1,1]} \cup \mathfrak{U}_{[1,1,2,1]} \cup \mathfrak{U}_{[1,1,1,2]} \cup \mathfrak{U}_{[1,1,1,1,1]}.$$

Using Lemmas 18, 21 and 23, we have the following lemma:

Lemma 24 *We have*

$$\mathfrak{U}_5 \rightsquigarrow \mathfrak{U}_{[5]} \cup \mathfrak{U}_{[4,1]} \cup \mathfrak{U}_{[3,2]} \cup \mathfrak{U}_{[2,3]} \cup \mathfrak{U}_{[1,4]}$$
$$\cup\, [\![\mathfrak{U}_{[3]},\, I_2]\!] \cup [\![I_1,\, \mathfrak{U}_{[3]},\, I_1]\!] \cup [\![I_2,\, \mathfrak{U}_{[3]}]\!]$$
$$\cup\, \mathfrak{A}(2,3) \cup \mathfrak{A}(3,2)$$
$$\cup\, \mathfrak{H}_5$$
$$\cup\, \mathfrak{H}\mathfrak{H}_5{}^{\circ} \cup {}^{T}(\mathfrak{H}\mathfrak{H}_5{}^{\circ}).$$

6.7 *Exponential Matrices of* $\mathfrak{U}_5$

For simplicity, we frequently write $\mathrm{Hk}_{r,s}$ in place of $\mathrm{Hook}_{r,s}(\mathfrak{P})$.

We define a subset $\mathfrak{H}_4{}^{\varepsilon}$ of $\mathfrak{H}_4{}^{\circ}$ as

$$\mathfrak{H}_4{}^{\varepsilon} := \left\{ \begin{pmatrix} 1 & a & b & \varphi \\ 0 & 1 & 0 & b \\ 0 & 0 & 1 & a \\ 0 & 0 & 0 & 1 \end{pmatrix} \;\middle|\; \begin{array}{l} a, b, \varphi \in k[T], \\ a, b \text{ are linearly independent over} k \end{array} \right\}.$$

Lemma 25 *The following assertions (1)–(12) hold true:*

(1) $\mathfrak{U}^E_{[5]} \rightsquigarrow \mathfrak{J}^E_{[5]}$.

(2) $\mathfrak{U}^E_{[4,1]} \rightsquigarrow \mathfrak{J}^E_{[4]} \,\urcorner\, \mathrm{Hk}_{0,1}$.

(3) $\mathfrak{U}^E_{[3,2]} \rightsquigarrow (\mathfrak{J}^{0,0}_{[3,2]})^E \cup (\mathfrak{J}^{0,1}_{[3,2]})^E \cup (\mathfrak{J}^{1,0}_{[3,2]})^E \cup (\mathfrak{J}^{1,1}_{[3,2]})^E$
$\rightsquigarrow [\![I_1,\, \mathfrak{U}_3,\, I_1]\!]^E \cup \mathfrak{J}^E_{[1,4]} \cup (\mathfrak{J}^{1,0}_{[3,2]})^E \cup (\mathfrak{J}^{1,1}_{[3,2]})^E$.

(4) $\mathfrak{U}^E_{[2,3]} \rightsquigarrow (\mathfrak{J}^{0,0}_{[2,3]})^E \cup (\mathfrak{J}^{1,0}_{[2,3]})^E \cup (\mathfrak{J}^{0,1}_{[2,3]})^E \cup (\mathfrak{J}^{1,1}_{[2,3]})^E$
$\rightsquigarrow [\![I_1,\, \mathfrak{U}_3,\, I_1]\!]^E \cup \mathfrak{J}^E_{[4,1]} \cup (\mathfrak{J}^{0,1}_{[2,3]})^E \cup (\mathfrak{J}^{1,1}_{[2,3]})^E$.

(5) $\mathfrak{U}^E_{[1,4]} \rightsquigarrow \mathfrak{J}^E_{[4]} \,\urcorner\, \mathrm{Hk}_{1,0}$.

(6) $[\![\mathfrak{U}_{[3]},\, I_2]\!]^E \rightsquigarrow \mathfrak{J}^E_{[3]} \,\urcorner\, \mathrm{Hk}_{0,2}$.

(7) $[\![I_1,\, \mathfrak{U}_{[3]},\, I_1]\!]^E \rightsquigarrow \mathfrak{J}^E_{[3]} \,\urcorner\, \mathrm{Hk}_{1,1}$.

(8) $[\![I_2,\, \mathfrak{U}_{[3]}]\!]^E \rightsquigarrow \mathfrak{J}^E_{[3]} \,\urcorner\, \mathrm{Hk}_{2,0}$.

(9) $\mathfrak{H}_5^E \rightsquigarrow \left(\mathfrak{J}_{[3]}^E \Lsh \mathrm{Hk}_{0,2}\right) \cup \left(\mathfrak{J}_{[3]}^E \Lsh \mathrm{Hk}_{1,1}\right) \cup \left(\mathfrak{J}_{[3]}^E \Lsh \mathrm{Hk}_{2,0}\right)$
$\cup\, \mathfrak{A}(1,4)^E \cup \mathfrak{A}(2,3)^E \cup \mathfrak{A}(3,2)^E \cup \mathfrak{A}(4,1)^E$
$\cup\, (\mathfrak{H}_5{}^\circ)^E$
$\cup \left((\mathfrak{H}_4{}^\circ)^E \Lsh \mathrm{Hk}_{0,1}\right) \cup \left((\mathfrak{H}_4{}^\circ)^E \Lsh \mathrm{Hk}_{1,0}\right).$

(10) If $p = 2$, then $(\mathfrak{H}_5{}^\circ)^E = \emptyset$.
(11) If $p = 2$, then $(\mathfrak{H}_4{}^\circ)^E \rightsquigarrow (\mathfrak{H}_4{}^\varepsilon)^E$.
(12) If $p = 2$, then $(\mathfrak{H}\mathfrak{H}_5{}^\circ)^E = \emptyset$ and $\left({}^\tau(\mathfrak{H}\mathfrak{H}_5{}^\circ)\right)^E = \emptyset$.

Proof (1) See Lemma 1.
(2) See Lemma 2.
(3) See Theorem 1 and Lemmas 6 and 8.
(4) Use the above assertion (3).
(5) See Lemma 3.
(6), (7), (8) are special cases of Lemma 16.
(9) Use Lemmas 17, 15, 18, 16.
(10) Suppose $(\mathfrak{H}_5{}^\circ)^E \neq \emptyset$. By Lemma 5, there exists a symmetric matrix S of $GL(3,k)$ whose all diagonal entries are zero. We can write S as

$$S = \begin{pmatrix} 0 & \lambda & \mu \\ \lambda & 0 & \nu \\ \mu & \nu & 0 \end{pmatrix} \qquad (\lambda, \mu, \nu \in k).$$

So, $\det(S) = 0$. This contradicts that S is regular.
(11) is clear.
(12) It is enough to show that $(\mathfrak{H}\mathfrak{H}_5{}^\circ)^E = \emptyset$. Suppose that there exists an element $A(T)$ of $(\mathfrak{H}\mathfrak{H}_5{}^\circ)^E$. Then we know from Lemma 5 that $A(T)$ has the following form:

$$A(T) = \left(\begin{array}{c|cc|c|c} 1 & x_1 & x_2 & z & w \\ \hline 0 & 1 & 0 & \lambda x_2 & \mu x_2 \\ \hline 0 & 0 & 1 & \lambda x_1 & \mu x_1 \\ \hline 0 & 0 & 0 & 1 & 0 \\ \hline 0 & 0 & 0 & 0 & 1 \end{array}\right) \qquad (\lambda, \mu \in k \setminus \{0\}).$$

Clearly, $(\lambda x_2, \lambda x_1)$ and $(\mu x_2, \mu x_1)$ are linearly dependent over k, which contradicts $A(T) \in \mathfrak{H}\mathfrak{H}_5{}^\circ$. □

Theorem 2 *We have*

$$\begin{aligned}\mathfrak{U}_5^E \rightsquigarrow\ & \mathfrak{J}_{[5]}^E \cup \left(\mathfrak{J}_{[4]}^E \Lsh \mathrm{Hk}_{0,1}\right) \cup \left(\mathfrak{J}_{[4]}^E \Lsh \mathrm{Hk}_{1,0}\right) \\ & \cup (\mathfrak{J}_{[3,2]}^{1,0})^E \cup (\mathfrak{J}_{[3,2]}^{1,1})^E \cup (\mathfrak{J}_{[2,3]}^{1,1})^E \cup (\mathfrak{J}_{[2,3]}^{0,1})^E \\ & \cup \left(\mathfrak{J}_{[3]}^E \Lsh \mathrm{Hk}_{0,2}\right) \cup \left(\mathfrak{J}_{[3]}^E \Lsh \mathrm{Hk}_{1,1}\right) \cup \left(\mathfrak{J}_{[3]}^E \Lsh \mathrm{Hk}_{2,0}\right) \\ & \cup \mathfrak{A}(1,4)^E \cup \mathfrak{A}(2,3)^E \cup \mathfrak{A}(3,2)^E \cup \mathfrak{A}(4,1)^E \\ & \cup (\mathfrak{H}_5{}^\circ)^E\end{aligned}$$

$$\cup \left((\mathfrak{H}_4{}^{\circ})^E \Lsh \mathrm{Hk}_{0,1} \right) \cup \left((\mathfrak{H}_4{}^{\circ})^E \Lsh \mathrm{Hk}_{1,0} \right)$$
$$\cup\ (\mathfrak{H}\mathfrak{H}_5{}^{\circ})^E \cup \left({}^{\tau}(\mathfrak{H}\mathfrak{H}_5{}^{\circ}) \right)^E.$$

Proof The proof follows from Lemma 24, Lemmas 1.10 and 1.11 in [7], and Lemma 25. □

Corollary 2 *The following assertions (1), (2), (3) hold true:*

(1) Assume $p = 2$. Then any exponential matrix of $SL(5, k[T])$ is equivalent to an exponential matrix belonging to at least one of the following 6 sets:

$$\mathfrak{A}(1,4)^E, \quad \mathfrak{A}(2,3)^E, \quad \mathfrak{A}(3,2)^E, \quad \mathfrak{A}(4,1)^E,$$
$$(\mathfrak{H}_4{}^{\varepsilon})^E \Lsh \mathrm{Hk}_{0,1}, \quad (\mathfrak{H}_4{}^{\varepsilon})^E \Lsh \mathrm{Hk}_{1,0}.$$

(2) Assume $p = 3$. Then any exponential matrix of $SL(5, k[T])$ is equivalent to an exponential matrix belonging to at least one of the following 16 sets:

$$(\mathfrak{J}^{1,0}_{[3,2]})^E, \quad (\mathfrak{J}^{1,1}_{[3,2]})^E, \quad (\mathfrak{J}^{1,1}_{[2,3]})^E, \quad (\mathfrak{J}^{0,1}_{[2,3]})^E,$$
$$\mathfrak{J}^E_{[3]} \Lsh \mathrm{Hk}_{0,2}, \quad \mathfrak{J}^E_{[3]} \Lsh \mathrm{Hk}_{1,1}, \quad \mathfrak{J}^E_{[3]} \Lsh \mathrm{Hk}_{2,0},$$
$$\mathfrak{A}(1,4)^E, \quad \mathfrak{A}(2,3)^E, \quad \mathfrak{A}(3,2)^E, \quad \mathfrak{A}(4,1)^E,$$
$$(\mathfrak{H}_5{}^{\circ})^E,$$
$$(\mathfrak{H}_4{}^{\circ})^E \Lsh \mathrm{Hk}_{0,1}, \quad (\mathfrak{H}_4{}^{\circ})^E \Lsh \mathrm{Hk}_{1,0},$$
$$(\mathfrak{H}\mathfrak{H}_5{}^{\circ})^E, \quad \left({}^{\tau}(\mathfrak{H}\mathfrak{H}_5{}^{\circ}) \right)^E.$$

(3) Assume $p \geq 5$. Then any exponential matrix of $SL(5, k[T])$ is equivalent to an exponential matrix belonging to at least one of the following 19 sets:

$$\mathfrak{J}^E_{[5]},$$
$$\mathfrak{J}^E_{[4]} \Lsh \mathrm{Hk}_{0,1}, \quad \mathfrak{J}^E_{[4]} \Lsh \mathrm{Hk}_{1,0},$$
$$(\mathfrak{J}^{1,0}_{[3,2]})^E, \quad (\mathfrak{J}^{1,1}_{[3,2]})^E, \quad (\mathfrak{J}^{1,1}_{[2,3]})^E, \quad (\mathfrak{J}^{0,1}_{[2,3]})^E,$$
$$\mathfrak{J}^E_{[3]} \Lsh \mathrm{Hk}_{0,2}, \quad \mathfrak{J}^E_{[3]} \Lsh \mathrm{Hk}_{1,1}, \quad \mathfrak{J}^E_{[3]} \Lsh \mathrm{Hk}_{2,0},$$
$$\mathfrak{A}(1,4)^E, \quad \mathfrak{A}(2,3)^E, \quad \mathfrak{A}(3,2)^E, \quad \mathfrak{A}(4,1)^E,$$
$$(\mathfrak{H}_5{}^{\circ})^E,$$
$$(\mathfrak{H}_4{}^{\circ})^E \Lsh \mathrm{Hk}_{0,1}, \quad (\mathfrak{H}_4{}^{\circ})^E \Lsh \mathrm{Hk}_{1,0},$$
$$(\mathfrak{H}\mathfrak{H}_5{}^{\circ})^E, \quad \left({}^{\tau}(\mathfrak{H}\mathfrak{H}_5{}^{\circ}) \right)^E.$$

In the following, we describe the above classes appearing in Theorem 2 and in Corollary 2 for user-friendliness:

$\mathfrak{J}^{E}_{[5]}$	$\begin{pmatrix} 1 & a & \frac{1}{2}a^2+b & \frac{1}{6}a^3+ab+c & \frac{1}{24}a^4+\frac{1}{2}a^2b+\frac{1}{2}b^2+ac+d \\ 0 & 1 & a & \frac{1}{2}a^2+b & \frac{1}{6}a^3+ab+c \\ 0 & 0 & 1 & a & \frac{1}{2}a^2+b \\ 0 & 0 & 0 & 1 & a \\ 0 & 0 & 0 & 0 & 1 \end{pmatrix}$ $(\, a, b, c, d \in \mathfrak{P}, \quad a \neq 0\,)$
$\mathfrak{J}^{E}_{[4]} \Lsh \mathrm{Hk}_{0,1}$	$\left(\begin{array}{cccc\|c} 1 & a & \frac{1}{2}a^2+b & \frac{1}{6}a^3+ab+c & d \\ 0 & 1 & a & \frac{1}{2}a^2+b & 0 \\ 0 & 0 & 1 & a & 0 \\ 0 & 0 & 0 & 1 & 0 \\ \hline 0 & 0 & 0 & 0 & 1 \end{array}\right)$ $(\, a, b, c, d \in \mathfrak{P}, \quad a \neq 0\,)$
$\mathfrak{J}^{E}_{[4]} \Lsh \mathrm{Hk}_{1,0}$	$\left(\begin{array}{c\|cccc} 1 & 0 & 0 & 0 & d \\ \hline 0 & 1 & a & \frac{1}{2}a^2+b & \frac{1}{6}a^3+ab+c \\ 0 & 0 & 1 & a & \frac{1}{2}a^2+b \\ 0 & 0 & 0 & 1 & a \\ 0 & 0 & 0 & 0 & 1 \end{array}\right)$ $(\, a, b, c, d \in \mathfrak{P}, \quad a \neq 0\,)$
$(\mathfrak{J}^{1,0}_{[3,2]})^{E}$	$\left(\begin{array}{ccc\|cc} 1 & a & \frac{1}{2}a^2+b & c & \frac{\lambda}{2}a^2+\mu ac+\frac{\nu}{2}c^2+d \\ 0 & 1 & a & 0 & \lambda a+\mu c \\ 0 & 0 & 1 & 0 & 0 \\ \hline 0 & 0 & 0 & 1 & \mu a+\nu c \\ 0 & 0 & 0 & 0 & 1 \end{array}\right)$ $\begin{pmatrix} a, b, c, d \in \mathfrak{P}, \quad a, c \text{ are linearly independent over } k, \\ \lambda, \mu, \nu \in k \end{pmatrix}$
$(\mathfrak{J}^{1,1}_{[3,2]})^{E}$	$\left(\begin{array}{ccc\|cc} 1 & a & \frac{1}{2}a^2+b & \lambda c & \frac{1}{6}a^3+ab+\frac{\lambda}{2}c^2+d \\ 0 & 1 & a & 0 & \frac{1}{2}a^2+b \\ 0 & 0 & 1 & 0 & a \\ \hline 0 & 0 & 0 & 1 & c \\ 0 & 0 & 0 & 0 & 1 \end{array}\right)$ $\begin{pmatrix} a, b, c, d \in \mathfrak{P}, \quad a, c \text{ are linearly independent over } k, \\ \lambda \in k \setminus \{0\} \end{pmatrix}$

<table>
<tr><td>$(\mathfrak{J}^{1,1}_{[2,3]})^E$</td><td>$$\begin{pmatrix}\begin{array}{cc|ccc} 1 & c & a & \frac{1}{2}a^2+b & \frac{1}{6}a^3+ab+\frac{\lambda}{2}c^2+d \\ 0 & 1 & 0 & 0 & \lambda c \\ \hline 0 & 0 & 1 & a & \frac{1}{2}a^2+b \\ 0 & 0 & 0 & 1 & a \\ 0 & 0 & 0 & 0 & 1 \end{array}\end{pmatrix}$$ $$\begin{pmatrix} a, b, c, d \in \mathfrak{P}, \quad a, c \text{ are linearly independent over } k, \\ \lambda \in k \setminus \{0\} \end{pmatrix}$$</td></tr>
<tr><td>$(\mathfrak{J}^{0,1}_{[2,3]})^E$</td><td>$$\begin{pmatrix}\begin{array}{cc|ccc} 1 & \mu a+\nu c & 0 & \lambda a+\mu c & \frac{\lambda}{2}a^2+\mu ac+\frac{\nu}{2}c^2+d \\ 0 & 1 & 0 & 0 & c \\ \hline 0 & 0 & 1 & a & \frac{1}{2}a^2+b \\ 0 & 0 & 0 & 1 & a \\ 0 & 0 & 0 & 0 & 1 \end{array}\end{pmatrix}$$ $$\begin{pmatrix} a, b, c, d \in \mathfrak{P}, \quad a, c \text{ are linearly independent over } k, \\ \lambda, \mu, \nu \in k \end{pmatrix}$$</td></tr>
<tr><td>$\mathfrak{J}^E_{[3]} \urcorner \mathrm{Hk}_{0,2}$</td><td>$$\begin{pmatrix}\begin{array}{ccc|cc} 1 & a & \frac{1}{2}a^2+b & c & d \\ 0 & 1 & a & 0 & 0 \\ 0 & 0 & 1 & 0 & 0 \\ \hline 0 & 0 & 0 & 1 & 0 \\ 0 & 0 & 0 & 0 & 1 \end{array}\end{pmatrix}$$ $$(a, b, c, d \in \mathfrak{P}, \quad a \neq 0)$$</td></tr>
<tr><td>$\mathfrak{J}^E_{[3]} \urcorner \mathrm{Hk}_{1,1}$</td><td>$$\begin{pmatrix}\begin{array}{c|ccc|c} 1 & 0 & 0 & c & d \\ \hline 0 & 1 & a & \frac{1}{2}a^2+b & e \\ 0 & 0 & 1 & a & 0 \\ 0 & 0 & 0 & 1 & 0 \\ \hline 0 & 0 & 0 & 0 & 1 \end{array}\end{pmatrix}$$ $$(a, b, c, d, e \in \mathfrak{P}, \quad a \neq 0)$$</td></tr>
<tr><td>$\mathfrak{J}^E_{[3]} \urcorner \mathrm{Hk}_{2,0}$</td><td>$$\begin{pmatrix}\begin{array}{cc|ccc} 1 & 0 & 0 & 0 & d \\ 0 & 1 & 0 & 0 & c \\ \hline 0 & 0 & 1 & a & \frac{1}{2}a^2+b \\ 0 & 0 & 0 & 1 & a \\ 0 & 0 & 0 & 0 & 1 \end{array}\end{pmatrix}$$ $$(a, b, c, d \in \mathfrak{P}, \quad a \neq 0)$$</td></tr>
<tr><td>$\mathfrak{A}(1, 4)^E$</td><td>$$\begin{pmatrix}\begin{array}{c|cccc} 1 & a & b & c & d \\ \hline 0 & 1 & 0 & 0 & 0 \\ 0 & 0 & 1 & 0 & 0 \\ 0 & 0 & 0 & 1 & 0 \\ 0 & 0 & 0 & 0 & 1 \end{array}\end{pmatrix}$$ $$(a, b, c, d \in \mathfrak{P})$$</td></tr>
</table>

$\mathfrak{A}(2,3)^E$	$\left(\begin{array}{cc\|ccc} 1 & 0 & a & b & c \\ 0 & 1 & d & e & f \\ \hline 0 & 0 & 1 & 0 & 0 \\ 0 & 0 & 0 & 1 & 0 \\ 0 & 0 & 0 & 0 & 1 \end{array}\right)$ $(a, b, c, d, e, f \in \mathfrak{P})$
$\mathfrak{A}(3,2)^E$	$\left(\begin{array}{ccc\|cc} 1 & 0 & 0 & a & b \\ 0 & 1 & 0 & c & d \\ 0 & 0 & 1 & e & f \\ \hline 0 & 0 & 0 & 1 & 0 \\ 0 & 0 & 0 & 0 & 1 \end{array}\right)$ $(a, b, c, d, e, f \in \mathfrak{P})$
$\mathfrak{A}(4,1)^E$	$\left(\begin{array}{cccc\|c} 1 & 0 & 0 & 0 & a \\ 0 & 1 & 0 & 0 & b \\ 0 & 0 & 1 & 0 & c \\ 0 & 0 & 0 & 1 & d \\ \hline 0 & 0 & 0 & 0 & 1 \end{array}\right)$ $(a, b, c, d \in \mathfrak{P})$
$(\mathfrak{H}_5{}^{\circ})^E$	$\left(\begin{array}{c\|c\|c} 1 & (a,b,c) & \frac{1}{2}(a,b,c)\cdot S\cdot {}^t(a,b,c)+d \\ \hline 0 & I_3 & S\cdot {}^t(a,b,c) \\ \hline 0 & 0 & 1 \end{array}\right)$ $\left(\begin{array}{l} a, b, c, d \in \mathfrak{P}, \quad a, b, c \text{ are linearly independent over } k, \\ S=(s_{i,j}) \text{ is a symmetric matrix of } GL(3,k) \end{array}\right)$
$(\mathfrak{H}_4{}^{\circ})^E \,\urcorner\, \mathrm{Hk}_{0,1}$	$\left(\begin{array}{cccc\|c} 1 & a & b & \frac{\lambda}{2}a^2+\mu ab+\frac{\nu}{2}b^2+c & d \\ 0 & 1 & 0 & \lambda a+\mu b & 0 \\ 0 & 0 & 1 & \mu a+\nu b & 0 \\ 0 & 0 & 0 & 1 & 0 \\ \hline 0 & 0 & 0 & 0 & 1 \end{array}\right)$ $\left(\begin{array}{l} a, b, c, d \in \mathfrak{P}, \quad a, b \text{ are linearly independent over } k, \\ \lambda, \mu, \nu \in k, \quad \lambda\nu-\mu^2 \neq 0 \end{array}\right)$
$(\mathfrak{H}_4{}^{\varepsilon})^E \,\urcorner\, \mathrm{Hk}_{0,1}$	$\left(\begin{array}{cccc\|c} 1 & a & b & ab+c & d \\ 0 & 1 & 0 & b & 0 \\ 0 & 0 & 1 & a & 0 \\ 0 & 0 & 0 & 1 & 0 \\ \hline 0 & 0 & 0 & 0 & 1 \end{array}\right)$ $(a, b, c, d \in \mathfrak{P}, \quad a, b \text{ are linearly independent over } k)$

$(\mathfrak{H}_4{}^{\circ})^E \urcorner \mathrm{Hk}_{1,0}$	$\left(\begin{array}{c\|cccc} 1 & 0 & 0 & 0 & d \\ \hline 0 & 1 & a & b & \frac{\lambda}{2}a^2+\mu ab+\frac{\nu}{2}b^2+c \\ 0 & 0 & 1 & 0 & \lambda a+\mu b \\ 0 & 0 & 0 & 1 & \mu a+\nu b \\ 0 & 0 & 0 & 0 & 1 \end{array}\right)$ $\left(\begin{array}{l} a,b,c,d \in \mathfrak{P}, \quad a,b \text{ are linearly independent over } k, \\ \lambda,\mu,\nu \in k, \quad \lambda\nu-\mu^2 \neq 0 \end{array}\right)$
$(\mathfrak{H}_4{}^{\varepsilon})^E \urcorner \mathrm{Hk}_{1,0}$	$\left(\begin{array}{c\|cccc} 1 & 0 & 0 & 0 & d \\ \hline 0 & 1 & a & b & ab+c \\ 0 & 0 & 1 & 0 & b \\ 0 & 0 & 0 & 1 & a \\ 0 & 0 & 0 & 0 & 1 \end{array}\right)$ $(a,b,c,d \in \mathfrak{P}, \quad a,b \text{ are linearly independent over } k)$
$(\mathfrak{H}\mathfrak{H}_5{}^{\circ})^E$	$\left(\begin{array}{c\|cc\|cc} 1 & a & b & \frac{\lambda}{2}a^2+\mu ab+\frac{\nu}{2}b^2+c & \frac{\lambda'}{2}a^2+\mu' ab+\frac{\nu'}{2}b^2+d \\ \hline 0 & 1 & 0 & \lambda a+\mu b & \lambda' a+\mu' b \\ 0 & 0 & 1 & \mu a+\nu b & \mu' a+\nu' b \\ \hline 0 & 0 & 0 & 1 & 0 \\ 0 & 0 & 0 & 0 & 1 \end{array}\right)$ $\left(\begin{array}{l} a,b,c,d \in \mathfrak{P}, \quad a,b \text{ are linearly independent over } k, \\ \lambda,\mu,\nu \in k, \quad \lambda\nu-\mu^2 \neq 0, \\ \lambda',\mu',\nu' \in k, \quad \lambda'\nu'-\mu'^2 \neq 0, \\ (\lambda,\mu,\nu),(\lambda',\mu',\nu') \text{ are linearly independent over } k \end{array}\right)$
$\left({}^{\tau}(\mathfrak{H}\mathfrak{H}_5{}^{\circ})\right)^E$	$\left(\begin{array}{cc\|cc\|c} 1 & 0 & \mu' a+\nu' b & \lambda' a+\mu' b & \frac{\lambda'}{2}a^2+\mu' ab+\frac{\nu'}{2}b^2+d \\ 0 & 1 & \mu a+\nu b & \lambda a+\mu b & \frac{\lambda}{2}a^2+\mu ab+\frac{\nu}{2}b^2+c \\ \hline 0 & 0 & 1 & 0 & b \\ 0 & 0 & 0 & 1 & a \\ \hline 0 & 0 & 0 & 0 & 1 \end{array}\right)$ $\left(\begin{array}{l} a,b,c,d \in \mathfrak{P}, \quad a,b \text{ are linearly independent over } k, \\ \lambda,\mu,\nu \in k, \quad \lambda\nu-\mu^2 \neq 0, \\ \lambda',\mu',\nu' \in k, \quad \lambda'\nu'-\mu'^2 \neq 0, \\ (\lambda,\mu,\nu),(\lambda',\mu',\nu') \text{ are linearly independent over } k \end{array}\right)$

7 Exponential Algebras

7.1 Basics

An associative k-algebra $\mathfrak{e}$ is said to be *exponential* if $\mathfrak{e}$ satisfies the following conditions (1) and (2):

(1) $\mathfrak{e}$ is commutative, i.e., for all $x, y \in \mathfrak{e}$, $xy = yx$.
(2) For all $x \in \mathfrak{e}$, $x^p = 0$.

Clearly, $\mathrm{Mat}(n, k)$ is an associative k-algebra. Let $\mathfrak{e}$ be a k-subalgebra of $\mathrm{Mat}(n, k)$. If $\mathfrak{e}$ is exponential, $\mathfrak{e}$ can yield exponential matrices.

Lemma 26 *The following assertions (1) and (2) hold true:*

(1) Let $\mathfrak{e}$ be a k-subalgebra of $\mathrm{Mat}(n, k)$. *Assume that $\mathfrak{e}$ is exponential. Let $N_0, \ldots, N_r \in \mathfrak{e}$. Then the matrix $A(T) = \prod_{i=0}^{r} \mathrm{Exp}(T^{p^i} N_i)$ is an exponential matrix of $SL(n, k[T])$.*

(2) Let $A(T)$ be an exponential matrix of $SL(n, k[T])$, and express $A(T)$ as $A(T) = \prod_{i=0}^{r} \mathrm{Exp}(T^{p^i} N_i)$, where $N_i \in \mathrm{Mat}(n, k)$ for $0 \le i \le r$. Let $\mathfrak{e} := \sum_{i=0}^{r} k\, N_i$ be the k-subalgebra of $\mathrm{Mat}(n, k)$. *Then $\mathfrak{e}$ is exponential.*

Proof The proofs of assertions (1) and (2) are straightforward (cf. [7, Lemma 1.3]). □

Let n, r, s be integers satisfying $n \ge 2$ and $r, s \ge 0$. Let $A \in \mathrm{Mat}(n, k)$ and $\alpha \in \mathrm{Hook}_{r,s}(k)$. We can write A and α as

$$A = \begin{pmatrix} A_{1,1} & A_{1,2} \\ A_{2,1} & A_{2,2} \end{pmatrix}, \qquad \alpha = \begin{pmatrix} \alpha_{1,1} & \alpha_{1,2} \\ 0 & \alpha_{2,2} \end{pmatrix},$$

where the sizes of $A_{1,1}, A_{1,2}, A_{2,1}, A_{2,2}$ are $1 \times (n-1)$, 1×1, $(n-1) \times (n-1)$, $(n-1) \times 1$, respectively, and the sizes of the submatrices $\alpha_{1,1}, \alpha_{1,2}, \alpha_{2,2}$ of α are $r \times 1$, $r \times s$, $1 \times s$, respectively. We define a matrix $A \Lsh' \alpha$ of $\mathrm{Mat}(r + n + s, k)$ as

$$A \Lsh' \alpha := \begin{pmatrix} O & O & \alpha_{1,1} & \alpha_{1,2} \\ O & A_{1,1} & A_{1,2} & \alpha_{2,2} \\ O & A_{2,1} & A_{2,2} & O \\ O & O & O & O \end{pmatrix}.$$

For a subset $\mathfrak{S}$ of $\mathrm{Mat}(n, k)$ and a subset $\mathfrak{T}$ of $\mathrm{Hook}_{r,s}(k)$, we let $\mathfrak{S} \Lsh' \mathfrak{T}$ be the subset of $\mathrm{Mat}(r + n + s, k)$ defined by

$$\mathfrak{S} \Lsh' \mathfrak{T} := \{A \Lsh' \alpha \in \mathrm{Mat}(r + n + s, k) \mid A \in \mathfrak{S} \text{ and } \alpha \in \mathfrak{T}\}.$$

Let $\mathfrak{u}_n$ be the subset of all upper triangular matrices of $\mathrm{Mat}(n, k)$ whose diagonal entries are zero. Clearly, $\mathfrak{u}_n$ becomes an associative k-algebra. If $2 \le n \le p$, we know that $x^p = 0$ for all $x \in \mathfrak{u}_n$. If $n \ge 3$, $\mathfrak{u}_n$ is not commutative. We have ${}^{\tau}\mathfrak{u}_n = \mathfrak{u}_n$.

Given a k-subalgebra of $\mathfrak{u}_n$ $(n \geq 2)$ which is exponential, we can construct other exponential algebras, as follows:

Lemma 27 *Let $\mathfrak{e}$ be a k-subalgebra of $\mathfrak{u}_n$ $(n \geq 2)$. Assume that $\mathfrak{e}$ is exponential. Let $r, s \geq 0$ be non-negative integers. Then the following assertions (1) and (2) hold true:*

(1) The k-subalgebra $\mathfrak{e} \urcorner' \mathrm{Hook}_{r,s}(k)$ of $\mathfrak{u}_{r+n+s}$ is exponential.
(2) The k-subalgebra ${}^{\top}\mathfrak{e}$ of $\mathfrak{u}_n$ is also exponential.

Proof The proofs of assertions (1) and (2) are straightforward. □

In the following Sects. 7.2–7.9, we give examples of exponential algebras.

7.2 $\mathfrak{j}_{[n]}$

Let $n \geq 2$ be an integer. We denote by $\mathfrak{j}_{[n]}$ the set of all matrices A of $\mathfrak{u}_n$ with the form $A = J(0, \alpha_1, \ldots, \alpha_{n-1})$, where $\alpha_i \in k$ $(1 \leq i \leq n-1)$. Clearly, for any $n \geq 2$, $\mathfrak{j}_{[n]}$ is a k-subalgebra of $\mathfrak{u}_n$, and for any $2 \leq n \leq p$, $\mathfrak{j}_{[n]}$ is exponential.

7.3 $\mathfrak{j}^{1,0}_{[n,2]}(S)$

Let $n \geq 2$ be an integer. For any symmetric matrix

$$S = \begin{pmatrix} \lambda & \mu \\ \mu & \nu \end{pmatrix} \in \mathrm{Mat}(2, k),$$

we denote by $\mathfrak{j}^{1,0}_{[n,2]}(S)$ the set of all matrices A of $\mathfrak{u}_{n+2}$ with the following form:

$$A = \left(\begin{array}{c|c|c} J(0, \alpha_1, \ldots, \alpha_{n-1}) & \alpha_n\, e_1 & {}^{t}(\gamma, \lambda\,\alpha_1 + \mu\,\alpha_n, 0, \ldots, 0) \\ \hline 0 & 0 & \mu\,\alpha_1 + \nu\,\alpha_n \\ \hline 0 & 0 & 0 \end{array}\right),$$

where $\alpha_i \in k$ $(1 \leq i \leq n)$ and $\gamma \in k$. Clearly, for any $n \geq 2$, $\mathfrak{j}^{1,0}_{[n,2]}(S)$ is a k-subalgebra of $\mathfrak{u}_{n+2}$, and for any $2 \leq n \leq p$, $\mathfrak{j}^{1,0}_{[n,2]}(S)$ is exponential.

7.4 $\mathfrak{j}^{1,1}_{[n,2]}(\lambda)$

Let $n \geq 2$ be an integer. For any $\lambda \in k \setminus \{0\}$, we denote by $\mathfrak{j}^{1,1}_{[n,2]}(\lambda)$ the set of all matrices A of $\mathfrak{u}_{n+2}$ with the following form:

$$A = \left(\begin{array}{c|c|c} J(0, \alpha_1, \ldots, \alpha_{n-1}) & \lambda\beta e_1 & {}^t(\gamma, \alpha_{n-1}, \ldots, \alpha_1) \\ \hline 0 & 0 & \beta \\ \hline 0 & 0 & 0 \end{array}\right),$$

where $\alpha_i \in k$ $(1 \le i \le n-1)$ and $\beta, \gamma \in k$. Clearly, for any $n \ge 2$, $\mathfrak{j}^{1,1}_{[n,2]}(\lambda)$ is a k-subalgebra of $\mathfrak{u}_{n+2}$, and for any $2 \le n \le p$, $\mathfrak{j}^{1,1}_{[n,2]}(\lambda)$ is exponential.

7.5 $\mathfrak{j}^{0,1}_{[2,n]}(S)$

Let $n \ge 2$ be an integer. For any symmetric matrix S of $\mathrm{Mat}(2, k)$, we let $\mathfrak{j}^{0,1}_{[2,n]}(S) := {}^{\tau}\big(\mathfrak{j}^{1,0}_{[n,2]}(S)\big)$. Clearly, for any $n \ge 2$, $\mathfrak{j}^{0,1}_{[2,n]}(S)$ is a k-subalgebra of $\mathfrak{u}_{n+2}$, and for any $2 \le n \le p$, $\mathfrak{j}^{0,1}_{[n,2]}(\lambda)$ is exponential (see Lemma 27).

7.6 $\mathfrak{j}^{1,1}_{[2,n]}(\lambda)$

Let $n \ge 2$ be an integer. For any $\lambda \in k \setminus \{0\}$, we let $\mathfrak{j}^{1,1}_{[2,n]}(\lambda) := {}^{\tau}\big(\mathfrak{j}^{1,1}_{[n,2]}(\lambda)\big)$. Clearly, for any $n \ge 2$, $\mathfrak{j}^{1,1}_{[2,n]}(\lambda)$ is a k-subalgebra of $\mathfrak{u}_{n+2}$, and for any $2 \le n \le p$, $\mathfrak{j}^{1,1}_{[2,n]}(\lambda)$ is exponential (see Lemma 27).

7.7 $\mathfrak{a}(i_1, i_2, i_3)$

For integers i_1, i_2, i_3 with $i_1 \ge 1$, $i_2 \ge 0$ and $i_3 \ge 1$, we define a subset $\mathfrak{a}(i_1, i_2, i_3)$ of $\mathfrak{u}_{i_1+i_2+i_3}$ as

$$\mathfrak{a}(i_1, i_2, i_3) := \left\{ \left(\begin{array}{c|c|c} O_{i_1} & O_{i_1,i_2} & \alpha \\ \hline O_{i_2,i_1} & O_{i_2} & O_{i_2,i_3} \\ \hline O_{i_3,i_1} & O_{i_3,i_2} & O_{i_3} \end{array}\right) \in \mathrm{Mat}(i_1+i_2+i_3, k) \;\middle|\; \alpha \in \mathrm{Mat}_{i_1,i_3}(k) \right\}.$$

We frequently use the notation $\mathfrak{a}(i_1, i_3)$ instead of $\mathfrak{a}(i_1, i_2, i_3)$ if we can understand the value of $i_1 + i_2 + i_3$ from the context. Clearly, $\mathfrak{a}(i_1, i_2, i_3)$ is a k-subalgebra of $\mathfrak{u}_{i_1+i_2+i_3}$, which is exponential.

7.8 $\mathfrak{h}_{m+2}(S)$

Let $m \geq 2$ be an integer and let S be a symmetric matrix of $\mathrm{Mat}(m, k)$. If $p = 2$, we assume that all diagonal entries of S are zero. Let $\mathfrak{h}_{m+2}(S)$ be the set consisting of all matrices N of $\mathfrak{u}_{m+2}$ satisfying the following conditions (1) and (2):

(1) $N = \eta(a_1, \ldots, a_{2m+1}) - I_{m+2}$ for some $a_1, \ldots, a_{2m+1} \in k$.
(2) $(a_{m+1}, \ldots, a_{2m}) = (a_1, \ldots, a_m)S$.

Clearly, $\mathfrak{h}_{m+2}(S)$ is a k-subalgebra of $\mathfrak{u}_{m+2}$, which is exponential.

7.9 $\mathfrak{hh}_{m+3}(S_1, S_2)$

Let $m \geq 2$ be an integer and let S_1, S_2 be symmetric matrices of $\mathrm{Mat}(m, k)$. If $p = 2$, we assume that all diagonal entries of S_i $(i = 1, 2)$ are zero. Let $\mathfrak{hh}_{m+3}(S_1, S_2)$ be the set consisting of all matrices N of $\mathfrak{u}_{m+3}$ satisfying both conditions

$$N_{(m+2,m+2)} \in \mathfrak{h}_{m+2}(S_2) \quad \text{and} \quad N_{(m+3,m+3)} \in \mathfrak{h}_{m+2}(S_1).$$

Clearly, $\mathfrak{hh}_{m+3}(S_1, S_2)$ is a k-subalgebra of $\mathfrak{u}_{m+3}$, which is exponential.

8 Exponential Algebras and Modular Representations of Elementary Abelian p-Groups

For any k-subalgebra $\mathfrak{e}$ of $\mathrm{Mat}(n, k)$, we can define a subset $\exp(\mathfrak{e})$ of $\mathrm{Mat}(n, k[T])$ as

$$\exp(\mathfrak{e}) := \left\{ \prod_{i=0}^{r} \mathrm{Exp}(T^{p^i} N_i) \;\middle|\; r \geq 0, \;\; N_i \in \mathfrak{e} \text{ for all } 0 \leq i \leq r \right\}.$$

If $\mathfrak{e}$ is exponential, $\exp(\mathfrak{e})$ is a subset of $\mathcal{E}(n, k[T])$.

Lemma 28 *Let $\mathfrak{e}$ be a k-subalgebra of $\mathfrak{u}_n$. Assume that $\mathfrak{e}$ is exponential. Let $r, s \geq 0$ be non-negative integers. Then the following assertions (1) and (2) hold true:*

(1) $\exp(\mathfrak{e}) \,\Lsh\, \mathrm{Hook}_{r,s}(\mathfrak{P}) = \exp(\mathfrak{e} \,\Lsh'\, \mathrm{Hook}_{r,s}(k))$.
(2) ${}^{\tau}\exp(\mathfrak{e}) = \exp({}^{\tau}\mathfrak{e})$.

Proof The proofs of assertions (1) and (2) are straightforward. □

Lemma 29 *The following assertions (1)–(9) hold true:*

(1) $\mathfrak{J}^{E}_{[n]} \subset \exp(\mathfrak{j}_{[n]})$.

(2) $(\mathfrak{J}^{1,0}_{[n,2]})^E \subset \bigcup_{S\in \mathrm{Sym}(2,k)} \exp\big(\mathfrak{j}^{1,0}_{[n,2]}(S)\big)$.

(3) $(\mathfrak{J}^{1,1}_{[n,2]})^E \subset \bigcup_{\lambda\in k\setminus\{0\}} \exp\big(\mathfrak{j}^{1,1}_{[n,2]}(\lambda)\big)$.

(4) $(\mathfrak{J}^{0,1}_{[2,n]})^E \subset \bigcup_{S\in \mathrm{Sym}(2,k)} \exp\big(\mathfrak{j}^{0,1}_{[2,n]}(S)\big)$.

(5) $(\mathfrak{J}^{1,1}_{[2,n]})^E \subset \bigcup_{\lambda\in k\setminus\{0\}} \exp\big(\mathfrak{j}^{1,1}_{[2,n]}(\lambda)\big)$.

(6) $\mathfrak{A}(i_1, i_2, i_3)^E \subset \exp\big(\mathfrak{a}(i_1, i_2, i_3)\big)$.

(7) If $p = 2$, *then* $(\mathfrak{H}_4{}^{\varepsilon})^E \subset \exp\big(\mathfrak{h}_4(J_2)\big)$, *where* $J_2 := \begin{pmatrix} 0 & 1 \\ 1 & 0 \end{pmatrix}$.

(8) If $p \geq 3$, *then* $(\mathfrak{H}_{m+2}{}^{\circ})^E \subset \bigcup_{S\in GLS(m,k)} \exp\big(\mathfrak{h}_{m+2}(S)\big)$.

(9) If $p \geq 3$, *then* $(\mathfrak{H}\mathfrak{H}_5{}^{\circ})^E \subset \bigcup_{S,S'\in GLS(2,k)} \exp\big(\mathfrak{h}\mathfrak{h}_5(S, S')\big)$.

Proof (1) See [7, Lemma 1.17].

(2) For any $A(T) \in (\mathfrak{J}^{1,0}_{[n,2]})^E$, we can express $A(T)$ as

$$A(T) = \mathrm{Exp}\left(\left(\begin{array}{c|c|c} J(0, \alpha_1, \ldots, \alpha_{n-1}) & \alpha_n\, e_1 & {}^t(\gamma, \lambda\,\alpha_1 + \mu\,\alpha_n, 0, \ldots, 0) \\ \hline 0 & 0 & \mu\,\alpha_1 + \nu\,\alpha_n \\ \hline 0 & 0 & 0 \end{array}\right)\right)$$

for some $\lambda, \mu, \nu \in k$ and $\alpha_1, \ldots, \alpha_n, \gamma \in \mathfrak{P}$. Let

$$S = \begin{pmatrix} \lambda & \mu \\ \mu & \nu \end{pmatrix}.$$

Thus $A(T) \in \exp\big(\mathfrak{j}^{1,0}_{[n,2]}(S)\big)$.

(3) For any $A(T) \in (\mathfrak{J}^{1,1}_{[n,2]})^E$, we can express $A(T)$ as

$$A(T) = \mathrm{Exp}\left(\left(\begin{array}{c|c|c} J(0, \alpha_1, \ldots, \alpha_{n-1}) & \lambda\,\beta\, e_1 & {}^t(\gamma, \alpha_{n-1}, \ldots, \alpha_1) \\ \hline 0 & 0 & \beta \\ \hline 0 & 0 & 0 \end{array}\right)\right)$$

for some $\lambda \in k \setminus \{0\}$ and $\alpha_1, \ldots, \alpha_{n-1}, \beta, \gamma \in \mathfrak{P}$. Thus $A(T) \in \exp\big(\mathfrak{j}^{1,1}_{[n,2]}(\lambda)\big)$.

(4) Use the above (2) and Lemma 28.

(5) Use the above (3) and Lemma 28.

The proofs of (6) and (7) are straightforward.

(8) For any $A(T) \in (\mathfrak{H}_{m+2}{}^{\circ})^E$, we can express $A(T)$ as

$$A(T) = \mathrm{Exp}\left(\eta\big(a_1, \ldots, a_m, (a_1, \ldots, a_m)S, a_{2m+1}\big) - I_{m+2}\right)$$

for some $S \in GLS(m, k)$ and $a_1, \ldots, a_m, a_{2m+1} \in \mathfrak{P}$. Thus $A(T) \in \exp\big(\mathfrak{h}_{m+2}(S)\big)$.

(9) The proof is similar to the proof of the above (8). □

Let $r, n \geq 1$ be integers. Given a modular representation $\varrho : (\mathbb{Z}/p\mathbb{Z})^r \to GL(n, k)$, we let $N_i := \varrho(e_i) - I_n$ $(1 \leq i \leq r)$ and then let

$$A_\varrho(T) := \prod_{i=0}^{r-1} \mathrm{Exp}(T^{p^i} N_{i+1}) \in SL(n, k[T]).$$

Clearly, $A_\varrho(T)$ is an exponential matrix of $SL(n, k[T])$.

Lemma 30 *Let $\varrho : (\mathbb{Z}/p\mathbb{Z})^r \to GL(n, k)$ be a modular representation and let $\mathfrak{e}$ be a k-subalgebra of $\mathrm{Mat}(n, k)$ satisfying $P^{-1} A_\varrho(T) P \in \exp(\mathfrak{e})$ for some $P \in GL(n, k)$. Assume that $\mathfrak{e}$ is exponential. Then $P^{-1}\varrho(e_i)P - I_n \in \mathfrak{e}$ for all $1 \leq i \leq r$.*

Proof By the definition of $A_\varrho(T)$, we have $A_\varrho(T) = \prod_{i=0}^{r-1} \mathrm{Exp}(T^{p^i} N_{i+1}) \in SL(n, k[T])$, where $N_i = \varrho(e_i) - I_n$ $(1 \leq i \leq r)$. Since $P^{-1} A_\varrho(T) P \in \exp(\mathfrak{e})$, we know that $P^{-1} N_i P \in \mathfrak{e}$ for all $1 \leq i \leq r$. □

9 Five-Dimensional Modular Representations of Elementary Abelian p-Groups

We frequently use $\mathrm{hk}_{r,s}$ instead of $\mathrm{Hook}_{r,s}(k)$.

The following theorem gives an overlapping classification of five-dimensional modular representations of elementary abelian p-groups.

Theorem 3 *Let $r \geq 1$ be an integer. Let $\varrho : (\mathbb{Z}/p\mathbb{Z})^r \to GL(5, k)$ be a modular representation. Then the following assertions (1), (2), (3) hold true:*

(1) If $p = 2$, then ρ is equivalent to a modular representation $\varrho' : (\mathbb{Z}/p\mathbb{Z})^r \to GL(5, k)$ such that the set $\{\varrho'(e_i) - I_5 \mid 1 \leq i \leq r\}$ is a subset of at least one of the following 6 exponential algebras:

$$\mathfrak{a}_{1,4}, \quad \mathfrak{a}_{2,3}, \quad \mathfrak{a}_{3,2}, \quad \mathfrak{a}_{4,1}, \quad \mathfrak{h}_4(J_2) \Lsh' \mathrm{hk}_{0,1}, \quad \mathfrak{h}_4(J_2) \Lsh' \mathrm{hk}_{1,0},$$

where

$$J_2 := \begin{pmatrix} 0 & 1 \\ 1 & 0 \end{pmatrix}.$$

(2) If $p = 3$, then ρ is equivalent to a modular representation $\varrho' : (\mathbb{Z}/p\mathbb{Z})^r \to GL(5, k)$ such that the set $\{\varrho'(e_i) - I_5 \mid 1 \leq i \leq r\}$ is a subset of at least one of the following exponential algebras:

$$\mathfrak{j}_{[3,2]}^{1,0}(S)\ \big(S \in \mathrm{Sym}(2,k)\big), \quad \mathfrak{j}_{[3,2]}^{1,1}(\lambda)\ (\lambda \in k\backslash\{0\}),$$
$$\mathfrak{j}_{[2,3]}^{1,1}(\lambda)\ (\lambda \in k\backslash\{0\}), \quad \mathfrak{j}_{[2,3]}^{0,1}(S)\ \big(S \in \mathrm{Sym}(2,k)\big),$$
$$\mathfrak{j}_{[3]} \Lsh' \mathrm{hk}_{0,2}, \quad \mathfrak{j}_{[3]} \Lsh' \mathrm{hk}_{1,1}, \quad \mathfrak{j}_{[3]} \Lsh' \mathrm{hk}_{2,0},$$
$$\mathfrak{a}_{1,4}, \quad \mathfrak{a}_{2,3}, \quad \mathfrak{a}_{3,2}, \quad \mathfrak{a}_{4,1},$$
$$\mathfrak{h}_5(S)\ \big(S \in GLS(3,k)\big),$$
$$\mathfrak{h}_4(S) \Lsh' \mathrm{hk}_{0,1}\ \big(S \in GLS(2,k)\big), \quad \mathfrak{h}_4(S) \Lsh' \mathrm{hk}_{1,0}\ \big(S \in GLS(2,k)\big),$$
$$\mathfrak{hh}_5(S,S')\ \big(S,S' \in GLS(2,k)\big), \quad {}^{T}\big(\mathfrak{hh}_5(S,S')\big)\ \big(S,S' \in GLS(2,k)\big).$$

(3) *If $p \geq 5$, then ρ is equivalent to a modular representation $\varrho' : (\mathbb{Z}/p\mathbb{Z})^r \to GL(5,k)$ such that the set $\{\varrho'(e_i) - I_5 \mid 1 \leq i \leq r\}$ is a subset of at least one of the following exponential algebras:*

$$\mathfrak{j}_{[5]},$$
$$\mathfrak{j}_{[4]} \Lsh' \mathrm{hk}_{0,1}, \quad \mathfrak{j}_{[4]} \Lsh' \mathrm{hk}_{1,0},$$
$$\mathfrak{j}_{[3,2]}^{1,0}(S)\ \big(S \in \mathrm{Sym}(2,k)\big), \quad \mathfrak{j}_{[3,2]}^{1,1}(\lambda)\ (\lambda \in k\backslash\{0\}),$$
$$\mathfrak{j}_{[2,3]}^{1,1}(\lambda)\ (\lambda \in k\backslash\{0\}), \quad \mathfrak{j}_{[2,3]}^{0,1}(S)\ \big(S \in \mathrm{Sym}(2,k)\big),$$
$$\mathfrak{j}_{[3]} \Lsh' \mathrm{hk}_{0,2}, \quad \mathfrak{j}_{[3]} \Lsh' \mathrm{hk}_{1,1}, \quad \mathfrak{j}_{[3]} \Lsh' \mathrm{hk}_{2,0},$$
$$\mathfrak{a}_{1,4}, \quad \mathfrak{a}_{2,3}, \quad \mathfrak{a}_{3,2}, \quad \mathfrak{a}_{4,1},$$
$$\mathfrak{h}_5(S)\ \big(S \in GLS(3,k)\big),$$
$$\mathfrak{h}_4(S) \Lsh' \mathrm{hk}_{0,1}\ \big(S \in GLS(2,k)\big), \quad \mathfrak{h}_4(S) \Lsh' \mathrm{hk}_{1,0}\ \big(S \in GLS(2,k)\big),$$
$$\mathfrak{hh}_5(S,S')\ \big(S,S' \in GLS(2,k)\big), \quad {}^{T}\big(\mathfrak{hh}_5(S,S')\big)\ \big(S,S' \in GLS(2,k)\big).$$

Proof The proofs of assertions (1), (2), (3) follow from Corollary 2 and Lemmas 29, 28, 30. □

Acknowledgements The author would like to thank the referee for careful reading of the article.

References

1. Fauntleroy, A.: On Weitzenböck's theorem in positive characteristic. Proc. Am. Math. Soc. **64**(2), 209–213 (1977)
2. Miyanishi, M.: Curves on Rational and Unirational Surfaces. Tata Institute of Fundamental Research Lectures on Mathematics and Physics, vol. 60. Tata Institute of Fundamental Research, Bombay (1978)
3. Seshadri, C.S.: On a theorem of Weitzenböck in invariant theory. J. Math. Kyoto Univ. **1**, 403–409 (1961/1962)
4. Tanimoto, R.: An algorithm for computing the kernel of a locally finite iterative higher derivation. J. Pure Appl. Algebra **212**(10), 2284–2297 (2008)
5. Tanimoto, R.: Representations of $\mathbb{G}_a$ of Codimension Two, Affine Algebraic Geometry, pp. 279–284. World Scientific Publishing, Hackensack (2013)
6. Tanimoto, R.: A note on the Weitzenböck problem in dimension four. Commun. Algebra **46**(2), 588–596 (2018)

7. Tanimoto, R.: Exponential matrices. Linear Algebra Appl. **572**, 153–251 (2019)
8. Tyc, A.: An elementary proof of the Weitzenböck theorem. Colloq. Math. **78**(1), 123–132 (1998)
9. Weitzenböck, R.: Über die invarianten von linearen gruppen. Acta Math. **58**(1), 231–293 (1932)
10. Zariski, O.: Interprétations algébrico-géométriques du quatorzième problème de Hilbert. Bull. Sci. Math. **78**(2), 155–168 (1954)

Mathieu-Zhao Spaces and the Jacobian Conjecture

Arno van den Essen

Abstract In this paper we define the notion of a Mathieu-Zhao space, give various examples of this concept and use the framework of these Mathieu-Zhao spaces to describe a chain of challenging conjectures, all implying the Jacobian Conjecture.

Keywords Mathieu subspace · Ideal theory · Jacobian Conjecture

1 Introduction

Since many years I have been studying the Jacobian Conjecture and various problems and conjectures related to it. Most of the conjectures implying the Jacobian Conjecture turned out to be false. This led me to formulate the following statement, which I stated publicely at various occasions:

If you have a conjecture which implies the Jacobian Conjecture, and which is not equivalent to it, then you can be sure that your conjecture is wrong.

When around 2009 Wenhua Zhao came up with several conjectures which are much more general than the Jacobian Conjecture I was convinced that they all were false and started to look for counter examples. However my research led me to discover the contrary, more evidence in favour of these conjectures. Moreover, Zhao developed a framework in which all these new conjectures can be formulated, by introducing the concept of a *Mathieu subspace* of a ring. In 2014, during an International Conference on Polynomial Automorphisms and the Jacobian Conjecture held in Tianjin, China, I changed the name Mathieu subspace into *Mathieu-Zhao space* (*MZ-space* for short), obviously to honor Wenhua Zhao for his great contributions.

The aim of this paper is to give a first introduction to MZ-spaces and to show their importance for our understanding of the Jacobian Conjecture. For readers who

A. van den Essen (✉)
Institute of Mathematics Astrophysics and Particle Physics,
Radboud University Nijmegen, Nijmegen, The Netherlands
e-mail: essen@math.ru.nl

S. Kuroda et al. (eds.), *Polynomial Rings and Affine Algebraic Geometry*,
Springer Proceedings in Mathematics & Statistics 319,
https://doi.org/10.1007/978-3-030-42136-6_14

want to know more about these spaces and want to read the proofs of the theorems mentioned, I refer to the upcoming book *New results on Polynomial Automorphisms and the Jacobian Conjecture*, to be published at Birkhäuser, in cooperation with Tony Crachiola and Shigeru Kuroda.

2 Mathieu-Zhao Spaces: Examples

Throughout this section A is a ring containing 1. The ring need not to be commutative. *What is a Mathieu-Zhao space?*

Roughly speaking, a Mathieu-Zhao space (MZ-space) of a ring A is a generalization of an ideal in A. Recall that a (left)-ideal I in A is a (non-empty) additive subset of A such that

$$a \in I, b \in A \Rightarrow ba \in I.$$

In other words, I is stable under multiplication by elements of A.

To generalize this concept we introduce a useful notation: let $M \subseteq A$ be an additive subset and $a \in A$. We say that $a^\infty \in M$ if there exists an $N \in \mathbb{N}$ such that $a^m \in M$ for all $m \geq N$. Often we write more informally $a^m \in M$ *for all large* m. More generally if $b, a \in A$ we say that $ba^\infty \in M$ if there exists an $N \in \mathbb{N}$ such that $ba^m \in M$ for all $m \geq N$. Now we are ready to formulate:

Definition An additive subset $M \subseteq A$ is a (left)-Mathieu-Zhao space of A if the following implication holds

$$a^\infty \in M,\ b \in A \Rightarrow ba^\infty \in M.$$

In a similar way one can define a right MZ-space and a two-sided MZ-space. Since in almost all examples in this paper the ring will be commutative, the above definition is sufficient for our purposes.

Obviously a left ideal $I \subseteq A$ is an example of an MZ-space of A. Now let's look at some more interesting examples.

Example 1 Let $A = M_n(\mathbb{C})$ and $M = \{C \in A \mid Tr\, C = 0\}$. Then M is a left MZ-space of A.

Proof Let $C^\infty \in M$. So there exists an $N \in \mathbb{N}$ such that $C^m \in M$ for all $m \geq N$. Hence $Tr\, C^m = 0$, for all $m \geq N$. Let $\lambda_1, \cdots, \lambda_n \in \mathbb{C}$ be the eigenvalues of C. Then $\lambda_1^m + \cdots + \lambda_n^m = 0$, for all $m \geq N$. It is well-known that this implies $\lambda_1 = \cdots = \lambda_n = 0$. Consequently C is nilpotent. So $C^n = 0$.

Now let $B \in A$. Then obviously $BC^m = 0$, for all $m \geq n$, whence $Tr\, BC^m = 0$, for all $m \geq n$, i.e. $BC^m \in M$, for all $m \geq n$. So $BC^\infty \in M$. Hence $C^\infty \in M$ implies that $BC^\infty \in M$, i.e. M is an MZ-space of A.

Remark 1 The argument given above also shows that if $C^\infty \in M$ and $B_1, B_2 \in A$, then $B_1 C^\infty B_2 \in M$, which implies that M is a two-sided MZ-space of A. On the other hand A is a simple ring, which means that A has no two-sided ideals other than A and $\{0\}$.

Remark 2 It was shown by Zhao in [18] that M is the only codimension one (left) MZ-space of A (and hence also the only two-sided MZ-space of A).

Example 2 Let A be a commutative ring and $v : A \to \mathbb{R} \cup \{\infty\}$ an order function i.e.

(i) $v(a+b) \geq min(v(a), v(b))$, for all $a, b \in A$.
(ii) $v(ab) = v(a) + v(b)$, for all $a, b \in A$.
(iii) $v(a) = \infty$ if and only if $a = 0$. Let $c > 0$. Put $M_c = \{a \in A \mid v(a) > c\}$. Then M_c is an MZ-space of A.

Proof If $a^\infty \in M_c$, then there exists an $N \in \mathbb{N}$ such that $v(a^m) > c$ for all $m \geq N$. Hence $mv(a) > c$, for all $m \geq N$. In particular $v(a) > 0$.

Now let $b \in A$, say $v(b) = t$. Then

$$v(ba^m) = v(b) + mv(a) = t + mv(a) > 0$$

for all large m, since $v(a) > 0$. So $v(ba^m) > c$, for all large m, i.e. $ba^\infty \in M_c$. Hence $a^\infty \in M_c$ implies that $ba^\infty \in M_c$. So M_c is an MZ-space of A.

In both examples we computed the set of $a \in A$ such that $a^\infty \in M$. This set deserves a name

Definition Let $M \subseteq A$ be any subset. Then

$$r(M) := \{a \in A \mid a^\infty \in M\}$$

i.e. $r(M)$ is the set of $a \in A$ such that $a^m \in M$ for all large m. This set is called the radical of M.

Some remarks concerning the radical are in order.

Remark 3 The radical is a very bizarre set. In some sense it resembles the Mandelbrot set. It is very hard to find some structure of the radical, except in some special cases. For example, even if M is an additive set its radical $r(M)$ need not to be additive as well: in Example 1 the radical of M is equal to the set of all nilpotent matrices. If $n > 1$ this set is not additive.

Remark 4 Once we can compute the radical of a given set, it is often rather easy to prove that this set is an MZ-space or not. The problem however is that in most cases it is *extremely difficult* to compute the radical. To illustrate this statement we give some challenging problems for the reader to try.

Problem 1 Let $A = \mathbb{C}[t_1, \cdots, t_n]$ and

$$M = \{f \in A \mid \int_0^1 \cdots \int_0^1 f(t)dt_1 \cdots dt_n = 0\}.$$

Is $r(M) = \{0\}$? In other words, does $\int_0^1 \cdots \int_0^1 f(t)^m dt_1 \cdots dt_n = 0$ for all large m imply that $f = 0$?

The answer is yes if $n = 1$ (see [9]). However the case $n \geq 2$ remains open.

Problem 2 Let $A = \mathbb{C}[t_1, \cdots, t_n]$ and

$$M = \{f \in A \mid \int_0^\infty \cdots \int_0^\infty f(t)e^{-(t_1+\cdots+t_n)}dt_1 \cdots dt_n = 0\}.$$

Is $r(M) = \{0\}$?

Again the answer is yes if $n = 1$ (see [6], Theorem 4.9), but the case $n \geq 2$ remains open.

Problem 3 Let $c_1, \cdots, c_n \in \mathbb{C}$, linearly independent over $\mathbb{Q}$,
$A = \mathbb{C}[t_1, \cdots, t_n, t_1^{-1}, \cdots, t_n^{-1}]$ and $D := \sum_i c_i t_i \partial_i : A \to A$.
Let $M := Im\, D$. Find a new proof of the following theorem (see [5]):

Theorem (Duistermaat-van der Kallen)

$$f \in r(M) \Leftrightarrow 0 \notin New(f)$$

($New(f)$ denotes the Newton polygon of f).

3 Background and History

In this section we very briefly give some historical background leading to the introduction of the notion of a Mathieu-Zhao space. In other words we consider the question

Where do MZ-spaces come from?

The starting point was the formulation of the Jacobian Conjecture in [10] by Ott-Heinrich Keller in 1939:

Jacobian Conjecture. *If* $F : \mathbb{C}^n \to \mathbb{C}^n$ is a polynomial map such that $det\, JF \in \mathbb{C}^*$, then F is invertible.

The first progress appeared in 1982 with the work of Bass, Connell and Wright and independently Yagzhev. They proved the following remarkable theorem (see [1, 12]):

Theorem 1 *It suffices to investigate the Jacobian Conjecture for all $n \geq 1$ and all F of the form*

$$(x_1 + H_1, \cdots, x_n + H_n)$$

where each H_i is either homogeneous of degree 3 or $H_i = 0$. Furthermore, the condition $\det JF \in \mathbb{C}^*$ is equivalent to JH is nilpotent.

Thirteen years later Mathieu in [11] made is very general conjecture and showed that his conjecture *implies* the Jacobian Conjecture. His conjecture is the following:

Mathieu's Conjecture. *Let G be a compact connected real Lie group with Haar measure σ. Let $f : G \to \mathbb{C}$ be a G-finite function such that*

$$\int_G f^m \, d\sigma = 0, \quad \text{for all } m \geq 1.$$

Then for every G-finite function $g : G \to \mathbb{C}$ we have

$$\int_G g f^m \, d\sigma = 0, \quad \text{for all large } m.$$

A function $f : G \to \mathbb{C}$ is called *G-finite* if the $\mathbb{C}$-vector space generated by the orbit $G \cdot f$ is finite dimensional.

Then in [3] Michiel de Bondt and the author improved upon the theorem of Bass, Connell, Wright and Yagzhev. Namely we showed the following:

Theorem 2 *It suffices to investigate the Jacobian Conjecture for all $n \geq 1$ and all maps of the form*

$$(x_1 + f_{x_1}, \cdots, x_n + f_{x_n})$$

where f is a homogeneous polynomial of degree 4 and f_{x_i} denotes the partial derivative of f with respect to x_i.

4 New Conjectures

Around the time we proved this result, Wenhua Zhao was working on new inversion formulas for polynomial mappings. When he heard about our result he used it to investigate the inverse of the gradient maps described in Theorem 2. This led him to the discovery of the following surprising conjecture, which he showed to be *equivalent* to the Jacobian Conjecture [13]. More precisely, the Jacobian Conjecture for all $n \geq 1$ is equivalent to the following statement:

Vanishing Conjecture VC(n). *Let $\Delta := \partial_{x_1}^2 + \cdots + \partial_{x_n}^2$ be the Laplace operator. If $f \in \mathbb{C}[x_1, \cdots, x_n]$ is homogeneous of degree ≥ 3, then $\Delta^m f^m = 0$ for all $m \geq 1$, implies that $\Delta^m f^{m+1} = 0$ for all large m.*

The structure of this reformulation of the Jacobian Conjecture resembles the statement of Mathieu's Conjecture. To make this resemblance even stronger Zhao and the author showed in [8] that the Jacobian Conjecture is also equivalent to the following statement for all $n \geq 1$:

Vanishing Conjecture (2nd form). *If $f \in \mathbb{C}[x_1, \cdots, x_n]$ is homogeneous of degree ≥ 3, then $\Delta^m f^m = 0$, for all $m \geq 1$, implies that for every $g \in \mathbb{C}[x_1, \cdots, x_n]$ we have $\Delta^m(gf^m) = 0$, for all large m.*

This conjecture led Zhao to the question: *what is special about the operator Δ?*

Investigating this question he found out that if the Vanishing Conjecture holds for Δ, then it also holds for all *quadratic homogeneous* differential operators with constant coefficients. This kind of considerations finally led him in [14] to formulate a stronger conjecture, where he replaced the operator Δ by *any* differential operator with constant coefficients and f by *any* polynomial (not necessarily) homogeneous. More precisely:

Generalized Vanishing Conjecture (GVC(n)). *Let $\Lambda \in \mathbb{C}[\partial_{x_1}, \cdots, \partial_{x_n}]$ and $f \in \mathbb{C}[x_1, \cdots, x_n]$ such that*

$$\Lambda^m f^m = 0 \text{ for all } m \geq 1.$$

Then for every $g \in \mathbb{C}[x_1, \cdots, x_n]$ we have

$$\Lambda^m(gf^m) = 0 \text{ for all large } m.$$

Remark 5 We can view the Generalized Vanishing Conjecture as a *family of Jacobian Conjectures*, one conjecture for each Λ, since the Jacobian Conjecture is just the case $\Lambda = \Delta$, as seen above.

Remark 6 If $n = 1$ the Generalized Vanishing Conjecture is true. If $n \geq 2$ several special cases of the conjecture have been proved. Furthermore de Bondt proved in [2] the following interesting result:

Theorem 3 *It suffices to investigate the Generalized Vanishing Conjecture for all $n \geq 1$ and all operators of the form*

$$\Lambda = \partial_{x_1}^{e_1} + \cdots + \partial_{x_n}^{e_n}, \text{ with all } e_i \geq 1.$$

Studying the Generalized Vanishing Conjecture, in particular observing that the differential operator Λ does not commute with the multiplication operator "multiplying by f", made Zhao to introduce n new variables $\zeta_1, \cdots, \zeta_n$ and to consider the $2n$-variable polynomial ring $\mathbb{C}[\zeta, x] := \mathbb{C}[\zeta_1, \cdots, \zeta_n, x_1, \cdots, x_n]$. Furthermore he defined the following $\mathbb{C}$-linear map

$$\mathcal{E}_n : \mathbb{C}[\zeta, x] \to \mathbb{C}[x] \text{ by } \mathcal{E}_n(\zeta^a x^b) = \partial^a(x^b).$$

For simplicity we will write $\mathcal{E}$ instead of $\mathcal{E}_n$. With these notations Zhao introduced in [15] the following conjecture:

Special Image Conjecture (SIC(n)). *Let $f \in \mathbb{C}[\zeta, x]$ be such that*

$$f^m \in ker\,\mathcal{E} \text{ for all } m \geq 1.$$

Then for all $g \in \mathbb{C}[\zeta, x]$ we have

$$gf^m \in ker\,\mathcal{E} \text{ for all large } m$$

i.e. $ker\,\mathcal{E}$ is an MZ-space of $\mathbb{C}[\zeta, x]$.

Remark 7 The name (Special) Image Conjecture comes from the fact that is was shown in [15] that

$$ker\,\mathcal{E} = \sum_{i=1}^{n}(\partial_{x_i} - \zeta_i)\mathbb{C}[\zeta, x]$$

The word "special" is added because the conjecture stated above is a special case of an even more general conjecture (which we don't need in this paper).

The importance of the Special Image Conjecture, in relation to the Jacobian Conjecture, stems from the following theorem (see [15]):

Theorem 4 *SIC(n)⇒GVC(n).*

Consequently we get the following chain of implications:

$$SIC \Rightarrow GVC \Rightarrow VC \Leftrightarrow JC$$

In particular we see that the Special Image Conjecture implies the Jacobian Conjecture, ie. $SIC \Rightarrow JC$.

To give the reader some idea of a proof of this result, we sketch a direct proof (avoiding the intermediate implications via GVC and VC). Furthermore this proof will give some insight concerning the conditions $f^\infty \in ker\,\mathcal{E}$ and $gf \in \ker\,\mathcal{E}$.

In order to prove the Jacobian Conjecture it suffices to show (as we have seen above) that if $F = x - H = (x_1 - H_1(x), \cdots, x_n - H_n(x))$, where each H_i is homogeneous of the same degree ≥ 2 (or $H_i = 0$) and $det\,JF \in \mathbb{C}^*$, then F is invertible. The following two theorems were obtained by Zhao in [17]. Let

$$f := \zeta_1 H_1(x) + \cdots + \zeta_n H_n(x) \in \mathbb{C}[\zeta, x].$$

Theorem 5 *$f^\infty \in ker\,\mathcal{E}$ if and only if $det\,JF \in \mathbb{C}^*$.*

Theorem 6 *For every $g \in \mathbb{C}[\zeta, x]\big[gf^\infty \in ker\,\mathcal{E}\big]$ if and only if F is invertible.*

Clearly both theorems imply

Corollary *SIC ⇒ JC.*

To conclude this paper we will describe a conjecture, which is even more general than the Special Image Conjecture. The results below can be found in the paper [4], due to Derksen, Zhao and the author.

Recall that above we defined the $\mathbb{C}$-linear map $\mathcal{E}_n : \mathbb{C}[\zeta, x] \to \mathbb{C}[x]$ by the formula $\mathcal{E}_n(\zeta^a x^b) = \partial^a(x^b)$. Using this map we define $\mathcal{F}_n : \mathbb{C}[\zeta, x] \to \mathbb{C}$ by

$$\mathcal{F}_n(f) := \mathcal{E}_n(f)_{|x=0}.$$

Then we have the following result:

Proposition *If ker* $\mathcal{F}_n$ *is an MZ-space of* $\mathbb{C}[\zeta, x]$*, then ker* $\mathcal{E}_n$ *is an MZ-space of* $\mathbb{C}[\zeta, x]$.

In order to formulate the last conjecture of this paper we give a reformulation of the statement *ker* $\mathcal{F}_n$ *is an MZ-space of* $\mathbb{C}[\zeta, x]$. First we define for each $f \in \mathbb{C}[x_1, \cdots, x_n]$

$$\mathbb{E}(f) := \int_{\mathbb{R}^n} e^{-(x_1^2+\cdots+x_n^2)} f(x)\, dx.$$

Gaussian Moments Conjecture (GMC(n). *Let* $f \in \mathbb{C}[x_1, \cdots, x_n]$ *be such that*

$$\mathbb{E}(f^m) = 0 \text{ for all } m \geq 1.$$

Then for all $g \in \mathbb{C}[x_1, \cdots, x_n]$ *we have that*

$$\mathbb{E}(gf^m) = 0 \text{ for all large } m,$$

i.e. ker $\mathbb{E}$ *is an MZ-space of* $\mathbb{C}[x_1, \cdots, x_n]$.

The importance of this conjecture stems from the following result:

Theorem 7 *ker* $\mathcal{F}_n$ *is an MZ-space of* $\mathbb{C}[\zeta, x]$ *if and only if* $GMC(2n)$ *holds.*

Corollary $GMC(2n)$ *implies* $SIC(n)$.

Summarizing all results described above, we get the following chain of implications

$$GMC \Rightarrow SIC \Rightarrow GVC \Rightarrow VC \Leftrightarrow JC.$$

References

1. Bass, H., Connell, E., Wright, D.: The jacobian conjecture: reduction of degree and formal expansion of the inverse. Bull. Am. Math. Soc.(NS) **7**, 287–330. [MR 83k:14028]
2. de Bondt, M.: A few remarks on the generalized vanishing conjecture. Arch. Math (Basel) **100**(6), 533–538 (2013)

3. de Bondt, M., van den Essen, A.: A reduction of the jacobian conjecture to the symmetric case. Proc. Am. Math. Soc. **133**(8), 2201–2205 (2005). [MR2138860]
4. Derksen, H., van den Essen, A., Zhao, W.: The gaussian moments conjecture and the jacobian conjecture. Israel J. of Math. **219**, 917–928 (2017). arxiv.org/pdf/1506.05192.pdf
5. Duistermaat, J., van der Kallen, W.: Constant terms in powers of a Laurent polynomial. Indag. Math. (N.S.) **9**(2), 221–231 (1998)
6. van den Essen, A., Wright, D., Zhao, W.: On the image conjecture. J. Alg. **340**(1), 211–224 (2011). See also arXiv:1008.3962 [math.RA]
7. van den Essen, A.: An introduction to Mathieu subspaces. Lectures delivered at the Chern Institute of Mathematics. Tianjin, China (July 2014)
8. van den Essen, A., Zhao, W.: Two results on homogeneous Hessian nilpotent polynomials. J. Pure Appl. Algebra **212**(10), 2190–2193 (2008). See also http://arxiv.org/abs/0704.1690v1
9. Francoise, J.P., Pakovich, F., Yomdin, Y., Zhao, W.: Moment vanishing problem and positivity: some examples. Bull. des Sci. Math. **135**(1), 10–32 (2011). Also available at:http://www.math. bgu.ac.ilakovich/Publications/FPYZ.pdf
10. Keller, O.H.: Ganze Cremona-Transformationen. Monatsh. Math. Phys. **47**, 299–306 (1939)
11. Mathieu, O.: Some conjectures about invariant theory and their applications. Algèbre Non Commutative, Groupes Quantiques et Invariants (Reims, 1995). Séminaries et Congrès, vol. 2, pp. 263–279 Société Mathématique de France (1997)
12. Yagzhev, A.V.: On Keller's problem. Siberian Math. J. **21**, 747–754 (1980)
13. Zhao, W.: Hessian nilpotent polynomials and the Jacobian conjecture. Trans. Am. Math. Soc. **359**(1), 294–274 (2007). See also math.CV/0409534
14. Zhao, W.: A vanishing conjecture on differential operators with constant coefficients. Acta Math. Vietnam. **32**(3), 259–286 (2007). See also arXiv:0704.1691v2
15. Zhao, W.: Images of commuting differential operators of order one with constant leading coefficients. J. Algebra **324**(2), 231–247 (2010). See also arXiv:0902.0210 [math.CV]
16. Zhao, W.: Generalizations of the image conjecture and the Mathieu conjecture. J. Pure Appl. Algebra **214**, 1200–1216 (2010). See also arXiv:0902.0212 [math.CV]
17. Zhao, W.: New proofs for the Abhyankar-Gujar inversion formula and equivalence of the Jacobian conjecture and the vanishing conjecture. Proc. AMS **139**,(9), 3141–3154 (2011)
18. Zhao, W.: Mathieu subspaces of associative algebras. J. Algebra **350**(1), 245–272 (2012). See also arXiv:1005.4260

The manufacturer's authorised representative in the EU is Springer Nature Customer Service Centre GmbH, Europaplatz 3, 69115 Heidelberg, Germany. If you have any concerns regarding our products, please contact ProductSafety@springernature.com

Printed and bound by CPI Group (UK) Ltd, Croydon, CR0 4YY
05/10/2026
02228120-0011